Gender and Rural Globalization

Gender and Rural Globalization

International Perspectives on Gender and Rural Development

Edited by

Bettina B. Bock

Associate Professor for Rural Sociology, Wageningen University, Wageningen, The Netherlands; Professor for Population Decline and Quality of Life, Groningen University, Groningen, The Netherlands

and

Sally Shortall

Duke of Northumberland Chair of Rural Economy, Newcastle University, Newcastle upon Tyne, UK

CABI

CABI is a trading name of CAB International

CABI
Nosworthy Way
Wallingford
Oxfordshire OX10 8DE
UK

Tel: +44 (0)1491 832111
Fax: +44 (0)1491 833508
E-mail: info@cabi.org
Website: www.cabi.org

CABI
745 Atlantic Avenue
8th Floor
Boston, MA 02111
USA

Tel: +1 (617)682-9015
E-mail: cabi-nao@cabi.org

British Library Cataloguing-in-Publication Data

A catalogue record for this book is available from the British Library.

Library of Congress Cataloging-in-Publication Data

Names: Bock, Bettina B. (Bettina Barbara), 1960- author. | Shortall, S., author.
Title: Gender and rural globalization : international perspectives on gender and rural development / Bettina B. Bock and Sally Shortall.
Description: Boston, MA : CABI, [2017] | Includes bibliographical references and index.
Identifiers: LCCN 2017023474 (print) | LCCN 2017024012 (ebook) | ISBN 9781780648422 (ePub) | ISBN 9781780646268 (ePDF) | ISBN 9781780646251 (hbk: alk. paper)
Subjects: LCSH: Women in rural development. | Sociology, Rural. | Sex role.
Classification: LCC HQ1240 (ebook) | LCC HQ1240 .B635 2017 (print) | DDC 305.3--dc23
LC record available at https://lccn.loc.gov/2017023474

ISBN: 978 1 78064 625 1 (hardback)
978 1 78064 626 8 (e-book)
978 1 78064 842 2 (e-pub)

Commissioning editor: David Hemming
Editorial assistant: Emma McCann
Production editor: Alan Worth

Typeset by AMA DataSet Ltd, Preston, UK.
Printed and bound in the UK by CPI Group (UK) Ltd, Croydon, CR0 4YY.

Contents

Contributors

Monika Agarwal, C 802, Olive Heights, Sector 56, Gurgaon, India 122011. E-mail: monikka.agarwal@gmail.com

Margaret Alston, Department of Social Work, Monash University, Caulfield Campus, Dandenong Road, Melbourne, VIC 3145, Australia. E-mail: margaret.alston@monash.edu

Bettina B. Bock, Department of Social Sciences, Chairgroup Rural Sociology, Wageningen University, Hollandseweg 1, 6706 KN Wageningen, The Netherlands. E-mail: bettina.bock@wur.nl

Berit Brandth, Department of Sociology and Political Science, NTNU, NO-7491 Trondheim, Norway. E-mail: berit.brandth@svt.ntnu.no

Angela Calvo, Department of Agricultural, Forestry and Food Sciences (DISAFA), ULF Economia, Turin University, Largo Paolo Braccini 2, 10095 Grugliasco (TO), Italy. E-mail: angela.calvo@unito.it

Megan Canning, The University of Edinburgh, Room 3.16, Chrystal Macmillan Building, 15a George Square, Edinburgh EH8 9LD, UK. E-mail: s1264507@sms.ed.ac.uk

Anne Cassidy, UNESCO Child and Family Research Centre, School of Sociology and Politics, NUI, Galway, Republic of Ireland. E-mail: anne.cassidy@nuigalway.ie

Francesca Alice Centrone, Department of Agricultural, Forestry and Food Sciences (DISAFA), and Research Centre for Women's and Gender Studies (CIRSDe), Turin University, Largo Paolo Braccini 2, 10095 Grugliasco (TO), Italy. E-mail: alicentrone@gmail.com

Majda Černič Istenič, Biotechnical Faculty, University of Ljubljana, Jamnikarjeva 101, SI-1000 Ljubljana, Slovenia. E-mail: majda.cernic.istenic@bf.uni-lj.si

Chrysanthi Charatsari, School of Agriculture, Aristotle University of Thessaloniki, Egnatia Street 148A, 54621 Thessaloniki, Greece. E-mail: chcharat@agro.auth.gr

Jessica Duncan, Rural Sociology, Wageningen University, Hollandseweg 1, 6706 KN Wageningen, The Netherlands. E-mail: jessica.duncan@wur.nl

Elisabeth Garner, 341 E. Waring Avenue, State College, PA 16801, USA. E-mail: elg154@psu.edu

Donatella Greco, University of Trieste, Piazzale Europa 1, 34127 Trieste, Italy. E-mail: grecod83@yahoo.it

Wendy Harcourt, International Institute of Social Studies, Erasmus University, PO Box 29776, 2502 LT The Hague, The Netherlands. E-mail: harcourt@iss.nl

Marit S. Haugen, Centre for Rural Research, University Centre Dragvoll, NO-7491 Trondheim, Norway. E-mail: marit.haugen@bygdeforskning.no

Susie Jacobs, Reader in Comparative Sociology, Department of Sociology, Manchester Metropolitan University, Manton Building, Manchester M15 6LL, UK. E-mail: s.jacobs@mmu.ac.uk

Lori McVay, Department of Sociology, Global Studies, and Criminal Justice, Spring Arbor University, 106 E. Main Street, Spring Arbor, MI 49283, USA. E-mail: lori.ann.mcvay@gmail.com

Ilona Matysiak, Maria Grzegorzewska University (Akademia Pedagogiki Specjalnej im. Marii Grzegorzewskiej), Szczęśliwicka 40, 02-353 Warsaw, Poland. E-mail: ilona.matysiak@gmail.com

Peter Mollinga, SOAS University of London, Thornhaugh Street, Russell Square, London WC1H 0XG, UK. E-mail: pm35@soas.ac.uk

Janet Momsen, Emerita Professor of Geography, University of California Davis, Davis, California, USA; Research Associate in International Gender Studies, Lady Margaret Hall, Norham Gardens, Oxford OX2 6QA, UK. E-mail: jdmomsen@ucdavis.edu

Angela Mosso, Department of Agricultural, Forestry and Food Sciences (DISAFA), ULF Economia, Turin University, Largo Paolo Braccini 2, 10095 Grugliasco (TO), Italy. E-mail: angela.mosso@unito.it

Caoimhe Ni Dhonaill, Queens University Belfast, 17 Hopefield Avenue, Antrim Road, Belfast BT15 5AP, UK. E-mail: c.nidhonaill@qub.ac.uk

Nargiza Nizamedinkhodjayeva, CGIAR Research Program on Water, Land and Ecosystem, International Water Management Institute (IWMI), IWMI HQ, PO Box 2075, 127 Sunil Mawatha, Pelawatte, Battaramulla, Colombo, Sri Lanka 10120. E-mail: nargiza00@hotmail.com

Mark Riley, Department of Geography and Planning, University of Liverpool, 413 Roxby Building, Liverpool L69 7ZT, UK. E-mail: mark.riley@liverpool.ac.uk

Carolyn Sachs, 111-C Armsby, Pennsylvania State University, University Park, PA 16802, USA. E-mail: xyl@psu.edu

Heather Sangster, Department of Geography and Planning, University of Liverpool, 610 Roxby Building, Liverpool L69 7ZT, UK. E-mail: Heather.Sangster@liverpool.ac.uk

Sally Shortall, Duke of Northumberland Chair of Rural Economy, Centre for Rural Economy, School of Agriculture, Food and Rural Development, Newcastle University, Agriculture Building, Newcastle upon Tyne NE1 7RU, UK. E-mail: sally.shortall@newcastle.ac.uk

Susanne Stenbacka, Department of Social and Economic Geography, Uppsala University, Box 513, S-751 20 Uppsala, Sweden. E-mail: susanne.stenbacka@kultgeog.uu.se

Margreet van der Burg, Department of Social Sciences, Wageningen University, Hollandseweg 1, 6706 KN Wageningen, The Netherlands. E-mail: margreet.vanderburg@wur.nl

Chiara Zanetti, University of Trieste, Piazzale Europa 1, 34127 Trieste, Italy. E-mail: zanetti_chiara@yahoo.it

1 Gender and Rural Globalization: An Introduction to International Perspectives on Gender and Rural Development

B.B. Bock[1,2,*] and S. Shortall[3]

[1]*Wageningen University, Wageningen, The Netherlands;* [2]*Groningen University, Groningen, The Netherlands;* [3]*Newcastle University, Newcastle upon Tyne, UK*

Introduction

In the Introduction to the book *Rural Gender Relations* in 2006 we discussed how globalization and liberalization of agriculture irrevocably changed rural life and with it rural gender relations. A lot has happened since 2006; yet we could very well repeat the same sentence introducing this book ten years later. Globalization is, after all, still ongoing and the liberalization of agricultural policies is still a topical issue today. Actually, it is the accelerating process of globalization and the sociological theories that elaborate on it, which convinced us of the need for reflection and a new collection on rural gender research.

Giddens defines globalization as the 'growing interdependence between different peoples, regions and countries in the world as social and economic relationships come to stretch worldwide' (Giddens, 2001, p. 690). In particular, he refers to the interchange of products and capital through global trade, the interchange of people through global travel and migration, as well as the interchange of culture, ideas and world views. Globalization, moreover, affects politics in terms of expanding collaboration as well as interdependency. There is increasing awareness that we are facing challenges that may only be addressed through collaboration (e.g. climate change) (Beck, 2016). On the other hand, we see that the fast travel of information through information and communications technology may 'globalize' local political events in an unforeseen manner and interrelate places at great geographical distance. Globalization does not stop at the border of the rural, as Woods (2007) has demonstrated. After all, farmers in many rural areas deliver agricultural products to the world market and are, hence, closely involved in global trade and through their organizations in global negotiations. The same is true for other 'global production sectors' such as forestry and fisheries, energy, biofuel, minerals, as well as the non-agricultural rural entrepreneurs and rural residents more generally. They all are increasingly involved in worldwide connections and networks of collaboration, not least because many rural areas are receiving 'newcomers' from all over the world.

* Corresponding author e-mail: bettina.bock@wur.nl

This idea of the interconnected world in which the travel of information is unbound has inspired Manuel Castells's theory of the network society in which he elaborates on the importance of electronically processed information networks to organize social structures and relations (Castells, 2000). These networks constitute a means by which information is exchanged and relations are established. As these networks are virtual and unbound by geographical location, traditional spatial barriers, such as remote location, lose their importance. It is the extent to which networks criss-cross and interconnect that defines their power and not their geographical location. In theory, rural peripheral locations may be as included in the network society as central and urban places. In practice, the so-called 'connectivity' of rural residents is often lower because, among other factors, of the limited accessibility of high-speed Internet (Salemink *et al.*, 2016).

Urry (2007) and Cresswell (2010) elaborate on the movement of people, ideas and things that go along with the network society and contribute to globalization. They define the current time as an era of mobility and stress the importance of understanding humans as travellers. According to Urry, modern mobility includes the ongoing corporeal travelling of people and objects between places; the pluri-local senses of belonging to, and engagements in, places that create translocal social networks; and the imaginative, virtual and communicative 'travelling' that vivifies and invigorates those networks. The mobility turn is relevant for rural studies in several ways: it opens our eyes to the fact that rural residents are far less sedentary or immobile than often assumed (Milbourne and Kitchen, 2014), yet also underlines the huge disadvantage of rural areas when mobility is structurally limited through lack of infrastructure (Bell and Osti, 2010).

Taking account of mobility has consequences for theoretical frameworks as well as research methodologies. This is forcefully expressed in what Beck and Sznaider (2010, p. 382) call 'the cosmopolitization of reality', in which not only cross-border communication and collaboration increases but also the material and immaterial interdependence among social actors across national borders. These borders become more fluent and pervious as a result and may challenge commonly known differences between nations and/or localities. Turning to Europe, we have already seen that rural–urban distinctions are increasingly difficult to identify for gender relations (Shortall, 2014). This calls for multi-sited and comparative research or a cosmopolitan perspective, following Beck's idea of 'the cosmopolitan turn' (Beck and Sznaider, 2010). Its explicit aim is to go beyond the usual isolation of social matters as local or national phenomena and their traditional separation as a concern of so-called 'developed or developing' countries, and to understand how they are co-created in a global interplay of forces. Beck therefore argues we should let go of the usual methodological nationalism and localism to understand the construction of social relations. In terms of rural gender relations, this means that we should look beyond rural places when trying to understand their construction and functioning; also, separate discussions of Southern and Northern rural gender relations make less and less sense in an increasingly connected world.

Globalization and Rural Gender Studies

Globalization matters greatly for rural gender studies in our view. Globalization offers chances as it promotes the inclusion of even remote rural areas in the global network society. As a result many of the traditional structural and cultural barriers may no longer hamper rural women's empowerment and, hence, promote more equal gender relations. The increasing material and virtual mobility and connectivity may, in addition, promote collaboration of men and women across rural–urban and national borders. It also includes risks of course when the rural becomes part of a global field of action and with it competition and conflict at a global level. Higher levels of mobility render travelling affordable also for rural residents, yet also enable outmigration – in particular of young women as we have seen (e.g. Johansson, 2016) – as well as the entry of newcomers from all over the world. Both may change the existing structure of rural societies. In the following we discuss how such developments are taken up in research. We reflect upon the development of rural gender studies as described in the previous book, the relevance of the then defined research themes and the emergence of new topics and methodologies. Thereafter we introduce the structure and content of the current book.

In the Introduction to *Rural Gender Relations* (2006) we argued that rural gender studies started with research that revealed women's engagement in farming, which was continuously overlooked. Gender research of that time also identified how development policies and projects that ignored women's productive role, undermined women's position in society and created unequal gender relations (Bock, 2006). In the course of the 1980s rural gender researchers were among those who criticized the dominant idea of modernization as development, as it supported the profits of capitalist firms and countries, contributed to social inequality more generally, reinforced patriarchy and with it the dependency of women. At the same time research revealed the important role that women played in the maintenance of family farms and the elaboration of alternative pathways towards organic or multifunctional agriculture, part-time farming and rural entrepreneurship (e.g. agritourism).

Rural gender research is strongly influenced by mainstream gender studies in terms of theoretical and methodological development as well as research topics. This is clearly reflected in the growing attention to differences among women and the flexibility of gender identities in terms of dominant definitions of rurality and sexuality. In the following years this topic has been further developed, also theoretically, which is mirrored in increasing attention to lesbian, gay, bisexual, transgender, intersex and questioning (LGBTIQ) identities and queer theory (Keller, 2015). Rural gender researchers also followed up on the mobility turn with more research into the effects of globalization and the increased mobility of women and men on rural gender relations.

Based on this we may argue that rural gender studies in the global North and South followed roughly the following development path, when developing research on the following topics:

1. The visibility of women in agriculture and rural development.
2. Critique of (rural) development and patriarchy.
3. Struggle for empowerment and recognition of women's agency.
4. The acknowledgement of heterogeneity and flexible gender identities.
5. Mobility, globalization and transnationalism.

Rural gender studies started with research into women in agriculture and development and proceeded from there. The development is, however, not linear as most of the themes remain relevant despite the emergence of new ones, as we will demonstrate by giving just a few examples. Gender and agriculture is still a very topical issue; it is however studied in a different way, taking account of the theoretical and methodological insights gained, and of the relevance of mobility and globalization. New research looks, for instance, at how the increasing involvement of women as paid labourers in the expanding plantation agriculture in the global South affects their position (Canning, Chapter 21, this volume), into women's and men's involvement in agriculture as migrant labour or how women's access to agricultural innovation could be improved (FAO, 2016). There is also still research going on which critically evaluates the attempts of rural development policy to involve rural women in targeted projects, which more often than not reinforce traditional rural identities and relations (Shortall and Bock, 2015). In the global South gender is still an important focus in mainstream development projects, even though several researchers have a rather critical view of it. Already in 2007 Cornwall *et al.* reflected critically on their own role in mainstreaming gender. In a recent article Cornwall (2016) elaborates how current attempts to empower women have lost their critical stance by focusing too much on the economic involvement of women. In doing so we have, in her view, foregone the earlier focus on supporting women's agency, which renders any attempt to transform gender relations futile.

One interesting development is the growing interest in heterogeneous and flexible gender identities. This is evident in the types of questions being asked in research about rural masculinities and in the acceptance of LGBTIQ orientations and identities in rural areas in different parts of the world (Abelson, 2016). It often confirms earlier research that pointed to the conservative character of rural areas as well as the limited inclusion and integration of those who are considered different (Cloke, 2007). Finally, there is an expanding body of research into rural mobility and migration. Such research examines the arrival of labour migrants in Europe who contribute to the revitalization of rural areas, the ambivalent reception of refugees as well as the continuous trend of population decline in (remote) rural areas throughout the world, the dreams of those who leave as well as the experiences of those who stay behind (Bock *et al.*, 2016). In addition, several researchers have meticulously detailed the mobility of rural residents and 'the rural' in Europe, addressing the supposed immobility of the rural in this era of mobility (see Chapter 2, this volume).

This Book

The developments discussed above inspired us to edit a book that explores how rural gender relations are changing in a globalizing world. It analyses the development of rural gender relations in specific places around the world and looks at the effects of the increasing connectivity and mobility of people across places. The book is not geographically or geo-politically organized but integrates experiences across the globe through the discussion of four themes which reflect the above listed key themes in rural gender research: agriculture, international development, gender identities and mobility. Each theme is introduced with a paper that gives an overview of the state-of-the-art in that specific thematic area and integrates the case studies that follow. The contributors present empirical work from the global North and South, and more particularly Sweden, Norway, Northern Ireland, Republic of Ireland, the UK, Poland, Greece, Italy, Slovenia, Uzbekistan, India, Africa, Asia and Latin America, Australia, New Zealand, Canada and the USA.

The first section (Chapters 2–6) examines how mobility affects men and women in rural areas. It explores gender differences in mobility patterns and analyses how mobility affects rural gender identities and relations. The section opens with Chapter 2, an overview in which Bettina Bock discusses the development of mobility research and the attention given to the impact mobility has on gender relations and vice versa. She then introduces the chapters that contribute to this section. Chapter 3 by Donatella Greco and Chiara Zachetti reports on the experiences of Ukrainian women who work in rural areas of Italy as domestic caregivers. In doing so they cover the inadequacy of the Italian welfare system, which no longer supports the availability of sufficient services in rural areas, particularly for (elderly) citizens with limited mobility. In Chapter 4, Nargiza Nizamedinkhodjayeva, Bettina Bock and Peter Mollinga discuss how labour migration in rural Uzbekistan, which increased enormously in the aftermath of its independence, affects the livelihood of those who stay behind and how the importance of the economic contribution of labour migrants impacts on household relations in terms of gender and generation. In Chapter 5, Jessica Duncan and Monika Agrawal explore how land grabbing hinders the mobility of pastoralist communities in Gujarat in India and how the restricted mobility affects their livelihood, personal well-being and gender relations. Chapter 6 by Janet Momsen compares how two opposing migration streams – tourist immigration and labour emigration – play out in two regions with very different histories of migration: the Caribbean islands and the Yucatán Peninsula in Mexico. She demonstrates how migration may strengthen the position of women through their increased involvement in gainful employment as well as the interaction with other lifestyles, norms and values.

The second section (Chapters 7–11) focuses on agricultural change and the development of agricultural and rural policies, the response of individuals within farm households and the implications for gender relations in rural areas. It starts in Chapter 7 with an overview of the development of research on gender and agriculture by Sally Shortall and an introduction of the chapters in this section. In Chapter 8, Margaret Alston presents a longitudinal study of farm families' struggles with the impact of climate change in Australia. She discusses how farm families' strategies to cope with drought and the reduced farm income affect gender relations and put pressure on farm families' well-being. In Chapter 9, Susanne Stenbacka follows a group of female entrants into farming and analyses the values that motivate them to start farming. Based on her study she concludes that the new female entrants differ from men in terms of their economic aspirations and aspired quality of life. They are active and ambitious farmers yet shape farming in their own way, different to the hitherto dominant masculine worldview. Chapter 10 by Majda Černič Istenič and Chrysanthi Charatsari studies and compares agricultural education and extension programmes in Slovenia and Greece in order to understand to what extent women participate in such programmes and if and how participation fosters their empowerment as farm women. In Chapter 11, the final one of this section, Anne Cassidy looks at the views of young women and men who are not the heirs to the farm and their role in contributing to the smooth succession and maintenance of Irish family farms. She unravels how Irish farm youths' socialization into traditional gendered position teaches, in particular women, not to expect to inherit yet to support their brother's takeover of the family property without objecting to sacrificing inheritance rights.

The third section (Chapters 12–17) focuses on the construction of identities and the changes occurring in the definition of rural femininity and masculinity as a result of rural transformations. Sally Shortall opens the section in Chapter 12 by presenting an overview of research on rurality and gender identity. She then introduces the chapters of this section. In Chapter 13, Lori McVay examines the identity of women leaders in Northern Ireland in order to understand how these women managed to become leaders, against all traditional odds, and which events and life choices shaped their identity. In doing so she elaborates how growing up in a rural context may nurture women's confidence and readiness to lead in the sense that taking on responsibility is considered part of a rural lifestyle. Chapter 14 by Marit Haugen and Berit Brandth studies how divorce affects the gendered identities in farming. Divorce has been considered particularly destructive in farming, where the unity of farm and family is perceived as the basis of its functioning and continuity. Their chapter demonstrates how divorce dismantles not only the socio-economic unit but also attacks the personal identity of farm women and men – as being a farmer is inseparably part of the personal identity of both. Losing the farm family undermines men's identity as farmers, whereas women – who generally leave the farm – have to let go of their farm identity altogether. How important farming is for the identity of men emerges too from Chapter 15 by Mark Riley and Heather Sangster, who study how retirement affects the male farming identity of farmers. They point at the central role of farm labour, of being physically involved in work on the farm. Proof of bodily strength and continued involvement in the farm are crucial for farm men's sense of self even if farm management is actually transferred. The importance of work as part of men's identity after retirement is also a theme of Chapter 16 in which Caoimhe Ni Dhonaill compares two different initiatives to tackle the social exclusion of elderly men and women in rural and urban areas. She demonstrates how men are often more vulnerable to loneliness when they lose work and how difficult it is for initiatives to involve men if there is no room for a post-work masculine space that allows men to reconfirm their masculine identity, for instance through do-it-yourself, wood carving and metal work. Chapter 17, the final chapter of this section, looks at the identity of female village representatives in Poland. Ilona Matysiak's key question is whether accessing the position of village representative promotes women's position in politics, which had traditionally been a male position of privilege. Based on her research she demonstrates how women's entry is facilitated by the devaluation of this position and does not lead to a gain in political power even though it promotes women's participation in village affairs.

The fourth section (Chapters 18–23) examines the role of international development policies in advancing women's wellbeing in the less developed parts of the world and some of the unintended consequences of such interventions. It starts with an overview in Chapter 18, in which Bettina Bock and Margreet van der Burg discuss the development of research in this field in terms of important topics as well as theories and methodologies. They underline the recurrence of gender equality on the international political agenda as an important precondition for sustainable development and food security in the light of an evergrowing world population. Their chapter closes with the introduction of the contributing chapters, all of which look at the changes of rural gender relations from the viewpoint of international relations and international development politics. Chapter 19 by Carolyn Sachs and Elisabeth Garner reviews recent literature on the feminization and masculinization of agriculture resulting from the reorganization of the agrifood system in both the global North and South contexts. They demonstrate how new developments in family farming and industrial agriculture, as well as the emergence of new alternative forms of farming, go along with shifts in gender relations and identities which may or may not provide opportunities for the empowerment of women. In Chapter 20, Wendy Harcourt presents evidence of glocal networking among rural women when presenting case studies in rural India, Tanzania and Italy from a feminist political ecology perspective. In doing so she stresses the importance of interaction and collaboration across place and between different groups in order to effectively promote equitable and sustainable development across the globe. Chapter 21 by Megan Canning looks at labour conditions in the sugarcane industry in Malawi and discusses how wage work in plantation agriculture redefines traditional gender roles and relations, realizing gain and loss at the same time. Gaining income improves women's position within the household, while at the same time women are prone to exploitation and sexual intimidation at their working place on the plantation. In Chapter 22, Francesca Alice Centrone, Bettina Bock, Angela Mosso and Angela Calvo present their experiences with the development of novel gender indicators and their usefulness in the light of the increased importance of accountability. Gender indicators are most often used at the international level. Gender indicators may, however, be useful also at regional or even project level by assessing the extent to which projects and policies do actually contribute to more equal gender relations and empower women in terms of, for instance, access to health, education, governance, knowledge and technology. In Chapter 23, the final chapter of this section, Susie Jacobs reviews current trends towards the formalization of land tenure practices and compares the effects of individual or joint land titling on the position of women in different countries across the globe. She concludes that titling may strengthen women's right to land, although the increased privatization that goes along with it may also undermine the right of access of the most vulnerable groups, especially when it is the state or a corporate business that acquires formerly common land.

The book closes in Chapter 24 with conclusions and reflections on the position of gender in rural research agendas and in rural academia more generally.

References

Abelson, M.J. (2016) 'You aren't from around here': race, masculinity, and rural transgender men. *Gender, Place & Culture* 23(11), 1535–1546. Available at: http://dx.doi.org/10.1080/0966369X.2016.1219324 (accessed 4 March 2017).

Beck, U. (2016) *The Metamorphosis of the World*. Wiley, Chichester, UK.

Beck, U. and Sznaider, N. (2010) Unpacking cosmopolitanism for the social sciences: a research agenda. *British Journal of Sociology* 61(Suppl. 1), 381–403. Available at: http://dx.doi.org/10.1111/j.1468-4446.2009.01250.x (accessed 4 March 2017).

Bell, M.M. and Osti, G. (2010) Mobilities and ruralities: an introduction. *Sociologia Ruralis* 50(3), 199–204. Available at: http://dx.doi.org/10.1111/j.1467-9523.2010.00518.x (accessed 4 March 2017).

Bock, B.B. (2006) Gender and migration: an overview. In: Bock, B.B. and Shortall, S. (eds) *Rural Gender Relations: Issues and Case Studies*. CAB International, Wallingford, UK, pp. 155–164.

Bock, B.B., Osti, G. and Ventura, F. (2016) Rural migration and new patterns of exclusion and integration in Europe. In: Shucksmith, M. and Brown, D. (eds) *Routledge International Handbook of Rural Studies*. Routledge, London and New York, pp. 71–84.

Castells, M. (2000) Materials for an exploratory theory of the network society. *British Journal of Sociology* 51(1), 5–25. Available at: http://dx.doi.org/10.1111/j.1468-4446.2000.00005.x (accessed 4 March 2017).

Cloke, P. (ed.) (1997) *Contested Countryside Cultures*. Routledge, Abingdon, UK and New York.

Cornwall, A. (2016) Women's empowerment: what works? *Journal of International Development* 28(3), 342–359. Available at: http://dx.doi.org/10.1002/jid.3210 (accessed 4 March 2017).

Cornwall, A., Harrison, E. and Whitehead, A. (2007) Gender myths and feminist fables: the struggle for interpretive power in gender and development. *Development and Change* 38(1), 1–20. Available at: http://dx.doi.org/10.1111/j.1467-7660.2007.00400.x (accessed 4 March 2017).

Cresswell, T. (2010) Towards a politics of mobility. *Environment and Planning D: Society and Space* 28(1), 17–31. Available at: http://dx.doi.org/10.1068/d11407 (accessed 4 March 2017).

FAO (2016) Transforming gender relations in agriculture through women's empowerment. Summary of the online discussion no. 130. Global Forum on Food Security and Nutrition. Available at: http://www.fao.org/fsnforum/activities/discussions/gender_childcare_nutrition (accessed 31 October 2016).

Giddens, A. (2001) *Sociology*, 4th edn. Polity Press and Blackwell Publishing Ltd, Cambridge and Oxford, UK.

Johansson, M. (2016) Young women and rural exodus – Swedish experiences. *Journal of Rural Studies* 43, 291–300. Available at: http://dx.doi.org/10.1016/j.jrurstud.2015.04.002 (accessed 4 March 2017).

Keller, J.C. (2015) Rural queer theory. In: Pini, B., Brandth, B. and Little, J. (eds) *Feminisms and Ruralities*. Lexington Books, London, pp. 107–118.

Milbourne, P. and Kitchen, L. (2014) Rural mobilities: connecting movement and fixity in rural places. *Journal of Rural Studies* 34, 326–336. Available at: http://dx.doi.org/10.1016/j.jrurstud.2014.01.004 (accessed 4 March 2017).

Salemink, K., Strijker, D. and Bosworth, G. (2017) Rural development in the digital age: a systematic literature review on unequal ICT availability, adoption, and use in rural areas. *Journal of Rural Studies* (in press). Available at: http://dx.doi.org/10.1016/j.jrurstud.2015.09.001 (accessed 4 March 2017).

Shortall, S. (2014) Gender mainstreaming or strategic essentialism?: how to achieve rural gender equality. In: Pini, B., Brandth, B. and Little, J. (eds) *Feminisms and Ruralities*. Lexington Books, London, pp. 69–80.

Shortall, S. and Bock, B.B. (2015) Rural women in Europe: the impact of place and culture on gender mainstreaming the European Rural Development Programme. *Gender, Place & Culture* 22(5), 662–669. Available at: http://dx.doi.org/10.1080/0966369X.2014.917819 (accessed 4 March 2017).

Urry, J. (2007) *Mobilities*. Polity Press, Cambridge, UK.

Woods, M. (2007) Engaging the global countryside: globalization, hybridity and the reconstitution of rural place. *Progress in Human Geography* 31(4), 485–507. Available at: http://dx.doi.org/10.1177/0309132507079503 (accessed 4 March 2017).

2 Gender and Mobility

B.B. Bock[1,2,*]

[1]Wageningen University, Wageningen, The Netherlands; [2]Groningen University, Groningen, The Netherlands

Introduction

Famous sociologists such as Castells (2000) and Urry (2007) describe the present time as the century of mobility. With this description, they refer to the fact that falling transportation costs and the new possibilities offered by information and communications technology have enabled people to travel more and spend their lives in different places (even at the same time, i.e. virtually) while remaining in contact with friends and family across the world. Cresswell (2010) even describes modern (wo)men as nomads who are constantly on the move.

That 'travelling' has increased appears self-evident – mirrored, for instance, by ever-increasing numbers of tourists, and air passengers in particular.[1] Nevertheless, the discourse of modern mobility meets with contestation regarding its overly optimistic connotations and its conflation of all types of movements, travellers and means of transport. Bauman (2000) emphasizes that access to mobility is unequal and that social inequality is reflected in (relative) immobility. Braidotti *et al.* (2013) warn that free travel is a privilege of the elite. Many current travellers are forced to travel, and in doing so they must overcome many barriers because their mobility is perceived as undesirable and limited by the elite.

Without denying the pertinence of mobility and its societal and scientific relevance, their critique is significant. With the experiences of 2015 in mind, and the arrival of several hundred thousand migrants and refugees on the coasts of Greece and Italy, nobody can any longer deny the significance of mobility in our society. We could even speak of 'mobilization' as reflected in the high numbers of people who move house or live in multiple places, commute to work, travel for fun, migrate on a regular or continuous basis for work or are on the run by force, driven by war and violence, natural disasters and poverty. To understand the meaning and social significance of (im)mobility, it is, however, essential to distinguish carefully among travellers and explore their different travel experiences. We need to study who is (not) travelling, for what reasons and with what consequences. Who receives support or opposition from whom? We must also unravel the significance of important markers of social difference such as gender, race/ethnicity, class and nationality. In doing so, it is important, also, to account for the geographical context and the type of 'borders' that travellers are crossing, as it is the movement from marginal to affluent places that is politically most contested. The walls in Texas, Spain, Hungary and Jerusalem are just some examples of countries blocking the entry of

* Corresponding author e-mail: bettina.bock@wur.nl

'others'; referring to the 'Fortress of Europe' is another well-known illustration (see, for instance, Casas-Corters *et al.*, 2015). As Hanson (2010, p. 8) explains, 'it should be impossible to think about mobility without simultaneously considering social, cultural and geographical context – the specifics of place, time and people'. In the context of this book, it is rural–urban borders that are of particular interest. Moving from rural to urban areas is often seen as proof of progress and modernization, yet such movement is also feared as the engine of rural decline and marginalization. Movements in the opposite direction are benevolently interpreted as supporting rural development (Brown and Glasgow, 2008) yet feared when counter-urbanization leads to the gentrification of rural areas (Scott, 2011; Gkartzios and Ziebarth, 2016). As discussed in the following sections, social characteristics such as gender, race and identity take on particular significance in the context of rural migration and mobility.

The significance of mobilization is also reflected in the fact that the concept of mobility is gaining ground within migration theories through its attention to the continuous character of modern movements. Yet, one might also argue that it is migration theorists who have identified the importance of ongoing movement and, hence, mobility. This is seen, for instance, in the body of work on transnationalism, which has already stressed for years that migration does not cross-cut but rather interlinks peoples across place (Levitt and Jaworsky, 2007; Vertovec, 2009). Another strength of mobility theory is that it includes all types of voyages and movements, whereas traditional migration theory clearly distinguished 'migration' from any other movement (Brown, 2012). Launching our discussion from the more inclusive concept of mobility – as opposed to migration – is advantageous, as it is often difficult to clearly define travel or the traveller as they are reflected in the heated political debate over the distinction between (economic) migrants and 'real' refugees. In practice, current research often combines both bodies of work, mobility and migration. This is also the case for scholars interested in the gender-specific aspects of migration and mobility.

This chapter provides a brief overview of the development of the field of gender and migration and mobility. It then focuses, in particular, on research on mobility and migration in the rural context and the research questions that have gained the most attention in recent years. It closes with a presentation of the chapters that comprise this section of the book (Chapters 3–6).

The Development of the Field

Gender and migration is, by now, an established body of knowledge and a long-standing topic of research (see, for instance, Passerini *et al.*, 2007; Oso and Ribas-Mateos, 2013). Gender-specific research has challenged the traditional notion of men as active migrants and women as their passive partners who may stay behind and wait for their return, accompany them and travel under their protection, or follow them once they have paved the way.

The concentration on male migrants may also be explained by the traditional interpretation of migration as an economic behaviour and a desire to better one's economic circumstances by migrating to places with more and better paid employment and opportunities for social mobility (Anthias, 2000). Limited attention to women as economic agents thus explains their neglect by studies of economic migrants. This has changed fundamentally in recent years. Women's economic agency is increasingly acknowledged, as is women's wish to migrate for work (Anthias and Lazaridis, 2000). Various studies document young women migrating in search of higher education and promising employment in cities and/or affluent countries (Ni Laoire, 2001; Donkersloot, 2012; Kuhmonen *et al.*, 2016; Tufor *et al.*, 2016). It is, however, also mature women, married and/or with children, who, by migrating to work for shorter or longer periods, sustain their families and the successful upbringing of their children

(Greco and Zanetti, Chapter 3, this volume). Most famous are probably Philippine women, who support not only their families but also the national economy through the remittances they earn through domestic labour in Europe and the Middle East (Ehrenreich and Hochschild, 2003). Similar cases may be found elsewhere (Marchand, 2009), including in Europe, where women from Poland, Bulgaria and Ukraine accept work as resident caregivers of elderly people in Germany, Spain and Italy (Marchetti and Venturini, 2014; Farris, 2015; Greco and Zanetti, Chapter 3, this volume). Such studies emphasize that migrants' labour markets are clearly segregated along lines of class. In general, migrants tend to find work that native workers are unwilling to do – often physically challenging and poorly paid, with flexible or no contracts, in the shadow economy (Farris, 2015). Sometimes their work is under the exploitative circumstances of near-slavery (Corrado, 2012; Bock *et al.*, 2016). The sectors in which migrants typically work are, in addition, clearly segmented by gender, with women overly present in health care, services and domestic labour, and increasingly agriculture, while men dominate agriculture, heavy industry and mining, transport and construction (Farris, 2015).

There are, however, other reasons to move from place to place besides employment or education. Searching for a better quality of life is one of them. This may be prompted by the desire for a safer or healthier environment, a place with more or better services, or a place offering more freedom to live according to one's own identity, norms and values. In the last case, escaping from social exclusion in rural places is a strong motive that has been reported for young (rural) women (Dahlström, 1996), migrants and refugees (Colvin, 2017), and members of lesbian, gay, bisexual, transgender and intersex communities (Manalansan, 2006; Abelson, 2016). The pleasure of being mobile is also in itself a motivation to move, as Hanson (2010) explains. This is reflected especially by young men and women who leave rural areas to lead a more adventurous life either elsewhere or in continuous mobility, with mobility and migration becoming an ever more popular rite of passage to adulthood (White, 2010; Farrugia, 2016). And then, there are numerous cases where moving is induced by love and friendship (Sunanta and Angeles, 2013). In this context it is important to note the increasing attention given to the phenomenon of transnational families and households and the fact that even intimate relations form configurations that cut across locations, keeping people emotionally and socio-economically connected (Mazzucato, 2008; Pantea, 2012; Sunanta and Angeles, 2013). It is here that the concept of transnationality is particularly important for explaining the multi-sitedness of belonging and identification and the emergence of a mobile network of social relations that is no longer dependent on sedentary settlement (Gustafson, 2001).

In 2003, Ehrenreich and Hochschild (2003, p. 5) stressed the 'feminization of migration', indicating that women migrants are outnumbering men. Recent numbers confirm this expectation; female migrants outnumbered males in several European countries in 2010 and their figures have, overall, been less affected by the crisis in Western Europe than has been the case for migrant men (Farris, 2015). Yet, there are so many travellers other than migrants, which makes it difficult to compare the mobility of men and women. What is, however, evident is that not only migration but also mobility is a gendered phenomenon: 'Decisions to migrate are made within a larger context of gendered interactions and expectations between individuals and within families and institutions' (Donato *et al.*, 2006, p. 12). The latter includes 'state regulations of the mobile' (Donato *et al.*, 2006, p. 21), such as governmental migration policies that, in Africa for instance, intentionally restricted women's migration until the 1950s and 1960s to prevent African urbanization and maintain reproduction of the rural labour force (Wright, 1995, p. 784). Gender relations are shaping patterns of migration and mobility, and both are shaping gender relations (Donato *et al.*, 2006; Hanson, 2010). This is clearly seen in studies that report on female migrants gaining recognition and bargaining power because of their economic contributions (Tufor *et al.*, 2016) and in studies that reveal how the domestic division of labour may be revised with men taking over 'feminine' responsibilities, at least temporarily (Choi, 2016). Hanson (2010, p. 9) notes that although 'mobility/immobility stand at the core of traditional gender ideologies', the understanding of this phenomenon has often been too simplistic, equating mobility with empowerment and masculinity and immobility with subordination and femininity. Whereas mobility may indeed be empowering, it is important, in her view, to account for rootedness as another important source of power, as it promotes the development of solid social networks. She also points to the fact that mobility may be forced and may result from a subordinate position, and she calls for 'detailed understandings of the gendered meanings and power relations embedded in various forms of mobility/immobility in various social and geographic contexts' (Hanson, 2010, p. 11).

Therefore, it is important to account for gender's interaction with other markers of social differentiation when studying mobility and migration. How this is reflected in the rural context is discussed next.

Gender, Migration and Mobility in the Rural Context

The genderedness of mobility and migration takes on a particular significance in the rural context because of the interesting interaction between the assumed differences in mobility between urban and rural areas and between men and women. Both rurality and femininity are traditionally connected to immobility, or at least to mobility constraints. Rising levels of migration and mobility thus challenge traditional notions of rurality and rural–urban distinctions, as well as rural gender identities and relations.

In Europe, Bell and Osti (2010) initiated the debate on the supposed immobility of the rural. While underlining that access to transport is essential for rural residents and for the inclusion and development of rural areas, the authors stress the high numbers of rural residents who increasingly must travel to regional centres to reach basic facilities and services. Since then, several studies have unravelled how mobile the rural actually are, materially as commuters, visitors and sojourners; this is in addition to newcomers and incoming, outgoing and returning migrants (Halfacree, 2012; Goodwin-Hawkins, 2014; Milbourne and Kitchen, 2014; Argent, 2016).

Mobility is, thus, an important issue for rural areas in various ways. First of all, it concerns the needs and opportunities of rural residents in terms of everyday mobility, such as commuting to work, school and basic services. European research reports the increased reliance on private means of transport resulting from cuts in public budgets and related reductions in public transport. It is those without a driver's licence and a car who are most disadvantaged in this regard, including the very young and old, women and those with low levels of income (Osti, 2010).

A second important topic concerns the outmigration of, particularly, young and well-educated residents, initiating a vicious cycle of rural marginalization and decline (Argent, 2016). This is a long-standing subject that has also been discussed repeatedly in the context of the rural exodus and flight from the land and ongoing urbanization of the rural population (Argent, 2016). The emigration of young rural women has particular significance in the European context, as it is perceived as a major threat to local development in terms of declining birth rates and motivation for young rural men to follow. The resulting masculinization of rural areas is interpreted as a reflection of local decline. As early as 2001, Ni Laoire emphasized that the ability to leave rural areas may be unequally distributed, with women being better equipped to leave because of their relatively higher levels of education and because they face lower expectations to stay compared with rural men, who are expected to maintain the family business, especially in farming. Dahlström (1996) stressed that rural areas have little to offer to young women, as both employment and recreational opportunities are more attuned to men's interests. Similar motivations of especially female adolescents to leave the rural areas and move to the city have been confirmed in recent research in several countries across Europe, including Sweden (Johansson, 2016; Rauhut and Littke, 2016), Hungary (Timár and Velkey, 2016), East Germany (Leibert, 2016), Spain, Ireland and Austria, as well as Australia and Canada (Leibert and Wiest, 2016).

Also particularly interesting is research that explores how mobility and emigration affect family life and the division of household responsibilities. Following the example of the famous research of Ehrenreich (Ehrenreich and Hochschild, 2003) and Lutz (Lutz and Pallenga-Möllenbeck, 2012) on the chain of care and love, there are several recent studies on how, when leaving to earn money, European women reorganize the care of their children by offering paid care to their employer elsewhere (Pantea, 2012; Tyldum, 2015). Also, European women tend to redistribute care work across generations, with other women, often their mothers (-in-law), stepping in, while redistribution across gender lines is a less frequent and often temporary solution (King and Vullnetari, 2009). Similar research is now beginning in Africa (Dimova *et al.*, 2015). Also interesting is research that discusses the experience of transnational families and households across the globe (Baldassar and Merla, 2014; Parella, 2016). Whereas many of the studies cited above examine the experiences of migrating women, there are also a number of studies exploring the situation of women left behind in rural areas (Gartaula *et al.*, 2012; Wu and Ye, 2016).

A third important topic concerns rural immigration, which has a long history of research, particularly in the global North in terms of counter-urbanization, lifestyle migration and the influx of newcomers and retirees (Berry, 2000; Halfacree, 2012; Argent, 2016). Counter-urbanization is most frequent in rural areas close to major cities and addresses the movement of middle-class families to rural areas in search of a better quality of life. While contributing to the economic development of these areas, such movement may also render housing unaffordable for rural residents (Shucksmith, 2000), endanger the rural culture and lifestyle, as well as affect the conservation of rural landscapes. In the 1990s, counter-urbanization was also discussed in gender research that stressed the constraints women experienced on their professional careers as a result of very traditional rural gender relations and ideologies (Little and Austin, 1996). Similar results have been found in studies looking at the adoption of more traditionalist gender roles among those moving back to rural areas following their dissatisfaction with urban work and consumption patterns (Wilbur, 2014). More recent research underlines that gender relations and ideologies are changing in rural areas and becoming more similar to urban gender relations (Bock, 2010).

The influx of international labour migrants and refugees into rural areas is a new topical issue in European rural research. These studies stress the difficulties migrants and refugees face in finding their way and successfully integrating into rural areas – difficulties due to the shortage of relevant services and facilities and the fact that they are often perceived as extremely different by the rural population, who frequently meets the arrival of 'strangers' with suspicion and anxiety (De Lima, 2012; Hedberg *et al.*, 2012; McAreavey, 2012). The latter had already been discussed in the late 1990s in the context of rural residents' difficulties in meeting the 'other'. At that time, this was also discussed in terms of gender and the idea that rural women were perceived as needing the protection of incoming strangers (Cloke, 1997). Recent studies on immigration may address gender issues in terms of differences between the opportunities and experiences of male and female migrants and the effect of novel experiences on gender relations in local and immigrant households (see, for instance, Petrou, 2012; Vullnetari, 2012; Farris, 2015; Maliepaard and Alba, 2016).

The gender-specific effects of migrant labour in agriculture deserve separate attention. These have been discussed in depth in the context of plantation labour in the global South and the USA, where the fair treatment of migrants in general and women in particular is increasingly discussed as an element of fair trade certification (Terstappen *et al.*, 2013). Whereas the poor treatment of migrants in agriculture is certainly an issue in Europe too (Bock *et al.*, 2016), there is no research so far that examines its gender-specific aspects.

The Chapters

The chapters presented in this section build on the legacy of earlier research into gender and rural migration, such as that sketched above, and contribute to the further development of research into the gender-specific dimensions of rural mobilization. They explore the significance of mobility for rural gender relations in very different contexts that cross-cut countries and continents. In doing so, they illustrate the gender-specific aspects and effects of increasing levels of mobility in the country of origin as well as the country of reception. However, they also discuss the effects of mobility constraints.

In Chapter 3, Donatella Greco and Chiara Zanetti present a case that elucidates the effects of increased international immigration into rural areas in Europe. Their chapter examines the experiences of Ukrainian women who play an increasingly important role in providing care for the elderly in rural areas and, in doing so, repair the inadequacy of the Italian welfare system, which no longer supports the availability of sufficient services in rural areas, particularly for (elderly) citizens with limited mobility. Many women are circular migrants who move between Italy and Ukraine and aim to support their families by providing paid care to the family members of others elsewhere. Greco and Zanetti provide an overview of recent research looking into the significance of this phenomenon and the experiences of these women, who welcome the opportunity to provide for their families at home yet also suffer from loneliness and isolation, especially in remote rural areas, and are often forced to accept precarious employment conditions.

Chapter 4 by Nargiza Nizamedinkhodjayeva, Bettina Bock and Peter Mollinga presents an example of how labour migration affects those staying behind and may change household relations in terms of gender and generation as a result of changing economic bargaining power as well as experiences of independence obtained abroad. In particular, they report on how increasing levels of male outmigration for work, which is necessary to make a living in the rural areas of Uzbekistan in the aftermath of the dissolution of the Soviet Union, have affected the division of labour and dominant norms and values in the extended rural households of Uzbekistan in terms of gender and generation. Again, the changes are not clearly in favour of or detrimental to gender equality. On the one hand, post-socialist transitions in the organization of the country have led to the impoverishment of many rural households and have weakened the position of women as traditional gender norms were reintroduced. At the same time, migration has enabled individual men and women to gain personal income and, with that, some level of independence from the usual patriarchal organization of extended households.

In Chapter 5, Jessica Duncan and Monika Agrawal explore the gender-specific effects of land grabbing on the social practices of pastoralist communities in Gujarat, India; in doing so they present an example of the unequal distribution of mobility. In this particular case, the mobilization of assets (land) by powerful groups results in the immobilization of those with less power. The chapter describes in considerable detail how restricted mobility affects the lives of pastoralists and the organization of gender relations. After presenting three cases of the state-sanctioned transfer of pastoralists' traditional land access rights to corporations, they reconstruct a typical day in the lives of the women who live in the affected areas. In doing so, they demonstrate how the loss of livestock brought on by restricted access to grazing land affects daily routines and rhythms and produces new economic relations within families. The forced immobility of pastoralists affects not only their livelihoods but also has gender-specific effects on their social practices and emotional well-being.

Janet Momsen discusses the effects of migration on rural areas in an historic perspective in Chapter 6. Moreover, she compares how two opposing migration streams – tourist immigration and labour emigration – play out in two regions with very different histories of migration: the Caribbean islands and the Yucatán Peninsula in Mexico. She demonstrates how migration influences the livelihoods of rural households and the position of women. Whereas the outmigration of male farmers has led to the decline of plantation agriculture and commercial crop production in general, farming is often continued as a safety net by those who remain behind. It is often women who continue farming to generate income in their own right and to avoid being fully dependent on migrants' remittances. At the same time, they want to safeguard a place for migrants to return. Some specialize in the production of food and, for example, marijuana for the tourism industry; others sell handicrafts to tourists. In doing so, engagement in the tourist industry also becomes an alternative to outmigration. Momsen concludes that earning income of their own adds to women's empowerment and creates a new interaction with tourists and urban lifestyles.

Note

[1]On the evolution of air passengers in Europe, see, for example: Eurostat (2016) Air passenger transport – monthly statistics. Available at: http://ec.europa.eu/eurostat/statistics-explained/index.php/Air_passenger_transport_-_monthly_statistics (accessed 7 September 2016).

References

Abelson, M.J. (2016) 'You aren't from around here': race, masculinity, and rural transgender men. *Gender, Place & Culture* 23(11), 1535–1546. Available at: http://dx.doi.org/10.1080/0966369X.2016.1219324 (accessed 4 March 2017).

Anthias, F. (2000) Metaphors of home: gendering new migrations to Southern Europe. In: Anthias, F. and Lazaridis, G. (eds) *Gender and Migration in Southern Europe: Women on the Move*. Berg, Oxford, UK, pp. 15–47.

Anthias, F. and Lazaridis, G. (2000) Introduction: women on the move in Southern Europe. In: Anthias, F. and Lazaridis, G. (eds) *Gender and Migration in Southern Europe: Women on the Move*. Berg, Oxford, UK, pp. 1–13.

Argent, N. (2016) Demographic change: beyond the urban–rural divide. In: Shucksmith, M. and Brown, D. (eds) *Routledge International Handbook of Rural Studies*. Routledge, Abingdon, UK and New York, pp. 29–35.

Baldassar, L. and Merla, L. (2014) *Transnational Families, Migration and the Circulation of Care: Understanding Mobility and Absence in Family Life*. Routledge, Abingdon, UK and New York.

Bauman, Z. (2000) *Liquid Modernity*. Polity Press, Cambridge, UK.

Bell, M.M. and Osti, G. (2010) Mobilities and ruralities: an introduction. *Sociologia Ruralis* 50(3), 199–204. Available at: http://dx.doi.org/10.1111/j.1467-9523.2010.00518.x (accessed 4 March 2017).

Berry, E.H. (2000) Review essay: rural sociology, migration and community change. *Rural Sociology* 65(4), 658–667.

Bock, B.B. (2010) *Personal and Social Development of Women in Rural Areas of Europe. IB/B/AGRI/IC/2010_089, PE 438.608*. Directorate General for Internal Policies, Policy Department B: Structural and Cohesion Policies, Agriculture and Rural Development, Brussels. Available at:

http://www.europarl.europa.eu/RegData/etudes/note/join/2010/438608/IPOL-AGRI_NT(2010)438608_EN.pdf (accessed 2 February 2017).

Bock, B.B., Osti, G. and Ventura, F. (2016) Rural migration and new patterns of exclusion and integration in Europe. In: Shucksmith, M. and Brown, D. (eds) *Routledge International Handbook of Rural Studies*. Routledge, Abingdon, UK and New York, pp. 71–84.

Braidotti, R., Hanafin, P. and Blagaard, B. (2013) *After Cosmopolitanism*. Routledge, New York and London.

Brown, D. (2012) How 'mobilities' are restructuring the rural–urban periphery. Keynote address at the *Agriculture in an Urbanizing Society Conference*, Wageningen University, Wageningen, The Netherlands, 1–4 April 2012.

Brown, D.L. and Glasgow, N. (2008) *Rural Retirement Migration*. Springer, Dordrecht, The Netherlands and New York.

Casas-Cortes, M., Cobarrubias, S. and Pickles, P. (2015) Riding routes and itinerant borders: autonomy of migration and border externalization. *Antipode* 47(4), 894–914.

Castells, M. (2000) Materials for an exploratory theory of the network society. *British Journal of Sociology* 51(1), 5–25. Available at: http://dx.doi.org/10.1111/j.1468-4446.2000.00005.x (accessed 4 March 2017).

Choi, S.Y.P. (2016) Gendered pragmatism and subaltern masculinity in China: peasant men's responses to their wives' labor migration. *American Behavioral Scientist* 60(5.6), 565–582. Available at: http://dx.doi.org/10.1177/0002764216632832 (accessed 4 March 2017).

Cloke, P. (ed.) (1997) *Contested Countryside Cultures*. Routledge, Abingdon, UK and New York.

Colvin, N. (2017) 'Really really different different': rurality, regional schools and refugees. *Race Ethnicity and Education* 20, 225–239. Available at: http://dx.doi.org/10.1080/13613324.2015.1110302 (accessed 4 March 2017).

Corrado, A. (2012) Ruralità differenziate e migrazioni del sud Italia. *Agriregionieuropa* 8(28), 72–74.

Cresswell, T. (2010) Towards a politics of mobility. *Environment and Planning D: Society and Space* 28(1), 17–31. Available at: http://dx.doi.org/10.1068/d11407 (accessed 4 March 2017).

Dahlström, M. (1996) Young women in a male periphery – experiences from the Scandinavian North. *Journal of Rural Studies* 12(3), 259–271. Available at: http://dx.doi.org/10.1016/0743-0167(96)00018-6 (accessed 4 March 2017).

De Lima, P. (2012) Boundary crossing. In: Hedberg, C. and De Carmo, E. (eds) *Translocal Ruralism*. Springer, Dordrecht, The Netherlands and New York, pp. 203–217.

Dimova M., Hough, C., Kyaa, K. and Manji, A. (2015) Intimacy and inequality: local care chains and paid childcare in Kenya. *Feminist Legal Studies* 23(2), 167–179. Available at: http://dx.doi.org/10.1007/s10691-015-9284-6 (accessed 4 March 2017).

Donato, K.A., Gabaccia, D., Holdaway, J., Manalansan, M.F. IV and Pessar, P.R. (2006) A glass half-full? Gender in migration studies. *International Migration Review* 40(1), 3–26. Available at: http://dx.doi.org/10.1111/j.1747-7379.2006.00001.x (accessed 4 March 2017).

Donkersloot, R. (2012) Gendered and generational experiences of place and power in the rural Irish landscape. *Gender, Place & Culture* 19(5), 578–599. Available at: http://dx.doi.org/10.1080/0966369X.2011.610095 (accessed 4 March 2017).

Ehrenreich, B. and Hochschild, A.R. (2003) Introduction. In: Ehrenreich, B. and Hochschild, A.R. (eds) *Global Woman: Nannies, Maids, and Sex Workers in the New Economy*. Granta Books, London, pp. 1–14.

Farris, S.R. (2015) Migrants' regular army of labour: gender dimensions of the impact of the global economic crisis on migrant labour in Western Europe. *The Sociological Review* 63(1), 121–143. Available at: http://dx.doi.org/10.1111/1467-954X.12185 (accessed 4 March 2017).

Farrugia, D. (2016) The mobility imperative for rural youth: the structural, symbolic and non-representational dimensions rural youth mobilities. *Journal of Youth Studies* 19(6), 836–861. Available at: http://dx.doi.org/10.1080/13676261.2015.1112886 (accessed 4 March 2017).

Gartaula, H.M., Visser, L. and Niehof, A. (2012) Socio-cultural dispositions and wellbeing of the women left behind: a case of migrant households in Nepal. *Social Indicators Research* 108(3), 401–420. Available at: http://dx.doi.org/10.1007/s11205-011-9883-9 (accessed 4 March 2017).

Gkartzios, M. and Ziebarth, A.C. (2016) Housing: a lens to rural inequalities. In: Shucksmith, M. and Brown, D. (eds) *Routledge International Handbook of Rural Studies*. Routledge, Abingdon, UK and New York, pp. 495–508.

Goodwin-Hawkins, B. (2014) Mobilities and the English village: moving beyond fixity in rural West Yorkshire. *Sociologia Ruralis* 55(2), 167–181. Available at: http://dx.doi.org/10.1111/soru.12043 (accessed 4 March 2017).

Gustafson, P. (2001) Roots and routes: exploring the relationship between place attachment and mobility. *Environment and Behaviour* 33(5), 667–686. Available at: http://dx.doi.org/10.1177/00139160121973188 (accessed 4 March 2017).

Halfacree, K. (2012) Heterolocal identities? Counter-urbanisation, second homes, and rural consumption in the era of mobilities. *Population, Space and Place* 18(2), 209–224. Available at: http://dx.doi.org/10.1002/psp.665 (accessed 4 March 2017).

Hanson, S. (2010) Gender and mobility: new approaches for informing sustainability. *Gender, Place & Culture* 17(1), 5–23. Available at: http://dx.doi.org/10.1080/09663690903498225 (accessed 4 March 2017).

Hedberg, G., Forsberg, G. and Najib, A. (2012) When the world goes rural. In: Hedberg, C. and do Carmo, R.M. (eds) *Translocal Ruralism*. Springer, Dordrecht, The Netherlands and New York, pp. 125–142.

Johansson, M. (2016) Young women and rural exodus – Swedish experiences. *Journal of Rural Studies* 43, 291–300. Available at: http://dx.doi.org/10.1016/j.jrurstud.2015.04.002 (accessed 4 March 2017).

King, R. and Vullnetari, J. (2009) The intersections of gender and generation in Albanian migration, remittances and transnational care. *Geografiska Annaler: Series B, Human Geography* 91(1), 19–38. Available at: http://dx.doi.org/10.1111/j.1468-0467.2009.00304.x (accessed 4 March 2017).

Kuhmonen, T., Kuhmonen, I. and Luoto, L. (2016) How do rural areas profile in the future dreams by Finnish youth? *Journal of Rural Studies* 44, 89–100. Available at: http://dx.doi.org/10.1016/j.jrurstud.2016.01.010 (accessed 4 March 2017).

Leibert, T. (2016) She leaves, he stays? Sex-selective migration in rural East Germany. *Journal of Rural Studies* 43, 267–279. Available at: http://dx.doi.org/10.1016/j.jrurstud.2015.06.004 (accessed 4 March 2017).

Leibert, T. and Wiest, K. (2016) The interplay of gender and migration in Europe's remote and economically weak rural regions: introduction to a special issue. *Journal of Rural Studies* 43, 261–266. Available at: http://dx.doi.org/10.1016/j.jrurstud.2016.01.007 (accessed 4 March 2017).

Levitt, P. and Jaworsky, B.N. (2007) Transnational migration studies: past developments and future trends. *Annual Review of Sociology* 33(1), 129-156. Available at: http://dx.doi.org/10.1146/annurev.soc.33.040406.131816 (accessed 4 March 2017).

Little, J. and Austin, P. (1996) Women and the rural idyll. *Journal of Rural Studies* 12(2), 101–111. Available at: http://dx.doi.org/10.1016/0743-0167(96)00004-6 (accessed 4 March 2017).

Lutz, H. and Pallenga-Möllenbeck, E. (2012) Care workers, care drain, and care chains: reflections on care, migration, and citizenship. *Social Politics* 19(1), 15–37.

Maliepaard, M. and Alba, R. (2016) Cultural integration in the Muslim second generation in the Netherlands: the case of gender ideology. *International Migration Review* 50(1), 70–94. Available at: http://dx.doi.org/10.1111/imre.12118 (accessed 4 March 2017).

Manalansan, M.F. IV (2006) Queer intersections: sexuality and gender in migration studies. *International Migration Review* 40(1), 224–249. Available at: http://dx.doi.org/10.1111/j.1747-7379.2006.00009.x (accessed 4 March 2017).

Marchand, M.H. (2009) The future of gender and development after 9/11: insights from postcolonial feminism and transnationalism. *Third World Quarterly* 30(5), 921–935. Available at: http://dx.doi.org/10.1080/01436590902959149 (accessed 4 March 2017).

Marchetti S. and Venturini, A. (2014) Mothers and grandmothers on the move: labour mobility and the household strategies of Moldovan and Ukrainian migrant women in Italy. *International Migration* 52(5), 111–126.

Mazzucato, V. (2008) The double engagement: transnationalism and integration. Ghanaian migrants' lives between Ghana and the Netherlands. *Journal of Ethnic and Migration Studies* 34(2), 199–216.

McAreavey, R. (2012) Resistance or resilience? Tracking the pathway of recent arrivals to a 'new' rural destination. *Sociologia Ruralis* 52(4), 488–506. Available at: http://dx.doi.org/10.1111/j.1467-9523.2012.00573.x (accessed 4 March 2017).

Milbourne, P. and Kitchen, L. (2014) Rural mobilities: connecting movement and fixity in rural places. *Journal of Rural Studies* 34, 326–336. Available at: http://dx.doi.org/10.1016/j.jrurstud.2014.01.004 (accessed 4 March 2017).

Ni Laoire, C. (2001) A matter of life and death? Men, masculinities and staying 'behind' in rural Ireland. *Sociologia Ruralis* 41(2), 220–236. Available at: http://dx.doi.org/10.1111/1467-9523.00179 (accessed 4 March 2017).

Oso, L. and Ribas-Mateos, N. (2013) *The International Handbook on Gender, Migration and Transnationalism: Global and Development Perspective*. Edward Elgar, Cheltenham, UK and Northampton, Massachusetts.

Osti, G. (2010) Mobility demands and participation in remote rural areas. *Sociologia Ruralis* 50(3), 296–310. Available at: http://dx.doi.org/10.1111/j.1467-9523.2010.00517.x (accessed 4 March 2017).

Pantea, M.C. (2012) Grandmothers as main caregivers in the context of parental migration. *European Journal of Social Work* 15(1), 63–80. Available at: http://dx.doi.org/10.1080/13691457.2011.562069 (accessed 4 March 2017).

Parella, S. (2016) Introduction: the portability of care in an increasingly mobile world: chains, drains and circulation. *Papers* 101(2), 253–257. Available at: http://dx.doi.org/10.5565/rev/papers.2264 (accessed 4 March 2017).

Passerini, L., Lyon, D., Capussotti, E. and Laliotou, I. (2007) *Women Migrants from East to West: Gender, Mobility and Belonging in Contemporary Europe*. Berghahn Books, Oxford, UK and New York.

Petrou, M. (2012) Rural immigration, family farm modernisation and reactivation of traditional women's farming tasks in Greece: masculinities and feminities reconsidered. *South European Society and Politics* 17(4), 553–571. Available at: http://dx.doi.org/10.1080/13608746.2012.654617 (accessed 4 March 2017).

Rauhut, D. and Littke, H. (2016) 'A one way ticket to the city, please!' on young women leaving the Swedish peripheral region Västernorrland. *Journal of Rural Studies* 43, 301–310. Available at: http://dx.doi.org/10.1016/j.jrurstud.2015.05.003 (accessed 4 March 2017).

Scott, M. (2011) Exclusive countrysides: rural gentrification, consumer preferences and planning. *Planning Theory & Practice* 12(4), 593–635. Available at: http://dx.doi.org/10.1080/14649357.2011.626304 (accessed 4 March 2017).

Shucksmith, M. (2000) *Exclusive Countryside? Social Inclusion and Regeneration in Rural Areas*. Joseph Rowntree Foundation, York, UK.

Sunanta S. and Angeles, L.C. (2013) From rural life to transnational wife: agrarian transition, gender mobility, and intimate globalization in transnational marriages in northeast Thailand. *Gender, Place & Culture* 20(6), 699–717. Available at: http://dx.doi.org/10.1080/0966369X.2012.709827 (accessed 4 March 2017).

Terstappen V., Hanson, I. and McLaughlin, D. (2013) Gender, health, labor, and inequalities: a review of the fair and alternative trade literature. *Agriculture and Human Values* 30(1), 21–39. Available at: http://dx.doi.org/10.1007/s10460-012-9377-7 (accessed 4 March 2017).

Timár, J. and Velkey, G. (2016) The relevance of the political economic approach: the interpretations of the rural in the migration decision of young women and men in an economically backward region. *Journal of Rural Studies* 43, 311–322. Available at: http://dx.doi.org/10.1016/j.jrurstud.2015.11.012 (accessed 4 March 2017).

Tufor, T., Sato, C. and Niehof, A. (2016) Gender, households and reintegration: everyday lives of returned migrant women in rural northern Ghana. *Gender, Place & Culture* 23(10), 1480–1495. Available at: http://dx.doi.org/10.1080/0966369X.2016.1204999 (accessed 4 March 2017).

Tyldum, G. (2015) Motherhood, agency and sacrifice in narratives on female migration for care work. *Sociology* 49(1), 56–71. Available at: http://dx.doi.org/10.1177/0038038514555427 (accessed 4 March 2017).

Urry, J. (2007) *Mobilities*. Polity Press, Cambridge, UK.

Vertovec, S. (2009) *Transnationalism*. Routledge, London.

Vullnetari, J. (2012) Women and migration in Albania: a view from the village. *International Migration* 50(5), 169–188. Available at: http://dx.doi.org/10.1111/j.1468-2435.2009.00569.x (accessed 4 March 2017).

White, A. (2010) Young people and migration from contemporary Poland. *Journal of Youth Studies* 13(5), 565–580. Available at: http://dx.doi.org/10.1080/13676261.2010.487520 (accessed 4 March 2017).

Wilbur, A. (2014) Back-to-the-house? Gender, domesticity and (dis)empowerment among back-to-the-land migrants in northern Italy. *Journal of Rural Studies* 35, 1–10. Available at: http://dx.doi.org/10.1016/j.jrurstud.2014.03.005 (accessed 4 March 2017).

Wright, C. (1995) Gender awareness in migration theory: synthesizing actor and structure in Southern Africa. *Development and Change* 26(4), 771–791. Available at: http://dx.doi.org/10.1111/j.1467-7660.1995.tb00574.x (accessed 4 March 2017).

Wu, H. and Ye, J. (2016) Hollow lives: women left behind in rural China. *Journal of Agrarian Change* 16(1), 50–69. Available at: http://dx.doi.org/10.1111/joac.12089 (accessed 4 March 2017).

3 Women's Migration for Work: The Case of Ukrainian Caregivers in Rural Italy

D. Greco* and C. Zanetti
University of Trieste, Trieste, Italy

Introduction

The aim of this chapter[1] is to analyse the experience of Ukrainian women who migrate to Italy for work.[2] In recent decades, women's migrant labour has played an increasingly crucial role, particularly in elderly care. This phenomenon has been most evident in large cities, but over time it has also occurred in rural areas. This chapter focuses on migrant women from Ukraine, who are typically employed in the caregiving sector. In addition, it examines the specific features of the settlement of Ukrainian women in rural areas, their interaction with their country of origin and the emotional and practical consequences of their stay in Italy at the level of their transnational family ménage. Moreover, by reviewing the relevant literature, this chapter provides an overview of the research on Ukrainian migrant women in rural Italy.

As several authors have recently stated, the interface between gender and migration studies has been problematic for many years.[3] However, a new approach to the topic of migration was adopted in the 1970s and 1980s: a number of scholars contributed by conducting a series of studies that prioritized the analysis of the woman's perspectives in migration.

The chapter is structured as follows. The first section focuses on the main concepts concerning female migration to analyse the case of Ukrainian women migrants in Italy. An overview of the experience of these women is provided by drawing on relevant national and international literature. A brief survey of the official national statistical data is conducted to contextualize this migration phenomenon in the Italian setting. We then present an analysis of official statistics on Ukrainian women in Italian rural areas and a review of their migration experiences in greater detail; the push-and-pull factors that explain their migratory paths and their different features; and finally, their interaction with the local rural areas. This approach is used to describe the interaction between Ukrainian women and the Italian host society. Moreover, this analysis prompts a series of considerations about the Italian welfare system, which appears to be inadequate in satisfying the needs of the most fragile segment of the population (i.e. the elderly), especially in rural areas that lack services.

* Corresponding author e-mail: bettina.bock@wur.nl

 Gender and Rural Globalization (eds B.B. Bock and S. Shortall)

Background and Context: Ukrainian Women in International Migration Flows

According to Pollini and Scidà (2002, p. 16), migration flows are a significant element of social, economic and political changes. This means that their appearance changes over time and they acquire different characteristics, depending on the historical and social context in which they develop. In recent decades, one of the most striking aspects of international migrations has been the so-called *feminization of migration flows*, to which migrating Ukrainian women have certainly contributed. However, the claim that women have been the protagonists of international migration only in recent years should be nuanced, as a number of international scholars have noted (Grigg, 1977; Bock and Shortall, 2006). In fact, the role of women in migration flows was highlighted by Ravenstein in his 'Laws of migration' in 1876. Ravenstein's law no. 5 entails the 'predominance of females among short-distance migrants' (Lee, 1966; White, 2016): according to this statement, women are more inclined than men to short-radius migration. Indirectly, he underlines that women, like men, have always migrated throughout the history of international migration flows. What differentiates women's migration experiences from those of men is the shorter distances covered between the country of origin and the final destination. Ravenstein's assertion concerning women's migration certainly referred to society as he knew it. Today, we know that women's migration patterns develop in response to changes in the social and political international context (Portes and Sensenbrenner, 1993). The migration of Ukrainian women has changed over the years as well, with longer distances covered when such movement was permitted by national and international events such as the fall of the Berlin Wall, the dissolution of the former Soviet regime and, more generally, the end of the Cold War.

Ambrosini (2011) analysed the main characteristics of international migration flows and demonstrated the numerical predominance of men and their status as *first migrants*, who dominated and fostered migration flows even in the case of migration as a family strategy (travelling as, e.g., fathers, sons, husbands or brothers). However, as DeLaet stated (1999, p. 13), 'The invisibility of women in international migration scholarship does not correspond to the reality of international migration. Women migrate across international boundaries at approximately the same rate as men'. Indeed, it is generally agreed that the presence of women in migratory systems is not a very recent phenomenon but was already evident in the so-called Fordist phase of migration after the Second World War (Ambrosini, 2011). Studies that analyse migration from the gender perspective are, however, quite recent. Since the 1980s, thanks to the input of a number of European and American feminist scholars, research has focused on the role of women in international flows, aiming to show their central role, in addition to that of men, in international migration over the centuries (Kofman *et al.*, 2000; Pessar and Mahler, 2003; Vianello, 2014).

Furthermore, in the context of contemporary mobility, the male bias that once influenced academic studies on migration does not reflect the current nature of worldwide migration flows. According to the latest statistics on international migration provided by the United Nations (UN) Department of Economic and Social Affairs (UN-DESA), in 2013, approximately 232 million people in the world were migrants, 48% of whom were women (Caritas italiana and Migrantes, 2015). These data confirm that in the 'age of migrations' (Castels and Miller, 2012), one of the principal characteristics of international migration flows is their gender balance (Donato and Donna, 2015). More specifically, the number of female emigrants is higher in developed (51.6%) than in developing countries[4] (46.2%) (Caritas italiana and Migrantes, 2015, p. 47). Later in this chapter, closer attention will be paid to how the literature has highlighted the dependence of female migration on economic and social imbalances among countries (Sassen, 2000). This framework is particularly suited to the analysis of Ukrainian women migrants, who began to migrate more systemically after the collapse of the Soviet regime.

The gender approach to female migration highlights different types of female migrants: the largest category consists of those termed *dependants*. These women started to leave their countries of origin in the context of family reunification during the 1970s. As stated by Lahav (in Kofman, 2000, p. 66), in the 1980s, the number of women entering European states for family reunification purposes increased.[5] The dependant female migrant is often represented as a passive actor who undergoes the migratory event without any personal initiative and as part of a larger family project. Recent research, however, has questioned the supposed passiveness of women in both migration projects and the labour market. A number of studies have observed that women participate in the definition of migration routes and are active in the labour market, albeit as irregular workers who are involved in household or nursing jobs. As Farris (2010) stated, this pattern of female migration has been dominant in Europe – particularly in the UK, France and Italy – among Bangladeshi, North African and Moroccan communities, respectively.

Another group of female migrants in Europe consists of the so-called *forerunners*. These are generally women who leave their country of origin before their parents or husbands. They initiate the migration chain that enables other members of the family to follow. Petersen speaks in this regard of an innovative migration event that is often motivated by the search for better work and life conditions (Petersen in Pollini and Scidà, 2002, p. 73). According to Petersen, emigrating women can also be identified with the so-called *free migration* phenomenon, wherein the push factor is the personal desire to seek a better life and better working conditions.

Another characteristic of female migration flows is the role of informal networks. This feature is particularly evident in the migration flows to Italy from Eastern Europe or Eritrea, a former Italian colony. Moreover, Catholic institutions have often acted as mediators helping women settle in the destination country. In general, women who migrated to Europe in the 1970s and 1980s came from countries with strong Catholic traditions (such as South America, the Philippines and, after 1989, Eastern European countries).

According to Favaro and Colombo (1993), marital and parental roles are likely to be reversed when women are the first to migrate: in this case, women tend to assume the functions of the head of the family and the main breadwinner. In many cases, they assume this role while supporting their families at home, thereby managing a transnational family life and household. Using a transnational perspective as a new explanatory model appears to be particularly suited to the study of Ukrainian women's migration to Italy, as explained in the following sections.

Ukrainian Women in Italy: General Evidence in the Form of Statistics

Although the first immigrants arrived in Italy in the 1970s, the number of foreign citizens who were regularly present in the country became substantial only 20 years ago. Since then, the population has increased from 625,000 in 1991[6] to just over 1 million in 2001,[7] and just over 5 million in 2015 (Istat, 2016), when foreign citizens represented 8.3% of the total population. Currently, according to official national statistics, Romanians form the largest ethnic community living in Italy, followed by Albanians, Moroccans and Chinese. Ukrainians are the fifth largest ethnic group at the national level: on 1 January 2016, approximately 230,728 Ukrainians were living in Italy, i.e. 4.6% of the foreigners in the country.[8]

In general, the migration of Ukrainian women began in the early 1990s after the dissolution of the Soviet Union and the transition to capitalist economies in former Soviet member states. The economic and social instability experienced by these countries prompted many residents – even members of families belonging to the (lower) middle classes – to emigrate (Bianchini and Privitera, 2005). Female emigration flows grew significantly, especially in rural areas where the socio-economic situation was even more difficult (Vianello, 2009). Subtelny (2003) identified this period as the fourth phase of Ukrainian emigration, in which the push factor was related to a profound economic crisis affecting the national system as a consequence of political instability. A considerable number of Ukrainian migrants (men and women) moved to Russia, Germany and Poland. In the second half of the 1990s, Italy and Spain also attracted Ukrainian migrants, particularly women, who found work in the care system – in which, even if they were highly educated, they were usually employed as caregivers.

In Italy, immigration from the Ukraine has grown more significantly and rapidly than has immigration by other ethnic groups (Blangiardo, 2005): on 1 January 2003, there were 12,730 Ukrainian residents in Italy, which grew to 230,728 Ukrainian residents on 1 January 2015.[9] Nevertheless, female Ukrainian migrants have arrived somewhat unnoticed because of their relative invisibility compared with other migrants (Favaro and Tognetti Bordogna, 1991). First, the white skin colour of these women means that they are less frequently perceived as foreigners than are members of other ethnic groups. In addition, work in the domestic sector, which is often on full-time contracts, severely limits their possibilities for social interaction with the host society.

The increasing number of Ukrainian citizens in Italy has been confirmed by an analysis of residence permits (see Table 3.1): in 2002, there were 12,618 Ukrainians legally residing in Italy, whereas in 2016, their number increased to 240,141. However, as shown in Table 3.2, there has been a small reduction in the number of new residence permits for non-EU citizens in the last two years (–6.5%). This trend, which is stronger for Ukrainians than for all other foreigners present in Italy (–25.6%), is probably due to the reduction of employment opportunities in the Italian labour market resulting from the economic crisis. Moreover, the instability of the country of origin and the development of neighbouring countries in Eastern Europe as new emigration destinations have probably played a role.

Table 3.1. Residence permits in Italy (for Ukrainian citizens and total non-EU citizens), with percentage of females in years 2002, 2007 and 2016. (Data from http://dati.istat.it (accessed 21 December 2015).)

	Residence permits 01.01.2002		Residence permits 01.01.2007		Residence permits 01.01.2016	
	n	Of which women (%)	*n*	Of which women (%)	*n*	Of which women (%)
Ukrainians	12,618	78.8	118,524	83.2	240,141	79.2
Total non-EU citizens	1,448,392	47.2	2,414,972	50.4	3,931,133	48.7

Table 3.2. New residence permits in Italy (for Ukrainian citizens and total non-EU citizens), with percentage of females in years 2013 and 2015 and percentage variation 2013–2015. (Data from http://dati.istat.it (accessed 20 October 2016).)

	New residence permits in 2013		New residence permits in 2015		
	n	Of which women (%)	*n*	Of which women (%)	Variation 2013–2015 (%)
Ukrainians	14,162	78.7	10,543	58.6	–25.6
Total non-EU citizens	255,646	47.8	238.936	41.7	–6.5

The presence of Ukrainians in the Italian regions primarily reflects their variable demand for workers in the domestic work and home care services sector. However, as noted above, the role of chain migration is also important because it explains the high presence of Ukrainians in Campania and Lazio (see Table 3.3). Campania is a longstanding destination for Ukrainian migrants for two reasons: the first is the extent of the informal economy; the second is the presence of commercial relationships between Naples and the port of Odessa, which facilitated the first Ukrainian newcomers (Mazzacurati, 2005; Vianello, 2009). Moreover, as stated by Vianello (2009), the relationship with former Polish emigration flows can help explain the location of Ukrainian destinations in Italy: Ukrainian forerunners were initially supported by the Polish social networks that had already been established in Rome, Naples, Milan and Bologna. In particular, the connections between the Catholic (Italian and Polish) and the Orthodox churches enabled them to develop a wide and permeable network through which they provide support during the initial stages of the Ukrainian immigration.

When comparing the presence of migrants in urban and rural areas, the concentration of migrants in cities becomes evident, although it is significant in medium-sized towns and small villages (Ambrosini, 2011; Carrosio, 2012; Balbo, 2015). Over time, agricultural and small/medium manufacturing companies have attracted numerous foreign workers. According to Carrosio (2012), quantifying the presence of immigrants in the rural areas of Italy requires a consideration of municipalities with fewer than 5000

Table 3.3. Presence of Ukrainians and foreign citizens in Italian regions by gender on 1 January 2016. (Data from http://dati.istat.it (accessed 20 October 2016).)

	Ukrainians			Total foreigners			
Italian regions	*n*	%	Of which women (%)	*n*	%	Of which women (%)	% of Ukrainians among foreign citizens in Italy
Lombardia	51,490	22.3	78.6	1,149,011	22.9	51.3	4.5
Campania	42,403	18.4	75.6	232,214	4.6	52.7	18.3
Emilia-Romagna	31,757	13.8	80.5	533,479	10.6	53.4	6.0
Lazio	23,198	10.1	78.8	645,159	12.8	52.4	3.6
Veneto	16,507	7.2	80.9	497,921	9.9	52.8	3.3
Toscana	11,336	4.9	80.9	396,219	7.9	53.9	2.9
Piemonte	10,244	4.4	79.2	422,027	8.4	53.3	2.4
Calabria	6,458	2.8	73.5	96,889	1.9	51.4	6.7
Marche	5,632	2.4	79.5	140,341	2.8	55.0	4.0
Friuli Venezia Giulia	5,188	2.2	80.8	105,222	2.1	52.8	4.9
Umbria	4,961	2.2	79.5	96,875	1.9	56.3	5.1
Liguria	4,806	2.1	82.9	136,216	2.7	53.8	3.5
Trentino-Alto Adige	4,152	1.8	77.6	94,920	1.9	53.7	4.4
Abruzzo	3,833	1.7	78.7	86,363	1.7	54.8	4.4
Puglia	2,759	1.2	78.4	122,724	2.4	52.6	2.2
Sardegna	2,304	1.0	85.6	47,425	0.9	54.4	4.9
Sicilia	2,120	0.9	82.6	183,192	3.6	47.9	1.2
Basilicata	781	0.3	77.8	19,442	0.4	53.2	4.0
Molise	539	0.2	75.0	12,034	0.2	51.3	4.5
Valle d'Aosta	260	0.1	84.6	8,480	0.2	57.8	3.1
ITALY	230,728	100.0	78.8	5,026,153	100.0	52.6	4.6

inhabitants and a population density lower than 150 people/km^2. On the basis of this selection, there were 4306 municipalities with 6,402,540 inhabitants in 2015, among whom 373,191 were foreigners, for an overall incidence of 5.8%. In these areas, there were 14,031 Ukrainians, with a predominance of women (11,256, i.e. 80.2% of the Ukrainians resident in the rural municipalities of Italy). Hence, migration flows are also attracted to rural areas, even if they are not necessarily their first choice. Migrants rarely choose to move to rural areas; in most cases, the ethnic network and their job placement determine their settlement in those places. According to Morris (in Urry, 2007, p. 33), rural areas can represent metaphorical 'motels' in which foreigners live for a while during their migration journeys. Eventually, the unpredictable evolution of the migratory event, the decision to stay or move on to new destinations, will depend on the level of their social inclusion. However, the Italian countryside may offer fewer opportunities than urban areas, especially in terms of social equality. Working conditions are characterized by low wages and low social prestige, and housing conditions are often inappropriate (Osti, 2010; Bock *et al.*, 2016). This is certainly the case for immigrant workers who are engaged in agriculture, but it also describes the conditions of female workers employed in family services as domestic workers and caregivers, which is the typical scenario for female Ukrainian migrants in Italy. As Morokvasic underlines (1984, p. 886), '(. . .) women from the peripheral world represent a ready-made labour supply which is, at once, the most vulnerable, the most flexible and, at least in the beginning, the least demanding work force'. In particular, the low degree of job mobility, which is particularly marked in rural areas, makes caregivers more economically, socially and emotionally vulnerable. These aspects are analysed below when we focus on Ukrainian women's inclusion in the receiving society, the precariousness of their jobs and the impact of migration on the family residing in the countries of origin.

Social Integration of Ukrainian Caregivers in Rural Areas

Work is usually considered the first stage in the process of migrants' inclusion in the receiving society: this condition applies even more to women when they arrive in a new country. According to Ambrosini (2016), migrant women employed in the domestic care sector benefit from widespread social recognition, even in the absence of a regular stay permit or other formal authorizations. However, 'the tolerance related to work cannot be easily transferred outside the work sphere' (Ambrosini, 2016, p. 155). In fact, there is often some distrust towards live-in caregivers on the part of the host family, especially if distance makes it difficult for the family or relatives of the care recipient to check on the developing relationship between him/her and the caregiver. This appears to occur especially in the case of work with the elderly: the main worry is that the caregiver may take advantage (financially and emotionally) of the situation. Moreover, the elderly person may be reluctant to relate so intimately to a stranger (even more so if the caregiver is not Italian). These attitudes are common in both urban and rural areas. However, isolation and demographic decline in some marginal areas in Italy exacerbate these issues (Lucatelli *et al.*, 2012) due to a lack of services and depopulation. In these circumstances, hiring foreign caregivers may be the only option to help the elderly and prevent their placement in a retirement home. Moreover, the caregiver's cultural differences may have an impact on the well-being of the elderly care recipient. For example, eating habits and lifestyles can also mediate the relationship between the client and the foreign caregiver, possibly even more so in rural areas, where the traditions of the elderly are deep-rooted and part of their identity.

Newly arriving foreign caregivers are also aware of the marginality of rural areas: the blurred distinction between work and spare time/spaces limits the migrants' inclusion in the receiving society, thereby increasing the perception of temporariness that characterizes the migratory experience. Female caregivers, especially those from Eastern Europe, tend to make friends only with other migrant caregivers belonging to the same ethnic group (Da Roit and Facchini, 2010). Otelli (2011) describes how little Ukrainian migrant women identify themselves with the local society receiving them, given that their migration is mainly due to economic factors. An example of this attitude is provided by their use of public spaces: in fact, groups of Eastern European women typically spend their time off at the local public gardens. This behaviour is distinctive of female live-in caregivers who do not have personal space. The 'use' of these public spaces becomes less intense when these women convert their migratory path from a temporary to a long-term stay: this circumstance occurs, for example, when their work is changed into employment by the hour and, consequently, when accommodation with their 'patient' is no longer provided.

The same applies to Ukrainian women: the only opportunity for them to socialize is often to meet co-nationals in the public gardens of the cities or villages where they live. These meetings serve to affirm their national and social identity in the host society. An alternative may be attending mass at an Orthodox church. In fact, thanks to specific agreements with the local Catholic authorities, such churches can be located in small or medium-sized villages as well. Hence, the inability of Ukrainian women to recognize the place where they live as a source of identity is proof of their lack of entrenchment and integration in the receiving society.

For foreign women, who usually do not have a driver's license, this is especially true in rural areas, which are characterized by low levels of mobility (Kaufmann, 2005).[10] This aspect has been confirmed by a large body of research. In particular, a study conducted by Friuli Venezia Giulia (Azienda per i Servizi Sanitari n. 6, 2009) showed that in 2009, 75% of foreign caregivers did not have a driver's licence. This seriously limited their ability to commute to work, visit the doctor or go to the grocery store. Instead, they had to rely on public transport, with generally rather limited services in rural areas (Lucatelli *et al.*, 2012). Moreover, in these areas, immigrant caregivers find it difficult to obtain informal support (e.g. from friends and acquaintances of the same national group) because ethnic networks usually operate in urban areas. Hence, especially in rural areas, migrant women face a significant risk of loneliness. Migrant women living in urban areas can rely on the presence of special facilities for migrants provided by associations, non-governmental organizations or public offices as meeting places; at those sites, they can find both formal and informal support, such as information and counselling. However, as Magnani (2012, p. 181) noted in a research project carried out in an Italian Alpine valley, living in rural areas can offer foreign caregivers economic advantages because the costs of living tend to be lower than those in cities. In sum, rural areas are often the only places of work for Ukrainians.

Finally, as Morawaska highlights (2005), the social integration of immigrants into the receiving society is a multidimensional process developed across several levels, including perceptions, projects and initiatives that can encourage or discourage integration. At an individual level, the strategies of foreign caregivers are mainly aimed at the functional use of rural areas and developing the relationship with the native inhabitants. At a meso level, interactions between foreign caregivers and the local community are generally not problematic thanks to the strategic role that the former play in caring for the elderly. Migrant women's work is also acknowledged by national and regional policies through, for example, specific offices located in public job centres that monitor the caregivers' activities. This should help control the informal economy and train foreign caregivers in handling their relationships with the elderly, their families, and the health and social institutions.

Ukrainian Migration in Italy as a Development of the International Care Economy

The principal feature of Ukrainian migration in Italy is the feminization of migration flows: women represent 79.2% of all Ukrainians legally residing in Italy, a ratio that reflects the high demand for women as domestic workers and caregivers in Italy and Greece[11] (Hellermann, 2006). The case of Ukrainian women employed as caregivers should be viewed in the wider context of the international caregiving sector (Vianello, 2009, p. 77). Since the 1990s, the employment of migrant women in the domestic sector has increased in several Western and Asian countries (Momsen, 1999; Tam, 1999; Sassen, 2000; Vianello, 2014). Given this international economic context, the labour market participation of Ukrainian women in Italy is exemplary: among all foreigners in Italy, only 20% of Ukrainian women are unemployed, whereas up to 60% of the other foreign women are jobless.

The high demand for women in the domestic and care sector may be explained by the specific features of the Mediterranean welfare system (Sciortino, 2004; Ambrosini, 2013) and the weak institutional support that it gives to families. In Italy, women traditionally took care of their elderly parents and parents-in-law; placing an elderly person in a retirement home was unusual. With the growth of female employment, however, women have become less able and willing to take care of the elderly. However, public retirement homes in Italy tend to have long waiting lists and have difficulty meeting this demand in a timely manner. Moreover, especially if the elderly person is healthy but not completely self-sufficient, he/she is not often inclined to choose the retirement home but instead wants to stay at home with his/her family. New solutions must therefore be found, especially with the rapid ageing of the population. In recent years, an increasing number of families now rely on caregivers who provide care while living in with the elderly. This scenario is even more common in rural areas, where public welfare services have declined significantly. The increasing demand for domestic elderly care has prompted migrant women (including Ukrainians) to move into these areas and accept jobs in domestic care, particularly in caring for elderly and disabled persons.

In Italy, approximately 60% of domestic workers are foreigners, most of whom are employed as caregivers[12] (Pasquinelli and Rusmini, 2015). Moreover, this sector has developed greatly due to limited competition because few Italian women or women from other ethnic communities are willing to accept these jobs (Ambrosini, 2013). In fact, this work is not perceived as being very attractive because it is generally grossly underpaid or offers little legal protection. Generally, there are two types of contract: either for a live-in caregiver or for a caregiver who is paid by the hour. Both contracts have a number of disadvantages for women and may harm their quality of life. This situation is exacerbated, as explained below, for women living in rural areas that lack public and private services and offer few opportunities for social interaction and recreation, particularly with co-nationals.

The contracts often expect 24 hours of daily work because the woman usually cohabits with the care recipient. Therefore, the caregiver's spare time is strictly subordinate to the needs of the patient, who is often the employer as well. The general idea is that these women should be available for the patient around the clock and generally for the duration of the assisted person's life. This is especially the case if the latter is very sick and/or old. In fact, patients are often very old or not self-sufficient, or they suffer from a form of senile dementia. As underlined by Vianello (2009, p. 80), boredom and drudgery are two evident consequences of live-in care work. A day off is the only occasion when caregivers can go out and meet friends. All of these features give rise to a condition that some authors describe as a form of slavery (Andall and Sarti, 2004, p. 7). The difficulties highlighted are amplified in rural areas, where the lack of services makes it more difficult for migrant women to have and maintain a network of personal contacts outside the work context. Moreover, all of these aspects (arduousness of the work, lack of a private life, emotional involvement) generate constant turnover as women seek and find better job opportunities (Ambrosini, 2013). In the case of female caregivers in rural areas, changing jobs is also an opportunity to return to the urban context, where more jobs are available and their ethnic networks are stronger.

Hence, the ethnic network facilitates job seeking (Ambrosini, 2011) but does not necessarily promote social integration and emancipation. Certainly, the social capital accumulated through an extensive and pervasive female network along the migratory route makes it easier to find a job and accommodations; however, this is often a disadvantaged job as a live-in caregiver. In fact, at the beginning of the migratory experience, women seem to experience fewer difficulties than men, although women eventually find themselves trapped in 24-hour jobs. Of course, many other variables influence the emancipation of migrant caregiver women, particularly when their migration project assumes a long-term perspective. Once foreign caregivers have obtained a residence permit, for example, and no longer cohabit with their client, their contract is often switched to a flexible contract with hourly payment. The advantages of the latter are increased personal freedom and time for recreation, with more opportunities for social interaction. This form of employment is also less demanding, both emotionally and practically, but it has a number of drawbacks as well. It is economically less remunerative and requires caregivers to work more hours with several families to earn enough money to live in Italy and send remittances to their families. Furthermore, caregivers are often forced to perform many tasks in a limited number of working hours. In rural areas, another problem is the distance between places and the time spent on travel. From an emotional point of view, however, the women feel often less pressured because they have more free time to invest in their social lives. In the long term, if the individual migratory project evolves into permanent settlement in Italy, Ukrainian women try to improve their working conditions and are eventually employed in formal services as, for example, waitresses or caregivers (Vianello, 2009). These are usually younger women who want to invest in their futures and interact positively with the receiving society. The lack of ties of affection with the country of origin helps transform their migration project from transitional to permanent because they can see a future for themselves in the receiving society rather than in the society of origin.

However, most Ukrainian migration flows to Italy are characterized by the involvement of mature women. In fact, among the Ukrainians residing in Italy in 2014, the percentage of minors was extremely low (9.2% compared with 24.1% among all non-EU residents). Moreover, the number of Ukrainian women between 18 and 39 years of age was less than the average. By contrast, the proportion of people aged 45 years or older was significantly higher. In particular, the 45–49 years age group accounted for 11.7% (the average among non-EU was 8.0%), the 50–54 years age group for 15.2% (versus average of 5.9%) and the 55–59 age group for 12.5% (versus average of 3.8%), whereas the women over 60 years accounted for 11.0%, i.e. more than double the number among all foreigners (4.9%). Mature women leave the Ukraine to support their families, especially when their husbands are unemployed and their children are attending school or university. According to Marchetti and Venturini (2014, p. 111), 'emigration can be an important "investment" not only for young men and women, but also for middle-aged women'. These mature women choose to emigrate to support their children and grandchildren.

The Consequences in the Country of Origin for Migrating Ukrainian Women

Migration paths connect sending and host countries: in these places, migrants make plans, set goals and spend their lives travelling between different countries and cultures. The consequences are visible both in Italy, where those women have 'occupied' specific labour sectors, and in the Ukraine, where the forerunners' relatives (husbands and children in particular) suffer from being left behind. In addition, another connection between the host and the origin country consists of the remittances that represent an economic support for the Ukrainian economy.

In addition to the aforementioned economic disadvantages, the high concentration of Ukrainian women in caregiving activities gives rise to other negative effects: above all, the difficulty of reconciling work and family life (being a wife and mother in particular). Many of the Ukrainian migrant women in Italy may be considered to personify transnationalism, which is understood as a process whereby immigrants build a social environment that connects both the country of origin and that of arrival (Glick Shiller *et al.*, 1992 in Ambrosini, 2007). Whereas migrants used to break with their place of origin, the transnational migrant now tends to maintain intense relationships in both countries. In recent years, this has been supported by new technology (e.g. Skype), although the connection with the country of origin was nurtured in the past as well. An example is the shipment of presents from Italy back home (toys, food and technological devices), which is organized by small groups of migrants. They generally meet in specific places in the city or in the countryside to organize transportation. The transnational aspect of Ukrainian caregivers in Italy can also be connected to the dual role of migrant women: their culture of origin legitimates their migration and abandonment of the family as a means to accumulate money for children and support the family. However, maintaining frequent communication with their country of origin may be more difficult for women living in rural areas because of the lack of transportation and fast Internet services.

What Does Being a Good (Migrant) Mother and Wife Mean?

At this point, it may be useful to ask how transnational migration has influenced the social and personal identity of Ukrainian migrant women. This issue has been analysed from two perspectives: changes in the personal identity of Ukrainian women and the relationships with their children and in their marital life.

As stated, husbands who are left behind often find it difficult to accept the redefinition of roles within the couple. In general, the eldest daughter in Ukrainian families assumes the responsibilities of the mother if the latter becomes the breadwinner and migrates for work. If there is no daughter or if she is not considered old enough, an aunt or grandmother may take over; it is rather exceptional for the father to assume the mother's place. If he does so, this is generally only a temporary solution: when the migrant women return home, they usually resume responsibility for care of the home and family. Most frequently, the organization of maternal care follows the so-called *matrioška* family model (Poskanzer, 1995), which involves three generations of women in the same family: the maternal grandmother, the mother (the migrant woman) and the granddaughter. They support each other through circular substitution. This 'virtuous circle' based on intergenerational solidarity is itself part of the migration project and characterizes these transnational families. Once the migrant woman has returned home, she will take care of the mother and her grown-up daughter is likely to emigrate in turn, rewarding her family (Vianello, 2014).

Furthermore, the men who are left behind (i.e. those who remain in the Ukraine while the wife works abroad) often show an ambivalent attitude towards remittances: the money transfers often become the symbol of the man's loss of power in the family. This frustration may provoke the onset of aggressive behaviour (i.e. alcoholism) and an ambivalent feeling towards money that is considered less valuable because it has not been earned by the man. At worst, the man's sense of economic inferiority may lead to the breakdown of the marriage. The husband may decide that having extramarital affairs with other women is legitimate because they restore order and a 'balance' of social and sexual power in the couple. Immigrant women may develop new relationships, although less so in rural areas because there are generally few opportunities to meet other people compared with the city (Caponio and Colombo, 2011).

The relationship between migrant women and their children generally becomes more complex as well. The migratory event influences women's motherhood: to paraphrase Sayad's (2002) concept of 'double absence' in regard to the difficult changes in personal identity due to migration, for Ukrainian women, it is more appropriate to speak of a type of 'double presence'. Even if they live in Italy, they try to remain vividly present in their children's lives as well with the help of new technologies (Skype).

For a Ukrainian migrant woman, being a good mother means working abroad and earning enough money for her offspring. In fact, remittances represent the material contributions of migrant mothers to the raising of their children: the money sent by the mothers means that their children can live more comfortable lives. The love and affection of mothers is often also expressed by sending gifts: this helps a woman remain the good mother that she used to be. Even if she is far from her children, she tries to maintain emotional proximity with them (Akesson *et al.*, 2012; Vietti *et al.*, 2012). Their education is consciously delegated to other relatives. It thus follows that for migrant mothers, material aid represents the bond with their children, the means by which they can still be part of their children's lives and their educational projects (Boccagni, 2009). Many of these women, however, find it difficult to accept that their children are growing up far away and without them, even if they know that migration allows them to support their children (Solari, 2010). Ukrainians often leave their country when their children are still young: as stated previously, their migration project extends over a rather long period of time. This enables them to gain economic stability over time, but it also creates a great physical and emotional distance during their children's growth. These transnational family relationships are also nurtured by the women's periodic trips back home. Another strategy, which is used especially by Ukrainian women living in rural areas, is to find a compatriot (generally a trusted friend or relative) who can replace them in their job during the period of time spent at home (approximately from 3 to 4 months) (Vianello, 2014). This arrangement prevents them from losing their job in Italy and allows them to check on the situation at home. This mechanism is generally accepted by the Italian families for whom the Ukrainians work as caregivers, provided that a suitable replacement is arranged during the 'regular' caregiver's period of absence. As Morokvasic (2004) observed, this is how Ukrainian women balance their productive and reproductive functions, i.e. by stabilizing their lives between two countries. In this scenario, the concept of maternity should be redefined in light of the longer distances, which require women to adopt new strategies to maintain relationships with their families (especially children). Because of migration, Ukrainian women must find new ways to be good mothers. This condition, termed *transnational motherhood* (Hondagneu-Sotelo and Avila, 1997), has shown how migration can change family roles and how women need to be acknowledged as migrants to establish a new position in their relationships with their children.

For Ukrainian migrant women, the migration is generally a family decision and experience: the family decides who will leave and how to substitute for the person during her absence. According to the new economics theory of migration, it is the outcome of a familial diversification strategy in which everyone has a role in building economic stability (Ambrosini, 2001, 2007).

Conclusions

What emerges from this chapter on the migration of Ukrainian women to Italy can be analysed from different angles. The aforementioned official statistical data have shown the impact that migration flows may have on local areas. Although the number of Ukrainian women living in rural areas is not substantial (approximately 11,200 units), their presence has an impact on the social context in which they live. What makes their presence necessary is certainly the current organization of the ménage of Italian families, which increasingly require extra help in taking care of children, elderly and disabled family members. This is even more the case in the ageing and depopulating rural places that suffer from service decline as a result of public budget cuts and welfare state reforms. In these remote rural areas, Ukrainian migrant women play an invaluable role, particularly as cohabiting caregivers for elderly people. The women themselves, however, experience multiple difficulties related to their often precarious working conditions and the scant opportunities for social integration resulting from the lack of transport and technology for daily life purposes (e.g. Internet, Wi-Fi), which limits the opportunities to develop a solid network of social relationships.

Although the importance of Ukrainian women for Italian rural areas has been analysed above, the attitudes of those women towards their migration experiences are also worth considering. Migration entails a radical change in the lives of migrants, establishing new priorities and changing their relationships both with others and with their culture of origin. For some Ukrainian migrant women, migration is a pathway to more personal freedom and emancipation, which is, in part, also a result of women assuming the breadwinner role. However, this new condition implies a new definition of roles at both the macro and micro levels in society. At the macro level, the society of origin must accept that women are becoming increasingly emancipated. At the micro level, women and men need to redefine their roles in the family context and establish a new balance between productive and reproductive tasks. Many Ukrainian women migrants with children try to organize their absence 'at home' in collaboration with female family members and, in particular, with their mothers and sisters. In doing so, they develop a transnational family model supported through the substitution and circulation of maternal roles.

In conclusion, the analysis reveals the 'utilitarian' profile of Ukrainian caregivers living in Italian rural areas: according to Ambrosini and Cominelli (2004), their migration paths are time-limited because they are focused on returning to the country of origin. For this reason, their migration experiences are strictly linked to their working conditions, whereas the relational and affective aspects connected to their lives in the host country seem to be less relevant. However, the receiving rural society also adopts utilitarian attitudes towards these women because, as explained above, they compensate for the inadequate welfare system. It can therefore be said that there is a reciprocal utilitarian relationship between Italian rural areas and Ukrainian caregivers. This aspect is more visible among Ukrainian women than those of other migrant groups because the push factor in their migration experience is work, which consequently characterizes their integration into the Italian rural territory and society.

Notes

[1]Although both authors were equally involved in this chapter, Donatella Greco wrote the sections on background and context, Ukrainian women in Italy and what being a good (migrant) mother and wife means, while Chiara Zanetti wrote the sections on Ukrainian caregivers' social integration in rural areas, Ukrainian migration in Italy as a development of the international care economy and consequences in the country of origin for migrating Ukrainian women. Both authors wrote the introduction and conclusions.

[2]Although Ukrainian immigration to Italy is mainly for work purposes, due to the economic and political instability in Ukraine over the past two years, the official UN Refugee Agency (UNCHR) data show an increase in applications for international protection

by Ukrainian citizens (men and women). These data reveal that international migration is, in fact, influenced by political, economic and social factors that change over time.

[3]Some scholars, for example Berger and Mohr (1975), have recognized this 'prejudice'. They explain it on the basis of an overall simplification of the object of study (migration); however, for others (e.g. Handlin, 1951), this unbalanced situation never existed.

[4]It should be noted that the percentage of women migrating from developing countries has increased since 1990: in fact, in 1990, the UN statistics reported that the percentage of women migrants from developing countries was approximately 43% (Caritas italiana and Migrantes, 2015).

[5]Female migrants of this type have often been associated with a Muslim background: generally speaking, these women are passively involved in the migration process, although their migration is part of larger family projects.

[6]13th General Population and Housing Census (Istat, 1991).

[7]There were precisely 1,334,889 foreign residents in Italy at the 14th General Population and Housing Census (Istat, 2001).

[8]Their increase is due to both real migration flows and several general national immigration amnesties, which have led to the inclusion of irregular immigrants who were already present in Italy in municipal registers.

[9]Irregular migrants are excluded from these official statistics: this issue is particularly important when discussing women who work in the care system, which includes a large number of irregular migrants. In fact, it has been observed that in 2013, 56.5% of all caregivers were working without all of the legal requirements: 26.0% of them lacked both a residence permit and a regular contract, and a further 30.5% were employed without a contract, although they were resident in Italy with a regular permit (Irs and Soleterre, 2015).

[10]Until 2015, the validity of the Ukrainian driver's licence was not automatically recognized by the Italian Government, and this was a considerable obstacle to autonomy for the caregiver and the care recipient.

[11]Hellermann (2006) states that the destinations of migration flows of Ukrainian men and women differ. Whereas women tend to move to Greece and Italy, Portugal attracts Ukrainian men.

[12]In 2014, INPS (National Social Security Service) counted 898,429 domestic workers, 77.1% of whom were foreigners. Of these, 412,822 (59.6%) came from Eastern Europe: 225,709 (54.7%) were employed as caregivers and 187,103 (45.3%) as housekeepers (INPS, 2015).

References

Akesson, L., Carling, J. and Drotbohm, H. (2012) Mobility, moralities and motherhood: navigating the contingencies of Cape Verdean lives. *Journal of Ethnic and Migration Studies* 38(2), 237–260. Available at: http://dx.doi.org/10.1080/1369183X.2012.646420 (accessed 4 March 2017).

Ambrosini, M. (2001) The role of immigrants in the Italian labour market. *International Migration* 39(3), 61–83. Available at: http://dx.doi.org/10.1111/1468-2435.00156 (accessed 4 March 2017).

Ambrosini, M. (2007) Prospettive transnazionali, un nuovo modo di pensare le migrazioni? *Mondi Migranti* 2/2007, 43–90.

Ambrosini, M. (2011) *Sociologia delle migrazioni*. Il Mulino, Bologna, Italy.

Ambrosini, M. (2013) *Irregular Migration and Invisible Welfare*. Palgrave Macmillan, Basingstoke, UK.

Ambrosini, M. (2016) From 'illegality' to tolerance and beyond: irregular immigration as a selective and dynamic process. *International Migration* 54(2), 144–159. Available at: http://dx.doi.org/10.1111/imig.12214 (accessed 4 March 2017).

Ambrosini, M. and Cominelli, C. (eds) (2004) *Educare al futuro. Il contributo dei luoghi educativi extrascolastici nel territorio lombardo*. Osservatorio Regionale sull'integrazione e la multietnicità – Fondazione Ismu, Milan, Italy.

Andall, J. and Sarti, R. (2004) Le trasformazioni del servizio domestico in Italia: un'introduzione. *Polis* 28(1), 5–16. Available at: http://dx.doi.org/10.1424/12894 (accessed 4 March 2017).

Azienda per i Servizi Sanitari n. 6 (2009) Aver cura. Rapporto di ricerca. Available at: http://goo.gl/pg6tBM (accessed 2 December 2015).

Balbo, M. (2015) *Migrazioni e piccoli comuni*. Franco Angeli, Milan, Italy.

Berger, J. and Mohr, J. (1975) *A Seventh Man: The Story of a Migrant Worker in Europe*. Penguin Books, Harmondsworth, UK.

Bianchini, S. and Privitera, F. (2005) *Guida ai Paesi dell'Europa centro-orientale e balcanica. Annuario politico economico 2004*. Il Mulino, Bologna, Italy.

Blangiardo, G.C. (ed.) (2005) *L'immigrazione straniera in Lombardia. La quarta indagine regionale. Rapporto 2004.* Fondazione Ismu, Regione Lombardia, Osservatorio Regionale per l'integrazione e la multietnicità, Milan, Italy.

Boccagni, P. (2009) Come fare le madri da lontano? Percorsi, aspettative e pratiche della maternità transnazionale dall'Italia. *Mondi Migranti* 1/2009, 45–67. Available at: https://www.francoangeli.it/rivisite/Scheda_rivista.aspx?IDArticolo=36390 (accessed 6 July 2017).

Bock, B. and Shortall, S. (eds) (2006) *Rural Gender Relations: Issues and Case Studies.* CAB International, Wallingford, UK.

Bock, B., Osti, G. and Ventura, F. (2016) Rural migration and new patterns of exclusion and integration in Europe. In: Shucksmith, M. and Brown, D.L. (eds) *Routledge International Handbook of Rural Studies.* Routledge, New York.

Caponio, T. and Colombo, A. (2011) Migrazione, separazione coniugale e ruoli di genere. Il caso delle lavoratrici domestiche in Italia. *Polis* 25(3), 419–448.

Caritas italiana and Migrantes (ed.) (2015) *Migranti, attori di sviluppo: XXIV Rapporto Immigrazione Caritas e Migrantes 2014.* Tau Editrice SrL, Todi, Italy.

Carrosio, G. (2012) Un'analisi ecologica della presenza immigrata nell'Italia minore. In: Osti, G. and Ventura, F. (eds) *Vivere da stranieri in aree fragili. L'immigrazione internazionale nei comuni rurali italiani.* Liguori, Naples, Italy, pp. 9–22.

Castels, S. and Miller, M.J. (2012) *L'era delle migrazioni. Popoli in movimento nel mondo contemporaneo.* Odoya, Bologna, Italy.

Da Roit, B. and Facchini, C. (2010) *Anziani e badanti. Le differenti condizioni di chi è accudito e di chi accudisce.* Franco Angeli, Milan, Italy.

DeLaet, D. (1999) Introduction: the invisibility of women in scholarship on international migration. In: Kelson, G. and DeLaet, D. *Gender and Immigration.* New York University Press, New York, pp. 1–17.

Donato, K. and Donna, G. (2015) *Gender and International Migration.* Russell Sage Foundation, New York.

Farris, S. (2010) Interregional migration: the challenge for gender and development. *Development* 53(1), 98–104. Available at: http://dx.doi.org/10.1057/dev.2009.89 (accessed 4 March 2017).

Favaro, G. and Colombo, T. (1993) *I bambini della nostalgia.* Arnolodo Mondadori, Milan, Italy.

Favaro, G. and Tognetti Bordogna, M. (1991) *Donne dal mondo. Strategie migratorie al femminile.* Guerini Associati, Milan, Italy.

Grigg, D.B. (1977) E.G. Ravenstein and the 'laws of migration'. *Journal of Historical Geography* 3(1), 41–54.

Handlin, O. (1951) *The Uprooted: The Epic Story of the Great Migration that Made the American People.* Little, Brown and Company, Boston, Massachusetts.

Hellermann, C. (2006) Migrant alone: tackling social capital? Women from Eastern Europe in Portugal. *Ethnic and Racial Studies* 29(6), 1135–1152. Available at: http://dx.doi.org/10.1080/01419870600960396 (accessed 4 March 2017).

Hondagneu-Sotelo, P. and Avila, E. (1997) I'm here, but I'm not there. The meanings of Latina transnational motherhood. *Gender and Society* 11(5), 548–571.

INPS (2015) Data warehouse. Available at: http://www.inps.it (accessed 5 December 2015).

Irs and Soleterre (2015) Lavoro domestico e di cura. Pratiche e benchmarking per l'integrazione e la conciliazione della vita familiare e lavorativa. Available at: http://goo.gl/YdihES (accessed 2 December 2015).

Istat (1991) 13th General Population and Housing Census. Data warehouse. Available at: http://dati.istat.it/ (accessed 21 December 2015).

Istat (2001) 14th General Population and Housing Census. Data warehouse. Available at: http://dati.istat.it/ (accessed 21 December 2015).

Istat (2016) Data warehouse. Available at: http://dati.istat.it/ (accessed 20 October 2015).

Kaufmann, V. (2005) Mobilités et réversebilités: vers des sociétés plus fluides? *Cahiers internationaux de sociologie* 118(1), 119–135.

Kofman, E., Phizacklea, A., Raghuram, P. and Sales, R. (2000) *Gender and International Migration in Europe.* Routledge, London.

Lee, E. (1966) A theory of migration. *Demography* 3(1), 47–57.

Lucatelli, S., Peta, E.A. and Luisi, D. (2012) Immigrazione come risorsa: welfare, accesso e integrazione nelle aree interne della Calabria. In: Osti, G. and Ventura, F. (eds) *Vivere da stranieri in aree fragili. L'immigrazione internazionale nei comuni rurali italiani.* Liguori, Naples, Italy, pp. 119–135.

Magnani, N. (2012) L'immigrazione internazionale nelle valli alpine tra cambiamenti demografici, trasformazioni del turismo e conflitto centro-periferia: il caso della Val Rendena, provincia di Trento. In: Osti, G. and Ventura, F. (eds) *Vivere da stranieri in aree fragili. L'immigrazione internazionale nei comuni rurali italiani.* Liguori, Naples, Italy, pp. 169–187.

Marchetti, S. and Venturini, A. (2014) Mothers and grandmothers on the move: labor mobility and the household strategies of Moldovan and Ukrainian migrant women in Italy. *International Migration* 52(5), 111–126. Available at: http://dx.doi.org/10.1111/imig.12131 (accessed 4 March 2017).

Mazzacurati, C. (2005) Dal blat alla vendita del lavoro. Come sono cambiate colf e bandati ucraine a Padova. In: Caponio, T. and Colombo, A. (eds) *Migrazioni globali e integrazioni locali.* Il Mulino, Bologna, Italy, pp. 145–174.

Momsen, J.H. (ed.) (1999) *Gender, Migration and Domestic Service.* Routledge, Abingdon, UK.

Morawaska, E. (2005) Immigrati di ieri e di oggi in Europa e fuori: insediamento e integrazione. In: Caponio, T. and Colombo, A. (eds) *Migrazioni globali, integrazioni locali.* Il Mulino, Bologna, Italy, pp. 145–174.

Morokvasic, M. (1984) Birds of passage are also women. *International Migration Review* 28(4), 886–907. Available at: http://dx.doi.org/10.2307/2546066 (accessed 4 March 2017).

Morokvasic, M. (2004) Settled in mobility: engendering post-wall migration in Europe. *Feminist Review* 77(1), 7–25. Available at: http://dx.doi.org/10.1057/palgrave.fr.9400154 (accessed 4 March 2017).

Osti, G. (2010) Fenomeni migratori nelle campagne italiane. *Agriregionieuropa* 6(22). Available at: http://goo.gl/pgq24U (accessed 2 December 2015).

Otelli, E. (2011) Donne migranti ucraine e luoghi di aggregazione fra connazionali. *Antrocom Online Journal of Anthropology* 7(1), 53–61.

Pasquinelli. S. and Rusmini, G. (2015) I lavoratori stranieri nel lavoro domestico e di cura. In: Centro Studi e Ricerche Idos (eds) *Dossier Statistico Immigrazione 2015.* Idos edizioni, Rome, pp. 287–292.

Pessar, P.R. and Mahler, S.J. (2003) Transnational migration: bringing gender. *International Migration Review* 37(3), 812–846. Available at: http://dx.doi.org/10.1111/j.1747-7379.2003.tb00159.x (accessed 4 March 2017).

Pollini, G. and Scidà, S. (2002) *Sociologia delle migrazioni e della società multietnica.* Franco Angeli, Milan, Italy.

Portes, A. and Sensenbrenner, J. (1993) Embeddedness and immigration: notes on the social determinants of economic action. *American Journal of Sociology* 98(6), 1320–1350.

Poskanzer, A. (1995) The matryoshka: the three-generation soviet family in Israel. *Contemporary Family Therapy* 17(4), 413–427. Available at: http://dx.doi.org/10.1007/BF02249354 (accessed 4 March 2017).

Sassen, S. (2000) Women's burden: counter-geographies of globalization and the feminization of survival. *Journal of International Affairs* 53(2), 502–524.

Sayad, A. (2002) *La doppia assenza. Dalle illusioni dell'emigrato alle sofferenze dell'immigrato.* Raffaello Cortina Editore, Milan, Italy.

Sciortino, G. (2004) Immigration in a Mediterranean welfare state: the Italian experience in comparative perspective. *Journal of Comparative Policy Analysis* 6(2), 111–130.

Solari, C. (2010) Resource drain vs. constitutive circularity: comparing the gendered effects of post-Soviet migration patterns in Ukraine. *Anthropology of East Europe Review* 28(1), 215–238.

Subtelny, O. (2003) *Ukraine. A History.* University of Toronto Press, Toronto, Canada.

Tam, V.C.W. (1999) Foreign domestic helpers in Hong Kong and their role in childcare provision. In: Momsen, J.H. (ed.) *Gender, Migration and Domestic Service.* Routledge, Abingdon, UK, pp. 263–276.

Urry, J. (2007) *Mobilities.* Polity Press, Cambridge, UK.

Vianello, A.F. (2009) *Migrando sole. Legami transnazionali tra Ucraina e Italia.* Franco Angeli, Milan, Italy.

Vianello, A.F. (2014) *Genere e migrazioni. Prospettive di studio e di ricerca.* Guerini e Associati, Milan, Italy.

Vietti, F., Portis, L. and Ferrero, L. (2012) *Il paese delle badanti.* SEI, Turin, Italy.

White, M.J. (eds) (2016) *International Handbook of Migration and Population Distribution.* Springer Netherlands Publishing, Dordrecht, The Netherlands.

4 Gender, Migration and Rural Livelihoods in Uzbekistan in Times of Change

N. Nizamedinkhodjayeva[1,*], B.B. Bock[2,3] and P. Mollinga[4]

[1]International Water Management Institute, Colombo, Sri Lanka; [2]Wageningen University, Wageningen, The Netherlands; [3]Groningen University, Groningen, The Netherlands; [4]SOAS University of London, London, UK

Introduction

This chapter explores how rural gender relations have changed during the transition process that Uzbekistan entered after its independence in 1991, in the aftermath of the dissolution of the Soviet Union. It contributes to the growing body of work studying the gender-specific consequences of the fundamental socio-economic and political changes that accompanied the initial years of post-socialism. Earlier research has demonstrated that the closure of collective farms, state-owned industries and public services produced high unemployment figures, especially in rural areas where there was almost no alternative employment (Spoor, 2013). Rural women's employment opportunities were also threatened by the termination and/or privatization of social services such as childcare. As a result, women's economic independence was endangered (Gal and Kligman, 2012).

Little research about rural Uzbekistan has been published so far. Uzbekistan presents a particularly interesting case because its transition towards post-socialism is still 'in the making'. Even today, Uzbekistan's means of production – including land – remains state-owned. Furthermore, the state continues to almost completely manage agricultural production by defining production targets and controlling the market for most products. At the same time, the state transfers production rights to quasi-owners to boost production (Veldwisch and Bock, 2011). This results in novel processes of social differentiation and social inequality in rural areas (Veldwisch and Spoor, 2008). Gender inequality also results from the revival of traditional culture and religion, undermining some of the gains in gender equality produced by the socialist system – for example through education. Yet there are also signs of liberalization and growing independence among young women and men who – importantly – support rural households with incomes gained through migratory labour. This chapter explores this process in more detail.

The chapter is based on research we conducted in Khorezm, a rural province in the north-west of Uzbekistan. The fieldwork took place between 2008 and 2010 and explored, among other topics, the effects of migration on the organization and functioning of rural households. Since independence, more and more households in Khorezm have come to depend on seasonal migration as their primary or sole source of

* Corresponding author e-mail: nargiza00@hotmail.com

income. This has important effects on gender and generational relations and might undermine hitherto patriarchal household structures.

The chapter comprises six sections. The following section provides background information about Khorezm and reviews the existing literature about changing gender relations in post-socialist Uzbekistan. Then the theoretical framework that guided the research is introduced, as well as the research methodology. Next, the research findings are presented and discussed. The chapter closes with conclusions about the role of migration in the transformation of gender relations in rural Khorezm.

Background and Context: Transition and Gender Relations in Rural Khorezm

Khorezm province is located in the northwest of Uzbekistan on the lower reaches of the Amu Darya River. It borders the deserts of Turkmenistan, the autonomous republic of Karakalpakstan and the Bukhoro province of Uzbekistan. The weather conditions in the province are harsh: summers are extremely hot and winters are extremely cold, dry and windy. At times, the temperature can fall below −15°C in winter and surpass +40°C in summer (Conliffe, 2009; Fig. 4.1).

The province is home to approximately 1.5 million people. Approximately 80% of the Khorezm population is rural. Administratively, the province (*viloyat*) includes ten districts (*tumanlar*). The districts comprise villages (*qishloq*) and a district administrative capital. Each district capital is subordinate to the provincial capital of Khorezm, which is the city of Urganch.

Earlier research has described gender relations in Uzbekistan as patriarchal and, in general, patriarchal relations were indeed predominant in the Central Asian region before, during and after Uzbekistan was part of the Soviet Union (Kandiyoti, 2002; Kamp, 2005, 2006). The majority of the population is Muslim. The Soviet system improved women's education levels and increased their engagement in productive activities (Kamp, 2006), but it did not terminate the gender-based division of labour or women's

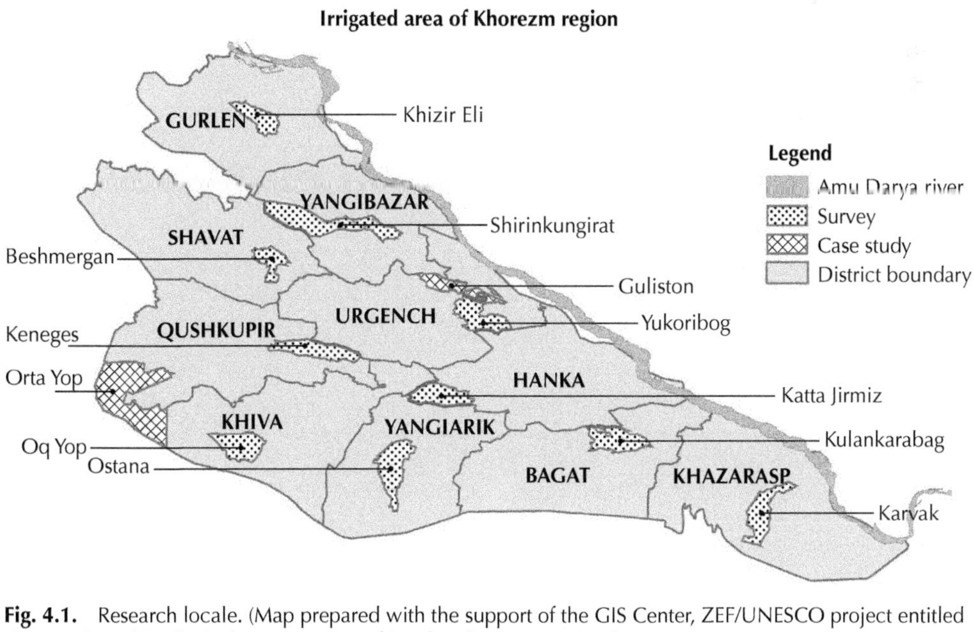

Fig. 4.1. Research locale. (Map prepared with the support of the GIS Center, ZEF/UNESCO project entitled 'Economic and Ecological Restructuring of Land and Water Use in Khorezm', 2009.)

economic dependence on men (Lazreg, 1999; Mee, 2001; Ruminska-Zimny, 2002; Kamp, 2005). Men had always been appointed as breadwinners and were responsible for physically demanding seasonal tasks at home (such as buying stocks for winter and performing construction and home repairs). Daily domestic work, childcare and subsistence agriculture were seen as the key responsibilities of women, regardless of their employment status (Ruminska-Zimny, 2002; Kamp, 2006). Women were generally occupied in so-called 'female jobs', including health, educational or social services (Pomfret and Anderson, 1997). According to Mee (2001) and Ruminska-Zimny (2002), they showed little interest in professional careers because of the double burden this would imply in terms of their productive and reproductive activities.

The post-Soviet transition made it very difficult for both men and women in Uzbekistan to secure their livelihoods (Ruminska-Zimny, 1997; Anderson and Pomfret, 2002; Kamp, 2004; IMF, 2008). This was especially the case in rural Khorezm – the focus of the present research – because of limited access to arable land and irrigation water and the paucity of non-agricultural income-generating opportunities (Kandiyoti, 2003; Müller, 2006). According to Kandiyoti (2003), poverty and unemployment in rural areas increased considerably as a result of the de-collectivization of agriculture, the closure of *kolkhozes* (Soviet collective farms) and the land reforms that went along with these changes. Land was not privatized and remained state property. It was, however, reallocated among individual farm managers, so-called *fermers*. They were supposed to organize agricultural production individually and in a more entrepreneurial manner than had been the case under the state-owned *kolkhozes*, even though state control on production and marketing remained very high. *Fermers* were, for instance, obliged to deliver certain quantities of products to the state for fixed prices. Therefore, they were not free to produce or sell whatever they wanted but had considerable freedom in their choice of labourers and how much they paid them. The de-collectivization of agriculture resulted not only in considerable unemployment among former *kolkhoz* labourers but also – with the subsequent land reform – in the concentration of access to land in the hands of a few. Eventually, 70% of the land in Khorezm was managed by approximately 5% of the rural population (Veldwisch, 2008; Veldwisch and Bock, 2011). The selection of *fermers* proceeded in a very unclear fashion, with nepotism and political favouritism playing an important role. Moreover, it clearly disadvantaged women: in 2010, only 14% of all *fermers* in Khorezm were women (Nizamedinkhodjayeva, 2013).

Those rural residents who did not become *fermers* would still have access to part of the arable land, as 30% of the land was distributed among rural households for private use, mainly for subsistence food production (Veldwisch, 2008). As so-called *dehqon* farmers, they had the right to two parcels of land: 0.12 ha for food production (called *qo'shimcha tomorka*) and 0.06–0.12 ha where they were allowed to build a house (so-called *uy tomorka*, or kitchen gardens). The quality of the land differed substantially and was defined to a large extent by the availability and accessibility of water for irrigation. The years of our fieldwork were very dry years, which made access to irrigation an enormous problem for all. The *tomorkas* were generally assigned to men as the supposed heads of households. Women had the right to use the land based on family/marriage relations (Nizamedinkhodjayeva, 2013).

The majority of rural households were home to extended families with three generations living together (Nizamedinkhodjayeva, 2013). Living in extended family households offered opportunities for mutual support but also implied submission to a patriarchal hierarchy that determined different rights, roles and responsibilities for every household member. As this chapter demonstrates, ownership of and access to assets, decision-making power and engagement in productive and reproductive activities differ by gender, generation and according to a person's position within the hierarchy of the household and the local community. Within this hierarchy, the youngest daughter-in-law comes last and the oldest man comes first.

In general, even extended families were unable to make a living using only the land and irrigation water they had at their disposal. Working for a *fermer*, however, was not sufficient either, as it was seasonal labour and generally poorly paid. Most households combined multiple income-generating activities, engaging in agricultural (on-farm), agriculture-related (off-farm) and non-agricultural (off-farm) activities (categorization by Ellis, 1998). Income diversification was an important livelihood strategy for two-thirds of rural households in Khorezm (Nizamedinkhodjayeva, 2013). However, non-agricultural employment was difficult to obtain and often poorly paid. Many households therefore turned to seasonal migration as an opportunity to generate cash income. In almost half of the surveyed households, one or more men were engaged in seasonal migration in 2009. Women would occasionally migrate as well, but generally not for long periods or whole seasons; they would travel to urban centres together with other women to sell products and return within a few days. It was the outmigration of (young) men that, importantly, affected the demography of the region as well as gender and generational relations, as demonstrated in the following sections.

Analytical Framework: Gender Relations and Change

Research tends to explain changing gender relations in different ways, emphasizing either structural or cultural processes. Duncan and Smith (2002) and others (Mandel, 2004; Korf and Oughton, 2006) stress the importance of cultural norms in reproducing existing gender relations and obstructing change. Reeves and Baden (2000) emphasize the importance of institutions, particularly the organization of decision making, whereas Breen and Cooke (2005) argue that it is the division of labour that matters most for changing or producing gender relations. Other researchers claim that unequal access to assets, technology, infrastructure and economic opportunities perpetuates gender inequalities (Ellis, 2000; Mukhopadhyay, 2008), including access to time- and labour-saving technologies, as well as basic service infrastructure such as childcare, public transportation and other public services (van Doorne-Huiskes and van Hoof, 1995). Over time, the emphasis placed on structural versus cultural or material versus immaterial factors has changed. Whereas structural factors and access to resources dominated the discussions of the 1970s and 1980s, immaterial and cultural factors gained prominence with the cultural turn (Jackson, 2001).

More recently, it has been acknowledged that all these factors interact and constitute each other. 'Norms do not float free: they are materialized in specific domains of social life and are often embedded in social institutions' (Pearse and Connell, 2016, p. 30). At the same time, norms change with changing structures and institutions, and with changing behaviour. Paraphrasing Butler, 'gender is done' – it is repeatedly constituted, renewed and changed through social interaction (Butler, 1988). An important question concerns the direction of change and its gendered effects. In feminist research, it is change towards women's empowerment that is desired – change that changes gender relations in such a way that women become less dependent on men and gain more autonomy, more say and more access to important assets. This should accompany changing gender norms. One prominent discussion has always concerned the extent to which empowerment requires substantial structural change.

In a recent publication, Andrea Cornwall (2016) reflects on how gender approaches were developed in international development research and how the focus on empowerment and political change prominent in the 1990s switched to more material and instrumental perspectives, as seen in currently dominant frameworks. The empowerment approach accounted for the importance of control over material resources; yet, at the same time, it underlined the need for 'control over ideology (beliefs, values and attitude)' (Sen, 1997, cited in Cornwall, 2016, p. 344). Cornwall stresses that this approach to transforming gender relations is becoming instrumentalized in the sense that improving women's access to the means of production is being recognized as important, as it may solve societal problems such as food insecurity. This is also presented as 'smart economics' (Chant and Sweetman, 2012; see Jacobs, Chapter 23, this volume). Similarly, the initial focus on empowerment as a change in relations and a collective endeavour is turned into empowerment as an individual achievement. In her view, 'the overemphasis on the material at the cost of the cultural' importantly limits the effect of gender projects (Cornwall, 2016, p. 354). From her perspective, it is critical to change the way women and men think, to 'engage critical consciousness, question taken for granted norms and make, in doing so, a vital contribution to shifting power relations' (Cornwall, 2016, p. 356). Others confirm the inability to separate material and immaterial change and argue that access to resources does not suffice in the absence of transformative change that explicates the current dimensions of 'injustice, (mal)-distribution and (mis)recognition' (Fraser, 2009, p. 107).

Connell and Pearse (2015) use the concept of a gender regime or gender order to indicate how the material and immaterial and structural and cultural interact in the institutionalization of gender relations at the level of organizations or societies and constitute relatively stable practices. Changing the gender order or gender regime opens up new trajectories and enables new practices. At the same time, new practices and material gender arrangements produce change at the level of the order or regime. Following Connell and Pearse (2015, p. 87), crisis tendencies play an important role because they 'undermine current patterns, and force change in the structure itself'.

Overall, research argues that it is difficult if not impossible to change gender relations unless one substantially changes the material and immaterial basis of a society, including its dominant norms and values, power-holding institutions and economic structure. Turning this argument around, however, one may also expect that once socio-economic and political institutions change substantially, changing gender relations comes within reach as well. The transition context is then a particularly interesting one, as it did indeed result in fundamental changes in the social, economic and political system and is a clear example of a crisis that produced change.

With independence, Uzbekistan entered a period of fundamental transition, with structural and cultural change occurring on several levels at the same time. The political and economic system changed when the country became independent and previously dominant socialist norms and values were questioned. Russia remained a friendly nation and the political system of Uzbekistan still contains 'socialist' characteristics, including a strong central government and centrally planned economy. At the same time, the economy was partly liberalized and partly privatized through the dismantling of state-owned firms. This change in socio-economic structures, as well as norms and values (and hence culture), also affected the rural households in Khorezm. Culture contains, among other characteristics, gendered norms of behaviour, such as those reflected in the gender-specific division of labour. Economic structure, spatial infrastructure and the provision of services define the job opportunities offered to women. In the case of Khorezm, it is not only the availability of childcare facilities that is important for supporting women's ability to combine family and paid work. Here, essential public services such as running water, electricity and gas often fail, which turns household chores into a labour-intensive, full-time business for some household members. Altogether, this situation encourages the reproduction of traditional gender relations, in which women depend on men economically, have little decision-making power and are subordinate to men in a way that is highly valued.

This situation may change in the case of male migration, when women need to replace men and take over jobs or make decisions in their absence. The impact of male migration on gender relations is, however, uncertain and ambivalent, and research has demonstrated that it may result in empowerment as well as in the reproduction of subordination (Palmary *et al.*, 2010).

The conceptual framework (see Fig. 4.2) presents the four different domains that we expect, in their interaction, to play a dominant role in the process of change: division of labour, access to assets, role in decision making and gender-specific norms and values. Change in one of the domains is expected to trigger change in the next and vice versa, all of which sets the gender regime in motion, transforming and reproducing gender relations. Migration, for instance, changes the division of labour, which in turn stimulates change in other domains. The presentation of results follows the structure of the framework, covering the different gendered elements in motion: division of labour, access to assets, decision-making power and gendered norms and values.

Methodology

The research methodology was designed using a multi-case study approach (Yin, 2002), with case studies at the local and household levels embedded in a regional case. Such an embedded case study approach incorporates units of analysis that are situated at different levels to look for consistent patterns of evidence across these units (Yin, 1984). This is particularly useful for studying the process of change taking place in rural Khorezm, as it enables capture of the change taking place in society at large, its effects on social relations and the dynamics of rural livelihoods in response to these changes. Applying a case study methodology is suitable because it provides insight into real-life situations and is especially useful for capturing the complexity of social phenomena (Marshall and Rossman, 1999; Bryman, 2004).

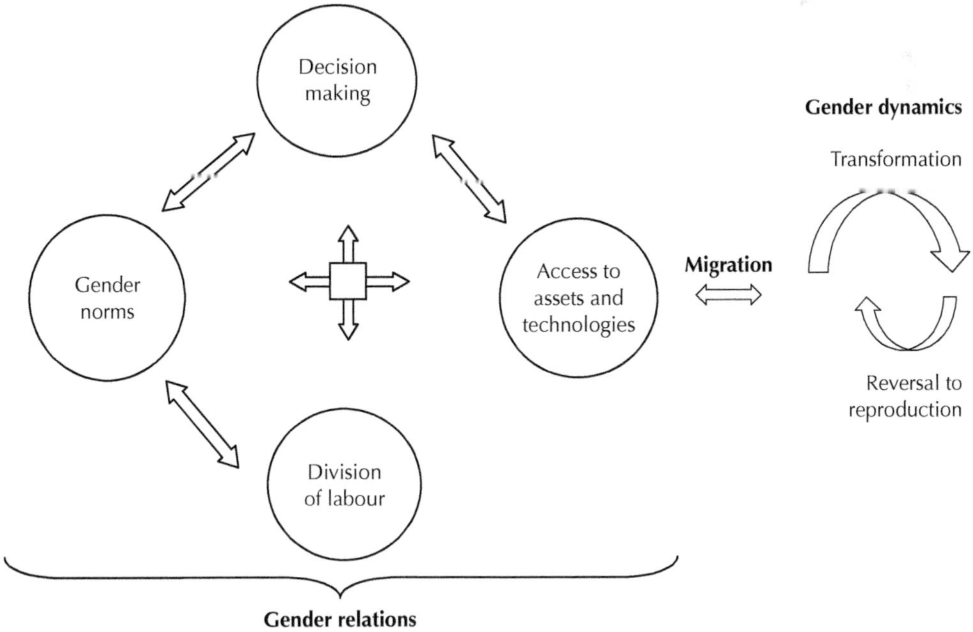

Fig. 4.2. Migration and change in gender relations. (Adapted from Nizamedinkhodjayeva, 2013.)

Within the case study multiple qualitative and quantitative data collection methods were combined, which allowed for 'triangulation, facilitation and complementarity' (Bryman, 2004). Triangulation enables cross-checking of findings based on data gathered in different ways, which strengthens the reliability and validity of the findings. Facilitation refers to the use of qualitative data to explain the results of quantitative analysis. Finally, qualitative data complement the understanding of phenomena that are difficult to capture with quantitative data and vice versa.

Figure 4.3 provides an overview of all the methods employed for each level of analysis: the community level, the household level and the level of the individual respondent. Methods such as statistical analysis of data, transect walks and mapping with a Geographic Information System (GIS) helped obtain a preliminary understanding of the organization of the community and levels of social differentiation. Case studies at the household level included (participant and non-participant) observation, formal and informal interviews, and open and semi-structured interviews with several household members, which allowed the functioning of households to be understood in terms of division of labour, decision-making power and dominant norms and values. Individual discussions, in combination with participant observation, provided insight into individual experiences and reasoning. Individual responses to surveys of smaller and larger scope allowed an understanding to be obtained of the frequency of behaviours and/or expectations, such as, for example, migration. Our case study of changing gender relations in Khorezm included four longitudinal case studies of rural households, which were followed and visited repeatedly from 2008 to 2010. In addition, household members were asked to keep track of their activities and experiences in research diaries. Moreover, so-called mini-surveys were organized about gender-specific norms and the division of labour ($n=42$) in 2008, about migration in 2008 and 2009 ($n=24$, 2008 and $n=30$, 2009), and a large survey of 300 respondents in 2009 about access to assets, activity patterns, division of labour and gendered roles and norms within the household as well as levels of livelihood security experiences throughout the years of transition. The survey data were analysed using the statistical software package IBM SPSS Statistics Version 22, and the transcripts of the open and semi-structured interviews were analysed with the help of the qualitative data analysis software ATLAS.ti7, version 7.0.71.

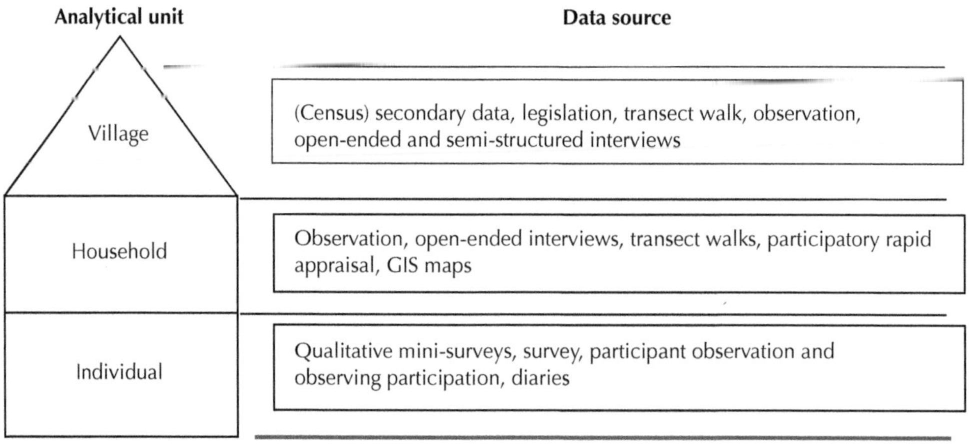

Fig. 4.3. Analytical units and data sources (GIS, geographic information system). (Adapted from Nizamedinkhodjayeva, 2013.)

The fieldwork took place between April 2008 and November 2010. It included three field visits as well as the continuous engagement of research assistants who collected data in the four case study households during the absence of the primary researcher in order to capture the seasonality of rural livelihoods. This was in addition to other data achieved through the diaries mentioned above.

Research Findings

Migration and its contribution to rural livelihoods in Khorezm

In approximately one-half of the rural households (44%) participating in our research, at least one person engaged in seasonal migration. Most of them were adult men, yet the migrants also included boys who were 16 years old. Only 6% of all migrant labourers were women. Women generally did not migrate alone. Those who migrated usually joined their husbands, leaving their children behind with their parents-in-law. Other women (generally older daughters-in-law) migrated together with their mothers-in-law, leaving the younger daughters-in-law to take care of the home and children. At migration destinations, these women usually worked as cleaners, dishwashers, salespersons or farm workers.

In general, women with small children did not migrate. Likewise, fathers generally did not travel outside Uzbekistan and did not leave home for long periods of time when there were infants at home:

> To earn, I find a job in other provinces of Uzbekistan. My child is less than 1-year old. I need to be at home at least once a month to do male jobs in the house. If I migrate outside Uzbekistan, it is impossible to be home every month.
>
> (Man 301-1; case study household, 2009)

Most migrants worked as unskilled labourers in construction (83%), agriculture, industry, trade or services (13%). Only 4% of migrants found jobs as skilled workers. The latter were generally men with 10 or more years of experience as migrants.

The duration of migration varied depending on the migrant's destination. Migration within Uzbekistan was generally short term and varied between 3 and 6 months. International migration generally lasted 5 months or longer. Most migrants migrated on a temporary basis; very few left their homes for longer than 4 years. There were two migration seasons per year. In the spring, men generally migrated outside Uzbekistan. At that time of year, the livelihood situation was often most desperate: winter stocks were depleted, spring fruits and vegetables were available only at the beginning of May, and very few job opportunities existed at home. Those who left in the second season, in May–June, generally engaged in short-term and short-distance migration within Uzbekistan. Both migration patterns were linked to the harvesting season in the place of origin. Most migrants returned home by September–October to prepare for the cold winters. Most weddings took place during these months as well.

Next to the migrants' skills, the destination and duration of migration also mattered for the remittances sent home. Those who worked outside Uzbekistan could earn about US$800–1500 per month whereas migrants within Uzbekistan could earn US$80–300 per month. Russia was the most common destination for migration (63% of migrants), followed by Kazakhstan (18%) and Tashkent city (17%). A third of migrants managed to send remittances home every month. Another third sent remittances every 2–4 months and on a regular basis. Approximately 42% of the migrants' households, however, could not rely on regular remittances and received remittances only occasionally, if at all. This was problematic because many households depended on remittances for their livelihoods. For about two-thirds of the households with migrants, these remittances contributed at least half of the household budget.

The costs of living were very high in Khorezm. Surviving in this region – where temperatures can drop to 20°C below zero in winter (e.g. winter 2008/09) and reach

50°C above zero in summer (e.g. summer 2008) – requires special preparations, such as the construction of thermostatic housing, accumulation of savings, preparation of food and wood stocks, and proper clothing. Many men decided to migrate because local incomes were generally insufficient for meeting these needs:

> It is very difficult to find a job with a reasonable income locally. Therefore, migration has considerably increased over the last 10 years. Soon everybody will leave for migration. No land, no factories, no employment. What can we do?
> (Man 328; mini-survey on migration, 2008)

> The money that can be earned here is not enough. If my husband works here, he earns too little money [in agriculture]. When he works outside the village, he sends money home and we do his work on the land.
> (Woman 412; open-ended interview, 2008)

Migration was generally seen as a panacea for many economic difficulties. Remittances could allow households to accumulate savings; pay for medical treatment; invest in agriculture (e.g. rent land and purchase livestock, fertilizers and even tractors); invest in housing and education; and acquire various assets (mini-survey on migration, 2008).

The first priority was generally to invest in housing (approximately 91% of the households with migrants), followed by investment in social obligations such as weddings, *beshik to'y* (celebrations of childbirth), *sunnat to'y* (celebrations of boys' circumcision) and the repayment of debts (approximately 81% and 4% of the households with migrants, respectively). About three-quarters of the households with migrants used their remittances to ensure food security. A quarter used remittances to cover the costs of medical treatment. And almost a quarter of the households invested their accumulated remittances in purchasing a car to generate income as taxi or lorry drivers; generally, owning a car was perceived as an indicator of success and upward social mobility.

Remittances were, therefore, a very important source of income in Khorezm. During the reporting period, migration was a very important means of securing the household's livelihood for almost half of households. For the other half, migration was not an option. Some could not rely on migration because there was only one man (or no man) in the household; in small households with many small children, there was a great demand for a man's presence and daily support. Others could not find jobs as migrants. Others decided against migration because they feared that it would harm their children and family life or because they wanted to avoid the many risks that go along with migration. Migrants generally worked under difficult conditions that were exacerbated by being unregistered and unprotected by legislation abroad.

Changing division of labour

Local Khorezmians distinguished between two types of households: *odatiy* (traditional) and *zamonaviy* (modern) households. Traditional households were patriarchal, with rights and responsibilities of household members depending on gender, age and household position. In the first type of household there was a clear division between female and male domains or spaces, with women generally responsible for child-rearing and domestic affairs and men responsible for production and all external affairs. Generation played a role as well, with the older generation having more decision making rights compared with the younger generation. In modern households, the division of labour and responsibilities was not as strictly organized. Here men and women could either work and make decisions together or take over each other's responsibilities in cases of absence. This occurred regularly in the case of migration.

Carrying out domestic activities required considerable effort and time in the context of Khorezm because of the poor infrastructure. For example, there was no running water in homes and bringing water from a well meant thawing the well in winter. Laundry was done by hand and generally in chilled water. Cooking and heating required collecting firewood. During the reporting period, the majority (162 out of 300 sample households; 54%) did not have regular access to gas and/or electricity. To reduce food costs, women processed and preserved vegetables and fruit for winter. In addition, there were no kindergartens with good conditions or food for children. The fulfilment of domestic tasks, therefore, was very time-consuming (requiring at least 12 h/day) and labour-intensive. Many men worked hard and in difficult conditions as well, but their workday usually ended when they came home. For (especially younger) women, the workday generally continued late into the night, with many household chores to finish (Table 4.1).

Women were also most engaged in agricultural activities, especially in production. Women's and men's efforts in agriculture differed in the duration, regularity and intensity of work. Women carried out most of the manual work and did so on a more regular basis compared with men (e.g., weeding, mulching). In general, women's work was unregistered and poorly paid (Table 4.2).

In agriculture, men generally performed the managerial or mechanized and thus relatively well-paid work (e.g., tractor driving), including seasonal labour that was physically demanding (e.g. fertilization) (Table 4.3).

When men migrated, women's responsibilities in agriculture increased. In the absence of men, women began to engage in livestock care and irrigation. These activities were previously perceived as typical male labour, not suitable for women:

> Before men started to migrate, heavy jobs were done by men, for example, irrigation of fields, transportation of goods. Now, women have to carry out all [agricultural tasks] when men are away.
>
> (Man 5; sample survey, 2009)

> It is very difficult and bad to be without men. I have to do all male work myself: I work on one hectare for a *fermer* and on *uy tomorka*. In addition, I do all the work at home.
>
> (Woman 324; mini-survey on migration, 2009)

The local discourses about 'what men should do' also started to change. For example, male migrants started to refer to their

Table 4.1. 'Female' household activities, Khorezm province, Uzbekistan, 2008–2010. (Data from self-completed diaries by members of case study households, 2008–2010.)

Domestic activity	Time spent	Household position (most input)
Cooking	2 h/day	Daughter-in-law
Bread-making	5 h/week	Mother-in-law
Pickling vegetables	3–4 h/day (during the season)	Daughter-in-law
Serving and cooking for guests	2–3 h/day	Daughter-in-law
Washing the dishes	2 h/day	Daughter-in-law
Buying food and other goods	Once weekly	Mother-in-law
Carrying purchased food from the market	Once weekly	Daughter-in-law
Cleaning	1 h/day	Daughter-in-law
Laundry	2 h/day	Daughter-in-law
Tailoring	4 h/day	Daughter-in-law
Childcare	Most of the time	Daughter-in-law
Caring for sick	When needed	Daughter-in-law
Water carrying	Once daily	Daughter-in-law
Working in the kitchen garden, milking cows and feeding poultry	3–4 h/day	Women
Labour for arranging community events	When needed	Younger women

Table 4.2. On-farm labour: gender divisions by month, Khorezm province, Uzbekistan, 2008–2010. (Data from self-completed diaries by members of case study households, 2008–2010.)

Month	Mainly male	Mainly female
January	Cleaning canals	
February	Soil leaching, making beds	
March	Second soil leaching, breaking beds	
April	Cotton planting using tractors, fertilization using machinery, covering furrows with plastic	Bringing seeds to the tractor, carrying bags of fertilizer to the tractor
May		Loosening hard soil, thinning crops
June	Making irrigation furrows with tractors, irrigation	Making beds using spades, weeding
July	Irrigation	Weeding, trimming
August	Irrigation	Weeding
September	Cotton transportation using a tractor	Cotton picking
October	Cotton transportation using a tractor	Cotton picking
November	Cotton transportation using a tractor	Cotton picking
December	No activity	No activity

Table 4.3. Engagement in on-farm activities by gender, Khorezm province, Uzbekistan, 2009. (Data from 2009 survey of 284 individuals with on-farm activities.)

	Men		Women	
On-farm activity	Count	%	Count	%
Managerial (*fermer*)	29	16	7	7
Technical work for a *fermer*	81	45	1	1
Labour for a *fermer*	72	40	94	92
Total individuals	182	64	102	36

work on *tomorka* as 'giving help' to women, whereas such work was formerly one of their responsibilities:

> Because of the migration of men, women and children have to do all the work in agriculture. In the past, men worked more on bog [*uy tomorka*] and *qo'shimcha tomorka* than now. Now men help when they come back home.
>
> (Man 371; sample mini-survey on perceptions about the gender-specific norms and the division of labour, 2008)

Overall, the findings suggest that men's seasonal migration challenges the status quo in the gendered division of labour. They also suggest that men's migration results in the feminization of agricultural labour and an increase in women's workloads. However, this feminization of agricultural labour did not go along with a transfer of property rights to land and control over income. Migration could also improve the situation of women. In some cases, remittances were used to invest in time- and labour saving equipment. This was generally the case in households with successful migrants or other well-off households, for example households of large-scale *fermers*, local authorities, or successful professionals such as surgeons. Such equipment often included washing machines and electric furnaces.

Changing access to assets

The decline of economic opportunities in rural areas reinforced women's dependence on men. Women's access to assets was generally assured through men. The only property a woman had for herself was her *sarpo* – a large chest, decorated with drawings or colourful stones and filled with wedding presents. After the wedding, it was customary for a woman to have a child within the first three years. If a woman could not have a child for health reasons, she could be divorced and sent back to the house of her parents: 'What can I do? I cannot have a child. Therefore, I was divorced. Now I rely on my dad. But, once he is not there, what then?' (woman 255; sample survey, 2009). However, the ability to earn an income through migration challenged generational relations and hierarchies. Migration enabled the younger generations to become economically independent from the older generations. Before the transition period, fathers were expected to provide housing for their sons; they paid for their sons' weddings and arranged their marriages. This all began to change when sons became economically independent and could decide for themselves whom to marry and where to live. Likewise, some young women could decide to study, even without the permission of their parents:

> When my husband and I migrated to earn, our daughter – without our permission – left for Tashkent to enter a university. We did not want her to study. We wanted her to marry. If there is education but there is no family, it is not good, especially for girls. She is already 28 years old and not yet married. She is about to get her Masters and she does not want to come back.
>
> (Woman 90; sample survey, 2009)

Challenging traditional norms, grown children also began to directly negotiate their concerns and interests with their parents and parents-in-law: 'Some daughters-in-law talk to their mothers-in-law now. Before, they could not say a word in her presence. This is because they are economically independent and live separately from their parents-in-law' (woman 15; sample survey, 2009). Overall, economic needs began to gradually transform the gender and generational structures in rural areas from 'traditional' to 'modern', although in 2010 most households were still traditionally organized.

Changing patterns of decision-making power

Economics and politics are always closely interlinked. The economic dependence of women on men restricted women's decision-making power in Khorezm. As explained above, decision-making power at the household level was divided into female and male spaces of responsibility and decision making (*o'rni* or 'places' in Uzbek), reflecting the division of labour into female and male domains. Women were generally responsible for household labour and (subsistence) agriculture, whereas men were supposed to take care of paid labour/production outside the context of the household. Decision-making power usually reflected this division.

The wife of the head of the household generally allocated different tasks to the younger women of the household. However, these allocated tasks were of different workloads, depending on the household position of the younger woman. In general, she would allocate easier tasks to her daughters and more difficult tasks to her daughters-in-law: 'My daughter will get tired when she is at her new home when she gets married herself' (woman 301-1; case study household, 2008). Apart from variations in workloads, different domestic activities had different rankings. For example, cleaning the dishes was an activity of the lowest-ranking woman that was generally done by the youngest daughter-in-law when there were several daughters-in-law in the household: 'Nobody will respect me here if I start washing the dishes instead of her' (woman 302-3; case study household, 2008). The youngest daughter-in-law typically had the heaviest workload. Any household member could 'tell' the youngest daughter-in-law what to do, and she in turn could 'request help' only from the youngest children in the family. An older daughter-in-law was one step higher than a younger daughter-in-law, especially when only the former had children.

Regarding activities outside the home, a household head (usually the oldest man or the man bringing the highest income) decided what every household member should or should not do: 'I quit my job because my husband did not allow me to work' (woman 195; sample survey, 2009).

> You know, I am actually trained as an accountant. My husband doesn't let me work though. Now I work for a *fermer* because the *fermer's* land is just across the street. Therefore, my husband allowed me to work there, but I get less money. [She was neither sad nor angry but had a smile on her face.]
>
> (Woman 303-1; case study household, 2008)

In other words, women from traditional households generally entered the labour market with the permission of men. In contrast to men, women generally never determined what men should do. Even when a woman had an income, the income of men was more esteemed than the woman's income, even if she earned more than a man (which was rare); and domestic activities remained an exclusively female responsibility:

> If a woman earns more than a man, she shall anyway behave as a woman. The man is the household head and the man shall not do female work (*x[h]otin kishini' ishi*). It humiliates his self-esteem. Even if a woman works all day long to earn money and earns more than a man, she shall do her female work at home.
>
> (Woman 302-1; case study household, 2008)

> If a woman earns more than a man, then there is more money in the family budget. But, it is not good for their relationship. If a woman is brought up well, she will behave correctly to her husband. I personally want a woman to do her female work. Earning money is male work.
>
> (Man 302; case study household, 2008)

Even when younger women worked and had an income, they generally did not control their income. The household budget was usually managed by an older woman (usually a mother-in-law). Men could take and use money from the household budget at any time, mainly because they contributed the most. Older women decided how to use money to meet ordinary household needs but sought permission from men if something special and expensive was to be bought. Younger women, and especially daughters-in-law, had no right to decide how to use the household budget, nor could they take even a small amount of pocket money without the approval of elders, even when they contributed considerably to the household budget. With age and over the course of their lives, men and women obtained more decision-making power. Young men generally had more decision-making power than young women. For example, a young husband could decide to migrate without consulting his young wife:

> My husband does not ask for my opinion or permission [to migrate]. For example, when he leaves home to earn money for a month, he just says that he needs a bag with his clothes.
>
> (Woman 302-4, 23 years old; case study household, 2008)

However, older men generally had more respect for older women. For example, a man who was more than 40 years old could not leave for migration without the permission of his wife, who was also more than 40 years old:

> My wife does not let me work far away. I left once and then she stopped eating and sleeping. When I called her she always cried, so I had to come back. I want to earn well, but I have to work here and do whatever I can do.
>
> (Man; mini-survey on migration labour, 2008)

Likewise, relatively older women could influence men's decisions to migrate to earn money:

> When men are unemployed and do not bring income, their wives say to them not to do female work but to do their own male work. This means that men shall earn money.
>
> (Man 304-2; case study household, 2008)

In most cases, older people had to agree that spouses and younger people could migrate. However, the distribution of decision-making powers was not static. It was challenged and gradually started to change due to the economic independence of the younger generation. Those sons who generated the most income in a household gradually gained bargaining and decision-making power within their households. They could considerably influence the decisions of the older men and women in the household. Because parents did not want to lose any contact with their now economically independent children, the former became flexible to the demands of the latter.

Changing gender norms and values

Some researchers suggest that cultural norms and gender ideals are central to the reinforcement of gender inequalities (Mandel, 2004; Korf and Oughton, 2006). In Khorezm, gender-specific norms and values were reinforced not only at the household level but also at the community level, through women's advisory committees and other channels:

> Our role is to explain to women how to be good home makers and how they shall behave in their new families when they are married. For example, we explain that they need to listen to and obey their husbands, be tolerant and not to argue.
> (Woman 430; open-ended interview, 2008)

In traditional households, daughters did not own any assets (whereas sons did) and, in addition, they were convinced by their parents not to continue their education after school:

> Why do girls need education? They sit and work at home. It is a man who shall provide for the family. Women are to take care of children and cook and clean the house. There is no need to be educated for this.
> (Man 431; open-ended interview, 2009)

> I wanted to study at a university but my dad said that a woman can live without education. Then, I stopped studying. My teachers were saying that I studied well and should continue my studies. But, I said that there was no point in it.
> (Woman 432; open-ended interview, 2009)

> I wanted to be a gymnastics teacher but my father said that instead I should learn how to tailor. Then, I went for tailoring courses.
> (Woman 429; open-ended interview, 2009)

As a result, many more men than women have graduated from universities. In this way, cultural norms have disadvantaged women in terms of their employment opportunities and income and have reinforced women's dependence on men. One woman explained that her father decided that she should not study at a university, her husband decided that she should not work aside from being a housewife and her employer preferred a male employee:

> Uzbeks usually do not want their wives to work outside of home. She should take care of her husband and make sure that he has clean clothes and is not hungry. And his responsibility is to provide for the family and bring money.
> (Woman 302; case study household, 2009)

> You employ a woman. It can take up to 6 months to train her. Then, she marries. Once married, first, she starts asking not to work overtime and not to work on Saturdays. Then, she takes a maternity leave or her husband decides that she shall not work.
> (Man 420, open-ended interview, 2008)

Adherence to cultural norms reinforced women's dependence on men and contributed to the reproduction of the traditional gendered division of labour, decision-making power and access to and control over assets in Khorezm.

Migration challenged these norms. Initially, people thought quite negatively about migration and the changes it induced. However, when successful migrants came back home with money and managed to build houses and to invest in local businesses, migration and the changes that went along with it became more accepted. In addition, men's and women's exposure to a context and culture of somewhat more (compared with Khorezm) egalitarian societies abroad started to change their views of their roles within the household.

The flexibility and change in cultural norms as a result of migration offers an opportunity to empower women. On the other hand, migration added to the workloads of many women and reconfirmed women's economic dependence on men and the remittances they sent home. This was also the case because income levels remained low in Korezhm and were insufficient for meeting household needs:

> We wait until my husband comes back. I hope he will come back in September. Then, we can prepare for winter. We need to buy coal, repair roof. Once he brings money, I will pickle vegetables for winter.
>
> (Woman 325; mini-survey on problem analysis, 2008)

Conclusions

This chapter set out to explore changing gender relations in rural Uzbekistan during the process of transition to post-socialism and independence after the dissolution of the Soviet Union. During this period, the transition process was interpreted as one of the major crisis tendencies that has the capacity to induce change by uprooting the institutionalized arrangements of gender relations at multiple levels as a result of the fundamental changes in the material structure as well as the incorporated gender ideologies or norms and values (Connell and Pearse, 2015; Pearse and Connell, 2016). In Uzbekistan, the shift towards a post-socialist society went along with changes in access to land and employment through the dismantling of state-owned firms and the dissolution of state-owned farms. At the same time, many public services that had been allocated through state employment and organized at the level of state firms vanished. Other public services, such as education, health services and transportation, deteriorated – especially in rural areas. The same was true for basic utilities such as running water and electricity. More generally, it can be said that transition led to the impoverishment of many rural and urban households and induced a process of social differentiation. The position of women was also weakened as a result of the reintroduction of traditional gender norms once the influence of the socialist ideal of gender equality evaporated. The chapter focused on changes in gender relations taking place in rural areas while elaborating, in particular, on the result of increasing levels of labour migration.

Interviews in rural households as well as among individual men and women underlined the importance of labour migration as a strategy to produce cash income. Importantly, nearly one out of two households depended on the remittances. As most migrants were men and often departed for several months, their absence induced shifts in the arrangements of household relations. Most of the rural households are home to extended families, with the oldest man acting as head of the household. He would regulate the division of labour, income and decision-making power in collaboration with his wife. Migration generally resulted in a repositioning of all members of the household and in a rearrangement of gender and generational relations and hierarchies.

On the one hand, men's migration added to women's workloads, as they had to take on many traditionally male tasks and duties. Migration did not change women's economic dependence on men, as the remittances men sent home were essential for the households. On the other hand, men's migration offered an opportunity for women to gain power and to redefine their position, as it was usually older women who acted as de facto household heads during men's absence. Changing roles and responsibilities in the division of labour supported a reorganization of decision making, which, in turn, gradually challenged existing gendered norms and values. This had the effect of changing gender relations in the long run, even if the transfer of responsibilities was initially on a temporary basis and everything returned to 'normal' when the men returned.

Migration, therefore, brings change as well as continuity; it brings empowerment as well as reproduction of dependence and subordination. The results presented here support our initial argument that it is difficult to change gender relations if societal change is limited to either norms and values or the division of labour. The results confirm the expectation that crisis tendencies induce change in existing relations through the combination of substantial changes in culture, politics and economics, which trigger changes in behaviour, norms and values, and social relations. Whereas it is still a question as to which of these changes taking place in rural Uzbekistan are most decisively changing gender relations, it is clear that as long as local income opportunities for both women and men are insufficient, there are few options to radically change household relations. It is, however, promising to watch the young generation adopt a stronger voice and more independence. For the young, migration is as an important gateway to empowerment and the creation of more equality between genders and generations. The transformative potential of migration may, therefore, unfold more clearly in the long term.

Acknowledgements

This research was conducted under the scope of the project entitled 'Economic and Ecological Restructuring of Land and Water Use in Khorezm', implemented by the agreement between the German Ministry of Education, the Centre for Development Research (ZEF), the United Nations Educational, Scientific and Cultural Organization (UNESCO) and the Ministry of Agriculture and Water Resources (MAWR) of Uzbekistan.

References

Anderson, K.H. and Pomfret, R. (2002) Relative living standards in new market economies: evidence from central Asian household surveys. *Journal of Comparative Economics* 30(4), 683–708. Available at: http://dx.doi.org/10.1006/jcec.2002.1803 (accessed 4 March 2017).

Breen, R. and Cooke, L. (2005) *The Persistence of the Gendered Division of Domestic Labour.* Oxford University Press, Oxford, UK.

Bryman, A. (2004) *Social Research Methods.* Oxford University Press, Oxford, UK.

Butler, J. (1988) Performative acts and gender constitution: an essay in phenomenology and feminist theory. *Theatre Journal* 40(4), 519–531.

Chant, S. and Sweetman, C. (2012) Fixing women or fixing the world? 'Smart economics', efficiency approaches and gender equality in development. *Gender & Development* 20(3), 517–529. Available at: http://dx.doi.org/10.1080/13552074.2012.731812 (accessed 4 March 2017).

Conliffe, A. (2009) The combined impacts of political and environmental change on rural livelihoods in the Aral Sea region of Uzbekistan. DPhil. thesis, University of Oxford, Oxford, UK.

Connell, R.W. and Pearse, R. (2015) *Gender: In World Perspective.* Wiley, Oxford, UK.

Cornwall A. (2016) Women's empowerment: what works? *Journal of International Development* 28(3), 342–359. Available at: http://dx.doi.org/10.1002/jid.3210 (accessed 4 March 2017).

Duncan, S. and Smith, D. (2002) *Geographies of Family Formations: Spatial Differences and Gender Cultures in Britain.* Royal Geographical Society and Institute of British Geographers, London.

Ellis, F. (1998) Household strategies and rural livelihoods diversification. *Journal of Development Studies* 35(1), 1–38. Available at: http://dx.doi.org/10.1080/00220389808422553 (accessed 4 March 2017).

Ellis, F. (2000) *Rural Livelihoods and Diversity in Developing Countries.* Oxford University Press, Oxford, UK.

Fraser, N. (2009) Feminism, capitalism and the cunning of history. *New Left Review* 56, 97–117.

Gal, S. and Kligman, G. (2012) *The Politics of Gender after Socialism: A Comparative Historical Essay.* Princeton University Press, Princeton, New Jersey.

IMF (2008) Republic of Uzbekistan: Poverty Reduction Strategy Paper. Available at: https://www.imf.org/external/pubs/ft/scr/2008/cr0834.pdf (accessed 10 August 2011).

Jackson, S. (2001) Why a materialist feminism is (still) possible – and necessary. *Women's Studies International Forum* 24(3–4), 283–293. Available at: http://dx.doi.org/10.1016/S0277-5395(01)00187-X (accessed 4 March 2017).

Kamp, M. (2004) Between women and the state: mahalla committees and social welfare in Uzbekistan. In: Luong, J. (ed.) *The Transformation of Central Asia: States and Societies from Soviet Rule to Independence*. Cornell University Press, Ithaca, New York, pp. 29–58.

Kamp, M. (2005) Gender ideals and income realities: discourses about labour and gender in Uzbekistan. *Nationalities Papers* 33(3), 403–422. Available at: http://dx.doi.org/10.1080/00905990500193287 (accessed 4 March 2017).

Kamp, M. (2006) *The New Woman in Uzbekistan: Islam, Modernity and Unveiling under Communism*. University of Washington Press, Seattle, Washington.

Kandiyoti, D. (2002) *Agrarian Reform, Gender and Land Rights in Uzbekistan. Social Policy and Development Programme Paper, 11*. United Nations Research Institute for Social Development, Geneva.

Kandiyoti, D. (2003) The cry for land: agrarian reform, gender and land rights in Uzbekistan. *Journal of Agrarian Change* 3(1–2), 225–256. Available at: http://dx.doi.org/10.1111/1471-0366.00055 (accessed 4 March 2017).

Korf, B. and Oughton. E. (2006) Rethinking the European countryside – can we learn from the South? *Journal of Rural Studies* 22(3), 278–289. Available at: http://dx.doi.org/10.1016/j.jrurstud.2005.09.005 (accessed 4 March 2017).

Lazreg, M. (1999) *Making the Transition Work for Women in Europe and Central Asia. Europe and Central Asia Gender and Development Series: Discussion Paper 411*. World Bank, Washington, DC.

Mandel, J. (2004) Mobility matters: women's livelihood strategies in Porto Novo, Benin. *Gender, Place & Culture* 11(2), 257–287. Available at: http://dx.doi.org/10.1080/0966369042000218482 (accessed 4 March 2017).

Marshall, C. and Rossman, G. (1999) *Designing Qualitative Research*. SAGE Publications Ltd, London.

Mee, W. (2001) *Women in the Republic of Uzbekistan. ADB Country Briefing Paper*. Asian Development Bank, Manila.

Mukhopadhyay, T. (2008) Whether monetary empowerment leads to gender empowerment or not: case studies in the context of micro credit programmes in India. *Global Academic Society Journal: Social Science Insight* 1(4), 18–37.

Müller, M. (2006) A general equilibrium approach to modelling water and land use reforms in Uzbekistan. PhD thesis, University of Bonn, Bonn.

Nizamedinkhodjayeva, N. (2013) Livelihood strategies: gender and generational specificities of rural livelihoods in transition. PhD thesis, Wageningen University, Wageningen.

Palmary I., Burman, E., Chantler, K. and Kiguwa, P. (eds) (2010) *Gender and Migration*. Zed Books, London and New York.

Pearse, R. and Connell, R. (2016) Gender norms and the economy: insights from social research. *Feminist Economics* 22(1), 30–53. Available at: http://dx.doi.org/10.1080/13545701.2015.1078485 (accessed 4 March 2017).

Pomfret, R. and Anderson, K.H. (1997) *Uzbekistan: Welfare Impact of Slow Transition, Working Paper 135*. United Nations University, World Institute for Development Economics Research, Finland. Available at: https://www.wider.unu.edu/sites/default/files/WP135.pdf (accessed 8 February 2017).

Reeves, H. and Baden, S. (2000) Gender and Development: Concepts and Definitions. Available at: http://www.bridge.ids.ac.uk/reports/re55.pdf (accessed 10 December 2011).

Ruminska-Zimny, E. (1997) Human Poverty in Transition Economies: Regional Overview for HDR 1997. Available at: https://ideas.repec.org/p/hdr/hdocpa/hdocpa-1997-03.html (accessed 10 September 2008).

Ruminska-Zimny, E. (2002) Gender Aspect of Changes in the Labour Markets in Transition Economies. Research Paper, United Nations Economic Commission for Europe. Available at: http://www.un.org/womenwatch/osagi/pdf/Ruminska-Zimny.FINAL.PDF (accessed 10 September 2008).

Spoor, M. (2013) Multidimensional social exclusion and the 'rural–urban divide' in Eastern Europe and Central Asia. *Sociologia Ruralis* 52(2), 139–157. Available at: http://dx.doi.org/10.1111/soru.12008 (accessed 4 March 2017).

van Doorne-Huiskes, A. and van Hoof, J. (1995) Gendered patterns in institutional constrains: an attempt at theoretical integration. In: van Doorne-Huiskes, A., van Hoof, J. and Roelofs, E. (eds) *Women and the European Labour Markets*. SAGE Publications Ltd, London, pp. 106–121.

Veldwisch, J. (2008) Cotton, rice & water: the transformation of agrarian relations, irrigation technology and water distribution in Khorezm, Uzbekistan. PhD thesis, ZEF, Bonn.

Veldwisch, G. and Bock, B. (2011) *Dehkans*, diversification and dependencies: rural transformation in post-Soviet Uzbekistan. *Journal of Agrarian Change* 11(4), 581–597. Available at: http://dx.doi.org/10.1111/j.1471-0366.2011.00327.x (accessed 4 March 2017).

Veldwisch, G. and Spoor, M. (2008) Contesting rural resources: emerging 'forms' of agrarian production in Uzbekistan. *Journal of Peasant Studies* 35(3), 424–451. Available at: http://dx.doi.org/10.1080/03066150802340420 (accessed 4 March 2017).

Yin, R. (1984) *Case Study Research: Design and Methods*, 1st ed. SAGE Publications, Inc., Beverly Hills, California.

Yin, R. (2002) *Case Study Research: Design and Methods*. SAGE Publications, Inc., Newbury Park, California.

5 'There is Dignity only with Livestock': Land Grabbing and the Changing Social Practices of Pastoralist Women in Gujarat, India

J. Duncan[1,*] and M. Agarwal[2]

[1]Wageningen University, Wageningen, The Netherlands; [2]Formerly NES India, International Land Coalition

Introduction

On 18 June 2013 over 10,000 men and women congregated in Gandhinagar, the capital of the Indian state of Gujarat, to express their dissent. Almost one year earlier the Government of Gujarat had declared a geographical area of over 50,000 ha, encompassing 44 villages, as the Mandal-Becharaji Special Investment Region (MB SIR) through a notification dated 24 September 2012. MB SIR was to be a hub for the automobile industry. The project was aligned with the Special Investment Region (SIR) Act of 2009 which aimed to develop global hubs of economic activity across the state. MB SIR was further designed to complement the national government's proposed Dedicated Freight Corridor between Delhi and Mumbai. The area of 150 km on both sides of the corridor, 38% of which is in Gujarat, was to be further developed as an Industrial Corridor. Yet throughout the planning process, the people and communities that lived in and relied on the land in question were not consulted, or even informed.

In May 2013, seven months after the official notification, the Government of Gujarat finally published a notice in a local newspaper, after civil society organizations launched a campaign against SIR on 7 May 2013. In a symbolic expression of dissent, a notice was painted outside affected villages stating that government officials, or any other official authorized under SIR, shall not enter the village for the purpose of survey or inspection (see Fig. 5.1).

After 100 days of unrelenting protests, signs of a victory for the people's movement emerged on 15 August 2013 when the Government of Gujarat gave notice of the withdrawal of 36 villages from MB SIR. As per media reports, towards the end of 2015, the state government decided to go ahead with only three out of the 14 SIRs originally announced in 2009. Although partial, this gave hope to people fighting for their right to land.

This conflict around land took place at a key moment in India's history. In 2013, the Right to Fair Compensation and Transparency in Land Acquisition, Rehabilitation and Resettlement (LARR) Act was passed

* Corresponding author e-mail: jessduncan@wur.nl

Fig. 5.1. Sign posted in the village of Hansalpur banning entry to those working on Special Investment Regions, February 2015. (Photo courtesy of J. Duncan.)

and came into force on 1 January 2014. Up to that point, land acquisition in India fell under the colonial British-era Land Acquisition Act of 1894 (Sud, 2014a,b). Land was, and remains, high on the social and political agenda.

Land is a fundamental resource that secures the livelihoods of many, especially the rural poor who often depend on access to and control over natural resources. Land provides food and shelter and is a key factor in economic growth (Deere and Leon, 1998; Deininger *et al.*, 2007; Sietchiping, 2010; Behrman *et al.*, 2012). Ensuring sustainable land use and eradicating food insecurity depends on how people and communities gain, maintain and control access to land. However, as the case of MB SIR illustrates, land access is increasingly subject to intensified competition (GRAIN, 2008; Hall, 2011; Harvey and Pilgrim, 2011; Behrman *et al.*, 2012). There is increasing concern that competition for food, energy and water will result in a land 'bottleneck' (Lambin and Meyfroidt, 2011). The resulting competition has been made visible through the rise of land grabbing or large-scale land acquisitions (Borras *et al.*, 2011; de Schutter, 2011; Margulis and Porter, 2013; Ykhanbai *et al.*, 2014).

Land grabbing has been defined as the 'purchase or lease of vast tracts of land by wealthier, food-secure nations and private investors from mostly poor, developing countries in order to produce crops for export' (Shepard and Mittal, 2009, p. 1). While foreign investments are a key aspect of large-scale land acquisition agreements, domestic governments and investors are also playing a major role. Echoing the MB SIR case, McMichael (2012, p. 682) notes that 'the term "grab" invokes a long history of violent enclosure of common lands to accommodate world capitalist expansion'. This amplified effort to privatize the commons is the key site of inquiry for this chapter.

Access to land and other natural resources is defined and regulated in societies through complex systems of tenure. Insecure tenure rights enhance vulnerability, conflict, food insecurity and poverty and can lead to increased environmental degradation and conflict (Sietchiping, 2010; FAO, 2012a). One group that is particularly vulnerable to the negative impacts of restricted access to land are pastoralists. Pastoralism is a socio-cultural and economic way of living that is reliant on the rearing of livestock and sustained through regional migration. Pastoralists around the world 'lack clear property rights because they occupy customary or tribal rangelands that are legally owned by the state, are controlled/owned by the pastoral community itself, or are claimed by other interest groups' (Behnke and Freudenberger, 2013, p. 1). As access to common rangeland and grazing land decreases, the livelihoods of pastoralists, and their livestock, come under increasing threat.

The aim of this book is to shed light on the fundamental impacts on the structure of agricultural life in rural areas and on urban–rural relations in a globalizing world from a gender perspective. In this chapter, we provide an alternative perspective in so far as we examine the opposite phenomenon. We are interested in understanding what happens when people whose livelihoods are dependent on, and cultures are defined by, mobility are no longer able to migrate as a result of restricted access to land brought about by processes associated with neoliberal globalization: specifically, land grabbing (McMichael, 2012).

To do this, the chapter begins by providing an introduction to pastoralism and land issues in Gujarat, India, with a particular focus on the evolving legal framework governing land in Gujarat. In the presentation of pastoralism in Gujarat, we deliberately focus on women in the recognition that 'analysts who fail to ask questions that specifically target women's tenure situation are unlikely to discover information about challenges women face in exercising or maintaining their rights' (Hannay and Scalise, 2014, p. 3). Subsequently the chapter presents an overview of the methods used in our fieldwork to collect data and to elaborate three cases of state-sanctioned land grabbing of lands traditionally used by pastoralists. Next, an analytic framework building on sociological theories of practices is introduced, which is used to analyse a typical day in the lives of the women who were interviewed. The objective here is to identify ways in which the loss of access to land, and correspondingly the loss of livestock, influences social practices. The chapter concludes by reflecting on the social, scientific and policy implications of the findings.

Background and Context: Pastoralism, Gender and Land in Gujarat, India

Pastoralism in Gujarat

Pastoralists represent endogamous social groups with professional specializations in animal husbandry (Sharma *et al.*, 2003), including livestock breeders, herders and dairy producers. In India, pastoralists share ethnicity with sedentary populations but form part of specific castes related to the rearing of livestock (Köller-Rollefson, 1994). This differs from Africa and the Middle East, where nomadic pastoralists tend to be represented by distinct ethnic groups, often organized along tribal lines. Pastoralism has historically been an important occupation in the semi-arid regions of Gujarat, India especially in Kutch, Saurashtra and North Gujarat. In an effort to build solidarity across these livestock communities (including *Rabari*, *Bharwad*, *Ahir*, *Charan*, *Gadhvi* and *Sindhi*), they have started to collectively identify as *Maldhari*: *mal* meaning both livestock and asset, and *dhari* meaning owner.

Each pastoralist community in Gujarat is unique, making it hard to draw generalizations. However, it is common that livestock management, including milking, administering medicine and watering, is shared work between women and men, although women take on the responsibility of caring for newly born and young animals (Rangnekar, 1994, p. 15). Women also have the responsibility of processing the milk and, traditionally, of selling milk and value-added dairy products (i.e. *ghee* (clarified butter) and *mava* (condensed milk)).

Most of the pastoralist communities in Gujarat have some form of permanent dwellings (traditionally known *nehdo* in Gujarati). Families will stay in the permanent dwellings while select members (men and women) will migrate with the animals when fodder is not locally available. Often those on migration will take the livestock of other members of their community, receiving money based on a fixed day rate per animal. Mobility is a strategy used by pastoralists to ensure access to grazing land and water for their herds. It is also a strategy that supports adaptation to different climatic regions. Pastoralist grazing systems adapt to fit the environment. However, to be able to maintain this approach to livestock rearing, they depend on access to common grazing lands and to the routes to bring them to these lands.

Land tenure

Land defines social status, political power and structures relationships (Agarwal, 1994, p. 2). Access to land and other natural resources is mediated by systems of land tenure. Tenure can be regarded as a social construct that defines the relationships between individuals and groups of individuals by which rights and obligations with respect to control and use of land are determined (Economic Commission for Africa, 2004, p. 21). In simple terms, land tenure systems determine who can use what resources for how long and under what conditions.

Land tenure is often understood as a 'bundle of rights' (Maxwell and Wiebe, 1999, p. 825) that enables and legitimizes groups or individuals to access land. 'Freehold', 'leasehold', 'statutory allocations' and 'customary systems' are the most common forms of land tenure (Economic Commission for Africa, 2004). Tenure systems include informal and unwritten customary rights of access to and use of land and natural resources, as well as more 'formal' arrangements (e.g. individual tilling, freehold and leasehold) mediated by written contracts, policies and laws. There are also secondary rights that include access to migratory routes and grazing lands. Insecure tenure rights enhance vulnerability, conflict, food insecurity and poverty and can lead to increased environmental degradation and conflict (Sietchiping, 2010; FAO, 2012a).

Pastoralists and land tenure

For centuries, pastoralists have had access to, and managed, rangelands through elaborate systems of land tenure based on customary rules and regulations, the authority of customary institutions and sustainable animal-rearing practices (Sharma *et al.*, 2003). Pastoralist tenure systems tend to be mixes of dynamic land-use regimes. These have served to regulate access to, and control over, pastoral territories. While farmers have relatively explicit systems of tenure, pastoralists tend to rely on more fluid systems. They have also maintained their livelihoods while fostering co-dependence and intricate economies with other communities (e.g. manure-for-grazing agreements with farmers; wool-for-cloth arrangements with weavers) (Duncan, 2013). However, these relationships and practices have all been eroding, in part due to the loss of access to grazing lands which have traditionally been held as part of the commons. 'The commons' refer to the cultural and natural resources held in common and not privately owned. Moreover, the commons are 'closely integrated to the social, economic and cultural identity of the communities which depend on them' (Sinha *et al.*, 2013, p. 1).

The legacy of Hardin's (1968) thesis on the 'tragedy of the commons', which upheld that common property resources shared by pastoralists led to overgrazing and environmental degradation, has led to decades of development policies aimed at privatizing these lands in the name of better land management (Fratkin, 1997). However, Hardin's thesis has been actively countered by social scientists who reject Hardin's assumption that communally held resources mean no restriction on use, and by natural scientists who have questioned the ecological evidence in the tragedy thesis (McCabe, 1990; Bromly, 1992; Behnke *et al.*, 1993; Turner, 1993; Ostrom *et al.*, 1999). Others have documented how pastoral systems can, and do, respond to restricted resources with cultural behaviours that include flexibility, mobility and diversity of species (Fratkin, 1997; see also Hjort, 1981; Coughenour *et al.*, 1985; Ellis and Swift, 1988). Growing evidence suggests that mobile pastoralism can be a sustainable form of production, resource management and land use (Blench, 2000; Rodriguez, 2008; ODI, 2009; Behnke and Freudenberger, 2013).

Today, pastoralists' access to common rangelands, upon which their livelihoods depend, is being restricted by new developments (urbanization) and expanded agricultural production. Through processes of privatization of the commons, previously communal land areas are shrinking. New technologies (i.e. irrigation) are, for example, pushing agricultural land into traditional grazing land. Access to the commons is further restricted by the development of new conservation zones and mining.

Shrinking common lands coupled with a lack of formal land rights puts pastoralists on the periphery of tenure rights, while migration patterns often result in restricted citizenship (Sietchiping, 2010; MARAG, 2012). This mirrors the situations of pastoralists the world over. As Nori *et al.* (2008, p. 18) note:

> Pastoral societies and people all over the world are experiencing processes that are redefining their territories and reshaping their resource utilisation patterns. Integration of pastoral economies into markets, changing migratory patterns and political processes of regional integration and decentralisation all carry threats and opportunities.

Women and land

Women face specific land tenure challenges that involve sensitive social and cultural practices as well as deeply rooted cultural norms and power structures. Just as gender relations are 'neither uniform across societies nor historically static' (Agarwal, 1994, p. 51), the particular land situations of women vary significantly depending on context (Hannay and Scalise, 2014, p. 2). However, women generally have access to less land, less secure land rights, and fewer rights to land they can access (Doss, 2013). Gender dimensions are frequently not captured in land-based assessments and data collection (UN-Habitat, 2012) and laws more broadly. Women can face fourfold vulnerability when it comes to securing land rights: (i) access to, ownership of and control of land; (ii) discrimination in decision making; (iii) relative income/cash poverty; and (iv) physical vulnerability (Pallas and Miggiano, 2011).

In India, only an estimated 10% of rural land is titled to women, whereas 83% of rural women provide agricultural labour (Kelkar, 2013, p. 2). Women's access to land depends on complex systems of laws, their enforcement, as well as customary and religious laws and practices. India has legislation protecting women's property rights; however, often traditions and customs serve to undermine or restrict these rights (Kelkar, 2013). International conventions protect women's rights, but may be at odds with national legislation and local practices and remain difficult to enforce. For example, in Gujarat, pasture land has traditionally been a community-managed common resource and could not be given away without consent. The International Labour Organization Convention 169 (Indigenous and Tribal Peoples Convention) and the United Nations Declaration on the Rights of Indigenous Peoples also recognize the customary right of communities to these lands. But, as the cases described below illustrate, this has not prevented the loss of rights and access to these lands. On the ground, perceptions of women's status affect the extent to which women can exercise their rights. In many cases, women are prevented from participating in decision making.

Governing land in Gujarat

When it comes to rights to land for pastoralists in Gujarat, the issue is rather complex. In an effort to protect agricultural land from urbanization or foreign interests after India gained independence in 1947, the Gujarat Revenue Act established that only farming families had the right to own agricultural land (popularly known as 'land to the tiller'). This effectively meant that any family that was tilling land in 1947 was eligible to hold tenure rights to that land. This also meant that most pastoralists were excluded from the right to own land zoned as agricultural since they did not till the land. However, it included lands upon which fodder was produced. The pastoralists in Gujarat, thus, have mainly depended on the commons including the village *gauchars* (the designated permanent pastures under village *panchayats*[1]), the common land under forests and the so-called 'wastelands'. The term 'wasteland' is a misnomer coined during colonial rule. Some of this land represents an important grazing area and source of livelihood for many, including livestock keepers and the landless in Gujarat. Yet, the term continues to be used to justify the transfer of land use to industrial purposes. At present, *gauchar* and wasteland access is declining as a result of large-scale conversion for industrialization. For example, in May 2005 the Government of Gujarat adopted a resolution for leasing 'wasteland' of up to 2000 acres (~810 ha) for a period of 20 years to corporations and larger-scale farmers.

The Gujarat Special Economic Zone (SEZ) Act 2004 and the SIR Act 2009 have also facilitated large-scale transfer of private agriculture and common land for industrial parks and investment regions. In the absence of a comprehensive policy framework, the governance of commons is weak in Gujarat and does not support pastoralists and their livestock. On top of rapid industrialization and classification challenges, it is estimated that 68% of the total geographical area of Gujarat is currently threatened by desertification (ISRO, 2007).

In India, land falls under the jurisdiction of the state-level government, but land acquisition is in the concurrent list of the constitution of India. This means that both the centre and state governments have legislative authority on issues related to the acquisition of land. The central LARR Act 2013 came into force on 1 January 2014. The Act seeks to clarify the government's commitment to securing a legal guarantee of the rights of individuals and families affected by the process of land acquisition, and strengthen the rights-based approach, while ensuring greater transparency in the land acquisition process. However, with a change in government, an amendment bill was introduced in 2015. This bill was set to reverse some of the social justice elements of the 2013 LARR Act and was strongly contested for being anti-farmer and too industry-friendly.

Given the dual-level jurisdiction over land, Indian states can amend the central act. The government of the state of Gujarat, for example, proposed the 'Right to Fair Compensation and Transparency in Land Acquisition, Rehabilitation and Resettlement (Gujarat Amendment) Bill 2016', which was introduced in the legislative assembly in 2016, to amend the central Right to Fair Compensation and Transparency in the LARR Act. The objective of this bill is to ease the stringent provisions of the central law, notably clauses of social impact assessment and consent of landlords for land acquisition for public purpose. The Bill was passed by the state government in the Assembly on 31 March 2016, amidst a walkout by Opposition members. Thus while the LARR was initially brought in to provide stronger legislation around land tenure and land rights for more vulnerable peoples, the Act is being systematically dismantled by the centre and state governments.

This is perhaps not surprising given that in 2015, the World Bank named the state of Gujarat as the most industry-friendly state in India. The ease of acquiring land, the result of several amendments in existing land policies, is a key reason for this. In the 1990s the state government of Gujarat began to introduce amendments to liberalize land markets. These amendments have gradually eased the restrictions on the conversion of agriculture land for non-agriculture purposes and also facilitated entry of non-farmers to buy, sell and mortgage agriculture land.

With an estimated population of 60.4 million, Gujarat is also home to 27.1 million livestock (Government of India, 2012). There is a state-level policy in Gujarat[2] that mandates each village maintain 16 ha (40 acres) of pasture land per 100 head of livestock in non-forest areas; and 8 ha per 100 head of livestock in forest areas. Following this policy, there is a huge deficit of *gauchar* land across the state. A 2010 report noted Gujarat required 39.56 lakh[3] hectares of grazing land required for cattle in the state, but that less than 8.5 lakh hectares were available. The lack of common grazing land is compounded by other factors, such as failure of the Government of Gujarat to update the official figure of 8.5 lakh hectares of pasture land since 1960 (Dave, 2011). There are more than 11,950 registered encroachments on *gauchar* land, of which majority are more than 5 years old (Dave, 2014). Since 2012, Gujarat has sold 116,000 m^2 of land for various purposes, leaving 424 villages without any pastoral land. Today, it is estimated that the state has just one-fifth of its required pastoral land (Mahapatra, 2012). It should be noted that the Supreme Court of India, in its judgement on 28 January 2011, directed all state governments to prepare schemes for the eviction of those occupying village commons and restore them to the community (Kaur, 2011). The Government of Gujarat has yet to comply with the Supreme Court order. Given these challenges, it is not surprising that land has emerged as one of the key socio-political issues in India.

Methodology

To better understand the impact of restricted access to grazing lands, fieldwork was undertaken in February 2015. Most families and livestock are in their villages in February, making it an appropriate time to conduct the research since it ensured access to families that had sedentarized their livestock practices, given up livestock keeping, or continued with livestock migration. This fieldwork built on several years of engagement with pastoralist communities. Further, after the primary data collection was over, when needed, additional visits were made to the villages to validate data and clarify questions that arose through the analysis.

The research sites were selected purposively using the following selection criteria. First, sites were selected on the basis of having pastoralist populations that were impacted by land grabbing. Second, the cases of land grabbing should have non-pastoralist actors (e.g. non-governmental organizations, food producer groups, legal support) engaged in them. This provided a way to triangulate the data acquired with the data collected by others. Third, the cases of land grabbing should be part of a formal legal dispute. This would mean that the villages had likely collected documents to defend their case that likely could be accessed as further data.

On the basis of these criteria, eight villages were selected. Focus groups were conducted in each of these villages. The number of women per focus group varied per village and also shifted over the course of the discussion, as women had to leave to attend to livestock or household chores and other women arrived late. The smallest focus group had five women and the largest counted 16 women. Overall, 87 women participated in the eight focus groups. This number excludes girls under 14 years old who were also present in the focus groups but rarely spoke, children and men, who were often present, but on the periphery. Focus groups were often conducted in a shaded location, near or in the house of a village leader. Strong social cohesion and a collective culture made one-on-one interviews not only impractical but also inappropriate.

Access to the community was gained via gatekeepers (local non-governmental organizations, pastoralist and social movement leaders, and development practitioners) and village leaders, who would inform the women of the research visit and encourage interested *Maldhari* women to participate. Upon arrival in a village, we first went to the home of one of the village leaders for a glass of water and sweet milky chai. Here we engaged in conversations that allowed us to begin to piece together the particular situation of the village. After about an hour, women would begin to arrive and were welcomed to sit in a circle on the ground. Sometimes blankets were spread out, other times cots were pulled up. Local facilitators (from pastoralist communities) guided the discussion and simultaneous translation was provided so the researchers could follow and ask additional questions.

The focus groups lasted from 45 min to 2 h. Chai was always served, usually by younger girls who were then encouraged to stay and listen. It was common for men to pass by, to stop and listen, and to provide input. We recognize that the presence of men may often impact women's willingness and ability to share. However, women in the villages visited play strong leadership roles and are used to speaking out. Most women did not show any apprehension when it came to disagreeing with the men. This was confirmed after the focus groups by some of the women in informal conversations as well as by the local facilitators.

The fieldwork also employed methods common to participatory rural appraisal, including participant observation, participatory resource mapping, power mapping, transect walks (and drives), and life histories. Furthermore, semi-structured key informant interviews ($n = 8$) were conducted to gain insight on land tenure challenges in India, and on women and land tenure, to support the triangulation of data. Interviews and focus groups were audio-recorded and two sets of field notes and field diaries were kept (one in English and one in Hindi/Gujarati). These were all used in data analysis.

After the fieldwork and initial analysis, three sites were selected to serve as sample case studies for this research (Fig. 5.2). These cases are presented in the following section.

Case Studies: Land Grabbing the Commons

Case 1: Zadwa and Sanghi Cement

Zadwa[4] was established over 500 years ago under the name of the Zadu fakir Zadwa. The region is marked by semi-nomadic, traditional grazing routes but facilitated sedentary forms of livestock rearing due to the relative abundance of natural resources and fodder for animals. Not surprisingly then, a group of *Maldhari* migrated from the north of Gujarat to this area 45 years ago to access the large grassland of Lakhpat. This group gained access to 100 ha of grassland. In total, 18 families were granted rights to access the land for farming. Alongside farming, most of the community partook in animal husbandry and continued to earn income from dairy. While the pastoralists moved to the region for its abundant grass, today they find themselves migrating again as much of the land has been encroached and the rest has been heavily impacted by dust from the nearby cement factory.

Sanghi Cement was first set up in the Zadwa region in the Lakhpat block of Kutch district in 1994. Through a legal land deal, Sanghi Cement Ltd acquired 1543 ha from the total area of 3043 ha. The land signed over to the company included grasslands, wastelands and lands on the perimeter of villages. At this time, 16 families sold a total of 48 ha of land to the company for 20,000 rupees (277 euros) per hectare per family.

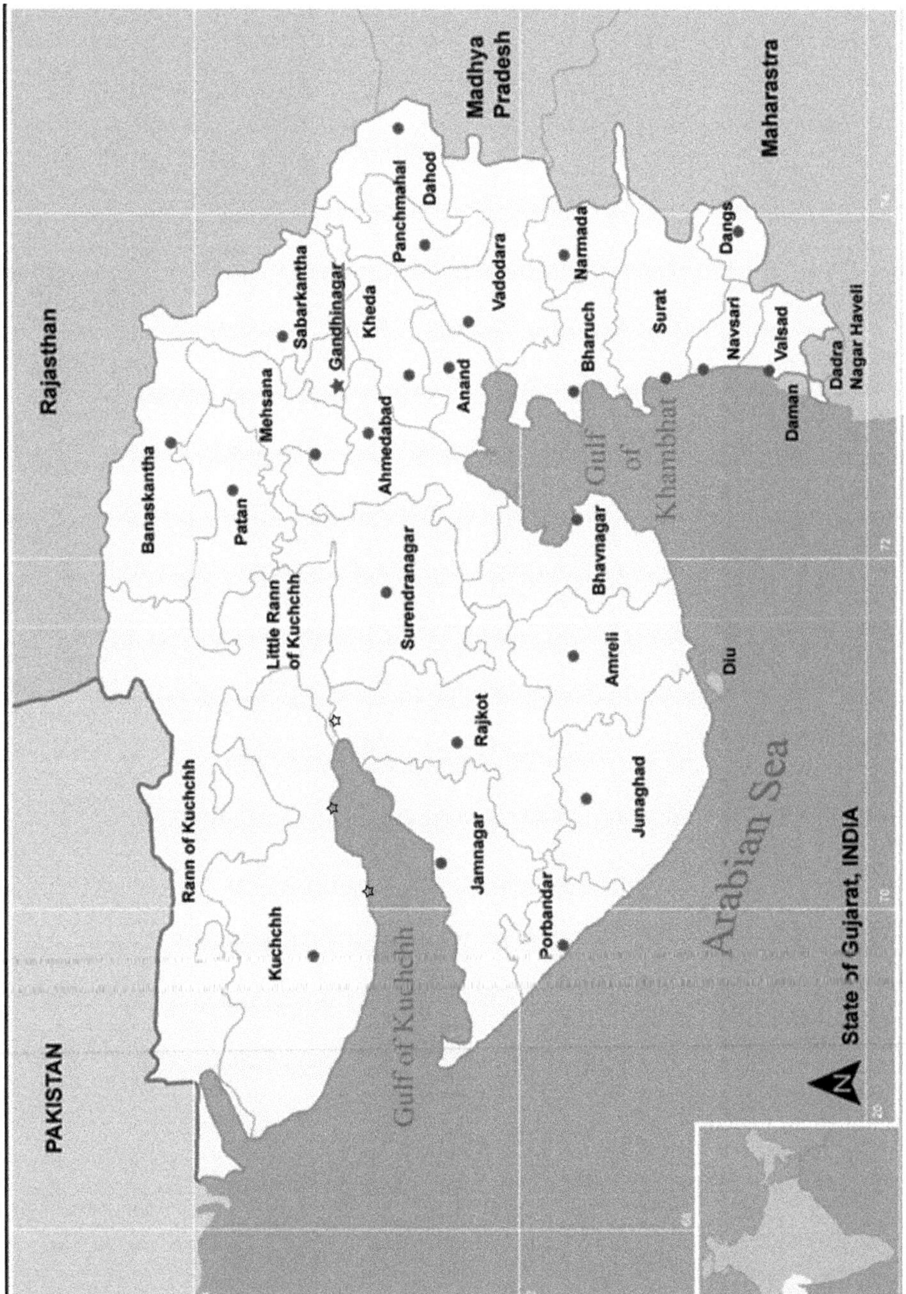

Fig. 5.2. Map of Gujarat state and districts, showing locations of the three case study villages (☆). (From https://commons.wikimedia.org/wiki/File:Map_Gujarat_state_and_districts.png.)

When the land was given to Sanghi Cement, there were agreements signed between the *gram panchayat* and the company. During interviews and discussions many villagers claimed that the *sarpanch* (elected head of a village government) signed the paper without their knowledge. Others told us that the public hearing was in fact held. During the public hearing, rules and promises for primary facilities were made. The agreement stated that villagers would get employment and the young boys would be given permanent jobs. Schools, a pre-school for children from economically weaker households, and a health centre were to be set up in the village. Further, drinking-water facilities were to be provided to the people as well as animals. Arrangement for fodder was to be made and a shelter for cows would be established in the village. Only a few of these promises materialized. Furthermore, Sanghi Cement has encroached on more common land than was defined in the agreement. Ongoing cases in the High Court are challenging this encroachment.

Case 2: Tunda Wandh and the Tata and Adani power plants

Tunda Wandh is a small village in Mundra block of Kutch district in Gujarat. Today, the village is literally sandwiched between two power plants owned by Tata and Adani. The power plants are located just outside the village boundary and overhead conveyor belts pass between the plants, effectively locking in the village (Fig. 5.3). The Adani power plant (commissioned in 2007) and the Tata power plant (commissioned in 2012) were built on village commons (*gauchar* and wasteland) and agriculture land.

The Adani power plant falls under the Adani Port and Special Economic Zone (APSEZ),[5] earlier known as the Mundra Port and SEZ Limited (MPSEZL). The coastal region of Mundra Taluka was targeted to become a new hub of industrialization and infrastructural projects after a large earthquake devastated the region in 2001. This was also India's first multi-product, port-based SEZ. It covers 3225.68 ha land, acquired from at least 14 villages. Besides the private port, the Adani Group also built a jetty and constructed roads, a private railway line, an airport, a power plant, warehouses and townships (Patel, 2013).

Adani has, over the years, leased an additional 7350 ha. This land was acquired primarily from 2005 onwards, from the government, in an area called Mundra in the Gulf of Kutch in Gujarat. Between 2005 and 2007, at least 1200 ha of grazing land were taken away from villagers. Under Indian law, land meant for grazing cattle can be used for something else only if it is in excess, defined by a specific formula (Bahree, 2014). Even then, the village chief

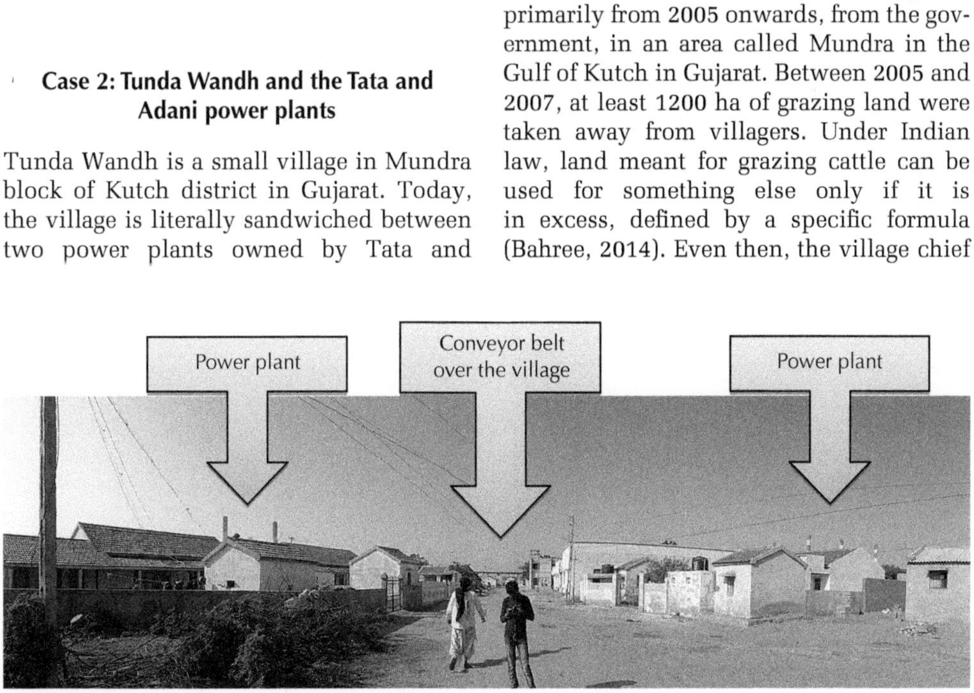

Fig. 5.3. Tunda Wandh village, sandwiched between two power stations, February 2015. (Photo courtesy of J. Duncan.)

has to give permission to take the land. Villagers in Adani's SEZ say their grazing land was signed away by earlier village chiefs without their knowledge. They have filed multiple cases in the Gujarat High Court to contest the government's actions, going back to 2005 and even earlier. Several cases are still pending. There have also been several protests and cases pending in Indian courts against the Adani SEZ on the grounds of the social and environmental impact on affected villages, particularly by fisherfolk and their association in the affected coastal areas. Several committees and teams have been constituted to study the impact on livelihoods and the environment.

Within this area, Tunda Wandh village is home to pastoralists who breed the Kharai camel, also known as the 'swimming camel of Kutch'. This camel is renowned for being able to swim up to 3 km in the sea and is adapted to a land and marine ecosystem. The number of Kharai camels has declined rapidly in the last two decades. For example, one family together owned 250 camels 15 years ago, but now each of the four brothers owns only one or two camels. One survey estimates that until 2003, the village had 1500 such camels (Shrivastava, 2013). There are now less than 250. Until recently the Kharai camel was not recognized as a separate breed. Experts believe now that it deserves special attention and care as it is estimated there are fewer than 10,000 Kharai camels left (*Times of India*, 2015). The Kharai camel feeds on mangroves. The residents of the village have reported that the mangroves are no longer accessible to camels.

In January 2014, the APSEZ faced a major setback when the Gujarat High Court ordered 12 out of 21 units in Mundra SEZ to shut down in response to a public interest litigation filed by some villagers who had claimed that construction activities at the SEZ were affecting their livelihood (*Economic Times*, 2014). However, in July 2014, the APSEZ received environment and coastal regulation zone clearance from the Ministry for Environment and Forests for an 8481 ha SEZ in Mundra (*Business Standard*, 2014).

Case 3: Becharaji-Hansalpur and the Maruti plant

> We tilled this land and our animals grazed on it for 60 years, yet we did not get ownership. How can Maruti get ownership?

Hansalpur is a small village in the Ahmedabad district of Gujarat. Hansalpur was one of the 44 villages under the proposed MB SIR which was partially withdrawn in August 2013. In 2012, the state government had allotted 647 acres (~262 ha) of 'wasteland' to Maruti Suzuki for its proposed plant. Of this land, 220 acres (89 ha) was being used for farming and the remaining as grazing area. This land had been used for at least 50 years by local members of the community. There is clear evidence that in 1954 the government allotted the 220 acres to 14 landless families, including some pastoralists. Since then, these families (now over 35) have been tilling this land and grazing animals on it. There is also evidence that these communities have paid taxes/fees on this land over this time period, further pointing to the fact that the government had recognized their right to access and use this land.

Yet, it appears that in 2007 title to this land was acquired silently by Gujarat Industrial Development Corporation (GIDC).[6] GIDC had acquired land from the village *panchayat* without any consultation or even notice. When the land rights were then transferred to Maruti there was again no public hearing. It appears that in 2008, during computerization of land records, this land was erroneously entered as fallow land instead of agriculture land and the names of the original 14 families were removed from the records. Thus, on paper, GIDC allocated 'wasteland' to Maruti. The families originally given right to the land filed a case to correct this error. However, while the matter was *sub judice*, Maruti went ahead and constructed a fence (see Fig. 5.4). The people and livestock of Hansalpur and neighbouring areas, who had always had access to this land, are now walled out. Moreover, out of 262 ha allotted to Maruti, the company maintains the right to sell 121 ha to the highest bidder. The affected families

Fig. 5.4. Maruti's fence outside the village of Hansalpur blocks access to common grazing lands, February 2015. (Photo courtesy of J. Duncan.)

filed a petition in the High Court. So far there have been more than 50 case hearings but government representatives have failed to attend the majority of them.

Summary of the cases

These cases read as extreme, but on paper, and in practice, they conform to deliberate policy programmes developed and advanced to enhance a very specific model of economic development. They raise important questions for which there are no simple answers: who wins and who loses? These land deals also include important trade-offs and demand analysis of costs versus benefits.

What is clear from our research is that in each case there have been immediate impacts on the lives of pastoralists. One major impact is the loss of access to common grazing land, or lands previously labelled as 'wasteland' despite being used by different communities. The loss of grazing land leads to the loss of livestock, as fodder banks and programmes are not able to supply adequate feed, nor do they often provide feed for smaller livestock such as sheep and goats. The loss of livestock removes the need to migrate and thus promotes a sedentary lifestyle. More than promoting a sedentary lifestyle, it challenges the pastoralists' sense of identity. To be *Maldhari* is to own livestock.

Many pastoralists continue to rear livestock in smaller numbers or change their herd composition to focus on cow and buffalo because there are stable and accessible markets for this milk. Others turn to agriculture, other look for jobs as labourers. This may provide more 'stable' employment for people but it also means changing from working for themselves and their community, to being employed by others. It is important to note here that pastoralists have historically had relatively steady incomes due to the strength and stability of government-supported village dairy cooperatives (Scholten and Basu, 2009; Duncan, 2013). With the factories also come 'strangers' and pastoralists (notably the women) are increasingly worried about their security.

In reflecting on the impact of these land grabs, it is not the change, so much as the speed of change that is striking. While cultures continuously evolve, the restricted access to grazing lands has had immediate repercussions. While pastoralists are keen adapters, they have always had the rangelands to support their adaptation strategies. In the villages where we undertook the research we found multiple cases of pastoralists, who identify as being livestock keepers, no longer able or willing to keep any livestock. While the question about who wins in these land deals cannot be answered, it is clear that, for the time being, the pastoralists we met are losing.

Analytic Framework: Theories of Social Practices

The objective of this chapter is to understand how the social practices of women pastoralists have been impacted by cases of land grabbing such as those described above. To do this, the chapter now makes use of theories of practice. Theories of practice are heterogeneous and are thus not presented here as a coherent theory but rather as a particular reading of an assembly of theoretical and analytic elements (Halkier, 2009, p. 3; see also Schatzki, 2001, 2012; Reckwitz, 2002; Warde, 2005).

A practice is understood to be a routinized behaviour of interconnected elements (Reckwitz, 2002): it is the routinized way in which bodies are moved, objects are handled, subjects are treated, things are described and the world is understood. These behaviours consist of several elements, interconnected to one other (Reckwitz, 2002, p. 250). Key characteristics of practice theory as identified by Reckwitz (2002, p. 249) include: 'forms of bodily activities, forms of mental activities, "things" and their use, a background knowledge in the form of understanding, know-how, states of emotion and motivational knowledge'.

To contextualize these elements and characteristics within pastoralist practices,[7] the chapter focuses on four sub-practices: bodies, mind, knowledge and things (Reckwitz, 2002; de Krom, 2014). Forms of bodily activities relate to movements of the body that are learned and repeated. Practices are performances of and through bodies. For pastoralists, such body activities include milking, activities associated with processing milk (stirring, tasting), transporting milk, inspecting animals, collecting water, sewing, cooking and migrating. Forms of mental activities recognize that while practices are routinized bodily performances, they are also mental activities, which means certain 'routinized ways of understanding the world, or desiring something, of knowing how to do something' (Reckwitz, 2002, p. 251). This can include desiring livestock, knowing migration routes, knowing how to milk and how to process it. Mental activities such as these are distinguished from knowledge, which goes deeper than simply 'knowing' to 'embrace ways of understanding, knowing how, ways of wanting and of feeling that are linked to each other within a practice' (Reckwitz, 2002, p. 253). For pastoralists, this knowledge extends to cultural traditions, rituals, animal husbandry and traditional indigenous knowledge, passed down from generation to generation. The final element relates to things: that is, the objects that form the necessary components of practices. For pastoralists, such things include milk jugs, ropes used to tie up animals when being milked, cots, blankets, traditional dress, items used to make handicrafts and cooking pots. In the analysis, the element of things has been extended to also include other materialities, including grazing land, roads, water, migration routes, markets, schools and veterinary services.

Theories of social practice are not without challenges or critiques. We acknowledge that the routinized nature of practices means that when carriers of new practices (i.e. new land tenure arrangements) enter into a practice the interconnectedness of the elements in routine behaviour can be taken for granted (Crivits and Paredis, 2013, p. 307). We also recognize that elements of practice cannot be easily transposed into empirical analysis, not least because they are deeply interconnected; and further that, because of this, these elements 'tend to be idealized, abstract and insufficiently attentive to the social processes involved in the creation and reproduction of practices' (Warde, 2005, p. 135). However, throughout the fieldwork and data collection, we found these elements helpful for categorizing the social practices of pastoralist women. Indeed, the heterogeneity associated with practice theory poses analytical challenges. Reckwitz (2002, p. 249) notes that a practice 'represents a pattern which can be filled out by a multitude of single and often unique actions reproducing the practice'. Given this, it is difficult to qualify clear descriptive categories. This analysis focuses on the four characteristics (i.e. body, mind, knowledge and things) of practices described above and groups them into practices before the land was grabbed and practices after the land was grabbed. We recognize that the events surrounding the land deal are not the only factors that are likely to shift practices. For example, climate change, urban migration, milk cooperatives, earthquakes and the rapid uptake of mobile phones all impact the social practices of women. However, our analysis focuses on the changes in routinized activities that can be directly linked to the land deals in question.

Warde (2005) identifies other challenges including: ways individuals engage in the same practices; how processes of diversification and multiplication of practices can impact established practices; and how implications of socio-economic and demographic composition of participants can impact routinized behaviours. Schatzki *et al.* (2001) present other limitations to practice approaches, including the implications of referring order to the field of practices so that order is understood as a feature of the field, and that aspects of the field are seen to be responsible for the establishment of the order. Schatzki (2001, p. 14) also challenges the tendency to reduce the scope and ordering power of reason, thereby 'abandoning the traditional conception of reason as an innate mental faculty and reconceptualising it as a practice phenomenon'. In contrast to Schatzki's concerns, we use practice theory explicitly because it moves away from claims to reason. Importantly, a practice approach allows analysts to avoid the normative and relativistic assumptions that often accompany research into livelihoods. Here, we do not seek to understand the logic in the change of practice, or assess how 'useful' or 'rational' these changes are for their survival. We want to understand what practices are changing and what new practices are emerging. From here we can reflect on implications.

The following recreates a 'typical day in the life of a woman *Maldhari*' living in a village impacted by land grab as a way of highlighting social practices. The quotes are taken from the various focus groups and interviews and thus respect the diversity across communities and cases, while also giving voice to the women who participated in this research. The names or locations of the people quoted are not disclosed to protect their identities. In this overview, we pay particular attention to the way in which social practices have changed as a result of loss of access to land and engagement in land-related protests in the three case study sites. Given the integration of livestock into the traditional (pre-land grab) social practices of pastoralists, the most obvious changes to social practices post-land deals are linked to the loss of livestock. However, we recognize that it is not always possible to attribute changes to social practices to the loss of access to land. We accept that other processes (e.g. introduction of new technologies, climate change) take place simultaneously and also rapidly change social practices. We try, to the best of our ability, to focus the analysis on changes that are clearly linked to the loss of access to land. Focusing on the pre- and post-land grab practices provides insight not only into how women pastoralists are impacted by land grabbing, but also how they respond.

A 'typical day'

Women have slept indoors with the children. The men have slept outside. Women used to sleep outside but are now too scared to sleep outside because of the influx of strangers who have come to work in the factories and settled in nearby villages. The women wake around 05:00 and dress in the dark. Women, especially from the *Rabari* community, wear long, heavy wool skirts and colourful tops that are bunched at the chest and tied at the back. The women also wear cotton or wool shawls, called *odhani*, over their heads. The women do not attach these shawls and are thus constantly readjusting them as they go about their daily activities. Women who have married into a village use the shawl to cover their face when an older man comes into view. They do this automatically, almost without thinking: drawing the shawl in front of their face, while they continue to talk, work and eat. They pull the shawl back when the man has left. If they are in their home community, they do not cover their face.

The *odhani* cost about 4000 rupees (55 euros), the women say. Due to the decline in the number of sheep, these woolen shawls are becoming more expensive. The older women explain that they prefer to wear wool shawls over cotton ones because they feel embarrassed to wear cotton, a much thinner material. Many of the pastoralist women (especially married and older women) we spoke to were not comfortable wearing non-traditional clothes. This causes tensions for some women who have now taken up work in factories where they are asked to wear cotton uniforms. As one respected female pastoralist leader remarked, '[we] sold off the livestock, now they ask 60-year-old women to wear uniforms'.

Once dressed, the women who still keep livestock walk to gather the animals and undertake the physical work of tying up the animals and separating them from their calves. The women, and often younger girls, easily identify the milking animals. The women make sounds to soothe the animals. They assume a deep squatting position and start to milk: one hand firmly gripped around the udders, thumb knuckles pressing down to extract the milk. Often the other hand is used to hold the milk jug, to pat the animal, or is also engaged in the practice of milking. The animals are untied and reunited with their young. The milk is then divided: milk for the family's consumption and milk to be sold. Most often the milk is carried to the location where it can be picked up for delivery to the local milk cooperatives; however, some *Maldhari* have arranged direct sales of their milk to neighbours, tea stands or sweet shops. Milk that is to be consumed by the family is boiled. This requires searching for firewood, lighting and tending to the fire, and watching to make sure that the milk does not boil over or burn.

The sons sleep while the daughters wake up and start to sweep and dust around the house. Younger women are sent to collect water. Collecting water is a key social practice. When a visitor arrives, it is customary to offer them a glass of water. These regions are semi-arid and water is a precious resource. In the case of Jadwa village, the water from a seasonal river has been diverted away from the village towards the cement mines, meaning the journey to collect water has increased. There is a well in the village for drinking but no water for livestock. The women explain that in dry periods, girls have to walk up to 5 km to collect water for the family. This impacts their ability to get to school on time, or at all. The trade-off represents a new worry for many of the older women that we spoke to.

Women are responsible for making all the meals for the family. In the morning they prepare breakfast and if they work outside the village, they also prepare lunch at this time. During our early morning walks around the villages, we watched the women cooking wheat roti. The women spent time in the morning mixing the dough from water and flour in stainless steel pans, rolling the prepared dough into balls, before rolling these out into perfect circles to be thrown on to a cast iron pan or clay dish laying in a fire. Once bubbles appear, the women flip the bread and then rotate them around the dish with their fingers as the roti start to brown and puff up. The roti are then taken off the fire and placed in a pile, ready to be eaten for breakfast or to be packed up and taken to the factories, fields or grazing lands for lunch.

While most of the women we met prepared wheat roti, traditional diets tended to centre around milk and millet. Rather than roti, rotla, a traditional flat bread made of millet, were the common breakfast food by pastoralists in Gujarat in the past. When we asked why they made roti instead of rotla, women tended to give two reasons. First, the roti take less time to cook and are easy to carry to the factory in a lunch box. Second, they said they had been influenced by workers from other states who bring roti in their lunch box. This points to a change in the practices of cooking, with implications for elements linked to cooking methods, utensils and the storage of food. In the case of Jadwa, earlier, pastoralists had access to plots of land that they either owned or worked on as agriculture labourers. Here they cultivated seasonal food and fodder crops such as millet, sorghum and moong (a type of lentil). This meant that village and families had usual access to grains, along with milk. Now the women buy wheat flour to make food. The older women expressed worry and concern about the nutritional quality of the wheat roti compared with the robust millet rotla they were raised on. But they also said in the villages that, due to dust and pollution from the cement factory, there is no point in continuing with agriculture.

In the villages we visited, people who had their own agricultural plots or fields, or fodder crops, complained that the pollution from the factory had negatively impacted their yields. In one village, a woman, laughing and pointing to one of us with pale skin, exclaimed: 'You will turn brown when you go outside' [because of the pollution]. While we did not measure the environmental impact of the factory on the fields, we noted in two case study locations that the air was heavy with dust and soot. Women showed us the amount of dust that collected in their homes; and how, despite daily cleaning, when they walked they left footprints in the dust that was constantly settling on the floor. They told us about the burden of additional cleaning.

After the breakfast and cleaning are complete, some women from the village tend to the livestock. One woman explained, 'before we were all busy with livestock'. They told how they used to have access to 200 acres (~80 ha) of grazing but the land was encroached by farmers. They are now left with 10 acres (4 ha) for grazing. Not enough to sustain their livestock. Some families opted to re-start the migration for grazing land, while others opted to give up the livestock and now work as labourers for the farmers who claimed that land.

The women with jobs in the factory start to walk to work around 08:00. They pick each other up at their homes along the way. Not all who walk to the factory have guaranteed employment. Some have work cards; others will walk upwards of an hour in the hope of receiving daily employment. The wages differ depending on the factory and the job but most women reported making between 110 and 200 rupees per day (1.50–2.80 euros). They are paid a daily wage but most receive the money at the end of the month. Work starts at 09:00 and continues to 17:00. They tell us that they get one hour for lunch between 13:00 and 14:00. The women are assigned mostly cleaning jobs. One of the women working as a cleaner at one of the factory's offices has the responsibility of cleaning 52 rooms per day. When she met with other women from her village in the evening she complained that her knees were sore, and they cracked in agreement as she sat down on the floor.

After the shifts end, women walk back to the village in groups. At about 19:00 the women start to prepare dinner. Around the village small fires are lit, and women take out pots and pans and start to prepare the food. Around this time, the boys and men who did not work at the factory come back to the village having spent the day guiding livestock to the grazing land. Just as it was with breakfast, the practices associated with cooking are changing. The rhythm of meals has shifted from a synchronicity with the milking, eating and breeding cycles of livestock, to following the schedule of the factories. After the meal, the women clear and wash up. They place all the dirty stainless steel dishes into a tub where they are washed and normally left to dry on an unused cot, where they quickly collect dust. The women get to sleep after 23:00. Each morning they wake up at about 05:00 to repeat the routines of the day before.

Research Findings

The recreation of a typical day of women in villages impacted by land grab has highlighted daily practices and related shifts in these practices brought about by changing access to land. One of the key implications, if not the most important implication, of land grabbing in the three case study areas has been the loss of access to grazing land for livestock. This has contributed to herd decline in each community and in turn had significant impacts on social practices. This section reflects on the impacts as they relate to changing rhythms or patterns or daily life, changing economic relations within the household, and the emergence of new worries.

New rhythms

As illustrated above, with the loss of animals, social practices which were so heavily dictated by the rhythm and needs of livestock have shifted. New social practices are developing around more formalized employment. This shift is marked by new practices of migrating, not with animals, but with the other women to the factories, fields and homes in peri-urban areas where some women (especially younger ones) are finding employment as domestic workers.

For those women who stay at home, new tensions between men and women were identified at the household level, notably around space and time which is being reorganized as a result of lost livestock. As one woman explained: ‘Before [we lost the livestock], men took the livestock and left the home early morning. Now they are at home during day, asking about everything.’ Another woman explained:

> Lots of land was available 15 years ago. Now it is not possible because of industry and land pollution. Land, grazing land, is not available. Before, men took care of women. Now there are no animals, no money and you see more violence. Men do not take care of women.

Women also explained that with the factory work they have lost their free time. Now they come home from the factories late in the evening. Men in the family often work later shifts and the women wait to serve them food. In the past, women used to have free time in the afternoon to partake in activities such as handicrafts.

Traditionally, women worked on heavily embroidered clothes, bedcovers and decorative pieces for homes. These were often exchanged as wedding gifts. There is now little time to continue this tradition.[8] The *Rabari* community is renowned for its embroidery skills. Indeed, there are key examples around this region of families losing livestock and turning to the production and sale of handicrafts, also opening up their villages for tourists. This is not an option for all, of course.

The limits of free time also impact family relations:

> Before [fixed working hours in factory and payment on number days worked in a month] we used to enjoy when relatives come to visit; now we feel they should leave as soon as possible. The relationships are not as ‘sweet’ now.

New economic relations

Women talked about how the practices associated with finance were changing, especially the way in which finances were managed. Traditionally income (from the sale of milk) was received, kept and managed by women. Now, income is often split depending on who sells milk and who works in the factories. Increasingly, men are taking financial decisions. This changes the family dynamics and power relations within the household, giving women a weaker position. One woman told us that now 'more labour work is available and this is a preference. Earn daily, spend daily. Men do not want livestock. Women have more responsibilities at home . . . earlier, men needed women. Now, men earn every day.' Another woman noted that, 'men want the high life, women think of family and the house'.

Women whose families had sold their livestock and/or land also reported that changes in the cash economy at the household level were linked to increased alcoholism (a taboo in these communities) and more money being spent on phones, motorbikes and other luxuries.

Instead of working in the traditional co-dependent ways with their husbands, and maintaining control over the family finances, new arrangements and practices are emerging which appear to enhance the burden of housework, decrease financial influence and increase vulnerability both by reducing the co-dependence between husbands and wives and by introducing a private space into previously communal and open societies (i.e. the introduction of houses with full walls and doors that close compared with the open-style traditional dwellings).

Worry

Throughout the day, women pastoralists in communities affected by land grab engage in the practice of worry: collective and individual worry. The women worry about the impact of the factory jobs on their health. They worry about the impact of the pollution on the crops, the remaining animals and on their own health. They worry about the newly diagnosed cases of cancer in the village and worry who will be next. They worry about the future of their culture and the future of their children. They worry about the changing dynamics in the household. They worry about the influx of people from outside the village, who come to work in the factories: 'earlier we did not fear anything, these days young girls cannot go out on their own.' They worry about alcohol entering into the communities. They worry for their livestock. This is not to suggest that there were not worries before. Our aim is not to measure the weight of worry, or even the impact, but to note that the practice of worrying, as it is performed through bodies and minds, and the translation of these worries into other practices, is evident across the case study sites.

With respect to worrying about the future, many of the women we spoke with, especially those over the age of 25 years, were struggling with deep ambivalence: with sadness and worry over the loss of their animals, their traditions, their culture and also hope for the future. They explained, with pride, that their daughters are getting an education. But older woman expressed concern: 'Young girls and boys go to school, so who will take care of livestock?'

When asked, older women overwhelmingly said that they preferred to raise livestock than to work in a factory. One woman who had lost her livestock and taken up work managing the livestock of a wealthier pastoralist family, told us: 'Before it was better. We owned livestock and could meet our expenses. I had sheep and sold *ghee* in the market. Now there is no market for sheep products.' Young girls tended to express the opposite desire. Laughing, one girl said, 'we don't like cow dung'. In another case site, the girls expressed similar sentiments: 'I do not want my husband to smell of cow dung'. And some middle-aged women agreed. One woman, whose family had joined with another to start a business renting farm equipment (i.e. tractors), explained that:

> with livestock there was more trouble. We had to walk a long way and we were vulnerable. There was fear that the livestock would be stolen. Now it is more peaceful. When there is no work, we just sit.

Yet, while these women no longer migrated with livestock by foot in search of grazing land, their families maintained herds of sheep and goats which came with them when the families embarked on a modern migration, driving to where their farm vehicles could be rented. One of the men takes the animals out to graze during the day, but morning and night, it is the women who milk the animals, prepare the food, take care of the children, and work on the elaborate traditional clothes and related handicrafts.

In one village, a woman explained: 'We migrated when we had smaller animals. We had to move. Now if we keep cows in the village, we can send our children to school. Our children will never raise livestock.' A man cut in: 'If there are three sons, hopefully one will keep livestock.' The woman continued: 'Our sons went to school. They have good jobs.'

Yet not everyone sees things that positively. In the vast majority of the families we spoke to who had given up their livestock, the men took up insecure positions as daily wage labourers. A community leader lamented that without livestock 'everyone will be beggars'. When asked what women can do to save livestock and pastoralism, the answers were the same: secure access to land. As one woman said: 'We need access to grazing land. We can do hard work, but we need access to grazing land.' One social activist working for over two decades in the region told us: 'The company will not feed them forever. Once construction is over, many of them will lose their temporary jobs.'

Another woman explained: 'We migrated, but the coming generation will not. It is up to God what happens to us, to our community. I do not have much hope.' When we asked what will happen in 10 years, one pastoralist woman leader said, '*Maldhari* will be wage labour'. A woman from another community explained: 'It will be a bad situation. We have no land, no livestock, no milk, no manure, no *Maldhari*.' When we asked what she hopes for the future of *Maldhari* communities, one woman told us that she hopes for 'dignity, for status, for livestock'. This echoes a common *Maldhari* saying *Maal chhe to mobho chhe*, meaning 'there is dignity only with livestock'.

It is important to note that this description of worry is not presented to evoke pity. It is presented to highlight the way in which changes to land access and the related loss of livestock, and in turn identity, is producing new routinized behaviours that include the mind, body, knowledge and the material world. We also note that alongside this worry, another social practice is emerging: a practice of resistance (Fig. 5.5). Here, the sub-elements of the practice become clear. The women are bringing their bodies to protests, to trainings and putting their bodies at risk of police violence, which is not uncommon at land demonstrations. They are thinking constantly about the value of land and their minds are dealing with the psychological threat of violence and the psychological issues associated with a loss of livelihood and culture. They are gaining and sharing new knowledge about their own rights, the court systems and resistance strategies. They are saving key papers, making use of mobile phones, designing and carrying protest signs. They are crucial to a growing resistance movement. As one woman noted: 'If men were enough, we would not have joined'.

Conclusions

The dynamics of development in an era of neoliberal globalization are complex. Land is at the centre of many of the dynamics and is increasingly a site of political tension and struggle. This is clear from the three cases presented in this chapter. However, our aim was not to focus explicitly on the dynamics of land grabbing in Gujarat, but rather to gain better insight into how this affects

Fig. 5.5. Women pastoralists defend their land right. (Photo courtesy of MARAG.)

women pastoralists by analysing their social practices.

Through a focus on changes in social practices we aimed to avoid some of the normative traps of a livelihood analysis. Practice theory aims to show how social beings, with their diverse motives and intentions, make and transform the world in which they live. It is a dialectic between social structure and human agency working back and forth in dynamic ways. Thus, the approach does not aim to apply a normative assessment to the actions of people. There are no evaluative criteria to assess whether the changes in practices are logical or rational. Understanding changes in practices sheds light on the challenges and adaptation strategies of, in this case, women pastoralists.

Our research shows that in the villages we visited, the land grab has led to the loss of livestock. This has led to shifts in social practices of pastoralist women, particularly around practices related to work and to domestic practices. As pastoralist societies adapt to the loss of livestock, the household dynamics are changing. Our research also illustrates the emergence of two new, interlinked social practices: deep worry and resistance. Yet it is not as easy as tallying up the wins and losses. Many of the women we spoke to struggle with the trade-offs of losing traditional practices but new routines are emerging: for example, children, including girls, have improved formal education.

What is clear is that changes in the routinized ordering of social practices of women pastoralists highlight that these land deals, which are being framed at the government level as good for development, are having impacts on some of Gujarat's most vulnerable communities. What is more, we fear the damage has already been done. The livestock has been lost, the quality of the land diminished and access to traditional grazing routes has been restricted. The knowledge and skills that make up the traditional social practices of pastoralists are being rapidly replaced and it is not clear that they can be re-routinized. For many of the women we spoke to, this is a true tragedy. It is a loss not only of their lifestyle, culture and heritage, but also their identity. Yet, as noted above, we acknowledge that a clear generational gap exists, as girls and younger women felt less tied to their cultural traditions.

Given the very selective sampling it is difficult to propose specific policy recommendations. However, it is accepted that effective land tenure governance is needed to ensure that the rights and access of people to land are respected (FAO, 2012b). India and the State of Gujarat are in the process of developing and reassembling land rights laws and regulations. All efforts should be taken to ensure that customary and traditional land rights are respected. Clear guidelines for the governance of land tenure have been developed and internationally endorsed and should be used by policy makers, business and civil society to ensure that rights are respected (FAO, 2012b). There are also gender-specific guidelines and technical tools to ensure that the rights of woman are taken into account (FAO, 2013; Hannay and Scalise, 2014). These should be adapted to the realities of pastoralist communities and applied where and when appropriate. Furthermore, more work is needed to understand the role and value of the commons in Gujarat. We recognize that the application of stronger governance mechanisms will prove a challenge given that, for example, despite a Supreme Court ruling to clear encroachments and restrict the sale of pastoral (*gauchar*) land, Gujarat has more than 12,000 cases related to illegal possession of pastoral lands (Dave, 2014).

Our hope is in the power of the people and in the emerging practices of resistance. This research affirms that pastoralists have a strong capacity to adapt while also developing strong networks to support knowledge sharing, dissent and protest. The organization of so-called 'pastoralist parliaments' and public demonstrations as well as the launch of more formal legal procedures is illustrative of enhanced civic engagement and renewed feelings of pride and identity in the face of immeasurable opposition.

Notes

[1]*Panchayat* refers to Panchayati Raj, a system of governance modelled on the traditional South Asian Panchata system. The system has three levels: *gram panchayat* (village level), *mandal parishad* or *block samiti* or *panchayat samiti* (block level) and *zila parishad* (district level).

[2]A notification signifies a public act by the government by way of publication, etc., whereby the concerned resolution is announced to the public in the form of a notification or order or rule.

[3]A lakh is a unit in the Indian numbering system equal to 100,000.

[4]Zadwa is also commonly referred to as Jadva.

[5]Industrial units within special economic zones get a waiver on import duty, excise and service tax for capital goods and raw materials procured. These zones have been promoted to attract foreign direct investment as well as to push exports.

[6]GIDC is the government nodal agency that acquires land to be given to industries. The primary purpose of GIDC is to promote small- and medium-scale industries and to promote public purposes.

[7]Here, 'pastoralist practices' is explicitly referring to how the practices employed by the pastoralists interviewed in Gujarat, India have been impacted by cases of land grabbing. The heterogeneity of practices in India and throughout the world is recognized.

[8]Note here that the focus on handicrafts is a contentious issue, with some *Maldhari* leaders going so far as to ban the practice in response to girls staying home from school to embroider for weddings. In discussions, women spoke of missing the practice of embroidering, notably the social element.

References

Agarwal, B. (1994) *A Field of One's Own: Gender and Land Rights in South Asia.* Cambridge University Press, Cambridge, UK.

Bahree, M. (2014) Doing big business in Modi's Gujarat. *Forbes Asia*, 12 March. Available at: https://www.forbes.com/sites/meghabahree/2014/03/12/doing-big-business-in-modisgujarat/#4fc9a2427e61 (accessed 11 November 2016).

Behnke, R. and Freudenberger, M.S. (2013) *Pastoral Land Rights and Resource Governance: Overview and Recommendations for Managing Conflicts and Strengthening Pastoralists' Rights.* US Agency for International Development, Washington, DC. Available at: https://www.land-links.org/issue-brief/pastoral-land-rights-and-resource-governance/ (accessed 6 July 2017).

Behnke, R., Scoones, I. and Kerven, C. (1993) *Range Ecology at Disequilibrium: New Models of Natural Variability and Pastoral Adaptation in African Savannas.* Overseas Development Institute and International Institute for Environment and Development, London.

Behrman, J., Meinzen-Dick, R. and Quisumbing, A. (2012) Gendered impacts of commercial pressure on land. *Journal of Peasant Studies* 39(1), 49–79.

Blench, R. (2000) *Extensive Pastoral Livestock Systems: Issues and Options for the Future. Prepared for FAO–Japan Cooperative Project 'Collection of Information on Animal Production and Health'.*

Overseas Development Institute, London. Available at: http://dlc.dlib.indiana.edu/dlc/bitstream/handle/10535/5313/Blench.pdf?sequence=1 (accessed 11 November 2016).

Borras, S.M., Hall, R., Scoones, I., White, I. and Wolford, W. (2011) Towards a better understanding of global land grabbing: an editorial introduction. *Journal of Peasant Studies* 38(2), 209–216. Available at: http://dx.doi.org/10.1080/03066150.2011.559005 (accessed 4 March 2017).

Bromly, D.W. (1992) *Making the Commons Work: Theory, Practice, and Policy.* ICS Press, San Francisco, California.

Business Standard (2014) Environment clearance for Adani's Mundra SEZ. *Business Standard*, 17 July. Available at: http://www.business-standard.com/article/companies/environment-clearance-for-adani-s-mundra-sez-114071600823_1.html (accessed 11 November 2016).

Coughenour, M.B., Ellis, J.E., Swift, D.M., Coppock, D.L., Galvin, K., McCabe, J.T. and Hart, T.C. (1985) Energy extraction and use in a nomadic pastoral ecosystem. *Science* 230(4726), 619–625. Available at: http://dx.doi.org/10.1126/science.230.4726.619 (accessed 4 March 2017).

Crivits, M. and Paredis, E. (2013) Designing an explanatory practice framework: local food systems as a case. *Journal of Consumer Culture* 13(3), 306–336. Available at: http://dx.doi.org/10.1177/1469540513484321 (accessed 4 March 2017).

Dave, K. (2011) Land of milk fast losing its pastures. *The Indian Express*, 6 March. Available at: http://archive.indianexpress.com/news/land-of-milk-fast-losing-its-pastures/758516/ (accessed 11 November 2016).

Dave, K. (2014) Over 11,000 encroachments on pastoral land in Gujarat. *The Times of India*, 2 April. Available at: http://timesofindia.indiatimes.com/city/ahmedabad/Over-11000-encroachments-on-pastoral-land-in-Gujarat/articleshow/33093379.cms (accessed 11 November 2016).

Deere, C.D. and Leon, M. (1998) Gender, land, and water: from reform to counter-reform in Latin America. *Agriculture and Human Values* 15(4), 375–386.

Deininger, K., Jin, S. and Nagaranjan, H.K. (2007) Land reforms, poverty reduction, and economic growth: evidence from India. *Journal of Development Studies* 45(4), 496–521. Available at: http://dx.doi.org/10.1080/00220380902725670 (accessed 4 March 2017).

de Krom, M.P.M.M. (2014) Governing animal–human relations in farming practices: a study of group housing of sows in the EU. *Sociologia Ruralis* 55(4), 417–437. Available at: http://dx.doi.org/10.1111/soru.12070 (accessed 4 March 2017).

de Schutter, O. (2011) How not to think of land-grabbing: three critiques of large-scale investments in farmland. *Journal of Peasant Studies* 38(2), 249–279. Available at: http://dx.doi.org/10.1080/03066150.2011.559008 (accessed 4 March 2017).

Doss, C. (2013) Intrahousehold bargaining and resource allocation in developing countries. *World Bank Research Observer* 28(1), 52–78.

Duncan, J. (2013) The White Revolution and reordering of relations among the pastoralists of Gujarat: a case for pastoralist policies. *Food Chain* 3(1–2), 89–103.

Economic Commission for Africa (2004) *Land Tenure Systems and Their Impacts on Food Security and Sustainable Development in Africa. ECA/SSD/05/09.* Economic Commission for Africa, Addis Ababa. Available at: https://www.unscn.org/web/archives_resources/files/Land_tenure_systems.pdf (accessed 4 February 2017).

Economic Times (2014) Adani Ports' Mundra special economic zone gets environment clearance. *The Economic Times*, 17 July. Available at: http://articles.economictimes.indiatimes.com/2014-07-17/news/51656810_1_the-sez-mundra-sez-environmental-clearance (accessed 11 November 2016).

Ellis, J.E. and Swift, D.M. (1988) Stability of African pastoral ecosystems: alternate paradigms and implications for development. *Journal of Range Management* 41(6), 450–459.

FAO (2012a) *More on Tenure Governance.* Food and Agriculture Organization of the United Nations, Rome. Available at: http://www.fao.org/fileadmin/user_upload/nr/land_tenure/pdf/More_on_Tenure_and_its_Governance_Eng.pdf (accessed 11 November 2016).

FAO (2012b) *Voluntary Guidelines for the Responsible Tenure of Land, Fisheries and Forests in the Context of National Food Security.* Food and Agriculture Organization of the United Nations, Rome. Available at: http://www.fao.org/docrep/016/i2801e/i2801e.pdf (accessed 4 February 2017).

FAO (2013) *Governing Land for Women and Men: A Technical Guide to Support the Achievement of Responsible Gender-Equitable Governance Land Tenure.* Food and Agriculture Organization of the United Nations, Rome. Available at: http://www.fao.org/docrep/017/i3114e/i3114e.pdf (accessed 11 November 2016).

Fratkin, E. (1997) Pastoralism: governance and development issues. *Annual Review of Anthropology* 26, 235–261.

Government of India (2012) *19th Livestock Census – 2012. All India Report. Government of India, New Delhi.* Available at: http://dahd.nic.in/sites/default/files/Livestock%20%205.pdf (accessed 4 February 2017).

GRAIN (2008) *Seized! The 2008 Land Grab for Food and Financial Security.* GRAIN, Barcelona. Available at: https://www.grain.org/article/entries/93-seized-the-2008-landgrab-for-food-and-financial-security (accessed 4 February 2017).

Halkier, B. (2009) A practice theoretical perspective on everyday dealings with environmental challenges of food consumption. *Anthropology of Food* (Suppl. 5), 1–19. Available at: http://aof.revues.org/6405 (accessed 11 November 2016).

Hall, D. (2011) Land grabs, land control, and Southeast Asian crop booms. *Journal of Peasant Studies* 38(4), 837–857. Available at: http://dx.doi.org/10.1080/03066150.2011.607706 (accessed 4 March 2017).

Hannay, L. and Scalise, E. (2014) Securing women's land tenure in northern Uganda: an empowerment approach. *Annual World Bank Conference on Land and Poverty*, Washington, DC, March 24–27, 2014. Available at: http://www.landesa.org/wp-content/uploads/Securing-Womens-Land-Tenure-in-Northern-Uganda-A-Women-First-Approach-Hannay-Scalise-March-2014.pdf (accessed 11 November 2016).

Hardin, G. (1968) The tragedy of the commons. *Science* 162(3859), 1243–1248. Available at: http://dx.doi.org/10.1126/science.162.3859.1243 (accessed 4 March 2017).

Harvey, M. and Pilgrim, S. (2011) The new competition for land: food, energy, and climate change. *Food Policy* 36(Suppl. 1), S40–S51. Available at: http://dx.doi.org/10.1016/j.foodpol.2010.11.009 (accessed 4 March 2017).

Hjort, A. (1981) A critique of 'ecological' models of pastoral land use. *Ethnos* 46(3–4), 171–189. Available at: http://dx.doi.org/10.1080/00141844.1981.9981220 (accessed 4 March 2017).

ISRO (2007) *Desertification and Land Degradation: Atlas of India.* Indian Space Research Organisation, Bangalore. Available at: http://www.isro.gov.in/desertification-and-land-degradation-atlas-released (accessed 11 November 2016).

Kaur, R. (2011) Return of village land. *Down to Earth*, 28 February. Available at: http://www.downtoearth.org.in/news/return-of-village-land-33020 (accessed 11 November 2016).

Kelkar, G. (2013) The fog of entitlement: women and land in India. *Annual World Bank Conference on Land and Poverty*, Washington, DC, April 8–11, 2013. Available at: http://www.landesa.org/wp-content/uploads/The-Fog-of-Entitlement-Women-and-Land-in-India-261-Kelkar.pdf (accessed 11 November 2016).

Köller-Rollefson, I. (1994) Pastoralism in western India from a comparative perspective: some comments. In: *A Collection of Papers from Gujarat and Rajasthan.* Overseas Development Institute, London, pp. 3–5. Available at: www.odi.org.uk/resources/docs/5405.pdf (accessed 11 November 2016).

Lambin, E.F. and Meyfroidt, P. (2011) Global land use change, economic globalization, and the looming land scarcity. *Proceedings of the National Academy of Sciences USA* 108(9), 3465–3472. Available at: http://dx.doi.org/10.1073/pnas.1100480108 (accessed 4 March 2017).

Mahapatra, R. (2012) Uncommon swing for commons. *Down to Earth*, 15 June. Available at: http://www.downtoearth.org.in/news/uncommon-swing-for-commons-38292 (accessed 7 February 2017).

MARAG (2012) MARAG (Maldhari Rural Action Network) homepage. Available at: http://www.marag.org.in/ (accessed 11 November 2016).

Margulis, M.E. and Porter, T. (2013) Governing the global land grab: multipolarity, ideas, and complexity in transnational governance. *Globalizations* 10(1), 65–86. Available at: http://dx.doi.org/10.1080/14747731.2013.760930 (accessed 4 March 2017).

Maxwell, D. and Wiebe, K. (1999) Land tenure and food security: exploring dynamic linkages. *Development and Change* 30(4), 825–849. Available at: http://dx.doi.org/10.1111/1467-7660.00139 (accessed 4 March 2017).

McCabe, J. (1990) Turkana pastoralism: a case against the tragedy of the commons. *Human Ecology* 18(1), 81–103. Available at: http://dx.doi.org/10.1007/BF00889073 (accessed 4 March 2017).

McMichael, P. (2012) The land grab and corporate food regime restructuring. *Journal of Peasant Studies* 39(3–4), 681–701. Available at: http://dx.doi.org/10.1080/03066150.2012.661369 (accessed 4 March 2017).

Nori, M., Taylor, M. and Sensi, A. (2008) *Browsing on Fences and Adaptation to Climate Change. Issue Paper No. 148.* International Land Coalition, Rome. Available at: http://pubs.iied.org/pdfs/12543IIED.pdf (accessed 11 November 2016).

ODI (2009) *Pastoralism and Climate Change: Enabling Adaptive Capacity. Synthesis Paper.* Overseas Development Institute, London. Available at: https://www.odi.org/sites/odi.org.uk/files/odi-assets/publications-opinion-files/4317.pdf (accessed 11 November 2016).

Ostrom, E., Burger, J., Field, C.B., Norgaard, R.B. and Policansky, D. (1999) Revisiting the commons: local lessons, global challenges. *Science* 284(5412), 278–282. Available at: http://dx.doi.org/10.1126/science.284.5412.278 (accessed 4 March 2017).

Pallas, S. and Miggiano, L. (2011) *Women's Legal Empowerment: Lessons Learned from Community-based Activities. ILC Briefing Note.* International Land Coalition, Rome. Available at: http://www.landcoalition.org/en/resources/womens-legal-empowerment-lessons-learned-community-based-activities (accessed 11 November 2016).

Patel, I. (2013) Impact of industrialization on rural community in Jharpara Village (Kutch District). *Network: Quarterly Publication of IRMA* 17(1&2), 7–10.

Rangnekar, S. (1994) Women pastoralists, indigenous knowledge and livestock production in northern Gujarat. In: *A Collection of Papers from Gujarat and Rajasthan.* Overseas Development Institute, London, pp. 15–16. Available at: https://www.odi.org/sites/odi.org.uk/files/odi-assets/publications-opinion-files/5405.pdf (accessed 11 November 2016).

Reckwitz, A. (2002) Toward a theory of social practices: a development in culturalist theorizing. *European Journal of Social Theory* 5(2), 243–263. Available at: http://dx.doi.org/10.1177/13684310222225432 (accessed 4 March 2017).

Rodriguez, L. (2008) *A Global Perspective on the Total Economic Value of Pastoralism: Global Synthesis Report based on Six Country Valuations.* World Initiative for Sustainable Pastoralism and International Union for Conservation of Nature, Nairobi. Available at: https://cmsdata.iucn.org/downloads/tev_report.pdf (accessed 11 November 2016).

Schatzki, T.R. (2001) Introduction: practice theory. In: Schatzki, T.R., Cetina, K.K. and von Savigny, E. (eds) *The Practice Turn in Contemporary Theory.* Routledge, London, pp. 10–23.

Schatzki, T.R. (2012) A primer on practices: theory and research. In: Higgs, J., Barnett, R., Billett, S., Hutchings, M. and Trede, F. (eds) *Practice-based Education.* Sense Publishers, Rotterdam, pp. 13–26.

Schatzki, T.R., Cetina, K.K. and von Savigny, E. (2001) *The Practice Turn in Contemporary Theory.* SAGE Publications, Inc., Thousand Oaks, California.

Scholten, B.A. and Basu, P. (2009) White counter-revolution? India's dairy cooperatives in a neoliberal era. *Human Geography* 2(1), 17–28.

Sharma, V.P., Köhler Rollefson, I. and Morton, J. (2003) *Pastoralism in India: A Scoping Study.* Indian Institute of Management, Ahmedabad and League for Pastoral Peoples, Ober-Ramstadt. Available at: http://www.dfid.gov.uk/r4d/PDF/outputs/ZC0181b.pdf (accessed 11 November 2016).

Shepard, D. and Mittal, A. (2009) *The Great Land Grab: Rush for World's Farmland Threatens Food Security for the Poor.* Oakland Institute, Oakland, California. Available at: https://www.oaklandinstitute.org/sites/oaklandinstitute.org/files/great-land-grab.pdf (accessed 7 February 2016).

Shrivastava, K.S. (2013) The sinking ship: A camel breed that can swim struggles for survival as its food source dwindles. *Down to Earth*, 15 April. Available at: http://www.downtoearth.org.in/coverage/the-sinking-ship-40705 (accessed 11 November 2016).

Sietchiping, R. (2010) Political economy of land access and control over natural and land-based resources: a means for conflicts prevention and management? In: *CAPRi Workshop on Collective Action, Property Rights, and Conflict in Natural Resource Management, Siem Reap, Cambodia.* CGIAR Systemwide Program on Collective Action and Property Rights (CAPRi), Washington, DC, pp. 1–12.

Sinha, H., Kumar, A. and Kumar, R. V. (2013) Introduction. In: Sinha, H. and Kumar, A. (eds) *Governance of Commons and Livelihood Security.* Xavier Institute of Social Science, Jharkhand, pp. 1–14.

Sud, N. (2014a) Governing India's land. *World Development* 60(1), 43–56. Available at: http://dx.doi.org/10.1016/j.worlddev.2014.03.015 (accessed 4 March 2017).

Sud, N. (2014b) The state in the era of India's sub-national regions: liberalization and land in Gujarat. *Geoforum* 51, 233–242. Available at: http://dx.doi.org/10.1016/j.geoforum.2013.06.002 (accessed 4 March 2017).

Times of India (2015) Kutch's 'Swimming Camels' recognized as separate breed. *The Times of India*, 19 April. Available at: http://timesofindia.indiatimes.com/city/ahmedabad/Kutchs-Swimming-Camels-recognized-as-separate-breed/articleshow/46973252.cms (accessed 11 November 2016).

Turner, M. (1993) Overstocking the range: a critical analysis of the environmental science of Sahelian pastoralism. *Economic Geography* 69(4), 402–421.

UN-Habitat (2012) *Gender Evaluation Criteria for Large-scale Land Tools. Brief 2/2012.* UN-Habitat, Nairobi. Available at: http://gltn.net/jdownloads/GLTN%20Documents/gec_brief_for_web.pdf (accessed 11 November 2016).

Warde, A. (2005) Consumption and theories of practice. *Journal of Consumer Culture* 5(2), 131–153. Available at: http://dx.doi.org/10.1177/1469540505053090 (accessed 4 March 2017).

Ykhanbai, H., Gard, R., Singh, A., Mioko, S., Enawgaw Benyene, C., *et al.* (2014) *Conservation and 'Land Grabbing' in Rangelands: Part of the Problem or Part of the Solution?* International Land Coalition, Rome. Available at: http://pubs.iied.org/pdfs/G03853.pdf (accessed 11 November 2016).

6 Gender and Rural Migration in Mexico and the Caribbean

J. Momsen*

University of California, Davis, USA and University of Oxford, Oxford, UK

Introduction

In the current era of globalism and transnationalism, with millions of people moving internationally, mostly from poor agriculture-based societies to rich industrialized countries, and roughly three-quarters of a billion migrating from rural to peri-urban and urban areas within the same country, small-scale farming is changing everywhere (Bhandari and Ghimire, 2016). This mobility turn (Lund, 2014) sees migration as a capability for different livelihood strategies leading to transformations in gendered responsibilities and practices. This chapter focuses on rural livelihood strategies and how these respond in the face of different types of migration and new environmental challenges within a feminist political ecology framework. It compares migration from rural areas in two regions with very different histories of migration: the island Caribbean and the Yucatán Peninsula, often known as the Mexican Caribbean. In both regions, tourism, a temporary form of in-migration, is having a widespread impact on rural areas (Torres and Momsen, 2011). Rural migration has often been thought to have opposing effects: loss of rural labour leading to land abandonment versus return migrants bringing capital and innovative ideas to renew local agriculture. This chapter examines both impacts within the greater Caribbean region where tourism and communal land tenure add additional complexity.

The Caribbean islands are populated by migrants from Europe, Africa and Asia, with the exception of a few indigenous peoples remaining in Dominica and some of the Greater Antilles. The Yucatan rural population is still largely Mayan, practising traditional *milpa* cultivation of maize and squash on communal land known as *ejidos*. Since the ending of slavery from 1838 onwards, Caribbean societies have developed small-scale farming side by side with the long-existing plantations, often on land held jointly by all members of a family, migrant or not (Besson, 1987). Many West Indians have turned to overseas migration for new livelihoods (Thomas-Hope, 1978), but this is a relatively new activity for the Mexican peasantry, especially in the Yucatán where current migration is mostly to regional tourist poles (Torres and Momsen, 2005a,b).

Gender analysis is critical to a study of migration and its influence on rural society and life chances. In general migrants have been men but young women are increasingly leaving independently. In terms of the impact of migration on agriculture much of the focus is on those left behind (Torres,

* Corresponding author e-mail: jdmomsen@ucdavis.edu

2015). Despite the increase in gender parity among migrants it is usually women who are left behind to care for children and to maintain the family home and farm, and so rural areas tend to have increasing numbers of female-headed households. Many of these 'left behind' groups become dependent on remittances from migrants and may give up farming or change their cropping patterns because of labour shortages. Or, more positively, migrants may provide an overseas market for local produce as in cheese from northern Mexico (Buechler, 2015) or tropical vegetables for migrant West Indians living in London and Toronto.

This chapter compares the impact of migration on rural households in the island Caribbean and in Mexico, mainly in the Yucatán, and the interaction of both outmigration and the temporary inward migration of tourists in both regions. The type of migration, whether transnational, national, permanent, temporary, circular or seasonal, affects the impact. The role of remittances from migrants varies in terms of amount, consistency and reliability as does their use by families left behind, whether for debt repayment, investment in agriculture, further migration or family well-being.

Recently attention has turned to climate change in the region (Dunn, 2013) and the relevance of a feminist political ecology approach. The impact of natural disasters such as hurricanes, or volcanic eruptions as in Montserrat, in precipitating migration, or through migrant short-term assistance providing much needed emergency aid to those left behind, is important. But changes such as rising sea levels, death of coral reefs, damage to infrastructure especially rural roads, drought and floods, and the expansion of mosquito-spread diseases such as dengue, chikungunya and zika and waterborne diseases such as cholera, affect both residents and tourists. Risk management needs to take into account gender aspects of climate change and its impacts on farmers. Thus gender needs to be understood as a critical variable in processes of ecological change and the pursuit of viable livelihood strategies (Elmhirst, 2015).

This chapter falls into several sections. First it looks at data sources and field data collection over time for information about small farming in the study region. Then it compares the historical background and contemporary economic situation of both study regions. It also looks at the interaction of migration and agriculture and the contribution of the various ethnic groups, indigenous and migrant, to the distinctive nature of agriculture. Then the chapter considers gender ideology and gender roles in Mexico and the Caribbean. Finally, it examines the gendered impact of migration from rural areas. Case studies based on fieldwork in both regions are used throughout.

Methodology

For the Caribbean, the chapter draws on over 50 years of fieldwork studying small farming in the Anglophone and Francophone islands of the region (Momsen, 1981, 1986, 1987, 2007, 2009; Besson and Momsen, 1987, 2007). Between 1963 and 2008, 11 random sample surveys of almost 1500 small farms in Barbados, St Lucia, St Vincent, Martinique, Montserrat and Trinidad were carried out. These farms specialized in sugarcane, banana, cocoa and vegetable production with a strong subsistence element. Focus groups were held in Jamaica, St Lucia and Dominica with vegetable farmers and Fairtrade banana producers (Momsen, 1986, 2009). Land use mapping of banana farms in Grenada, St Lucia, St Vincent and Dominica was also undertaken (Momsen, 1966).

In Mexico, the fieldwork has focused on the links between tourism, agriculture and migration. Field studies were conducted in the states of Quintana Roo and southern Yucatán (Momsen, 1999, 2003) and work with postgraduate students in several other parts of Mexico (Torres and Momsen, 2005a,b; Chambers and Momsen, 2007; Juarez Varela, 2014). Gender roles in agriculture were a major focus in the Mexican studies, with detailed questionnaire surveys on women's time use and in-depth studies of *ejidatarios* in Quintano Roo and rural migrants in shanty towns on the outskirts of Cancún (Torres and Momsen, 2011). Focus groups of women and men separately in several areas revealed that the apparent patriarchal dominance of agriculture was variable and often men and women cultivated different plots and grew different crops, as also noted by Lope-Alzina (2007). At the same time, the opening up of new opportunities for women to earn money in tourist areas as hotel workers, while men worked in construction, resulted in new gender roles and the growth of part-time farming and female-headed households.

Background and Context: The Historical Background to Migration in the Caribbean and Mexico

These regions were settled by Europeans in the 15th and 16th centuries. The Spanish conquered Mexico and the Dutch, French, English and Spanish settled on the islands of the Caribbean. This European colonization and the pre-Columbian movement of Amerindians from elsewhere in the Americas were followed in the Caribbean by the forced transfer of several million Africans as plantation slaves. The post-Emancipation arrival of indentured agricultural workers from Indonesia, India and Europe, and the more recent circum-Caribbean and international migration of West Indian workers counterbalanced by a seasonal flow of sun-seeking tourists from the global North, are all elements of the human flux in the region.

This chapter examines the interaction of migration and agriculture and the contribution of the various ethnic migrant groups to the distinctive nature of agriculture. Migration can influence the nature and success of agriculture through the new ideas, crops and cultivation methods introduced from different places and the experiences of individuals. On an immediate practical level, migrants can support agriculture by sending money earned overseas as remittances to non-migrant family members and may return to become farmers themselves.

When Europeans first settled the Caribbean they knew very little about tropical agriculture. Amerindians were brought from northern South America to Barbados to teach the settlers how to farm in this new environment and they introduced many crops, especially cassava, which soon became a staple in the islanders' diet (Higman, 2011). The processing of the cassava and its cooking were done by women according to Ligon (1657). There are similarities in the process of the making of cassava bread in seventeenth-century Barbados and the production of maize tortillas in Mexico today. Thus migration and gender roles had a major influence on agriculture from the earliest period of European settlement.

Sugarcane soon became the main crop in the Caribbean and was produced on plantations with labour provided by African slaves. From Africa came other food crops and wild medicinal plants grown using the traditional knowledge of the slaves (Watts, 1966). These crops were often preserved on the small plots cultivated by slaves and the knowledge of their use remained largely in the hands of women. Lope-Alzina (2007) and Chambers and Momsen (2007) both noted that the genetic diversity of traditional varieties of maize and squash is usually conserved *in situ* by women on Mexican small farms. In the 19th century indentured workers from India introduced new crops, especially rice and intensively grown vegetables, to Trinidad and Guyana. In Guyana it was noted that women did most of the farming of food while men worked on estates (Schuler, 2003). Peasant farmers known as *panyols* were also brought from Venezuela to central Trinidad in the mid-19th century to cultivate cocoa under a contract system (Momsen and Richardson, 2009).

Today sugarcane cultivation has ended in many Caribbean countries as it is no longer economically viable. Commercial production of henequen and chicle in the Yucatán has also disappeared. Cocoa plantations are being rehabilitated in Trinidad but the crop is now of relatively minor importance (Momsen and Richardson,

2009). The banana plantations that replaced sugarcane in the 1960s are also disappearing as the rules of the World Trade Organization make European subsidies for these export crops no longer possible. However, small-scale niche marketing of Fairtrade bananas for export has expanded in the Windward Islands and the Dominican Republic (Momsen, 2009). Local demand from the most recent seasonal migrants, tourists, for a wide range of foods has stimulated the production of new crops such as varieties of salad items, herbs and fruits in both the Caribbean and Yucatán (Torres and Momsen, 2011). Tourists also provide a local market for illegal crops such as marijuana widely grown by small farmers throughout the region.

Research Findings

Migrant remittances

There is some evidence that money earned overseas by West Indian migrants in the late 19th and the early part of the 20th century was invested in plots of land (Momsen, 1998). Men working on the building of the Panama Canal sent money home to women relatives. These remittances were often invested in land in the woman's own name so the number of small farms increased, especially in Barbados, as did the proportion owned by women (Momsen, 1998). However, the land available to women for purchase tended to be the most marginal in terms of both accessibility to market and quality of land. It is still true today that women's land tends to be the most isolated and French (1997, p. 316) points out that such problems as an unreliable water source, inadequate roads, lack of access to means of transport and primitive facilities in small rural communities, especially in the Windward Islands, make it difficult for women to combine farming with their domestic chores.

Between January 1906 and December 1910 Barbadians returning from Panama brought back with them a total of £102,456 which led to a rise in the price of small pieces of land to between £80 and £100 per acre while larger acreages were half the price (Momsen, 2007). The importance of the cash economy among these return migrants led to land being used as collateral to finance sugarcane growing and a loss of interest in food production. Farming for semi-subsistence was left in the hands of women as it still is largely today. In 1847, 19.5% of holdings were owned by women, rising to 21.7% in 1875, and between 1887 and 1895 the number of holdings owned by women almost doubled (Momsen, 1998). By the mid-1970s, based on questionnaire surveys of 1307 farmers in the eastern Caribbean, the proportion of female small farmers had reached 56% in Montserrat, 53% in Barbados, 43% in St Lucia and 36% in Martinique (Momsen, 1987).

Philpott (1973) points out that, in northern Montserrat, savings from migration to the USA in the 1920s and early 1930s and to the Netherlands Antilles in the 1940s enabled rural workers to buy up land when marginal estates were sold off in small lots and thus enhance their subsistence base and their financial independence. However, far from this land being used to increase the overall agricultural productivity of the island, it was often utilized as collateral for loans to finance further migration, as also occurred in nearby Nevis (Frucht, 1968). More recently, return migrants to Barbados are especially successful as small producers of the high-value crops demanded by tourists, reflecting a combination of investment capital earned overseas, a willingness to innovate, awareness of new crops and changing local demand, and, above, all access to land provided by the Barbados government to returning citizens (Torres and Momsen, 2011). Some migrants returning to Montserrat and Nevis put their savings into livestock because it brought a faster return for less labour than crop cultivation. Many of these new livestock owners had no land, with indiscriminate grazing by an increasing number of animals bringing both environmental degradation and a decline in cultivation, as wandering animals destroyed crops (Momsen, 1993). In

St Lucia, Nevis and Carriacou remittances are invested in new houses, often on family land, which are rented out until the migrant wishes to return home (Mills, 2007). The land surrounding these new houses is often farmed by relatives in the interim and later by the former migrants after their return. Environmental disasters such as hurricanes have a gendered impact in that men as livestock farmers may have to search for fodder, water and shelter for their animals, while women are more likely to be farming close to the home and so face problems of crop loss and care for small animals on the homestead.

Migrants are exposed to new ideas and experiences, so increasing their human capital, and it might be expected that the return migrant would become a leader in agricultural innovation and an agent of change but this occurs only to a limited extent (Faist, 2008). Employment overseas provides an opportunity to build up financial capital to be sent back to the migrants' home area as regular remittances or as a lump sum at the end of the migratory period. A lack of capital, it is generally acknowledged, is one of the major constraints on the development of small-scale agriculture in the Caribbean and Mexico. Thus it might be expected that the remittances of migrant labourers would be instrumental in stimulating greater productivity in Caribbean agriculture. Field evidence is fragmentary and variable.

In Mexico the indigenous peoples were forced to work on European *encomiendos* or large-scale holdings but the land reform following the 1917 Revolution established *ejidos*, communally held land holdings, and protected small farmers. Within the last few decades neoliberal ideas, the break-up of *ejidos* after 1991, the North American Free Trade Act of 1994 allowing import of cheaper products from the USA, and the reduction of subsidies to farmers have undermined traditional small-scale *milpa* agriculture in Mexico. Today Mexican farmers in the Yucatán living relatively close to tourist poles such as Cancún are also turning to production for the tourist market (Torres, 2011).

Gender, migration and agriculture

The improvement in accessibility over the last six decades in the study region has encouraged both migration and tourism while improved communications have made frequent interactions between home and overseas much easier for family members. Sex-specific migration tends to result in a feminization of agriculture in the region (Momsen, 1987). In work in Jamaica, Griffith (1988) noted women taking over as farm managers when male partners emigrated. He suggested that this usually led to a scaling down of cash crop production but an increase in women's technical farming knowledge. Remittances allow increased dependence on hired labour and also closer links with related farm households for mutual support, so increasing domestic group security. Ishemo (2005, 2009), also utilizing fieldwork in Jamaica, saw remittances as helping older farmers, both men and women, by enabling them to pay for inputs such as fertilizer, pesticide, irrigation and hired workers, as well as to acquire land, and in this way maintain productivity despite the physical limitations of age.

In my own field studies in Nevis, St Lucia and Montserrat in the eastern Caribbean (Momsen, 1986), analysis of data using principal components analysis indicated that in St Lucia farms run by former migrants were associated with more parcels of land and more tree crops than other farms and negatively associated with the production of subsistence root crops. Thus farms operated by former migrants were larger and more commercially oriented, showing capital investment in long-term crops and land. These farmers had fewer pest and marketing problems than other farmers, indicating a higher level of information from the agricultural extension services and the marketing agencies, helped by the generally higher level of education of returned migrant farmers compared with other farmers. Migrant farmers in Nevis were more likely than non-migrant producers to grow cotton for export rather than crops for local sale perhaps because of the higher capital investment needed and a more international outlook. In

Table 6.1. Gender differences in migration and remittances on small-scale farms in the eastern Caribbean. (Data from field surveys: Barbados 1987; St Lucia 1971; Nevis 1979; Montserrat 1973.)

	Barbados (*n*=128)		St Lucia (*n*=68)		Nevis (*n*=99)		Montserrat (*n*=66)	
	Men (%)	Women (%)	Men (%)	Women (%)	Men (%)	Women (%)	Men (%)	Women (%)
Total farmers	73	27	84	16	68	32	67	33
Returned migrants	43	7	47	20	67	28	46	9
Remittances: none received	94	83	28	55	39	22	50	45
Remittances providing over half of income	4	14	n/a	n/a	16	44	0	23

n/a, not applicable.

both Nevis and Montserrat women farmers were less likely to have migrated and to be more financially dependent on remittances than men (Table 6.1).

Return migration and family land

The ending of slavery gave the former slaves both the freedom to refuse plantation work and occupational choice for the first time. Shortage of land available for subsistence agriculture and the gradual ecological degradation of many of the smaller islands forced many West Indians to adopt 'livelihood mobility' (Richardson, 1983, p. 35) and seek new employment opportunities outside their birthplace.

Livelihood mobility dates back to the period following slave emancipations: 1804 when Haiti won independence; 1838 in the British colonial territories; 1848 in the Danish and French colonies; 1863 for the Dutch colonies with finally full emancipation in Cuba in 1886 (Williams, 1970). Landless ex-slaves – proto-peasants – were lured to labour-short areas by promises of free passage, higher wages and better living conditions. Between 1861 and 1871 more than 20,000 people, predominantly men, left Barbados for other British colonies – primarily Trinidad – but also for Dutch Suriname and St Croix, which was then Danish (Momsen, 1986). In the 1870s Nevis, a tiny island with a population of about 11,000, had a net emigration of almost 2000 (Richardson, 1983). Some migrants went to cut cane elsewhere but most went into construction projects such as building the Panama Canal (Newton, 1984).

Most Caribbean migrants leave intending to return at some point. Some are regular, seasonal agricultural labourers leaving home to cut cane in Cuba and the Dominican Republic in the 1920s and 1930s, later in St Kitts and Barbados and most recently in Florida. Rural Mexicans, both women and men, are also beginning to move to the USA and Canada as circular migrants to help in harvesting and other farm work (Becerril, 2016). Others have been temporary migrants travelling in response to short-term labour demand on specific projects such as the Panama Canal or the Madeira–Mamoré railway in southern Amazonia (Greenfield, 1983) or the Limón–San José railway in Costa Rica. With the completion of these projects most Latin American countries introduced restrictions on Antillean immigration. Peach (1968) has shown that West Indian migration to the UK was more closely related to labour demand in Britain than to population growth in the Caribbean and that flow was severely restricted in the early 1960s by new immigration laws. Others are forced to return by declining demand for their labour as occurred in the oil industry of the Netherlands Antilles in the 1950s and 1980s (Momsen, 1986).

In the 1980s, global recession encouraged return migration from the industrialized North, especially among those who left in the 1960s and had reached pensionable age (Momsen, 1987). In Mexico, where migration to North America is more recent, the recession of 2008 combined with stricter border controls both reduced initial migration northwards and limited circular migration (Torres, 2015). Such recessions have had severe impacts on the size of the remittances sent back to families of migrants. This was especially noticeable in Jamaica in the late 1970s and early 1980s (Momsen, 1987) and today in Mexico where currently remittances are over US$22.4 billion (Torres, 2015), a slight decline since 2008. In both countries remittances make up a significant part of foreign exchange earnings and provide a social safety net for many families.

Migration and land

Byron (2007) sees land as both a catalyst for and a product of migration. Often migrants leave other members of the family to cultivate the land while they are overseas, sending remittances to maintain the link. In this way they retain use of this land when they return, especially if it is settlement land as in Nevis (Byron, 2007) or family land as in St Lucia, Jamaica or Carriacou (Mills, 2007) or *ejido* land as in Mexico (Torres, 2011). All members of the family have rights to harvest from family land even if they personally have not cultivated it. Such land is often left in the hands of female members of the family to sustain them while other relatives migrate to work elsewhere. Mills argues that the existence of family land encourages return migration although this land is more often used for housing than for agricultural purposes.

In Mexico rights to *ejido* land can be lost if no member of the family is cultivating this land. Thus where the male head of household migrated the land was often put in the wife's name, which could lead to friction when the husband returned home, although it did provide security for the family (Radel *et al.*, 2012). In fieldwork in Yucatán we found examples of circular migration by different members of the family from farms in the south to the northern tourist poles, so that farming could continue on the *ejido* land and maintain family unity (Torres, 2011).

Many return migrants if they have land will cultivate it, as in Barbados, and may specialize in production for tourist hotels (Richardson-Ngwenya and Momsen, 2011). On the other hand, older migrants who return with pensions from working overseas will not go into farming but put their savings into a retirement home and may feel unwelcome in their country of birth (Sorhaindo and Pattullo, 2009).

The left behind

Often agriculture is most affected by the situation of those families left behind by migrants. In general women make up the majority of the 'left behind' so that migration imposes severe, economic, social and bodily costs on them, their children and other dependant family members (Torres, 2015). The bulk of remittances is consumed by basic subsistence needs such as housing and education. The impact on those left behind by migration can be positive or negative and vary considerably depending on the specificities of place. Children may experience immediate material gains, especially through gifts from parents overseas, but negative impacts have been reported such as increased juvenile delinquency and poor academic performance, especially when left in the care of grandmothers in the Caribbean. Absence of fathers may lead to increased child labour and further migration. Additional pressures on the families of migrants occur where migrants return home because illness or accidents make them unable to work overseas. They are usually unable to access health care, especially in the USA, and so return to be cared for by their 'left behind' families, thus increasing the burden on women (Juarez Varela, 2014).

Gender ideology and agriculture

Gender roles and female empowerment vary depending on existing gender ideologies and patriarchal structures. Boserup (1970) suggested statistics showed that in Latin America women played a very minor role in agriculture while in the Caribbean women were important in small farming and food production. West Indian women may be the decision makers on their own farm, they may market the produce of their own or other farms, or they may work as paid or unpaid agricultural labour. At various times individuals may take on all three roles. In the Anglophone and Francophone Caribbean women and men were brought from Africa and Asia to work on plantations as individuals, and so although society is patriarchal (Momsen, 2002) there is no stigma attached to women as independent farmers. On the other hand, in the Hispanic Caribbean and Mexico, with a strong tradition of *machismo*, some women may experience a new autonomy in the absence of partners, while others may be subject to greater surveillance, dependency and restriction (Juarez Varela, 2014). Social pressure from their community may force women to go to great lengths in order to feed their children in the absence of fathers. Rocheleau (2015) records that in the Dominican Republic a 'left behind' woman cultivated her land with the help of her sons at night so that the neighbours did not see. She felt judged and shunned but proud of her plot and the harvest it could yield. Another abandoned woman developed a relationship with a married man in order to support her children and obtain help with her farm and this was accepted by her community (Rocheleau, 2015).

In field studies in the Yucatán various adaptations of gender roles became apparent. In one *ejido* where male migration was rare and only of short duration, men said that women only worked in the home. However, when women were questioned separately, they reported that they often accompanied their husbands to work in the fields, especially at harvest time, and also cultivated a home garden (Momsen, 1999). In another study where the wife and children had moved to work in Cancún, the father and eldest son would return to their *milpa* to plant maize in order to maintain occupancy and so retain their rights as *ejidatarios*. They moved frequently between the farm and Cancún with several members of the family returning to help on the farm for short periods, but the lack of full-time attention to their land meant that their yields suffered from losses due to pests and diseases (Torres, 2011). Eventually almost the whole family moved back to the *ejido* as they found living in Cancún difficult and expensive. The husband then took up membership in the PROCAMPO programme, which provides direct financial transfers to those who cultivate and own land, and the wife began to sell farm produce in the nearby town. Another study area close by with many transnational male migrants had an increasing number of women receiving PROCAMPO payments as men transferred rights to these payments to their wives so that the household would continue receiving the benefits during the absence of husbands (Radel and Schmook, 2009). This change in turn led to increased spatial mobility for women as they had to travel to the municipal seat to collect the payments and also tended to give women greater control over household income (Radel and Schmook, 2009).

As Radel *et al.* (2012) note, in the Mexican male-dominated family farming system accompanied by strong gender norms against women's field labour participation, a variety of strategies is utilized such as cultivating less, planting pasture and/or hiring male workers to maintain gender ideologies in the face of male outmigration. Migration is one of the adaptive strategies to climate change as are planting later, utilizing a greater number of maize varieties and livelihood diversification, reflecting declines in precipitation and greater frequency of droughts since 1973 (Mardero *et al.*, 2015).

It is important to distinguish between labour and management. Women who could afford it would hire male labourers to carry out tasks in the field. As in the Caribbean, men were often not willing to work for women and it could be hard for women to find male labourers. In the Yucatán women were not allowed to supervise men in the field without another woman present, thus making management of workers difficult. In both areas, the jobs for which women always sought male labour were those involving heavy tasks such as land clearance and the use of chemicals on crops in the fields.

Conclusion

Both the island Caribbean and the Mexican Caribbean of the Yucatán Peninsula have been changed by the recent influx of rich tourists attracted by the region's beautiful sandy beaches and warm blue sea. Both areas have a long history of small subsistence farms interdigitated with large commercial plantations. Plantation agriculture has declined with the loss of overseas markets for their crops but some small farms have taken advantage of the new opportunities as tourists brought the market closer. Some have become specialized vegetable producers with contracts with specific hotels. On the other hand, some farmers have turned to marijuana production and profitable direct sales to individual tourists.

Migration has a generally negative effect on small-scale agriculture but there is field evidence that the specific nature of this impact varies with migrant experience in recipient countries, the age and physical capability of the migrant on return, land availability and local variations in rural investment opportunities. The existence of family-owned land in both Mexico and the Caribbean islands has provided a safety net for those left behind by migrants and has also offered a familiar place to return to for migrants, for retirement if not to cultivate.

Loss of male farmers may lead to a decline in commercial crop production but agrobiodiversity is often maintained as women left behind have greater influence over choice of food crop varieties. Intensity of production may decline because of shortage of labour and land may even be left idle, but women may gain greater rights to land. Rural women may also develop alternative livelihood strategies such as producing and marketing handicrafts and selling farm produce from their own and other farms, thus gaining income in their own right. Overall, the feminization of agriculture is not an automatic response to male migration although there may be an increase in women's empowerment. Climate change and environmental pressures may also have a gendered effect on changes in rural areas, exacerbated by migration and opportunities in the new tourism economy often more open to women. When mothers move to tourism poles they often are accompanied by children who benefit from new opportunities for education and become increasingly cosmopolitan. Thus the new interaction of countryside and beach in the region has brought urban and rural lifestyles closer, changed the nature of farming and gender roles, and, to some degree, reduced the need for long-term and long-distance migration among rural populations.

References

Becerril, O. (2016) Gendered policies, single mothers and transnational motherhood: Mexican female migrant farmworkers in Canada. In: Meghani, Z. (ed.) *Women Migrant Workers: Ethical, Political and Legal Problems*. Routledge, Abingdon, UK and New York, pp. 131–153.

Besson, J. (1987) A paradox in Caribbean attitudes to land. In: Besson, J. and Momsen, J. (eds) *Land and Development in the Caribbean*. Macmillan, London and Basingstoke, UK, pp. 13–45.

Besson, J. and Momsen, J. (eds) (1987) *Land and Development in the Caribbean*. Macmillan, London and Basingstoke, UK.

Besson, J. and Momsen, J. (eds) (2007) *Caribbean Land and Development Revisited*. Palgrave Macmillan, New York and Basingstoke, UK.

Bhandari, P. and Ghimire, D. (2016) Rural agricultural change and individual out-migration. *Rural Sociology* 81(4), 572–600. Available at: http://dx.doi.org/10.1111/ruso.12106 (accessed 4 March 2017).

Boserup, E. (1970) *Women's Role in Economic Development.* Allen and Unwin, London.

Buechler, S. (2015) Climate–water challenges and gendered adaptation strategies in Rayón, a riparian community in Sonora, Mexico. In: Buechler, S. and Hanson, A.-M. (eds) *A Political Ecology of Women, Water and Global Environmental Change.* Routledge, London and New York, pp. 99–117.

Byron, M. (2007) Collateral and achievement: land and Caribbean migration. In: Besson, J. and Momsen, J. (eds) *Caribbean Land and Development Revisited.* Palgrave Macmillan, New York and Basingstoke, UK, pp. 243–254.

Chambers, K.J. and Momsen, J.H. (2007) From the kitchen and the field: gender and maize diversity in the Bajío region of Mexico. *Singapore Journal of Tropical Geography* 28(1), 39–56.

Dunn, L. (ed.) (2013) *Gender, Climate Change and Disaster Risk Management. Working Paper Series No. 7.* Institute of Gender and Development Studies, Mona Unit, University of the West Indies; and Friedrich Ebert Stiftung, Jamaica and the Eastern Caribbean.

Elmhirst, R. (2015) Feminist political ecology. In: Coles, A., Gray, L. and Momsen, J. (eds) *The Routledge Handbook of Gender and Development.* Routledge, London and New York, pp. 58–66.

Faist, T. (2008) Migrants as transnational development agents: an inquiry into the newest round of the migration–development nexus. *Population, Space and Place* 14(1), 21–42. Available at: http://dx.doi.org/10.1002/psp.471 (accessed 4 March 2017).

French, J. (1997) Gender issues in Caribbean agriculture: methodology and process in the movement towards change. In: Leo-Rhynie, E., Bailey. B. and Barrow, C. (eds) *Gender: A Caribbean Multi-disciplinary Perspective.* Ian Randle, Jamaica and Oxford, UK; and James Currey, Oxford, UK, pp. 311–323.

Frucht, R. (1968) Emigration, remittances and social change: aspects of the social field of Nevis, West Indies. *Anthropologica* 10(2), 193–208.

Greenfield, S.M. (1983) Barbadians in the Brazilian Amazon. *Luso-Brazilian Review* XX(1), 44–64.

Griffith, D. (1988) Labor migration and changing peasant agriculture in Jamaica. In: Brierley, J.B. and Rubenstein, H. (eds) *Small Farming and Peasant Resources in the Caribbean. Manitoba Geographical Studies No. 10.* University of Manitoba, Winnipeg, Canada, pp. 101–117.

Higman, B.W. (2011) *A Concise History of the Caribbean.* Cambridge University Press, Cambridge, UK.

Ishemo, A. (2005) Population mobility and the sustainability of farming households: the Rio Grande Valley, Jamaica. Doctoral dissertation, University of the West Indies, Mona, Jamaica.

Ishemo, A. (2009) Migration and the small farming experience: the Rio Grande Valley, Jamaica. In: Thomas-Hope, E. (ed.) *Freedom and Constraint in Caribbean Migration and Diaspora.* Ian Randle, Kingston, Jamaica, pp. 280–283.

Juarez Varela, V.X. (2014) Dreams and miseries of the Mexican migration. Doctoral dissertation, University of California, Davis, California.

Ligon, R. (1657) *A True and Exact History of the Island of Barbados 1657.* Cass Library of West Indian Studies, London.

Lope-Alzina, D.G. (2007) Gendered production spaces and crop varietal selection: case study in Yucatán, Mexico. *Singapore Journal of Tropical Geography* 28(1), 21–38.

Lund, R. (2014) Gender, mobilities and livelihood transitions: an introduction, In: Lund, R., Kusakabe, K., Mishra Panda, S. and Wang, Y. (eds) *Gender, Mobilities, and Livelihood Transformations. Comparing Indigenous People in China, India and Laos.* Routledge, Abingdon, UK and New York, pp. 1–20.

Mardero, S.B., Schmook, C., Radel, Z., Christman, D., Millones, M., Nickl, E., Rogan, J. and Schneider, L. (2015) Smallholders' adaptations to droughts and climatic variability in southeastern Mexico. *Environmental Hazards* 14(4), 271–288. Available at: http://dx.doi.org/10.1080/17477891.2015.1058741 (accessed 4 March 2017).

Mills, B. (2007) 'Leave to come back': the importance of family land in a transnational Caribbean community. In: Besson, J. and Momsen, J. (eds) *Caribbean Land and Development Revisited.* Palgrave Macmillan, New York and Basingstoke, UK, pp. 233–242.

Momsen, J.D. (1966) *Report of a Banana Acreage Survey of the Windward Islands with Accompanying Maps.* Ordnance Survey, Directorate of Overseas Surveys, London.

Momsen, J.D. (1981) Women in small scale farming in the Caribbean. In: Horst, O. (ed.) *Papers in Latin American Geography in Honor of Lucia C. Harrison.* CLAG, Muncie, Indiana, pp. 44–56.

Momsen, J.D. (1986) Migration and rural development in the Caribbean. *Tijdscrift voor Economische en Sociale Geografie* 77(1), 50–58.

Momsen, J.H. (1987) The feminization of agriculture in the Caribbean. In: Townsend, J.G. and Momsen, J.H. (eds) *Geography of Gender in the Third World*. Hutchinson, London, pp. 344–347.

Momsen, J.H. (ed.) (1993) *Women and Change in the Caribbean: A Pan-Caribbean Perspective*. Ian Randle, Kingston; Indiana University Press, Bloomington and Indianapolis, Indiana; and James Currey, London.

Momsen, J.H. (1998) Gender ideology and land. In: Barrow, C. (ed.) *Caribbean Portraits: Essays on Gender Ideologies and Identities*. Ian Randle and Centre for Gender and Development Studies, University of the West Indies, Kingston, Jamaica, pp. 115–132.

Momsen, J. (1999) A sustainable tourism project in Mexico. In: Hemmati, M. (ed.) *Gender and Tourism: Women's Employment and Participation in Tourism*. UN Education and Development Forum, UK Committee, London, pp. 142–148.

Momsen, J. (2002) The double paradox. In: Mohammed, P. (ed.) *Gendered Realities. Essays in Caribbean Feminist Thought*. University of the West Indies Press and Centre for Gender and Development Studies, Kingston, Jamaica, pp. 44–55.

Momsen, J.H. (2003) Participatory development and indigenous communities in the Mexican Caribbean. In: Pugh, J. and Potter, R. (eds) *Participatory and Communicative Planning in the Caribbean*. Ashgate, Aldershot, UK, pp. 155–172.

Momsen, J.H. (2007) The waxing and waning of peasant agriculture in Barbados. In: Besson, J. and Momsen, J. (eds) *Caribbean Land and Development Revisited*. Palgrave Macmillan, New York, pp. 175–188.

Momsen, J.H. (2009) Fair trade versus unfair trade: the Windward Island banana industry. *Caribbean Geography* 15(2), 130–141.

Momsen, J.H. and Richardson, P. (2009) Caribbean cocoa: planting and production. In: Grivetti, L.E. and Shapiro, H.-Y. (eds) *Chocolate, History, Culture and Heritage*. Wiley, New York, pp. 481–491.

Newton, V. (1984) *The Silver Men: West Indian Labor Migration to Panama 1850–1914*. Institute of Social and Economic Research, University of the West Indies, Kingston, Jamaica.

Peach, C. (1968) *West Indian Migration to Britain: A Social Geography*. Oxford University Press, London.

Philpott, S.B. (1973) *West Indian Migration: The Montserrat Case. London School of Economics Monographs on Social Anthropology No. 47*. The Athlone Press, London.

Radel, C. and Schmook, B. (2009) Migration and gender: the case of a farming *ejido* in Calakmul, Mexico. *Yearbook of the Association of Pacific Coast Geographers* 7, 144–163. Available at: http://digitalcommons.usu.edu/cgi/viewcontent.cgi?article=1036&context=envs_facpub (accessed 4 March 2017).

Radel, C., Schmook, B., McEvoy, J., Méndez, C. and Petrzelka, P. (2012) Labour migration and gendered agricultural relations: the feminization of agriculture in the *ejidal* sector of Calakmul, Mexico. *Journal of Agrarian Change* 12(1), 98–119. Available at: http://dx.doi.org/10.1111/j.1471-0366.2011.00336.x (accessed 4 March 2017).

Richardson, B.C. (1983) *Caribbean Migrants*. University of Tennessee Press, Knoxville, Tennessee.

Richardson-Ngwenya, P. and Momsen, J. (2011) Tourism and agriculture in Barbados: changing relationships. In: Torres, R.M. and Momsen, J.H. (eds) *Tourism and Agriculture. New Geographies of Consumption, Production and Rural Restructuring*. Routledge, Abingdon, UK, pp. 139–148.

Rocheleau, D. (2015) A situated view of feminist political ecology. In: Harcourt, W. and Nelson, I.L. (eds) *Practising Feminist Political Ecologies: Moving Beyond the 'Green Economy'*. Zed Books Ltd, London, pp. 29–66.

Schuler, M. (2003) Liberated Africans in nineteenth-century Guyana. In: Moore, B.L., Higman, B.W., Campbell, C. and Bryan, P. (eds) *Slavery, Freedom and Gender. The Dynamics of Caribbean Society*. University of the West Indies Press, Kingston, Jamaica, pp. 133–157.

Sorhaindo, C. and Pattullo, P. (compilers) (2009) *Home Again. Stories of Migration and Return*. Papillote Press, London and Roseau, Dominica.

Thomas-Hope, E. (1978) The establishment of a migration tradition: British West Indian movements to the Hispanic Caribbean in the century after Emancipation. In: Clarke, C.G. (ed.) *Caribbean Social Relations. Monograph Series No. 8*. Centre for Latin American Studies, University of Liverpool, Liverpool, UK, pp. 66–81.

Torres, R.M. (2011) Life between the two *milpas*: tourism, agriculture and migration in the Yucatán. In: Torres, R.M. and Momsen, J.H. (eds) *Tourism and Agriculture: New Geographies of Consumption, Production and Rural Restructuring*. Routledge, London, pp. 47–71.

Torres, R.M. (2015) Gendered costs to the 'left behind': a challenge to the migration and development nexus. In: Coles, A., Grey. L. and Momsen, J. (eds) *The Routledge Handbook of Gender and Development*. Routledge, London and New York, pp. 365–374.

Torres, R. and Momsen, J. (2005a) Planned tourism development in Quintana Roo, Mexico: engine for regional development or prescription for inequitable growth? *Current Issues in Tourism* 8(4), 259–285. Available at: http://dx.doi.org/10.1080/13683500508668218 (accessed 4 March 2017).

Torres, R.M. and Momsen, J.D. (2005b) Gringolandia: the construction of a new tourist space in Mexico. *Annals of the Association of American Geographers* 95(2), 314–335. Available at: http://dx.doi.org/10.1111/j.1467-8306.2005.00462.x (accessed 4 March 2017).

Torres, R.M. and Momsen, J.H. (eds) (2011) *Tourism and Agriculture: New Geographies of Consumption, Production and Rural Restructuring*. Routledge, London.

Watts, D. (1966) *Man's Influence on the Vegetation of Barbados 1627 to 1800. Occasional Papers in Geography No. 4*. University of Hull, Hull, UK.

Williams, E. (1970) *From Columbus to Castro: The History of the Caribbean 1492–1969*. André Deutsch Limited, London.

7 Gender and Agriculture

S. Shortall*
Newcastle University, Newcastle upon Tyne, UK

Introduction

Reflecting on the gender and farming literature since our 2006 volume, continuity and change are key themes that emerge, as discussed in Chapters 8–11 in this section. The questions that concerned early research still concern us today, although how these questions are constructed has changed. The gendered nature of farm work (Reimer, 1986; Whatmore, 1991; Gasson, 1992; Haugen, 1998; Overbeek *et al.*, 1998; Shortall, 2010; Charatsari and Černič Istenič, 2016; Černič Istenič and Charatsari, Chapter 10, this volume), power relations within the farm family (Sachs, 1983; Bokemeier and Garkovich, 1987; Shortall, 1992; de Haan, 1994; Oldrup, 1999; Whatmore, 1991; Matysiak, 2015; Cassidy, Chapter 11, this volume), agency and resistance (O'Hara, 1998; Bock, 2004a; Hoggart, 2004; Pini *et al.*, 2014; Stenbacka, Chapter 9, this volume), the construction of gender identity (Bock, 2006; Brandth and Haugen, 2010; Shortall, 2016a,b), and a research theme that has considerably developed is the analysis of women's off-farm work and its implications for the construction of gender relations within the farm family (Gasson and Errington, 1993; Blekesaune, 1996; O'Hara, 1998; Oldrup, 1999; Brandth, 2002; Kelly and Shortall, 2002; Bock, 2010; Shortall, 2010; Alston, Chapter 8, this volume). Agriculture, and the position of women in agriculture and society generally, has transformed over the last 40 years. The lifting of the marriage ban and the greater availability of divorce and contraception have changed the lives of women dramatically. Gender segregation is still a pervasive feature of all labour markets and there are differences in men's and women's employment by sector, workplace and occupation. Nevertheless, women have become more permanently attached to the labour market and the presumption that women are carers first and workers second, is no longer true (Rubery and Rafferty, 2013). Women's lives on farms have also changed considerably, with women much more active in the labour market. However, social patterns persist and within agriculture the perception of farming as a male industry is hard to break, although there are of course variations to the rule (see Stenbacka, Chapter 9, this volume).

This chapter provides an update of the 2006 introduction to the section of women in farming and introduces the chapters that follow.

Women and Farm Work

An early and continuing focus of research for sociologists, geographers and economists is the 'invisibility' of farm women's

* Corresponding author e-mail: sally.shortall@newcastle.ac.uk

work and theoretical analyses of why it is so (Sachs, 1983; Gasson, 1992; Reimer, 1986; Little and Panelli, 2003). This body of work borrowed heavily from Marxist debates, particularly notions of petty commodity production and the separation of productive and reproductive work on the farm (Reimer, 1986). Feminist scholars argued that narrow definitions of productive farm work meant that much of the reproductive work carried out by farm women was unacknowledged (Bouquet, 1982; Whatmore, 1991; Brandth, 2002; Little and Panelli, 2003). Feminists identified the many ways in which women's farm work was essential to the farm business (Gasson, 1992). Attempts were made to unpack the 'black box' (Argent, 1999) of the family farm to illustrate the unequal gender relations within the family and the different statuses of work carried out by different family members. This research followed broader feminist trends by noting it was not the nature of women's work that led to lack of recognition, but rather their position within a patriarchal household (Oakley, 1974; Walby, 1990; Whatmore, 1991; Delphy and Leonard, 1992). Whatmore's theory of patriarchal gender relations remains the most sophisticated analysis to date of women's farm work (Whatmore, 1991). Her concept of domestic political ideology is developed from the recognition that home and work share the same location on a farm, and production, reproduction, family and economy must be analysed in an integrated rather than a fragmented fashion. Through this approach an understanding of the exploitation of women as farm housewives and unpaid farm labourers was advanced. In most of Western Europe, women are no longer as likely to work full-time on the farm, although there is considerable variation. Patterns vary by country and by size and type of holding. Couples make decisions on the basis of skills and ability to generate off-farm income, and women may work full-time on the farm if men can earn more off-farm (Shortall, 2010). The ability of agricultural statistics to reflect, and agricultural extension workers to value the farm work of women remains an issue of concern (Trauger *et al.*, 2008, 2010; Barbercheck *et al.*, 2009; Alston, Chapter 8, this volume; Černič Istenič and Charatsari, Chapter 10, this volume). Trauger *et al.* (2008, 2010) note that women fail to be seen as 'authentic' farmers, a concept borrowed by Černič Istenič and Charatsari in Chapter 10.

Farming and Power Relations

Early studies on power relations focused on relations within the farm family and the situations and circumstances that influenced women's involvement in farm decision making (Bokemeier and Garkovich, 1987; Oldrup, 1999). The reasons why men and women occupy different positions within the family farm was a central component of empirical and theoretical analysis. Of particular interest was that women enter and engage in farming through specific kinship relations, as wives, daughters, mothers and widows (Whatmore, 1994). This is an entrenched social structure that has changed little (Morell and Bock, 2008; Shortall, 2010). Given the patrilineal nature of land transfer from father to son, women marry into farming, and thus enter the occupation through marriage rather than through occupational choice (Shortall, 1999; Brandth, 2002). Their husband is already established as the 'farmer' and he is also (at least initially) the owner of the capital resources necessary to farm. This position impacts on the valuation of women's work and on their public place in farming (Trauger *et al.*, 2008, 2010; Barbercheck *et al.*, 2009). A great deal of subsequent research has examined the power relations embedded in the public representation of women. Research has examined the under-representation of women in farming organizations (Teather, 1996; Shortall, 2001; Pini, 2002; Alston, 2003). Other work has explored the stereotypical representations of men and women in the farming media which reinforce the conception of farm work as masculine (Duggan, 1987; Brandth, 1995; Brandth and Haugen, 2000; Liepins, 2000). Further gender-segregated space is evident in the provision of agricultural training. Most agricultural training is structured in a vocational way for those who will enter the occupation, so in many ways it is unsurprising that most agricultural programmes have a majority of male students. However, the socially constructed identities of women as home makers and farmers' wives means that they do not obtain a knowledge transfer appropriate to their farming roles. Women farmers are underserved in agricultural education and technical assistance (Shortall, 1996; Alston, 1998; Liepins and Schick, 1998; Trauger *et al.*, 2008). Women often view training groups and programmes as being for men and feel unwelcome and conspicuous in this space. Agriculture extension workers do not always see women as 'authentic' farmers because they do not occupy outdoor space and hence do not invite them to training initiatives or address programmes to their work (Teather, 1994; Barbercheck *et al.*, 2009; Trauger *et al.*, 2010). Here women's self-verification of not being the farmer is being institutionally reinforced by agricultural extension workers. It is remarkable that this gender-divided space persists. It is problematic, because increasingly off-farm employment to support the farm is decided between the couple, and educational levels and life cycle issues determine who will work on the farm and who will work off the farm (Benjamin and Kimhi, 2006; El-Osta *et al.*, 2008). Seeing men as the authentic farmer means the relevant person on the farm may not receive appropriate training. Černič Istenič and Charatsari take up this question in Chapter 10. Even though there are situations where women are land owners, the pervasiveness of male land ownership is a key component lending weight to the ideology that positions men at the heart of farming. In Chapter 11, Cassidy considers how siblings in the Irish family context facilitate the continuity of the patrilineal line by not questioning the patrilineal line of inheritance.

Agency and Resistance

The initial focus of research on farm women sought to illuminate women's farm work, which had previously been overlooked, and to understand the different gender and power relations within the farm family. It tended to present subordinate women and dominant men as static and homogeneous categories, and sought structural and causal explanations (Brandth, 2002; Berg, 2004). Research in the 1970s and 1980s has been described as occupying 'the rural women's subordination category' (Berg, 2004). More recently, choice, agency, resistance and the altering of gender identities over time have become more prominent in the research agenda, and this is the topic of the chapters by Stenbacka (Chapter 9) and, in a less obvious way, Alston (Chapter 8) in this book. Research has examined how women on farms are not simply accepting victims of patriarchal relations, but rather they are active agents, constructing and shaping their roles within farming. Public resistance, largely in the form of farm women's agricultural networks, has been examined in North America, Australasia and Europe (Mackenzie, 1992; Shortall, 1994; Teather, 1996; Panelli and Pini, 2005; Brandth *et al.*, 2014; Leach, 2014). Research noted how the emergence of these networks created a public space where women debated the nature of their farm work and its valuation, patriarchal gender relations on the farm and the patriarchal nature of agricultural institutions. Yet, debates remained conservative in terms of advocating a radical feminist position, reflecting the difficulties of such a position within a family farm business. As Brandth (2002) has noted, farm women's strained relationship to gender equality and feminism has continued to puzzle feminist researchers. Research has identified off-farm work, and the subsequent financial independence, as an expression of women's agency and resistance (Wozniak and Scholl, 1990; O'Hara, 1998). Brandth (2002) argues that earning their own money turns women into more self-assured, visible and autonomous subjects (p. 190). The growing emphasis on agency, choice and resistance is a necessary counterbalance to explanations that seemed to lean towards structural determinism. Hoggart (2004) cautions that there is still a tendency for research to focus on women's 'subordination' and there is more scope for 'celebratory explorations' of women in rural societies (p. 2). While there is merit in this assertion, it remains the case, in farming at least, that women's options in terms of resistance and choice are ones that have not greatly diminished the patriarchal nature of agricultural institutions. This is equally true for farm men. Men in rural areas are constrained by traditional masculine expectations (Campbell and Bell, 2000; Brandth and Haugen, 2005; Campbell *et al.*, 2006). In Chapter 11, Cassidy's research found that young men who had no interest in farming would nevertheless have taken on the role of farmer if they were the heir. The research debate has tended to present research as sitting in one of two opposing camps; on one side is research that focuses on 'patriarchy and the subordination', and on the other is research that focuses on the 'resistance and agency'. However, both structural constraints and strategies of resistance coexist. Women and men are agents who make choices and engage in both strategies of resistance and cooperation in farming and within the farm household. It is also the case that the patriarchal nature of farming and the farming industry persists despite resistance and a changing society. This is an area that requires further research.

Off-farm Employment

The extent to which off-farm earnings alter gender roles and positions within the farm family has been a research question of interest for some time; classifications or models of farm women have long included the category of 'women working off the farm' or 'women in paid work' (Whatmore, 1991; Haugen and Blekesaune, 1996; O'Hara, 1998). Women's unequal status within the farm family was seen as tied to their subordinate economic position in relation to the male breadwinner (Stebbing, 1984; Brandth, 2002; Little and Panelli, 2003). With the generally declining income of the agricultural industry, women's off-farm work and farm diversification work are increasingly subsidizing the farm, and women on farms are now more likely to be the primary breadwinner (Brandth, 2002; Kelly and Shortall, 2002). There are many national case studies across Europe showing that women are active in farm diversification activities (Bock, 2004a,b, 2010; Gorman, 2006; Brandth and Haugen, 2010, 2011). Interestingly, the way in which farm diversification develops often reinforces gender identities on the farm. Bock (2004a) argues that women undertake smaller-scale diversification activities. This is reflective of their more restricted access to capital, but also their desire to fit diversification activities around their other caring commitments and wishing to multi-task other domestic gender identity roles. In many cases the decision to work off-farm is motivated by the dire need for more income (Kelly and Shortall, 2002; Bock, 2004b). Other research found the decision to work off the farm and increase independent earnings to have been a positive choice linked to increased educational levels and women's greater attachment to the labour market (O'Hara, 1998; Bock, 2004a, 2010; Shortall, 2014, 2016a; Charatsari and Černič Istenič, 2016; Černič Istenič and Charatsari, Chapter 10, this volume). Regardless of the motivating factors, this represents a fundamental change in women's economic status within the family farm and could potentially have significant implications for gender relations. However, it is not necessarily increased resource contributions that lead to renegotiated domestic work and gender roles, but rather gender ideologies (West and Zimmerman, 1987; Layte, 1998; Shortall, 2005). Agrarian gender ideology is such that even though women may have an independent source of income off the farm, the fact of living on a farm means they continue to be positioned as farm women and traditional gender roles remain pervasive (Oldrup, 1999; Brandth, 2002). It is also the case that for farm women, an individualistic approach confuses the fact that women not only act as individuals but also as members of farm households (Wheelock and Oughton, 1996). Women's off-farm labour is often part of a farm household survival strategy to maintain the farm and men's occupation as the farmer. Any analysis of the likely impact of women's off-farm earnings on gender roles within the farm family must also take account of the historical context, power and established gender relations in the farm family. In addition, women's off-farm work can have implications for men's sense of identity. Recent research has shown that for men on farms, their identity is closely tied to their position as the breadwinner, and also to the power and privilege that has been associated with being a landowner. When the economic and social standing of their position is threatened, it has significant implications for men's mental health (Ni Laoire, 2001; Alston, 2006, 2012; Barlett, 2006; Alston and Kent, 2008; Price and Evans, 2009). The chapters that follow touch on these questions.

The Chapters

The following collection of chapters (Chapters 8–11) further contributes to the key debates on gender and farming within Western rural studies.

There are some common themes that emerge time and time again in this set of chapters and, interestingly, these are similar to the themes discussed in the 2006 volume (Bock and Shortall, 2006). One overarching theme is continuity and change. In different ways the chapters discuss the power and autonomy of women to act within a traditional and patriarchal industry in various ways. A number of chapters struggle with the difficulties of researching individual choice within the household unit, as people within the farm family act as individuals and as a collective. The chapters also situate their local or national analyses within the wider macro framework that impacts on micro socio-economic relations, climate change, globalization or changing EU legislation.

Margaret Alston's chapter (Chapter 8) provides a fascinating oversight of the impact of climate change in Australia on gender relations within farming families. Her methodological approach is interesting; she conducts a longitudinal study over 25 years using her own qualitative data sets. This is an interesting approach to qualitative analysis. Climate change has had devastating effects in Australia, with long difficult droughts, and it is predicted that Australia will become hotter and drier. Alston discusses how government responses to the drought led farm families to view the issue as a personal one rather than a more widespread global problem. It is in the individual, personalized responses that the profound gender implications emerge. Alston charts the devastating impact of drought, the uncertainty about when it will end, if it will end, and quotes one interviewee who talks of the 'waiting, waitingness'. The men and women she interviewed early in the process worried about each other's mental well-being and the toll the drought was taking on the other. Women's off-farm employment became crucial to the survival of the family farm, and many women worked away all week. Later in the process, Alston notes that women become less committed to the family farm ideology and have a more individualized response to the drought. She identifies climate change, government and farming bodies' actions and inactions around the interpretation of drought, and its impact on the farm family as the key drivers of the gendered outcomes of the drought. She notes that it feels Australian agriculture is in a weaker position going forward.

In Chapter 9, Susanne Stenbacka considers the values that drive women farmers in Sweden. She was involved in a study which considered why farmers continue to farm and what inspires young farmers to enter farming, and as part of this she considered the values of young women farmers. While the Swedish agricultural labour force is declining, the percentage of women who are part of that labour force is increasing, particularly among part-time farmers. There are policies targeting women to encourage their entry into farming. Of the regular agricultural force in Sweden, 35% are women, making it one of the highest percentages in Europe. She is interested in the lessening of gender segregation in Sweden and argues that the idea of hegemonic masculinity in Swedish agriculture is being challenged. After reviewing literature Stenbacka turns to consider the role of values. She notes that values can be individual and/or collective, and they are shaped by cultural values. They are of course not discrete, but rather intertwined. Values are essential to understanding motivations, as they guide the social actions we chose to take. Stenbacka conducted ten in-depth interviews with women farmers. She identifies the key themes of values motivating women farmers' actions: quality of life, financial goals and socio-economic aspirations. Each theme interacts with and offsets the other. She concludes that women farmers' values go beyond family-related motivations. Women's economic and strategic goals make them active farmers in their own right, thus questioning the dominant world view which sees farmers as men with women's role being a complementary one.

Why do women farmers participate in agricultural education programmes and extension courses? This is the question Majda Černič Istenič and Chrysanthi Charatsari explore in Chapter 10. They explore whether it is because of the characteristics of the programmes or other motivating factors. They compare Slovenia and Greece, two countries that have a similar agrarian structure but very different socio-political and economic contexts. They argue that the empowering force of education is well understood, as is its ability to foster participation in public and political life. They are interested then in the formal and informal educational provisions and the extent to which women participate. They provide an excellent overview of the literature on women and agricultural education. They note that in both developing and developed world contexts, women fare badly in accessing extension training. Extension providers have, in general, paid little attention to the role of women in agriculture. Men are seen as the farmers and women as the wives. Černič Istenič and Charatsari give an excellent overview of the literature, revisiting the thorny question of whether women-only programmes are progressive or regressive. They conducted a survey of farmers in Slovenia and Greece, and roughly 25% in each place were women farmers. They found that women seek agricultural training to counter invisibility, gain access to economic resources and specific knowledge. They also found that extension services are imbued with gender stereotypes, which limits their ability to reach women, as they continue to see men as the 'authentic' farmer. This is an issue that continues to require research and revised practices.

The British Isles are unusual in the European context in that children can legally be disinherited. This is the subject of Chapter 11 in which Anne Cassidy looks at the role of the non-heirs on Irish farms in securing the continuity of the family farm. They do this by accepting and not contesting the patrilineal line of inheritance. The very small amount of agricultural land that comes on the Irish[1] market demonstrates that inheritance is the main route of entry into farming. She uses Ulrick Beck's theory of individualization to examine the levels of acceptance and complicity in a system that is inherently unequal. Cassidy examines how family farm young people who will not inherit move easily between conflicting world views. She interviewed 15 young women and 15 young men at university. Previous research has shown that Irish farm families have always tended to progress to third-level education and succeed there. Through in-depth interviews she examines the relationship the family as a whole has with succession and how avoiding a dispute about the intergenerational transfer of the asset maintains its continuity. She explores the ways young boys and girls are socialized into traditional roles on the farm and into gendered expectations. While young women are encouraged to be financially independent and have successful careers, it is always clear that this is understood to be off the farm. While they are encouraged to drive and have their own cars, they are discouraged from driving tractors. Cassidy sees the paradox in these young women's high ambitions, alongside their acceptance of the inequality of land transfer. She recounts that young men have also been socialized to become the heir, regardless of their interest in the farm. In some cases their university education is encouraged in order to bring additional off-farm income to the farm. The young people she interviewed see themselves as part of the collective of the farm family. She argues that while they make individual choices about their future, these are within the structural constraints of ensuring the future of the family farm.

Conclusion

These chapters provide an important contribution to our understanding of gender and agriculture. Alston shows how men and women's roles and commitment to the farm family ideology have changed independently in a crisis situation. It is interesting that she notes that early in her career she would sometimes have interviewed farm couples together, but would no longer do so. Stenbacka's chapter shows changed patterns of women's participation in agriculture and considers the importance of their motivations. It is interesting that the EU is actively trying to encourage more women to come forward as new entrants to farming, although this policy will be translated through cultural patterns and ideologies. Černič Istenič and Charatsari revisit the thorny question of whether women-only agricultural programmes are a progressive or regressive step. Regardless, they conclude that in both the developed and developing world, women's lack of access to agricultural extension services persists as a social problem. Cassidy's chapter is a fascinating unpacking of relations within and between siblings in the Irish context, where it is legal to leave the entire farm to one child and disinherit the rest. Access to land and the subsequent inequalities it generates, remains a question of global importance.

Note

[1]Irish/Ireland here refers to the Republic of Ireland.

References

Alston, M. (1998) Farm women and their work: why is it not recognised? *Journal of Sociology* 34(1), 23–34. Available at: http://dx.doi.org/10.1177/144078339803400103 (accessed 4 March 2017).

Alston, M. (2000) *Breaking Through the Grass Ceiling: Women, Power and Leadership in Rural Organisations.* Harwood Publishers, London.

Alston, M. (2003) Women's representation in an Australian rural context. *Sociologia Ruralis* 43(4), 474–488. Available at: http://dx.doi.org/10.1046/j.1467-9523.2003.00256.x (accessed 4 March 2017).

Alston, M. (2006) The gendered impact of drought. In: Bock, B. and Shortall, S. (eds) *Rural Gender Relations: Issues and Case Studies.* CAB International, Wallingford, UK, pp. 204–220.

Alston, M. (2012) Rural male suicide in Australia. *Social Science & Medicine* 74(4), 515–522. Available at: http://dx.doi.org/10.1016/j.socscimed.2010.04.036 (accessed 4 March 2017).

Alston, M. and Kent, J. (2008) The big dry: the link between rural masculinities and poor health outcomes for farming men. *Journal of Sociology* 44(2), 133–147. Available at: http://dx.doi.org/10.1177/1440783308089166 (accessed 4 March 2017).

Argent, N. (1999) Inside the black box: dimensions of gender, generation and scale in the Australian rural restructuring process. *Journal of Rural Studies* 15(1), 1–15. Available at: http://dx.doi.org/10.1016/S0743-0167(98)00044-8 (accessed 4 March 2017).

Barbercheck, M., Brasier, K.J., Kiernan, N.E., Sachs, C., Trauger, A., Findeis, J., Stone, A. and Moist, L.S. (2009) Meeting the extension needs of women farmers. *Journal of Extension* 47(3), 1–11.

Barlett, P. (2006) Three visions of masculine success on American farms. In: Campbell, H., Bell, M.M. and Finney, M. (eds) *Country Boys: Masculinity and Rural Life.* Pennsylvania State University Press, University Park, Pennsylvania, pp. 47–65.

Benjamin, C. and Kimhi, A. (2006) Farm work, off-farm work, and hired farm labour: estimating a discrete-choice model of French farm couples' labour decisions. *European Review of Agricultural Economics* 33(2), 149-171. Available at: http://dx.doi.org/10.1093/erae/jbl002 (accessed 4 March 2017).

Berg, N.G. (2004) Discourses on rurality and gender in Norwegian rural studies. In: Goverde, H.J., de Haan, H. and Baylina, M. (eds) *Power and Gender in European Rural Development.* Ashgate, Aldershot, UK, pp. 127–144.

Blekesaune, A. (1996) *Family Farming in Norway. An Analysis of Structural Changes within Farm Households between 1975 and 1990. Report No. 6/96.* Centre for Rural Research, Trondheim, Norway.

Bock, B. (2004a) Fitting in and multi-tasking: Dutch farm women's strategies in rural entrepreneurship. *Sociologia Ruralis* 44(3), 245–261. Available at: http://dx.doi.org/10.1111/j.1467-9523.2004.00274.x (accessed 4 March 2017).

Bock, B. (2004b) It still matters where you live: rural women's employment throughout Europe. In: Buller, H. and Hoggart, K. (eds) *Women in the European Countryside.* Ashgate, Aldershot, UK, pp. 14–42.

Bock, B. (2006) Rurality and gender identity: an overview. In: Bock, B. and Shortall, S. (eds) *Rural Gender Relations: Issues and Case Studies.* CAB International, Wallingford, UK, pp. 279–287.

Bock, B. (2010) *Social Situation of Women in Rural Areas. IP/B/AGRI/IC/2010_089.* European Parliament, Brussels.

Bock, B. and Shortall, S. (eds) (2006) *Rural Gender Relations: Issues and Case Studies.* CAB International, Wallingford, UK.

Bokemeier, J, and Garkovich, L. (1987) Assessing the influence of farm women's self identity on task allocation and decision making. *Rural Sociology* 52(1), 13–36.

Bouquet, M. (1982) Production and reproduction of family farms in south-west England. *Sociologia Ruralis* 22(3), 227–244. Available at: http://dx.doi.org/10.1111/j.1467-9523.1982.tb01060.x (accessed 4 March 2017).

Brandth, B. (1995) Rural masculinity in transition – gender images in tractor advertisements. *Journal of Rural Studies* 11(2), 123–133. Available at: http://dx.doi.org/10.1016/0743-0167(95)00007-A (accessed 4 March 2017).

Brandth, B. (2002) Gender identity in European family farming: a literature review. *Sociologia Ruralis* 42(3), 181–201. Available at: http://dx.doi.org/10.1111/1467-9523.00210 (accessed 4 March 2017).

Brandth, B. and Haugen, M. (2000) From lumberjack to business manager. Masculinity in the Norwegian farming press. *Journal of Rural Studies* 16(3), 343–355. Available at: http://dx.doi.org/10.1016/S0743-0167(00)00002-4 (accessed 4 March 2017).

Brandth, B. and Haugen, M. (2005) Doing rural masculinity – from logging to outfield tourism. *Journal of Gender Studies* 14(1), 13–22. Available at: http://dx.doi.org/10.1080/0958923042000331452 (accessed 4 March 2017).

Brandth, B, and Haugen, M. (2010) Doing farm tourism: the intertwining practices of gender and work. *Signs* 35(2), 425–446. Available at: http://dx.doi.org/10.1086/605480 (accessed 4 March 2017).

Brandth, B. and Haugen, M. (2011) Farm diversification into tourism – implications for social identity? *Journal of Rural Studies* 27(1), 35–44. Available at: http://dx.doi.org/10.1016/j.jrurstud.2010.09.002 (accessed 4 March 2017).

Brandth, B., Follo, G. and Haugen, M.S. (2014) Paradoxes of a women's organisation in the forestry industry. In: Pini, B., Brandth, B. and Little, L. (eds) *Feminisms and Ruralities.* Lexington Books, London, pp. 57–69.

Campbell, H. and Bell, M. (2000) The question of masculinities. *Rural Sociology* 65(4), 532–546. Available at: http://dx.doi.org/10.1111/j.1549-0831.2000.tb00042.x (accessed 4 March 2017).

Campbell, H., Bell, M.M. and Finney, M. (eds) (2006) *Country Boys: Masculinity and Rural Life.* Pennsylvania State University Press, University Park, Pennsylvania.

Charatsari, C. and Černič Istenič, M. (2016) Gender, farming and rural social research: a relationship in flux. In: Shucksmith, M. and Brown, D.L. (eds) *Routledge International Handbook of Rural Studies.* Routledge/Falmer, London and New York, pp. 389–399.

de Haan, H. (1994) *In the Shadow of the Tree – Kinship, Property and Inheritance among Farm Families*. Het Spinhuis, The Hague.

Delphy, C. and Leonard, D. (1992) *Familiar Exploitation*. Polity Press, Cambridge, UK.

Duggan, C. (1987) Farming women or farmers' wives? Women in the farming press. In: Curtin, C., Jackson, P. and O'Connor, B. (eds) *Gender in Irish Society*. Galway University Press, Galway, Ireland, pp. 54–70.

El-Osta, H.S., Mishra, A.K. and Morehart, M.J. (2008) Off-farm labor participation decisions of married farm couples and the role of government payments. *Applied Economic Perspectives and Policy* 30(2), 311–332. Available at: http://dx.doi.org/10.1111/j.1467-9353.2008.00406.x (accessed 4 March 2017).

Gasson, R. (1992) Farmers' wives: their contribution to the farm business. *Journal of Agricultural Economics* 43(1), 74–87. Available at: http://dx.doi.org/10.1111/j.1477-9552.1992.tb00200.x (accessed 4 March 2017).

Gasson, R. and Errington, A. (1993) *The Farm Family Business*. CAB International, Wallingford, UK.

Gorman, M. (2006) Gender relations and livelihood strategies. In: Bock, B. and Shortall, S. (eds) *Rural Gender Relations: Issues and Case Studies*. CAB International, Wallingford, UK, pp. 27–47.

Haugen, M. (1998) The gendering of farming. The case of Norway. *European Journal of Women's Studies* 5(2), 133–153.

Haugen, M. and Blekesaune, A. (1996) *The Labour Situation of Farm Women in Norway. Report No. 4/96*. Centre for Rural Research, Trondheim, Norway.

Hoggart, K. (2004) Structures, cultures, personalities, places, policies: frameworks for uneven development. In: Buller, H. and Hoggart, K. (eds) *Women in the European Countryside*. Ashgate, Aldershot, UK, pp. 1–14.

Kelly, R. and Shortall, S. (2002) 'Farmers' wives': women who are off-farm breadwinners and the implications for on-farm gender relations. *Journal of Sociology* 38(4), 327–343. Available at: http://dx.doi.org/10.1177/144078302128756714 (accessed 4 March 2017).

Layte, R. (1998) Gendered equity: comparing explanations of women's satisfaction with the domestic division of labour. *Work, Employment and Society* 12(3), 511–532. Available at: http://dx.doi.org/10.1177/0950017098123006 (accessed 4 March 2017).

Leach, B. (2014) Feminist connections in and beyond the rural. In: Pini, B., Brandth, B. and Little, L. (eds) *Feminisms and Ruralities*. Lexington Books, London, pp. 81–95.

Liepins, R. (2000) Making men: the construction and representation of agriculture-based masculinities in Australia and New Zealand. *Rural Sociology* 65(3), 605–620. Available at: http://dx.doi.org/10.1111/j.1549-0831.2000.tb00046.x (accessed 4 March 2017).

Liepins, R. and Schick, R. (1998) Gender and education: towards a framework for a critical analysis of agricultural training. *Sociologia Ruralis* 38(3), 286–302. Available at: http://dx.doi.org/10.1111/1467-9523.00079 (accessed 4 March 2017).

Little, J. and Panelli, R. (2003) Gender research in rural geography. *Gender, Place & Culture* 10(3), 281–289. Available at: http://dx.doi.org/10.1080/0966369032000114046 (accessed 4 March 2017).

Mackenzie, F. (1992) 'The worse it got, the more we laughed': a discourse of resistance among farmers of Eastern Ontario. *Society and Space: Environment and Planning Development D* 10(6), 691–714. Available at: http://dx.doi.org/10.1068/d100691 (accessed 4 March 2017).

Matysiak, I. (2015) The feminization of governance in rural communities in Poland: the case of village representatives (*sołtys*). *Gender, Place & Culture* 22(5), 700–716. Available at: http://dx.doi.org/10.1080/0966369X.2013.879104 (accessed 4 March 2017).

Morell, A.I. and Bock, B. (eds) (2008) *Gender Regimes, Citizen Participation and Rural Restructuring. Research in Rural Sociology and Development Series, Vol. 13*. Elsevier, Oxford, UK and Amsterdam.

Ni Laoire, C. (2001) A matter of life and death? Men, masculinities and staying 'behind' in rural Ireland. *Sociologia Ruralis* 41(2), 220–236. Available at: http://dx.doi.org/10.1111/1467-9523.00179 (accessed 4 March 2017).

Oakley, A. (1974) *The Sociology of Housework*. Martin Robertson, Oxford, UK.

O'Hara, P. (1998) *Partners in Production? Women, Farm and Family in Ireland*. Berghahn Books, Oxford, UK and New York.

Oldrup, H. (1999) Women working off the farm: reconstructing gender identity in Danish agriculture. *Sociologia Ruralis* 39(3), 343–358. Available at: http://dx.doi.org/10.1111/1467-9523.00112 (accessed 4 March 2017).

Overbeek, G., Efstratoglou, S., Haugen, M.S. and Saraceno, E. (1998) *Labour Situation and Strategies of Farm Women in Diversified Rural Areas of Europe.* Office for Official Publications of the European Communities, Luxembourg.

Panelli, R. and Pini, B. (2005) 'This beats a cake stall!': farm women's shifting encounters with the Australian state. *Policy and Politics* 33(3), 489–503. Available at: http://dx.doi.org/10.1332/0305573054325710 (accessed 4 March 2017).

Pini, B. (2002) The exclusion of women from agri-political leadership: a case study of the Australian sugar industry. *Sociologia Ruralis* 42(1), 65–77. Available at: http://dx.doi.org/10.1111/1467-9523.00202 (accessed 4 March 2017).

Pini, B., Brandth, B. and Little, L. (eds) (2014) *Feminisms and Ruralities.* Lexington Books, London.

Price, L. and Evans, N. (2009) From stress to distress: conceptualizing the British family farming patriarchal way of life. *Journal of Rural Studies* 25(1), 1–11. Available at: http://dx.doi.org/10.1016/j.jrurstud.2008.03.008 (accessed 4 March 2017).

Reimer, B. (1986) Women as farm labour. *Rural Sociology* 51(2), 143–155.

Rubery, J. and Rafferty, A. (2013) Women and recession revisited. *Work, Employment and Society* 27(3), 414–432. Available at: http://dx.doi.org/10.1177/0950017012460314 (accessed 4 March 2017).

Sachs, C.E. (1983) *The Invisible Farmers, Women in Agricultural Production.* Rowman and Allanheld, Totowa, New Jersey.

Shortall, S. (1992) Power analysis and farm wives: an empirical study of the power relationships affecting women on Irish farms. *Sociologia Ruralis* 32(4), 431–451. Available at: http://dx.doi.org/10.1111/j.1467-9523.1992.tb00941.x (accessed 4 March 2017).

Shortall, S. (1994) Farm women's groups: farming or feminist or community groups, or new social movements? *Sociology* 28(1), 279–292. Available at: http://dx.doi.org/10.1177/0038038594028001017 (accessed 4 March 2017).

Shortall, S. (1996) Training to be farmers or wives? Agricultural training for women in Northern Ireland. *Sociologia Ruralis* 36(3), 269–286. Available at: http://dx.doi.org/10.1111/j.1467-9523.1996.tb00022.x (accessed 4 March 2017).

Shortall, S. (1999) *Gender and Power – Women and Farming.* Macmillan Press, London.

Shortall, S. (2001) Women in the field: women, farming and organisations. *Gender, Work & Organisation* 8(2), 164–182. Available at: http://dx.doi.org/10.1111/1468-0432.00127 (accessed 4 March 2017).

Shortall, S. (2005) Politics, gender and the farmyard. *Canadian Woman Studies* 24(1), 42–49.

Shortall, S. (2010) *Women Working on the Farm: How to Promote Their Contribution to the Development of Agriculture and Rural Areas in Europe. IP/B/AGRI/IC/2010_090.* European Parliament, Brussels.

Shortall S. (2014) Farming, identity and well-being: managing changing gender roles within Western European farm families. *Anthropological Notebooks* 20(3), 67–81.

Shortall, S. (2016a) Changing configurations of gender and rural society. In: Shucksmith, M. and Brown, D. (eds) *Routledge International Handbook of Rural Studies.* Routledge/Falmer, London and New York, pp. 347–357.

Shortall, S. (2016b) Gender and identity formation. In: Shucksmith, M. and Brown, D. (eds) *Routledge International Handbook of Rural Studies.* Routledge/Falmer, London and New York, pp. 400–412.

Teather, E. (1994) Contesting rurality: country women's social and political networks. In: Whatmore, S., Marsden, T. and Lowe, P. (eds) *Gender and Rurality.* David Fulton Publishers, London, pp. 31–50.

Teather, E. (1996) Farm women in Canada, New Zealand and Australia redefine their rurality. *Journal of Rural Studies* 12(1), 1–14. Available at: http://dx.doi.org/10.1016/0743-0167(95)00026-7 (accessed 4 March 2017).

Trauger, A., Sachs, C., Barbercheck, M., Kiernan, N.E., Brasier, K. and Findeis, J. (2008) Agricultural education: gender identity and knowledge exchange. *Journal of Rural Studies* 24(4), 432–439. Available at: http://dx.doi.org/10.1016/j.jrurstud.2008.03.007 (accessed 4 March 2017).

Trauger, A., Sachs, C., Barbercheck, M., Kiernan, N.E., Brasier, K. and Schwartzberg, A. (2010) The object of extension: agricultural education and authentic farmers in Pennsylvania. *Sociologia Ruralis* 50(2), 85–103. Available at: http://dx.doi.org/10.1111/j.1467-9523.2010.00507.x (accessed 4 March 2017).

Walby, S. (1990) *Theorizing Patriarchy.* Blackwell, Oxford, UK.
West, C. and Zimmerman, D. (1987) Doing gender. *Gender & Society* 1(2), 125–151. Available at: http://dx.doi.org/10.1177/0891243287001002002 (accessed 4 March 2017).
Whatmore, S. (1991) *Farming Women: Gender, Work and Family Enterprise.* Macmillan, London.
Whatmore, S. (1994) Theoretical achievements and challenges in European rural gender studies. In: van der Burg, M. and Endeveld, M. (eds) *Women on Family Farms: Gender Research, EC Policies and New Perspectives.* Circle for Rural European Studies, Wageningen University, Wageningen, The Netherlands, pp. 107–120.
Wheelock, J and Oughton, E. (1996) The household as a focus for research. *Journal of Economic Issues* 30(1), 143–159.
Wozniak, P. and Scholl, K. (1990) Employment motivations of farm spouses. *Lifestyles* 11(4), 321–342. Available at: http://dx.doi.org/10.1007/BF00987344 (accessed 4 March 2017).

8 The Genderness of Climate Change, Australia

M. Alston*

Monash University, Melbourne, Australia

Introduction

Climate changes resulting from increases in greenhouse gas emissions are impacting across the globe. As one of the driest continents on earth, Australia has been particularly affected by both slow-onset and catastrophic events, and, therefore, will continue to be at the forefront of climate change experiences and actions. In the last two decades, Australia has been subject to one of the longest and most severe droughts on record, an event that was accompanied by widespread dust storms and declining water stocks (Botterill and Fisher, 2003), and has experienced multiple catastrophic weather events including bush fires, floods and cyclones. What is already evident in Australia is that rural areas reliant on agricultural production are particularly affected by climate-induced weather challenges and that these events are making agricultural production more fraught with uncertainty. Further, as these climate events are predicted to become increasingly severe, rural areas and the people who live and work in these vulnerable regions will be most directly affected by the mitigation efforts adopted by national governments to address climate challenges (Molnar, 2010). This raises significant questions about the ongoing viability of Australian agriculture and, more particularly, the social relations that have shaped the lives of farming families over generations. While these relations have changed incrementally over time in response to wide-ranging and complex structural adjustments, they remain deeply rooted in patriarchy. How are farm families coping with this new challenge and what adjustments are being made? Do climate challenges threaten traditional patriarchal foundations?

This chapter adopts a longitudinal appraisal of research conducted over two decades to interrogate whether climate changes are impacting on gender relations in agriculture. It presents evidence that climate changes have had, and will continue to have, significant additional impacts on agricultural production and on the gender relations embedded in the family farm units that make up the largest proportion of Australian agricultural production units. By examining research findings relating to two major climate events that have occurred in Australia in the last 15 years, the catastrophic effects on the people and places affected are demonstrated and the 'genderedness' of these climate challenges is drawn out. These events – (i) a major 10-year drought in the late 1990s and early 2000s and (ii) the consequent declining water in the Murray–Darling Basin system that has

* Corresponding author e-mail: margaret.alston@monash.edu

led to water being withdrawn from some irrigation farms – have had critical impacts on agriculture, farm family units and the individuals affected. Before examining these events and their consequences, the next sections provide a brief overview of climate change predictions for Australia and the ongoing restructuring taking place in Australian agriculture.

Background and Context

Climate change impacts in Australia

Australia is predicted to become hotter and drier as our climate continues to change. There will be less rainfall, particularly across the southern parts of the continent, river flows will be lower (and hence the ongoing concerns regarding water allocated for irrigated agriculture), sea and air temperatures will rise and catastrophic climate events such as bushfires, heat waves, floods and cyclones will be more frequent and more intense (Garnaut, 2008; Department of the Environment, 2015). These predictions are borne out by the range of significant weather events over the last two decades. Devastating bushfires that spread across the state of Victoria in 2009 and Western Australia in 2011, floods that washed into central Brisbane in 2011 and cut off towns in New South Wales in 2011 and 2012, cyclones across the north of the country, including cyclones Yasi and Anthony in 2011, and Marcia, Nathan and Lam in 2015, the long 10-year drought that affected the entire continent and declining water in the Murray–Darling Basin river systems are all evidence of the increasing impacts of Australia's changing climate.

Yet, somewhat surprisingly, climate change is not necessarily an accepted fact in Australia, particularly in conservative political circles, including many farm representative bodies. As noted elsewhere (Alston and Whittenbury, 2012), the ongoing, vitriolic and highly politicized debate about whether climate change has anthropogenic causes or is part of a natural cycle of weather changes has had a particular impact on rural people. Despite catastrophic climate events and scientific predictions that they will become more frequent and intense, the Australian government under recently deposed conservative Prime Minister Tony Abbott has adopted an attitude of denial. In the lead-up to the Paris climate meeting in 2015, Abbott argued that 'coal is good for humanity' (Australian Broadcasting Commission, 2014) and, in determining that Australia would delay any emissions reduction targets, called the attempts of other nations to set carbon reduction targets 'airy-fairy' (*Sydney Morning Herald*, 2015a). His government halted support for renewable energy strategies including wind farms and solar power (*Sydney Morning Herald*, 2015b).

Disturbingly this distracting debate has led to rural people in general, and farming families in particular, viewing climate challenges as personal troubles rather than a more widespread and, therefore national and global, problem (Dunlap, 2010; Alston and Whittenbury, 2012). With such robust climate denial and aggressive opposition to mitigation efforts and renewable energy, it is inevitable that Australian farming families will view their own experiences of weather variability as private problems to be addressed within the production unit through reshaped livelihood strategies. This view has a number of consequences. First, there is no concerted industry or rural lobby for climate mitigation actions that are equitably shared across the nation; and second, farm families and rural people adopt individualized strategies to address their own interpretation and experiences of climate challenges. This is a critical point because, as outlined below, climate changes have significant gendered consequences in areas that are already shaped around gender differentiations. Thus individualized responses to climate changes have further profound gender implications. In rural Australia, this is manifest in a number of ways including during the climate event itself, the consequent reshaping of livelihood strategies in response to climate events, in the reallocation of work roles, the health issues that arise and the ongoing general well-being of individual family members. Before examining these in more detail, a close examination of Australian agriculture reveals that restructuring is ongoing and, therefore, that the challenges of climate change are additional to a long list of factors that are reshaping agricultural production in Australia.

Agriculture

Australian agriculture remains a critical industry across rural areas and, in many cases, is the dominant economic activity for many of the hundreds of rural towns across the vast continent. Nevertheless, Australian agriculture has been subject to ongoing restructuring for several decades (Gray and Lawrence, 2001), a restructuring that has reshaped the social relations of agriculture with significant gendered outcomes (Alston, 1995a, 2000, 2004; Alston and Whittenbury, 2012). From the 1980s to the 2011 Census a decline of approximately 40% in the number of farms from 240,000 to 135,000 was recorded, the equivalent of nearly 300 farming families leaving the industry each month during that period (Australian Bureau of Statistics, 2012). There are a number of factors associated with this decline quite apart from the drought, including globalization and the consequent exposure of farming families to world markets; reduced prices for produce; a cost-price squeeze; and an increase in the numbers of larger corporate farms. This restructuring has also resulted in critical social changes: an exodus of young people – and particularly young women – to the cities, greater reliance on income generated off-farm and changing gender roles.

While there has been a rise in large corporate farms in recent years, now accounting for 6% of Australian farms, agricultural production continues to be dominated by farm family units, the majority of which earn less than AU$100,000 per year (Australian Bureau of Statistics, 2012). It is important to note that the top 6% of farms earn over AU$1 million and are usually larger than family farms, suggesting that farm production units are far from homogeneous. Nevertheless, the research we have undertaken over the last two decades focuses on those farm family production units (approximately 70–80% of farms) that have made a modest living from farming and are classified as family farms – that is, they are largely reliant on family labour. Within these families, social relations are continuing to evolve. Historically Australian agriculture has been highly dependent on gendered social relations (see for example Lake, 1987) and this has continued to shape expectations affecting women and men in farm families. While the vast majority of landowners and identified farmers are men – a factor facilitated by patrilineal inheritance – significant changes are evident in the livelihood strategies adopted by farm families. Women are still largely viewed as responsible for care and household tasks, yet additional expectations both on- and off-farm have been added over the last several decades (Alston, 1995a, 2000; Alston and Whittenbury, 2012).

The critical need to access off-farm income to ensure the family's continuation in farming defines Australian agriculture. Particularly evident is the generation of off-farm income, most often undertaken by women. I have written extensively on this trend, noting that women's 'farm wife' role (Whatmore, 1991) has expanded to incorporate off-farm and on-farm work (Alston, 1995a, 2000; Alston and Whittenbury, 2012) and that their work has not necessarily given them significant power within the enterprise because it is viewed as an extension of their gender-determined role and is defined largely as a farm survival strategy (Alston and Whittenbury, 2012).

Despite definitional anomalies, the changes in Australian agriculture and the pressures exerted by drought and ongoing restructuring have had major gendered impacts, changing the economic significance of on- and off-farm work. Women and men on Australian farms continue to work long hours on and off the farm in order to stay in agriculture. Yet gender relations within the farm family units are undergoing major transformations as families adopt a diverse range of livelihood strategies to ensure their survival in farming (Alston, 1995a, 2000). The notion that climate changes are personal troubles to be addressed at farm level adds new and complex layers to the gender negotiations within farm households and a turning in of responsibility for these events. The neoliberal mantra adopted by successive governments fostering self-reliance and individual responsibility for one's own success adds a critical burden to farm family units.

Methodology

This chapter draws on a number of research projects conducted with Australian farm families from the early 1990s and throughout the period of the long-running drought – a slow-onset climate event. The drought in the late 1990s and early 2000s created significant trauma across Australia, not only on farms but also in the communities reliant on agriculture, and this created intense research interest. Studies on the social and gendered impacts of the drought were conducted throughout this period. These qualitative studies were conducted in small rural communities with farming families and community members, and relied on interviews and focus groups with purposive samples of people deeply affected by drought. These research projects are well documented (see for example Alston, 1995a, 1996; Alston *et al.*, 2004, Alston and Kent, 2006; Alston and Witney-Soanes, 2008; Alston *et al.*, 2010; Alston and Whittenbury, 2012) and indicate the ongoing changes in gender relations during a 25-year period of significant agricultural restructuring.

Current research continues our analysis of the changing gender dynamics in agriculture within the context of the Murray–Darling Basin, where an ongoing result of the drought and over-allocation of water has been the reduced access to water for irrigation purposes (Alston *et al.*, 2016). The restructuring of water use emerged during the drought years as a major consequence of reduced water in the river systems and fears that too much water was being allocated to irrigation. This had led to concerns that the ecological environment was eroding, that species were under threat and that salination was occurring at a rapid rate. The process of addressing water restructuring has been lengthy and highly politicized and has resulted in further significant changes in the areas affected. This has resulted in water licences being decoupled from land so that each can be sold separately. In many areas farms have been consolidated into larger holdings, many have had to revert to dryland farming, farm families have been displaced and several communities are affected by a critical loss of productivity. The government determination to return water from irrigators to the environment has created significant challenges for farm families in the affected regions and our current research examines how these issues are affecting farm families and communities (Alston and Whittenbury, 2012; Alston *et al.*, 2016).

This chapter undertakes a meta-analysis of these studies, drawing on the analysis of data conducted at the time of these research projects. Because these studies occurred frequently and over a long period, they provide a useful source of metadata. When viewed as a whole it is possible to see changes occurring in response to societal, industry and climate changes. While these changes are incremental over time, an analysis of data from the early 1990s to the current period reveals wide-ranging and easily detectable trends in gender relations. These events, the drought and water restructuring, represent a slow-onset climate incident followed by consequent changes in water allocation from the river systems. The chapter draws out the responses of people to these events, highlighting the changes in gender relations over the last two decades. In particular it focuses on changes in livelihoods, involuntary or voluntary displacement, health, care issues and commitment to farming.

Research Findings

Major changes in women's access to education and careers in the 1970s and 1980s, together with the wider global women's movement, had led Australia's farm women to a more critical appraisal of gender relations in the early 1990s and to women mobilizing around their own needs and rights, yet still with an ongoing commitment to a family farming ideology (Alston, 1995b). This resulted in the spectacular rise of the Australian Women in Agriculture movement in the 1990s (Liepins, 1996), a movement that had a number of successes in this period, including successfully lobbying for the establishment of rural women's policy units at national and state levels. These units provided, albeit for a brief time, a pathway to policy makers and helped to construct more gender-sensitive farm-related policies including in the area of drought support.

Perhaps not surprisingly, the decline of women's activism coincided with the drought years as women were drawn into greater involvement in farm work and income-generating activities. While the National Rural Women's Coalition, a body including the leaders of rural women's community organizations, is still operating, of concern is that there are now no remaining rural women's policy units. These were systematically disbanded under conservative governments at state and federal levels. Women's access to policy avenues, and the creation of gender-sensitive policies, has become significantly constrained by neoliberal governments that focus on agricultural sector productivity rather than territorial policies that might shape a more family-friendly environment and more critical attention to climate changes. What has been lost is a focus on policies that would assist women and their families to be better supported in their enhanced roles – particularly in areas such as childcare, disability and aged care, flexible working arrangements, transport and telecommunications policies.

Changing livelihood strategies

The lengthy drought was one of the most severe to affect Australia. The then Prime Minister, John Howard, speaking in 2006, described it as a one-in-1000-year event requiring plans to be drawn up to secure water supplies (*The Age*, 2006). Critical to understanding the impact of such an event is that there is never any insight into how long a drought will endure and no historical precedent that can signpost when it will end. Thus the slow creeping and devastating event was not easily addressed, and nor was it initially understood as a major climate change event. As each year unfolded there was hope that things must change and some years were better than others. Yet as each year brought lengthy dry conditions, farm families were forced to make major adjustments as debts mounted and no substantial rains came. As one farm woman explained in research conducted in 2003 (Alston *et al.*, 2004): 'It's like a waitingness or something. Not awkwardness, it's just a waiting, waiting, waitingness. Like everybody's holding their breath.'

Over time families turned inward, withdrawing from their communities. In the same study women spoke of the constraints affecting their lives, with little or no income and a policy environment that did not initially support the provision of welfare to farming families:

> We have no social life. We can't walk into a shop . . . we don't have fun like we used to . . . we can't afford to go anywhere, we can't afford to go to the movies, we just can't afford to do anything. Everything is too dear.
>
> (Farm woman)

Women spoke of reducing their purchases of essential items, not buying shampoo or hygiene products for example, not having any breaks away from the pressures of farming and limiting their visits to town to save on petrol – yet women remained committed to the family farming ideology in these critically tough times.

Meanwhile governments had embraced a neoliberal mantra and were slow to respond. Critically a major policy change occurred in the mid-1990s, when first the Labor government and then the Howard conservative government reinforced that drought was no longer to be considered a 'disaster' with the associated financial supports that would flow from such an event. From this point individual farm families were required to view drought as a normal cyclical event for which they must plan and be prepared (Botterill and Fisher, 2003). Facing the major drought crisis, rural Australians felt that governments, and with them the entire Australian community, were abandoning them. As the Deputy Prime Minister, and leader of the National Party at the time, John Anderson, noted in 1999, 'much of regional and rural Australia was feeling a deep sense of alienation, of being left behind, and of no longer being recognized and respected for the contribution being made to the nation' (Anderson, 1999). Yet as each new year unfolded, the do-nothing policy position became more untenable as widespread poverty spread across farming communities (Alston, 2007). One country newspaper reported on its front page that Australia's farm children were starving and that the Red Cross had had to step in to provide breakfast programmes for rural schoolchildren (*Albury Border Mail*, 2005).

Our own research (Alston *et al.*, 2004; Alston and Kent, 2006; Alston and Witney-Soanes, 2008) revealed similar stories of significant hardship. This work captured the impact of a neoliberal policy environment where the self-reliance mantra was biting deeply into the well-being of farm families. The counter political argument offered to justify this policy position was that farming families should not expect the community to assist them through normal weather cycles (Botterill and Fisher, 2003). The result was grim for many families. Service providers we interviewed noted the extraordinary stress on farm families and the particular issues for the aged and children. Stress was very evident within families and acknowledged by partners who worried deeply about their spouses. Women worried about the mental health of their husbands working alone in damaged, dry and dusty landscapes, with spiralling financial debt burdens resulting from the need to provide expensive feed and water for livestock. Men worried about their wives, particularly as complex taxation compliance changes had facilitated women being largely responsible for doing the farm books. One man in our 2003 research noted the impact on his wife:

> Oh apart from being . . . a little bit more worried than usual, my concern has been for my better half. She's been becoming a bit of a nervous wreck. Well she knows the books. . . . She doesn't sleep. . . . A fair bit of crying. . . . when she's crying I try to bite my tongue. Yeah, swallow the lump or two in my throat, just to really try and support her. But yeah I still get teary about it. . . . She's been fantastic – like I think other women probably would have broken more.
> (Farm man)

In the same study, many men articulated their individualized responses. This corroborated the notion that they experienced the drought as a personal failure rather than a major climate change event for which they should have been able to anticipate more community support: 'Emotionally the worst period of my life. . . . I feel very isolated. . . . I'm running out of resilience to keep taking the blows and keep moving on. . . . I carry the hurt inside' (farm man).

Policy changes did come, but very slowly. For example, in the period leading into Christmas 2002, the government announced that Exceptional Circumstances funding (a special category of welfare payment) would be available to families in areas that could prove they were experiencing a one-in-20-year exceptional event (Department of the Prime Minister and Cabinet, 2002). The process for proving the exceptional event was slow and cumbersome, and not always successful. This resulted in families having to make drastic decisions about their welfare long before the process was concluded. For those able to access exceptional circumstances funds, these were extremely welcome: 'I'm much more comfortable now I've got a bit of money to buy groceries . . . like any farm, if there's any money in the bank it doesn't get spent on the family' (farm woman, 2003 study).

Nevertheless, the process had been uneven in its area allocations necessitating the sourcing of off-farm income for many families, particularly by women who took work wherever they could find it (Alston *et al.*, 2004; Alston and Kent, 2006). Initially Exceptional Circumstances restrictions precluded families earning more than AU$10,000 off the farm and this had a direct impact on younger families, and families most in need of support, where women were working part-time. An outcry from women's groups and communities forced a change in the legislation and, as a result, the threshold was raised to AU$20,000 (see Alston, 2009 for a detailed discussion of this process). However, support funds remained difficult to access and were often underspent and not reaching their target groups, because government processes were so complex. This period where welfare support was restricted had a direct impact on gender role allocation and adherence to the family farming ideology.

Gender role allocation

In research conducted in the late 1990s and early 2000s (Alston, 2000; Alston *et al.*, 2004; Alston, 2006) we learnt of the way farm families were negotiating these gender changes. While there had been incremental shifts in work allocation from the 1980s, men had largely been responsible for the physical farm labour and women for household work. The notion that women were responsible for work inside the homestead fence and men beyond it, so noticeable in research conducted in the early 1990s (Alston, 1995a), was changing, and changing rapidly.

During subsequent decades, Australian farm women gradually took on more physical labour and, as financial compliance legislation added more arduous tasks to farm book keeping, women generally took over the task of doing the books. At the same time, through the drought years women were seeking work off-farm in large numbers, often taking work in regional centres at a distance from home, requiring them to live away during the week – an involuntary separation, and therefore a form of displacement, that became more common during these years. Their off-farm income became critical to the family remaining in farming and was viewed a farm survival strategy (Alston and Whittenbury, 2012).

This change in the way income was procured was made more remarkable by the lack of recognition from farm bodies and governments that Australian agriculture was now significantly reliant on women earning an income to keep it viable. It was a similar lack of acknowledgement and validation that had resulted in the Australian Women in Agriculture movement gaining momentum during the 1990s (Liepins, 1996). At that time women were deeply committed to farming and to the farm family ideology articulated so well by Brandth (2002). Yet as time went by, and the lack of public acknowledgement of women's economic contribution continued, our research (Alston and Whittenbury, 2012; Clarke, 2015) suggests that a significant further change occurred. What we are seeing in more recent research is a steep decline in women's commitment to the family farming ideology and the inherent hardships associated with supporting traditional gendered arrangements.

Respondents in research conducted in 2008–2009 (Alston and Whittenbury, 2012) were indicating dramatic changes in women's attitudes to agriculture. There are discernible gendered differences in the way women and men were responding to the ongoing pressures of agricultural sustainability. Men remained largely locked into a farm family ideology that valorizes masculine hegemony – an ideology that holds them in a commitment to the land and to the traditional practices that shape agriculture. Their inability to make this work successfully, either financially or socially, during the drought years was leading many into despair. Several women commented on their husbands' commitment to traditional roles and their eulogizing of family farming. In our 2008–2009 study Amy was one of many to make an ironic comment on the centrality of the land to her partner: 'I've always felt as though I'm living on the sacred site. The land, the land – the land is everything.'

In the same study, Robert represented the more traditional positioning of men. He had inherited his farm, had battled the drought and was determined to stay in agriculture. His wife had moved away from this view, leaving him perplexed that she no longer shared his passion nor his commitment to the hardship associated with family farming:

> [Wife] no longer does the books – she will not do the books. [she said] twenty-one years of good service, got the gold watch and said I've had enough of this. [Wife] is over it . . . she has given up on it because I know that she reckons we can't get through it. . . . She's just seen it every year . . . I'm not bitter that we're not going forward together but I'm still happy we're still together and as long as I understand and not be too torn because she's taken a different life . . .we're not sharing the same life. We're actually two lives in the one house and we couldn't be further apart.
>
> (Robert)

Robert tried to understand what had happened to his wife, who was now working night shifts as a nurse at the hospital, caring for their four children during the day, and who had abandoned her previously willing contribution to the farm:

> [My wife] is a lot . . . [angrier] than she used to be and I see a grumpy person now instead of the person who was the *fun-loving* housewife that she used to be. . . . it's not a *normal* life, you're living an alternative life. And she doesn't help on the farm anymore whereas she used to. . . . [before] I'd be pulling half a day out of her.
>
> (Robert)

In this research, and representing the significant shifts that we were observing, couples noted quite diverse gendered differences. While men yearned for the traditional farming family lifestyle, women expressed their scepticism and exhaustion with supporting their partners' ongoing dream. As one women noted in our later research conducted in 2011, she had moved to the city following several years of dry conditions and has worked some hours away from her family farm for several years. This has caused her to live a parallel life:

> If you want your marriage to continue then you have to start looking after yourself and almost live a parallel life for a while so that you're able to cope. He doesn't get it. He's just stuck up there by himself and it's all falling down around him like a deck of cards – that's how I see it. I really have a completely separate life down here.
>
> (Chloe)

This comment is significant as it mirrors that of Robert quoted above – and indicates that women and men articulate that they are now living their lives independently with different ideological foundations. Chloe went on to note despair that her husband did not understand her contribution to the farm because of his traditional view of the family farm:

> John doesn't have to support me at all . . . he just doesn't realize how good he's got it with me . . . he can't see the positive. He can only see the negative because it's just been so hard for so long and it's very hard. . . . The more I'm successful [in my career] the worse it is for him – it's that sense of failure. I'm sure he'd ultimately like me to be back there and just run all of that [household and books] for him, but I'd go nuts because I'd have no money . . . and he's not unintelligent, and he's not a bad farmer. But he's overburdened and he's not alone. . . . it's catastrophic what's happened to us . . . I'm so tired. . . I can't speak to John about it. . . . sometimes it gets overwhelming.
>
> (Chloe)

What our research studies reveal is this gradual shift away from a strong and shared commitment to family farming to a much more individualized and gendered response to the hardship of drought and declining income. This shift has not been without its consequences on the health and well-being of family farm members.

Impacts on health

It was the link between men's commitment to the hegemonic masculinity so inherent in the family farming ideology that led to further work examining the impacts on men's health. In 2008 we drew on this research to describe the deleterious impact of the drought on men's health and of their personal sense of responsibility for what was happening to their families and farms (Alston and Kent, 2008). We noted that because of patrilineal inheritance and family tradition, it appeared that men felt personally responsible for letting their ancestral family down. Hegemonic masculinity and a failure by governments to address climate challenges demonstrated by the drought meant that men were feeling the overwhelming burden of personal failure. In a later article (Alston, 2012) this was linked to the high male farmer suicide rates across Australia. The stoicism so evident in male farmers, and so potent in good times, was reducing their capacity to seek help and to address their mental health issues. What we were seeing was an extreme gendered response from men who viewed the drought and associated other challenges they were facing as personal failings. This view was reinforced in our research conducted with farm families affected by water restructuring (Alston and Whittenbury, 2012). The longevity of the drought and the impact of changing water policy were taking its toll on the health of farmers: 'It's just been too long, too long and that's what's crushing everybody's spirit' (male broad-acre farmer).

The health status of farm men throughout the drought was acknowledged as a significant issue by governments, farmer organizations and through initiatives undertaken by non-government organizations such as Beyond Blue. This organization conducted a nationwide campaign aimed at male farmers and designed to 'normalize' depression as an illness that required treatment. For a brief period from 2005 to 2007 the writer was a member of a mental health taskforce established by the NSW Farmers' representative body to advise governments and non-government agencies on ways to facilitate mental health services to farmers. Among other initiatives the group produced a mental health blueprint (NSW Farmers, 2007). These initiatives were useful in broadcasting the types of services available and how to access them.

While these efforts were laudatory and necessary, there was less of a focus on farm women's health. Throughout the drought, and in projects conducted across the eastern states of Australia, we heard from rural women who worried deeply about the health of their partners and families but ignored their own health crises. Women also spoke of having radical treatment alternatives such as mastectomies rather than chemotherapy in order to reduce their time away from the farm. The health issues for women were not being addressed in the same systematic way as those of men.

Meanwhile there was emerging evidence that another result of the drought disaster was a rise in violence against women (Whittenbury, 2012). This outcome mirrors what was evident in other climate change events across the world – the rise in gender-based violence in post-disaster sites is well documented (Enarson, 2006, 2009, 2012; Partenza, 2012). In our research, while women were reluctant to speak about their experiences of violence, service providers were not. They noted that the numbers of women presenting to services had risen and that there were periodic spikes in numbers relating to events such as when water allocation restrictions were announced (Whittenbury, 2012).

While health issues and gender-based violence were apparent, less evident were the services required to deal effectively with farm family members. Neoliberal policies relating to service provision – including user pays, privatization, and a demand- rather than a needs-based approach – had resulted in a steady withdrawal of services across the rural inland.

While rural financial counsellors were appointed and funded to assist families with financial matters, those staffing these services reported being ill-equipped to deal with the social issues presenting to them (Alston *et al.*, 2004).

Discussion: Shifting Boundaries

In summing up the gender changes we have observed over the last two decades, I would argue that the drought has significantly changed both the way many families operate as a production unit and the commitment that binds individual family members to that unit. There is evidence that women have moved away from a farm family ideology, and that women and men operate as individuals rather than as traditional farm families. There is however a marked gender difference in attitudes to the farming way of life consequent on the drought. Men are more likely to express a yearning for tradition, while women are more likely to have moved away from an identification with farming life, caught up in their own careers and trying to hold the many threads of family life together.

There is no doubt that men have a heavy investment in family farming. Thus the erosion of commitment to farming evident among some women during the drought was difficult for them to understand. In our more recent research we are assessing the gendered responses of women and men to the forced decisions about their future in irrigated agriculture. From 2008 farming families in irrigation areas have been given the option of selling their water for significant amounts of money. These offers have forced the decisions about the farming future to be squarely addressed within families. Not surprisingly these discussions have highlighted gendered differences. Many women are more likely to want to sell and move on – whether it be the water licences or the whole farming enterprise – and to take the money on offer. For many women, and for some men, this is a critical decision as it can be a way out of debt and hard work and can create a future that is more certain. Women, who are less anchored to the land, are more likely to accept this and to want to try something else. As one woman noted in our 2008 research: 'On the whole I think women are more ready to go back out into the current and see where it takes them' (Robin).

In our 2009 study (Alston, Whittenbury and Haynes, 2010), service providers such as Pam noted that women wanted to move on as much for themselves as for their husbands:

> It's breaking their heart to see what their husband is going through not so much that they've had enough [but] that they can't cope seeing him go through that anymore and it's like 'just get out, let's sell up and let's be townies'.
>
> (Pam)

The ground-breaking research of Clarke (2015) is highly significant in explaining the ongoing gender dynamics and shifting boundaries in gender negotiations among Australia's farm families. In 2012–2015 Clarke undertook research in the Mallee region of Victoria with farm families critically affected by both drought and water reallocations. She suggests that farm women and men leave farming mentally and physically at different times along a continuum of leaving. She corroborates our finding that there are gender differences that influence the leaving and in individuals' commitment to farming. This reinforces our view that family members are responding as individuals rather than as a unit (Alston and Whittenbury, 2012). Thus she notes that women may move into off-farm work, be still living on the farm with their partners, but be mentally disengaged to such a point that they can no longer be tagged as farming. This work explains why women disassociate themselves while still being part of a 'farm family' – viewing it more as a social rather than economic unit. Gender negotiations in these families have moved a long way from those described in our earlier research and it is far less likely that women would rally to form a 'women in agriculture' movement today.

Conclusion

There have been significant changes in gender relations in Australian agriculture over the last two decades. By taking a longitudinal view, and drawing on a number of studies conducted with farm families over the last two decades, it seems clear that there are discernible shifts in the way women and men conduct their relationships. There is no doubt that climate change is a significant driver of change, although not the only one. However, a failure of governments and farmer bodies to accept climate change as a significant threat to agriculture, and to advocate for a national approach to mitigation and adaptation efforts, leaves farm families addressing these issues as private troubles. This has had perhaps the most critical impact on gender relations, reinforcing the need for significant changes in livelihood strategies at the same time as agricultural production has become more uncertain and a less reliable income source. A critical farm survival strategy has been the adoption of off-farm work as a major source of income support and women are usually responsible for this income generation.

The theoretical advances that are evident from the meta-analysis conducted for this chapter relate to a consequent significant shift, particularly by women, away from a commitment to a farming family ideology. In the face of climate changes, and particularly where farm families are being offered inducements to sell their water entitlement back to the government, many viewed this as a ticket to a new, less stressful future. Many women have necessarily moved away for work and are now viewing themselves, not as farmers, but as workers in their off-farm occupation. Thus, arguably, the uncertainty created by climate changes has had a dramatic effect on the process of leaving farming – a process that falls along a continuum as described by Clarke (2015). While men remain deeply entrenched in the ideology of family farming, women have moved along the continuum and many have mentally, if not physically, exited farming.

As a consequence, Australian farming has become much more masculinized in a public sense. What is not acknowledged is that this masculinization of agriculture and of the decision-making bodies that foster and develop agricultural policy, overshadows the fact that the industry is highly dependent on women bringing in an income to support the façade. This ongoing lack of recognition for women and a failure to capture their needs through policies that support *family* farming will continue to have a profound effect on agricultural production in Australia.

There is no doubt that gender relations in agriculture continue to evolve amidst a host of pressures that require constant gender negotiations within family farm households in Australia. In this chapter I suggest that climate changes in Australia have added a significant pressure to these negotiations, exposing the genderedness of family farming, and have led to significant and often unexpected changes in the social relations of agriculture. Critical to the way these negotiations unfold is the failure by governments and industry bodies to first acknowledge climate change and second to address the impacts on agricultural production in any systematic way. The neoliberal policy environment that dominates Australian society in general and agriculture in particular trumpets a dogma of self-reliance and individual responsibility for one's own welfare. This exacerbates the lack of community responsiveness to climate challenges particularly in rural locations.

As noted in this chapter, this has led to farming families in general and farm men in particular viewing challenges such as the drought and water restructuring as personal troubles to be addressed at farm level. Because we have undertaken studies over a prolonged period, these reveal the evolving gendered responses and the divergence between farm men, many of whom are steeped in tradition and farm family ideology, and farm women who express a reluctance to continue to support the farm and family through off-farm work. Because this work of women, and the support it provides to agriculture, is largely unacknowledged, it is not supported by social policies including child and other care, and adequate transport and communications that might assist women to a better quality of life. Further, the burden of climate challenges is falling equally on the shoulders of men who view their lack of success as a personal failure. In our research we are seeing a confusion on the part of many men and a declining commitment on the part of women to farm family ideology. This indicates a less secure future for agriculture.

Meanwhile the empirical insights gained from this meta-analysis are equally significant. Over the course of the last two decades I have made significant shifts in the way I collect data. Early on, it would not have been unusual for me to interview farming couples together. Where I might interview women alone in my earlier studies, it was not unusual for their male partner to sit in on the interview, or to police the information in some way. My current research is very different. I interview farming couples separately and there is no attempt by either to control information. In fact, couples usually wish to give their own versions of their lives, which they see as different, each with their own validity, and they can be quite critical of their partners. I view this shift as a move away from a joint commitment to a farm family ideology towards a more individualistic, detraditionalized view. Nevertheless men still yearn for a return to the joint venture that was family farming, while women have generally moved to a post-farming future.

Arguably the methodology used to develop this chapter is contestable, given that data were collected for different purposes at different times in diverse areas. However I would note that I have been responsible for collecting these data over time, I am familiar with how they were gathered, I have been deeply embedded in all phases of the analysis and I can reflect on the patterns and changes over the entire period. For this reason I suggest it is indeed a very useful methodology.

If we are to address the challenges facing Australian agriculture, a number of factors must be resolved. First, it is necessary for governments and industry bodies to accept the challenges of climate change and to address these as a community and industry issue demanding cooperative attention to the future of agriculture. Displacement of families will continue in Australian agriculture – this is inevitable. However this can be eased by a more thoughtful attention to policies that support families to assess their situations in a supportive environment. Second, the need for social policies that support families in rural communities is long overdue. These policies must accept that rural women, like their urban counterparts, must work and that to do so they require satisfying work in their own communities, and support structures and flexible working conditions that will allow them to enjoy work and family life. Finally, it is critical to adopt gender equality as a centrepiece of family farming and to address the changing gender dynamics inherent in this mode of production, including inheritance practices, if we are to have viable industries based on family production units into the future.

References

The Age (2006) A one-in-1000-year drought. *The Age*, 8 November. Available at: http://www.theage.com.au/news/national/a-onein1000year-drought/2006/11/07/1162661662965.html (accessed 30 July 2015).

Albury Border Mail (2005) Our farm kids are starving. *Albury Border Mail*, 26 May.

Alston, M. (1995a) *Women on the Land: The Hidden Heart of Rural Australia.* University of NSW Press, Sydney, Australia.

Alston, M. (1995b) *Report from the National Rural Women's Forum.* Charles Sturt University, Wagga Wagga, Australia.

Alston, M. (1996) Backs to the wall: rural women make formidable activists. In: Lawrence, G., Lyons, K. and Momtaz, S. (eds) *Social Change in Rural Australia*. Rural Social and Economic Research Centre, UCQ, Rockhampton, Australia. pp. 77–84.

Alston, M. (2000) *Breaking Through the Grass Ceiling: Women, Power and Leadership in Rural Australia*. Harwood Publishers, London.

Alston, M. (2004) Who's down on the farm? Social aspects of Australian agriculture in the 21st century. *Agriculture and Human Values* 21(1), 37–46. Available at: http://dx.doi.org/10.1023/B:AHUM.0000014019.84085.59 (accessed 4 March 2017).

Alston, M. (2007) Maintaining vibrant rural communities. In: Swain, D. (ed.) *Redesigning Animal Agriculture*. CAB International, Wallingford, UK, pp. 18–29.

Alston, M. (2009) Drought policy in Australia: gender mainstreaming or gender blindness? *Gender, Place & Culture* 16(2), 139–154. Available at: http://dx.doi.org/10.1080/09663690902795738 (accessed 4 March 2017).

Alston, M. (2012) Rural male suicide in Australia. *Social Science & Medicine* 74(4), 515–522. Available at: http://dx.doi.org/10.1016/j.socscimed.2010.04.036 (accessed 4 March 2017).

Alston, M. and Kent, J. (2006) *Impact of Drought on Rural and Remote Education Access: A Report to DEST and Rural Education Fund of FRRR*. Centre for Rural Social Research, Wagga Wagga, Australia.

Alston, M. and Kent, J. (2008) The big dry: exacerbating the link between rural masculinities and poor health outcomes for rural men. *Australian Journal of Sociology* 44(2), 133–147. Available at: http://dx.doi.org/10.1177/1440783308089166 (accessed 4 March 2017).

Alston, M. and Whittenbury, K. (2012) Does climatic crisis in Australia's food bowl create a basis for change in agricultural gender relations? *Agriculture and Human Values* 30(1), 115–128. Available at: http://dx.doi.org/10.1007/s10460-012-9382-x (accessed 4 March 2017).

Alston, M. and Witney-Soanes, K. (2008) *Social Impacts of Drought and Declining Water Availability in the Murray–Darling Basin*. Institute of Land, Water and Society, Charles Sturt University, Wagga Wagga, Australia.

Alston, M., Kent, J. and Kent, A. (2004) *Social Impacts of Drought: Report to NSW Agriculture*. Centre for Rural Social Research, Charles Sturt University, Wagga Wagga, Australia.

Alston, M., Whittenbury, K. and Haynes, A. (2010) *The Social Impacts of Declining Water Availability and Ongoing Drought in the Murray–Darling Basin: Short Report*. Monash University Department of Social Work, Melbourne, Australia.

Alston, M., Whittenbury, K., Western, D. and Gosling, A. (2016) Water policy, trust and governance in the Murray–Darling Basin. *Australian Geographer* 47(1), 49–64.

Anderson, J. (1999) Speech to the Regional Australia Summit. Available at: https://infrastructure.gov.au/department/annual_report/1999_2000/RegionalAustraliaSummit.aspx (accessed 24 July 2015).

Australian Broadcasting Commission (2014) Coal 'good for humanity', Prime Minister Tony Abbott says at $3.9bn Queensland mine opening. *ABC News*, 13 October. Available at: http://www.abc.net.au/news/2014-10-13/coal-is-good-for-humanity-pm-tony-abbott-says/5810244 (accessed 2 May 2015).

Australian Bureau of Statistics (2012) *Australian Farmers and Farming*. ABS Catalogue No. 4102. Available at: http://www.abs.gov.au/AUSSTATS/abs@.nsf/Lookup/4102.0Main+Features10Dec+2012 (accessed 6 July 2015).

Botterill, L. and Fisher, M. (2003). Introduction. In: Botterill, L. and Fisher, M. (eds) *Beyond Drought, People, Policy and Perspectives*. CSIRO Publishing, Collingwood, Australia, pp. 1–8.

Brandth, B. (2002) Gender identity in European family farming: a literature review. *Sociologia Ruralis* 42(3), 181–200. Available at: http://dx.doi.org/10.1111/1467-9523.00210 (accessed 4 March 2017).

Clarke, J. (2015) A gender analysis of agricultural and rural restructuring in the Mallee region. PhD thesis, Monash University, Melbourne, Australia.

Department of the Environment (2015) Climate change in the future. Available at: http://www.environment.gov.au/climate-change/climate-science/climate-change-future (accessed 7 July 2015).

Department of the Prime Minister and Cabinet (2002) Transcript of the Prime Minister the Hon John Howard MP joint Press Conference with John Anderson and Warren Truss. Parliament House,

Canberra. Available at: http://pmtranscripts.pmc.gov.au/release/transcript-21773 (accessed 10 February 2017).

Dunlap, R.E. (2010) Comment: Climate change and rural sociology: broadening the research agenda. *Rural Sociology* 75(1), 17–27. Available at: http://dx.doi.org/10.1111/j.1549-0831.2009.00010.x (accessed 4 March 2017).

Enarson, E. (2006) Women and Girls Last?: Averting the Second Post-Katrina Disaster. Available at: http://understandingkatrina.ssrc.org/Enarson/ (accessed 14 March 2011).

Enarson, E. (2009) *Women, Gender and Disaster: Global Issues and Initiatives.* SAGE Publications, Inc., Thousand Oaks, California.

Enarson, E. (2012) *Women Confronting Natural Disaster: From Vulnerability to Resilience.* Lynne Reiner Publishers, Boulder, Colorado.

Garnaut, R. (2008) *The Garnaut Climate Change Review: Final Report.* Cambridge University Press, Melbourne, Australia.

Gray, I. and Lawrence, G. (2001) *A Future for Regional Australia: Escaping Global Misfortune.* Cambridge University Press, Cambridge, UK.

Lake, M. (1987) *The Limits of Hope: Soldier Settlement in Victoria 1915–1938.* Oxford University Press, Oxford, UK.

Liepins, R. (1996) Women in agriculture: a geography of Australian agricultural activism. PhD thesis, University of Melbourne, Melbourne, Australia.

Molnar, J.J. (2010) Climate change and social response: livelihoods, communities and the environment. *Rural Sociology* 75(1), 1–16. Available at: http://dx.doi.org/10.1111/j.1549-0831.2010.00011.x (accessed 4 March 2017).

NSW Farmers (2007). NSW Farmers Mental Health Network. Available at: http://www.aghealth.org.au/blueprint/ (accessed 25 July 2015).

Partenza, N. (2012) 'The way he tells it . . .': relationships after Black Saturday. Available at: http://www.whealth.com.au/documents/publications/whp-TheWayHeTellsIt.pdf (accessed 10 February 2017).

Sydney Morning Herald (2015a) Other countries 'airy-fairy' on climate change, says Tony Abbott, as Australia delays new emissions target announcement. *Sydney Morning Herald*, 13 July. Available at: http://www.smh.com.au/federal-politics/political-news/other-countries-airyfairy-on-climate-change-says-tony-abbott-as-australia-delays-new-emissions-target-announcement-20150713-giazve.html (accessed 15 July 2015).

Sydney Morning Herald (2015b) Government pulls the plug on household solar. *Sydney Morning Herald*, 13 July. Available at: http://www.smh.com.au/federal-politics/political-news/government-pulls-the-plug-on-household-solar-20150712-gian0u.html (accessed 15 July 2015).

Whatmore, S. (1991) *Farming Women: Gender, Work and Family Enterprise.* Macmillan, London.

Whittenbury, K. (2012) Climate change, women's health, wellbeing and experiences of gender based violence in Australia. In: Alston, M. and Whittenbury, K. (eds) *Research, Action and Policy: Addressing the Gendered Impacts of Climate Change.* Springer, New York, pp. 207–221.

9 Where Family, Farm and Society Intersect: Values of Women Farmers in Sweden

S. Stenbacka*
Uppsala University, Uppsala, Sweden

Introduction

This chapter is based on a study conducted within the scope of the 'Looking for farmers – young farmers' future strategies in a transforming sector' project.[1] One of the main questions guiding the project was that of what prompts farmers to persist in their occupation and/or inspires young agricultural students to enter it. As part of this overall aim, this chapter seeks to identify the values of contemporary women farmers and investigates how these values help them to actively develop occupational strategies.

Sweden's agricultural labour force is steadily shrinking. Only some 2% of the economically active population is currently engaged in farming (Statistics Sweden, 2015). Simultaneously, the share of women farmers has increased.[2] In 2013, women constituted 6–7% of full-time farmers in Sweden.[3] The proportion of women in part-time farming is higher. In 2013, 17–18% of part-time farmers were women (Swedish Board of Agriculture's Official Statistics of Sweden, 2015). Women farmers are slightly younger than their male counterparts. While 50% of the male farmers were aged 54 years or over in 2013, the corresponding figure for women farmers was 42%. In terms of gender distribution, there is a higher share of women employees than of farm owners. In 2013, 43% of the employees were women. This is due to both more women entering the sector and more men leaving it (Swedish Board of Agriculture's Official Statistics of Sweden, 2015). Figures for Europe (EU-28) show that in 2014, women constituted 35% of the regular agricultural labour force. The highest proportions of women in the labour force, close to 50% (Eurostat, 2014), were found in the Baltic states and Austria.

This trend towards a somewhat less gender-segregated sector is a point of departure for this chapter. Learning about women farmers' values enhances understanding of why they embark on or continue in farming, and how the sector is being transformed. Moreover, farmers' values intersect with their strategies and knowledge of these strategies is of interest to policy makers and consumers alike. Growing interest in food in both the public and the private sphere, and greater concern for factors affecting food production and quality, raises questions about farmers' values and how they actively influence entrepreneurial strategies. The current downward trend in the number of farmers and rise in the proportion of women farmers are intensifying the demand for knowledge of women farmers' motives and goals, as well as of how these

* Corresponding author e-mail: susanne.stenbacka@kultgeog.uu.se

intersect with the transformation of farming as a livelihood.

The research question is formulated as follows: 'How can women farmers' values be defined and how do these values relate to formulation of farming strategies?'

Background and Context: Gender and Farming in Sweden

The scope of research on farming and gender has expanded over the past three decades. Gender differences in farm ownership and farm employment have encouraged discussions about how the farming sector produces and reproduces gender relations, and research has emphasized the situation of women in family farming and gendered aspects of family labour (Brandth, 2002). The roles and importance of women have been accentuated in such respects as development of farm activities (Brandth, 2002) and financial contributions to make farm survival possible (Bock and Shortall, 2006). Nevertheless, contemporary views on farming see the occupation as first and foremost masculine and family farming as still the dominant form (Little, 2002; Pini, 2008).

This view has meant that women's role in agriculture is seen as more complementary. Women's entrepreneurship is integrated with the family and work on the farm (Caballé, 1999; Sattler Weber, 2007) and balancing work and family has been identified as a reason why women become agricultural entrepreneurs (Bock, 2004). Farm entrepreneurship includes integrating the identity of entrepreneur with those of mother and woman farmer (Sattler Weber, 2007). How women farmers approach the farming business and how they organize and value diverse aspects of their lives thus seem to follow a traditional path. Women in farm households are seen as more devoted to caring not only for the family but also for the animals, and therefore as special and possessing certain qualities important for business success (Farstad and Heggem, 2008). This view indicates that valuing women's work is combined with distinguishing between the types of task for which men and women are best suited. The same study also reveals male farmers' assumptions that women are relatively uninterested and unskilled in financial matters. This may be linked to women taking care of animals, since they are assumed to prioritize animal welfare over profit (Farstad and Heggem, 2008). While these results clarify contemporary dissociation of women's and men's capacities and engagements, they also prompt further investigation of how women farmers relate to such questions as whether it is feasible to maintain farm operations or how to create a solid and prosperous business.

Other studies have identified associations between women and financial motives. Their reasons for becoming agricultural entrepreneurs can be formulated in terms of financial independence and contributing to the family income (Petridou and Glaveli, 2008). Other motives may be: the need to secure an income; difficulty in finding employment; personal interest; and identified market demand (Warren-Smith and Jackson, 2004). Household concerns and finances are also apparent from a study of women working in agritourism. Their motives thus have both a market focus based on farm resources and a household focus, but when their chosen strategies lead to conflicts between market and household considerations the latter have priority (Engeset and Heggem, 2015). These results reflect traditional female activities, such as caring for others and fostering social relations, and women's view of their own financial impact as contributing to regular farm income; but the results also point to women's interest in market relations and attainment of individual life goals.

How female entrepreneurship and farming businesses develop is also connected to men's practices. One change observed is that men are increasingly involved in childcare and household labour (Miller, 2010). This affects women's opportunities in terms of gainful employment and entrepreneurship. A study of two generations of Norwegian farmers showed that the current generation of farmers is more involved in childcare in the domestic arena, while the earlier generation integrated childcare with introducing children to farm work. The movement of mothers into the non-agricultural labour force, combined with new 'fathering moralities' in society, contributes to this development (Brandth, 2016).

Restructuring processes identified recently indicate blurring of the boundaries between what men and women are expected to do in the farming industry (Silvasti, 2003; Price, 2012; Heggem, 2014b). The increasing contribution of women to farms and the agricultural industry that has been identified (Brandth, 2002; Bock and Shortall, 2006; Seuneke and Bock, 2015), and the transformation of gender identities and practices among farmers (Brandth, 2016), are among processes affecting this rural livelihood both materially and socially. Greater knowledge of the transformation processes, values and strategies involved is crucial for developing policies for agriculture and rural development alike. Agriculture needs to be reflected upon as a sector intertwined with the rural environment and with rural development in general. Gender identities under transformation intersect the past and the future, and the study described here aimed to provide insight into such dependencies.

The chapter proceeds as follows. The analytical framework section elaborates on the concept of 'values' and how it relates to farming and gender. This discussion forms a basis for the arguments that values are developed in relation to collective and individual goals and that this can be understood by recognizing the values as reflecting three levels: the individual, business and societal levels. All three are important for the viability of farms and for rural communities. The methodology section introduces the theme of farming and values, and describes the method and the respondents. The research findings section presents the results, and the following section contains a discussion of the results and their implications for research and policy formulation.

Analytical Framework: Farmers' Values and Strategies

'Values' may be defined as 'conceptions of the desirable that guide the way social actors (such as organizational leaders, policy-makers, individual persons) select actions, evaluate people and events, and explain their actions and evaluations' (Schwartz, 1999, p. 25). Values thus constitute trans-situational criteria or goals and active guiding principles in everyday life. Schwartz refers to underlying motivational structures and two dimensions that can represent the various value types. The first is openness to change versus conservation, reflecting the distinction between an individual's willingness to act independently and unwillingness to change. The second dimension, self-enhancement versus self-transcendence, reflects the distinction between values oriented towards self-interest and those oriented towards the welfare of others (Schwartz, 1992, 1994). Several researchers argue that people who prioritize self-transcendent values are more willing to engage in various forms of altruistic or cooperative behaviour (Rokeach, 1968; Stern *et al.*, 1995; Nordlund and Garvill, 2002).

Values relate to actions on an individual as well as a group level, and individual values are 'a product of shared culture and of unique personal experience' (Schwartz, 1999, p. 26). Cultural values represent shared abstract ideas about what is good, right and desirable in society (Williams, 1970 in Schwartz, 1999); this collective aspect means that social institutions or leaders can draw on them to influence people's behaviour and transmission of cultural values is an outcome of informal and formal socialization (Schwartz, 1999, p. 25). In recent years, greater interest in food and its production has placed farming and farms in the limelight and opinions from policy makers and consumers, for example, now reach farmers through public discourse. Institutions, which provide political, social and economic environments, underpin and thus influence human interaction (North, 1990). Accordingly, farmers' strategies involve reflections and negotiations covering a wide range of stakeholders and factors, such as consumers, veterinary surgeons, bankers, politicians, policy and subsidies. This in turn influences strategies for building legitimacy and creating a viable business.

Values, personal and collective, are linked to strategies and behaviour because of their close relationship with legitimacy and the social context. Collective values are thus mirrored in farmers' individual values, which are developed in interaction with the local social, political and economic context. Drawing on the concept of values oriented towards the pursuit of self-interest and values oriented towards the welfare of others, this work demonstrates how these values play an active part in the development of occupational strategies. Moreover, since values are always directed towards something, the distinction serves as a conceptual framework. As a way of making sense of the complex expressions of values in relation to farming, three levels of interest are defined: the individual, the business, and the society level. Farming practices also serve as an example of how these levels intersect.

Studies seeking to identify farmers' values have emphasized their association with the land and attachment to land, individual time management, profitability and societal goals. Björkhaug (2006) emphasizes the significance of five values characterizing a farmer habitus: working the land (boosting productivity); knowing the land (from an early age); owning the land (when it has been in the family for a long time); freedom (taking one's own initiatives and planning one's time); and hard work. These values are closely connected with succession and differ from those of 'farmers without a family connection': women farmers or farmers with an alternative career (previous or current).

Differences in values with regard to age and gender have been identified in a study on Finnish farmers (Niska *et al.*, 2012). There, profitability was valued more highly among younger farmers, while older farmers considered workers' equality, the common good of the nation and job creation to be more important. Women saw employee well-being as more important, as did male farmers who had diversified (Niska *et al.*, 2012, p. 460). The most popular value among the Finnish farmers was 'autonomy', followed by 'economy' and 'the community', while 'transgenerational continuity' was the least popular (Niska *et al.*, 2012, p. 462).

Two values often mentioned in relation to the context of farming are autonomy and the specific rural and agrarian lifestyle (Mayberry *et al.*, 2005). These values are sometimes related to a form of entrepreneurship known as 'ecological entrepreneurship' (Marsden and Smith, 2005). Ecological entrepreneurship is a process whereby 'key actors facilitate sustainable development in the countryside by a combination of fragmentation, specialisation and quality building strategies' (Marsden and Smith, 2005, p. 440). Marsden and Smith assert that ecological entrepreneurship is closely connected to sustainable rural development, including agriculture that strives for a lower impact on the environment, rural vitality and environmental well-being (Marsden and Smith, 2005). Ecological entrepreneurship thus reflects certain values, which are reflected in individual farmers' strategies. While ecological entrepreneurship has not been used as an organizational tool in the analysis, the definition is relevant to any attempt to understand and explain contemporary farming strategies.

Methodology

Keener interest in food and food production combined with a decreasing number of Swedish farmers give rise to questions about farmers' motives. An identified increase in women farmers, in particular, prompts study of motives and values among female farmers. By focusing on the reasoning and explanations of women farmers regarding their choice of occupation and how this choice intersects with other parts of their lives, the chapter highlights both contemporary processes of achieving sustainability in farms and how these processes are connected to the farms' rural localities and to society at large. The focus on women farmers also generates greater knowledge of the transformation of the agricultural sector and a local context (the rural).

This chapter builds upon semi-structured, in-depth interviews with ten women farmers in two counties. The interviewees represent a broad variety of farm ownership forms, type of farm (dairy, vegetables, etc.) and means of entry into farming. Help was initially sought from the Federation of Swedish Farmers to make the first contacts. Some of these farmers then provided further contacts; that is, 'snowball sampling' took place. These counties, Uppsala and Södermanland (and in one case Dalarna, close to the Uppland border), are located in central Sweden near (north and south of) the capital region that includes the city of Stockholm. This location affects market composition and opportunities as well as land prices and diversification options. Interviews took place on the farms, in kitchens, living rooms or gardens, and lasted for about 1.5–2 h. In four cases the women farmers were interviewed with their respective male partners, and in one case two cooperating women farmers were interviewed together. In these cases, the two informants sometimes confirmed each other's stories or experience, while sometimes opposing each other.

One observation is that the informants seemed to be used to discussing questions of the kinds raised. Another was their apparent awareness of each other's attitudes and opinions. Such pre-understandings or trained reflexive competencies entail a risk of producing repertoires and repeating what is already known, and the informants may encourage each other in this direction. One advantage, however, is that given the opportunity to develop their own thoughts and arguments on a familiar topic, the informants feel safe and perceive their own background and competence as a legitimate basis for expanding on a theme, expressing opinions and elaborating on their interpretations. This may be advantageous in individual as well as in group (or couple) interviews. The interviews were transcribed and analysed, partly by means of the research questions and interview guide and partly with *in vivo* coding; that is, themes emerged from the transcription itself.

For the women farmers interviewed, choosing farming and succession had not always been straightforward. Rather, they had started working in a different occupation – as a biomedical engineer, agricultural adviser or insurance salesperson, for example. In cases of succession, the opportunity to take over the family farm resulted in the woman negotiating with herself. For some it was a dream come true, for others more of an experiment. Women farmers who did not take over the family farm either started a new business or joined their partner's family farm. Some respondents owned the family farm jointly with the husband or several family members, while others had yet not taken over ownership. There are also examples of tenancy where the respondent leased the buildings and/or land while owning the animals. Access to machinery varied: while some owned all the machinery, others leased or hired machinery and the labour needed to operate it (Table 9.1).

This particular method was chosen to enable explanations and arguments to be generated about being a farmer and creating a profitable business, and to explore how farming as an occupation relates to an individual's family and societal relationships and aspirations. Its purpose is also to contribute to the discussion about transforming gender relations, taking women farmers' values and strategies as its point of departure. Niska *et al.* (2012) argue that supporting or rejecting a value in an interactive situation is not simply a reflection of the farmer's inner disposition, but an indication of what farming is all about. The interaction in an interview situation is thus well suited to an examination of farming as a practice in relation to society and rural transformation. Putting the essence of farming into words is also itself a means of changing society, through communicative acts and the way farm work is actually performed. In addition, farmers constitute a group that is relatively poorly represented in rural development projects (Shortall, 2008).

Table 9.1. Introducing the farmers.

Farmer's name	Family/ non-family succession	Age (years)	Main production	Ownership	Location (county)
Frida	Family	30	Crop and forest	Owns the farm	Uppsala
Sofie	Non-family	26	Dairy	Manages the farm for an investor	Södermanland
Cecilia	Family	34	Crop and meat	Tenancy	Södermanland
Sara	Family	38	Crop, forest, meat and tourism	Owns the farm	Södermanland
Karin	Non-family	30	Crop	Owns the farm	Uppsala
Margareta	Family	35	Dairy	Owns the farm	Dalarna
Maria	Non-family	35	Dairy	Owns the farm	Uppsala
Catherine	Family	28	Dairy	Parents still own the farm	Uppsala
Tina, Gabriela	Non-family	35 45	Dairy (including cheese)	Shared ownership	Uppsala

Research Findings

Women farmers' values

Everyday life involves continuous expressions of collective and individual interests. One problem is naturally that of distinguishing between the two. Entrepreneurial activities may, for example, be interpreted as individual interests. But it is also common knowledge that entrepreneurs often take their families into consideration in relation to the business, their employees, the environment, the local community or other, related aspects.

During the interviews, no questions directly pertaining to values were asked. Rather, the discussion related to how to achieve a sustainable and prosperous farming enterprise and what lends meaning to the occupation and the way of life to which it contributes. Values were thus discernible in the interviewees' explanation of motives for their behaviour. In this case study, values have been identified through analysis of the interview transcriptions. The analytical procedure involved trying to identify key concepts and central themes, rather than imposing a predetermined theoretical framework. Discussions on motives for embarking on farm entrepreneurship and persisting in farming activities included relating the farm to individual goals and the family, the farm business and its development and earning capacity, and the farmer's role in society. One observation is that the strategies and values identified, emerging from the interview content, partly support earlier findings and partly lead to new insights into women's values and strategies in relation to entrepreneurship. This is not to say that the values identified are new among women (such a comparative approach would require a different methodological procedure), but rather that an ongoing transformation affects how individuals express themselves and also how their narratives are interpreted.

The analysis builds upon the interviews with ten women farmers and the quotations are strategically chosen to exemplify the themes emerging in the material as a whole. The values identified can be divided into three themes: quality of life, financial goals and socio-ecological aspirations. These values reflect the three different levels at which farming is perceived: the individual and societal levels. These levels intersect and impact each other: for example, the financial situation at the farm affects quality of life, and how far socio-ecological goals are met may affect finances. The correspondence between the values identified and the three levels also illustrates how individual and collective interests intersect, and how farming as an occupation involves placing individual lives and personal goals in a broader environmental context. Schwartz's definition of values may be useful in this analysis. 'Conceptions of the desirable', expressed by the farmers in terms of, for example, being able to supply health-giving food or contributing to sustainable agriculture, are at work when they formulate strategies and behaviour, and guide their evaluation of the situation (Schwartz, 1999, p. 25).

Quality-of-life values include freedom and autonomy, work–life balance or 'love–work balance' (see Grubbström *et al.*, 2014) and enjoyment of life. These values are connected to the other two sets of values, being mutual prerequisites to some extent. Sound finances and good relations with the local community are included in quality-of-life aspects.

Values relating to financial goals include ensuring that nothing goes to waste. Using resources optimally intersects with a desire to make a profit and also communicate to other farmers that a certain strategy is profitable. This strategy of visualizing success involves a competitive attitude.

Related to financial values are *socio-ecological aspirations*, since they include economizing and nurturing (animals, for example) combined with a wish to include consumers and the community in farming practice. Ecological and social values are expressed as an interest in multifunctionalism or diversification, which involves a wider range of tasks, such as selling directly to consumers or offering tourism, leisure activities and care services. These values also place the agricultural business in a community context.

Quality-of-life values

Tina and Gabriela work together on a milk farm. They have a herd of 18 Swedish highland cows and produce cheese from the milk. The two women used to be neighbours living on a farm near Stockholm. One interest they share is horses – they also describe themselves as 'horsey types' – but neither comes from a farming family. Tina's husband worked on a pig farm while she worked at an agricultural college. Gabriela's background is in insurance, developing new products, and subsequent employment as a sales representative at an alarm system company. However, her educational background was as an agronomist. Tina, who went to agricultural college, also worked with cattle and had an idea about starting a micro-dairy.

> Tina: 'I didn't want several million kronor in bank loans and hundreds of cows. I wanted a niche business. . . . I looked at free-range pigs, deer breeding and things but thought a dairy would be good. I attended a course on operational leadership. My partner and I took the course together, started to do the calculations and drafted a financial plan. And over a piece of cheese (!) I told Gabriela about my plans.'

> Gabriela: 'And I thought, wow! This sounds fun, I want to get involved!'

> Tina: 'And that was something I'd thought about from the start: that you should do this with someone. 'Cause I want to be able to go on holiday, and it's great to be able to swap ideas with someone other than the person you're living with – who perhaps is getting bored listening to you.'

Tina expresses financial caution: she did not want to borrow too much money from the start. She also describes efforts to achieve personal goals, such as finding a balance between work and leisure (including holidays), and developing a sustainable business within a partnership. The commitment involved in running a farm may put a great deal of strain on a relationship and she stresses the importance of involving another, equally committed partner. Her strategy also included educating herself in order to feel secure and be well equipped to run a business.

'Quality of life' thus also includes family life and the 'love–work balance'. Such values have previously been shown to explain women's agricultural entrepreneurship and the work strategies developed (Heather *et al.*, 2005), which is also evident here, if not consistently. One woman, Sara, mentions that being at home, closer to her children, is an advantage. This applies to both herself and her husband: 'Even if we work a lot, we're also close to our children. That's worth a lot.' For Sara, farming is about working together. One driver is the desire to combine their interests and skills. The focus of the farm has emerged from ad hoc decisions, including reflections and negotiations regarding their life situation in general. 'Working together and living like this has been a driver.' The farming business means long hours, and she has an ever-present feeling of neglecting the children and possibly not giving them the same level of attention and kind of childhood as other children in terms of travel and related experience, for example. However, Sara finds comfort in her closeness to her children and the scope for integrating work and childcare, enabling her to spend more time with them in their day-to-day lives.

Margareta, who took over her family farm with her husband, explains their rationale for diversifying and introducing cheese-making. Her husband had to drop his off-farm work and planned to start making cheese, but they concluded that with three young children, travelling to local markets every weekend would be neither easy nor desirable. For Cecilia, the situation is different. She describes her husband as 'an urbanite who prefers asphalt to mud'. She also describes how he has to take care of the children and the domestic chores when he gets home from work as she is busy on the farm in the evenings. He is not interested in the business, so her strategy is directed more towards making him feel comfortable in the farm environment. A common feature of all these cases is that entrepreneurial strategies are developed with consideration for other aspects of life. The women's male partners – partners in life and sometimes also business – are acknowledged in these negotiations, which result in adaptation on both sides.

Freedom and autonomy are values often associated with farming (Björkhaug, 2006) and this is confirmed in the present study.

> You get the opportunity to work with something you enjoy – many people do, but we also work for ourselves and there's an opportunity to be outside and be active, to not be shut in. . . . The dairy is fun, we produce something. . . . And people like what we produce.

This short statement contains several hints about values such as autonomy ('work for ourselves'), being in the rural physical environment ('outside and be active') and satisfaction in producing something people want. She also expresses a sense of freedom in combination with routines and being on site at the farm. Autonomy may also concern deciding when to be sociable and showcase your business (for example, at farmers' markets or by inviting visitors to the farm), and when to immerse yourself in daily routines without needing to adjust to scrutiny by others, which Sara describes as follows: 'It's good not always having to be sociable, that you can be on your own. I've had colleagues, but you work by yourself in a way. No need to be . . . superficially nice 24 hours a day.'

Cecilia expresses similar feelings. She has been approached by a neighbour who wants to spend time on the farm and help her out of interest in farming, but her reaction to this has been 'If she comes round once in a while, that's enough – I can't stand having someone here all the time!' It should be added that this need to be on their own does not prevent the farmers from being sociable and positive towards, for example, meeting customers or working with other farmers, or from envisioning the option of employing a team.

Financial values: thrift and money-making

For Tina and Gabriela, the dairy farm was a project they wanted to work on with a business partner rather than in the context of a family business. Support and understanding from other family members are important in terms of their living situation but unnecessary for running the business. They found a farm for sale in Uppland, a dairy farm shut down three years earlier. Relatively early on, they decided to buy Swedish highland cattle. Their calculation included 12 cows, and at the time of the interview they had 18. The breed is at risk of extinction and these animals are seen as special by many people who remember them from their childhood. They are also 'cute', which has proved important in marketing terms. Another advantage is that their milk contains a special protein that enables more cheese to be produced per litre of milk. Therefore, their motives for focusing on this specific niche were multifaceted and included financial aspects, marketing options and emotional content related to the cows' appearance and significance. Media interest has been growing – as has interest in the products – and has led to frequent interviews and documentaries. However, the explanation given by Tina and Gabriela for their success is 'luck' and that they only do 'what we find enjoyable'.

Sara also reflects upon the challenges involved in making a profit. She speaks in terms of economizing and ensuring that nothing goes to waste:

> And that's what I find exciting, making use of each part of the cow we're slaughtering so we can get maximum payment for every cut of meat – because we sold the right cut as entrecôte and made sausages from another.

Alternating between explanations that indicate fortuitous success and well-founded observations about finances, quality and marketing options indicates an entrepreneurial spirit that encompasses both job satisfaction and financial goals. In describing their entrepreneurial practices, the women do not highlight their proficiency or expertise to any great extent; instead, they play down these aspects by talking about feelings and luck. Nevertheless, entrepreneurship involves aspiring to earn money, and decisions are motivated not solely by feelings but also potentially by financial considerations.

> Tina: 'From the start we didn't want to grow. We wanted to stick to ten to 12 cows. But now we want to expand.'
>
> Interviewer: 'What happened?'
>
> Tina: 'It's also hard to say no. When regular customers call, you can't say you have no products. You want to meet their requirements and can see financial advantages in expanding the business. Perhaps you get greedy.'

Another reason for expanding was that Tina's husband wanted to work in agriculture at the farm. This also posed a financial challenge, and expansion was a way to make this possible. Various factors thus contribute to business development. The women also explain their success in terms of people's growing interest in local produce and of starting their business at the right time.

> Tina: 'It's very popular – all nice restaurants, all good . . .'
>
> Gabriela: '. . . gourmet shops . . .'
>
> Tina: '. . . gourmet shops but also ordinary shops . . .'
>
> Gabriela: '. . . want Swedish . . .'
>
> Tina: '. . . that is, ICA and shops like that . . . they want organic and it has to be . . .'
>
> Gabriela: '. . . Swedish and produced locally . . .'
>
> Tina: 'Yes, exactly, so most of them are desperately searching for their own farm.'

Karin and Peter have settled on Peter's family farm. Karin talks both of the importance of Peter's skills and how their skills are complementary. She describes the future in terms of a great number of plans they will probably have to put on hold for a while. The schedule has been changed, but the courage to plan and persevere is Peter's: 'He's braver – I'm more of a realist.' She continues: 'He's also good at making calculations to determine whether it's a safe bet or not. And he's more of an entrepreneur.' When asked about her own qualifications, Karin responds: 'I'm much better with animals. He's more into arable farming.'

One of their business ideas was to sell meat packages consisting of several different cuts of meat from the same animal – an idea Karin picked up at her former job on a farm in another region. These packages can be delivered to customers, or customers can visit the farm to see the animals and find out more about the production process. While she describes Peter as more daring and having more of an entrepreneurial spirit, she also reveals that the meat packages were her idea. She does not associate this with being entrepreneurial, perhaps because of her view that she and her husband complement each other and have different roles in the company. However, in terms of taking care of the children and shouldering domestic responsibilities unrelated to the business, she assumes that they have similar roles.

When a woman starts a business as an independent farmer, her entrepreneurial identity may be strong. But explicitly entrepreneurial strategies are offset by such motives as 'having fun', balancing work and child care, and being able to pursue leisure activities such as horse riding. These are aspects also related to quality-of-life motives and goals. It seems that going into business – whether this is self-initiated and independent or arises from being a farmer's life partner or spouse – may affect people's views of themselves as entrepreneurs and of their own roles in the business. 'Traditional' gendered identities and attitudes prevail in cases where the male partner is the main farmer. A woman farmer may be active in drawing up business strategies, but still not consider herself an entrepreneur. Nevertheless, entrepreneurial identities may be observed among the women farmers studied, such as being responsible for developing the business, branching out into new activities or crops, and communicating with customers.

Socio-ecological values

Transformation in the farming industry is sometimes referred to as 'multifunctionality'. This is not a recent phenomenon; farmers have always supplemented their income from such sources as forestry, hunting, fishing and off-farm work (Björkhaug and Blekesaune, 2008). More recent developments, however, are the increased elements of service and care, along with the wider range of activities (van der Ploeg *et al.*, 2002). Farmers included in the study express an interest in multifunctionalism or diversification, involving a broad range of functions, such as selling directly to consumers, offering tourism and leisure activities, and providing care services. The awareness and interpretation of this context have socio-ecological key signatures (van der Ploeg *et al.*, 2002).

The values identified indicate that women farmers' strategies and values may be interpreted in terms of what Marsden and Smith (2005, p. 440) term 'ecological entrepreneurship', in which ecological and social values intersect. Sara explains that she and her partner sell their meat directly to consumers and that meeting consumers is an important aspect of their business. It gives them the opportunity to provide information about their products and find out 'who is eating our animals'. Their farm is multifunctional: as well as producing crops and meat, they run a B&B and a small shop, inviting consumers and visitors to buy their products and come face to face with the animals – 'allowing the children to feel the lambs' wool between their fingers'. Ecological values are intertwined in the entrepreneurial strategies. One goal is to make the farm function as a closed-loop ecosystem: 'I enjoy finding a way to close the loop and actually sell the meat to nearby consumers, who will also gain an understanding of the origin of their food.'

Consumers are also active in shaping the development of a farm business. For Sara, consumer requests have led to her decision to start breeding Angus and Hereford cows. These are not traditionally bred in Sweden but are in demand because of their high-quality meat. The decision was underpinned by financial, ecological and animal-welfare reasoning:

> Angus and Hereford, they aren't Swedish . . . but they're quite small (beef) cattle and they're hardy, because we don't want them to eat soya – they should eat grass. . . . they should be good for preserving the landscape, and they should give birth easily.

Socio-ecological values also include relations with the community. Margareta explains how they have striven to establish good relations with their neighbours and the local community by, for example, inviting people to the farm and showing them how they want to develop it:

> We thought we should get everyone involved by showing them what we were up to because there was going to be some noise and traffic during the construction period. And when we'd finished building, we invited everyone to a 'pub evening' in our staff room. Lots of people came and it was great fun.

Neighbours and colleagues are local resources that need nurturing, since they can be important during hard times. When Margareta's father fell ill, he was unable to be there as planned during the busiest period of the year: 'Everything was chaotic.' However, she goes on to say that:

> People just showed up – it was fantastic. They stood there in our yard and said, 'We thought we'd help out today,' and then you realize how many valuable people there are that you didn't know about. Fantastic.

Starting or growing a farming business may be linked with goals that exert an impact outside the farm as well, such as educating consumers or creating job opportunities. McGeheee *et al.* (2007) argue that women are more motivated than men to become tourism entrepreneurs because of their higher interest in the issues concerned. In Nebraska, the motive for starting a farmers' market was a desire to preserve the local community. Entrepreneurship was embedded in the local context: the closure of a local church and declining income from agriculture and the women's former business activities (Sattler Weber, 2007).

The women farmers included in this study refer to entrepreneurial strategies and goals. The interviewees' paths into farming varied. Their farms are of differing sizes, with various areas of focus. These women's farms have represented an opportunity for them to concentrate their efforts on something they care about and consider important, as well as a catalyst in their search for challenges and coherence. This coherence involves the farm, the family and the community, or 'overlapping social, financial and personal motives' (Caballé, 1999).

Discussion and Conclusion

One means of understanding the ongoing transformation of the agricultural sector and of gender relations – manifested in, for example, larger elements of tourism and care services (van der Ploeg *et al.*, 2002) and farmers' quest to engage with consumers – is investigation of farmers' values. In this chapter, women farmers' values have been identified to clarify entrepreneurial motives and how agricultural entrepreneurship is linked to individual and community contexts.

'Multifunctionality', a key concept in discussion of agricultural change, refers not only to various products and services generated in the farming industry but also to diversification and rural entrepreneurship (Rönningen *et al.*, 2012). Several scholars have argued that agriculture and farming are developing into a more diversified labour market, open to men and women with varying interests and skills. From Heggem (2014a), it appears that a transformation of the sector allows women to participate more, one reason being that this diversified entrepreneurship is more in line with 'traditional' women's skills. It leads to processes involving emphasis on femininities, but also to transformation of gendered identities among both men and women.

In Sweden, the share of women farmers is increasing. Identifying the values of contemporary women farmers and investigating how far these values are instrumental in their development of occupational strategies produce knowledge that is crucial regarding how female (and male) entrepreneurs should be viewed and approached by, for example, politicians and agricultural advisors. The present results support earlier findings that hegemonic masculinity related to farming and agriculture is being challenged. It may be added that women farmers' values relating to quality of life and finances include more than family-related reasons, and that their economic and strategic goals extend beyond being supplementary stakeholders in the system.

The research question guiding this study is formulated as: 'How can women farmers' values be defined and how do the values relate to the formulation of farming strategies?' 'Values' are defined as trans-situational criteria or goals that are active guiding principles in everyday life (Schwartz, 1999). The values identified in this case study can be divided into three themes: quality of life, financial goals and socio-ecological aspirations. Returning to Schwartz (1992, 1994) and his 'first dimension', openness to change versus conservation, the strategies developed are changing the existing mode of operation rather than reproducing it, as Seuneke and Bock (2015) have also identified. However, the results presented here cannot be used to assert that women farmers' values are changing. Rather, they show that how values are expressed in the context of contemporary transformations may change our conceptions of gender relations and of women's roles and self-awareness. Moreover, the farmers in this study seem to reflect a mix of values oriented towards self-interest and values oriented towards the welfare of others, in line with the 'second dimension' (Schwartz, 1992, 1994). The welfare of others has so far primarily been discussed in terms of family welfare, while social and environmental resilience are themes of the material referred to here.

The experiences described by the women farmers shed light on the ambiguities involved in explaining values and motives. Societal gender structures such as expectations and individuals' own aspirations and desires in relation to their businesses are visible in language and self-perception. A simultaneous link with individual and collective goals (Schwartz, 1992, 1994) is evident from the farmers' motives for starting and/or staying in business. Achieving a 'love–work balance' may be an individual consideration, while socio-ecological values intersect with ideas about what is desirable from a community perspective.

Farming is an activity integrated into the local rural context, as well as into a broader community context involving other producers and consumers. Such intentions are, moreover, found in this study. They are expressed in communicative strategies involving the local rural community and the wider consumer community. Interaction with the community – customers, neighbours and local residents – is integrated into entrepreneurship. Communicating what farming is about and putting farming into practice constitute steps towards changing society (Shortall, 2008) and the farmers interviewed express such aspirations explicitly in terms of communicating how food is produced and how consumers can be more active. From a gender perspective, communicating what farming is about may include intentionally or unintentionally informing others how gendered practices are being transformed. One expectation of this study is that the results will contribute to a deeper understanding of transformation of gender relations in the farm sector and the rural geographical context.

The farmers studied value the financial gains that a farm business can generate. Their financial values concern maximizing profit, economizing and ensuring that nothing goes to waste. The study results thus show that financial values are not synonymous with a positive attitude towards policies promoting restructuring of the sector in terms of bigger units and higher profits. Farm finances are intertwined with other values in that a farm business also develops in relation to farmers' intentions regarding their families, 'love–work balance' and relations with the community – values oriented towards self-interest and community welfare. A desire to make an impact on society and the food production industry may also be included in a discussion of strategy formulation. Farmers want to be part of the rural agricultural landscape as well as the social scene. Such aspirations may be interpreted as involving increases in local dependencies, collaborations or common interests. It means that agriculture needs to be reflected upon as a sector intertwined with the rural environment and rural development in general. Further research could aim to shed light on the effect on farmers' roles in social networks and on rural resilience.

Notes

[1]The project was funded by the Swedish Farmers' Foundation for Agricultural Research (project no. H0946336).

[2]In Sweden, policy has been developed to encourage women's entrepreneurship. It stresses that both men and women should be encouraged to start businesses. The Federation of Swedish Farmers (Lantbrukarnas Riksförbund; LRF) states the aim that gender equality should permeate the whole organization and has established the LRF Gender Equality Academy. The purpose of this think tank is to develop ideas, initiate research and generate opinion, in order to boost knowledge of men's and women's scope for starting and running businesses in agriculture, forestry and related industries (LRF, 2017).

[3]A 'farmer' is defined here as a person who is financially and legally responsible for agricultural business operations.

References

Björkhaug, H. (2006) Sustainable agriculture in the Norwegian farmers' context: exploring farming habitus and practice on the Norwegian agricultural field. *International Journal of Environmental, Cultural, Economic and Social Sustainability* 4(2), 123–131.

Björkhaug, H. and Blekesaune, A. (2008) Gender and work in Norwegian family farm businesses. *Sociologia Ruralis* 48(2), 152–165. Available at: http://dx.doi.org/10.1111/j.1467-9523.2008.00456.x (accessed 4 March 2017).

Bock, B.B. (2004) Fitting in and multi-tasking: Dutch farm women's strategies in rural entrepreneurship. *Sociologia Ruralis* 44(3), 245–260. Available at: http://dx.doi.org/10.1111/j.1467-9523.2004.00274.x (accessed 4 March 2017).

Bock, B.B. and Shortall, S. (eds) (2006) *Rural Gender Relations: Issues and Case Studies.* CAB International, Wallingford, UK.

Brandth, B. (2002) Gender identity in European family farming: a literature review. *Sociologia Ruralis* 42(3), 181–200. Available at: http://dx.doi.org/10.1111/1467-9523.00210 (accessed 4 March 2017).

Brandth, B. (2016) Rural masculinities and fathering practices. *Gender, Place & Culture* 23(3), 435–450. Available at: http://dx.doi.org/10.1080/0966369X.2015.1013454 (accessed 4 March 2017).

Caballé, A. (1999) Farm tourism in Spain: a gender perspective. *GeoJournal* 48(3), 245–252. Available at: http://dx.doi.org/10.1023/A:1007044128883 (accessed 4 March 2017).

Engeset, A.B. and Heggem, R. (2015) Strategies in Norwegian farm tourism: product development, challenges, and solutions. *Scandinavian Journal of Hospitality and Tourism* 15(1–2), 122–137. Available at: http://dx.doi.org/10.1080/15022250.2015.1005332 (accessed 4 March 2017).

Eurostat (2014) Farm Structure Statistics. Available at: http://ec.europa.eu/eurostat/statistics-explained/index.php/Farm_structure_statistics (accessed 9 February 2016).

Farstad, M. and Heggem, R. (2008) Like barn leker best?: om mannlige bønders holdninger til kvinner i landbruket (Birds of a feather? Male farmers' attitudes to women in agriculture). In: Almås, R., Haugen, M.S., Rye, J.F. and Villa, M. (eds) *Den nye bygda* (*The New Village*). Tapir Akademisk forlag, Trondheim, Norway, pp. 85–102.

Grubbström, A., Stenbacka, S. and Joosse, S. (2014) Balancing family traditions and business: gendered strategies for achieving future resilience among agricultural students. *Journal of Rural Studies* 35, 152–161. Available at: http://dx.doi.org/10.1016/j.jrurstud.2014.05.003 (accessed 4 March 2017).

Heather, B., Skillen, L., Young, J. and Vladicka, T. (2005) Women's gendered identities and the restructuring of rural Alberta. *Sociologia Ruralis* 45(1–2), 86–97. Available at: http://dx.doi.org/10.1111/j.1467-9523.2005.00292.x (accessed 4 March 2017).

Heggem, R. (2014a) Den nye «bonden» – mellom endring og kontinuitet. Empiriske studier av rekruttering til norske landbrukseiendommer sett i lys av det multifunksjonelle landbruket (The new 'farmer' – between change and continuity. Empirical studies of recruitment to Norwegian farm estates in the light of multifunctional land use). PhD thesis, Norwegian University of Science and Technology (NTNU), Trondheim, Norway.

Heggem, R. (2014b) Exclusion and inclusion of women in Norwegian agriculture: exploring different outcomes of the 'tractor gene'. *Journal of Rural Studies* 34, 263–271. Available at: http://dx.doi.org/10.1016/j.jrurstud.2014.03.002 (accessed 4 March 2017).

Little, J. (2002) Rural geography: rural gender identity and the performance of masculinity and femininity in the countryside. *Progress in Human Geography* 26(5), 665–670. Available at: http://dx.doi.org/10.1191/0309132502ph394pr (accessed 4 March 2017).

LRF (2017) Om LRF. Uppdrag, vision och värdegrund. Jämställdhet (Mission, vision and values. Gender equality). Available at: https://www.lrf.se/om-lrf/uppdrag-vision-och-vardegrund/jamstalldhet/jamstalldhetsakademin/ (accessed 12 February 2017).

Marsden, T. and Smith, E. (2005) Ecological entrepreneurship: sustainable development in local communities through quality food production and local branding. *Geoforum* 36(4), 440–451. Available at: http://dx.doi.org/10.1016/j.geoforum.2004.07.008 (accessed 4 March 2017).

Mayberry, D., Crase, L. and Gullifer, C. (2005) Categorising farming values as economic, conservation and lifestyle. *Journal of Economic Psychology* 26(1), 59–72. Available at: http://dx.doi.org/10.1016/j.joep.2003.10.001 (accessed 4 March 2017).

McGehee, N., Kyungmi, K. and Jennings, G. (2007) Gender and motivation for agri-tourism entrepreneurship. *Tourism Management* 28(1), 280–289. Available at: http://dx.doi.org/10.1016/j.tourman.2005.12.022 (accessed 4 March 2017).

Miller, T. (2010) *Making Sense of Fatherhood: Gender, Caring and Work.* Cambridge University Press, Cambridge, UK.

Niska, M., Vesala, H.T. and Vesala, K.M. (2012) Peasantry and entrepreneurship as frames for farming: reflections on farmers' values and agricultural policy discourses. *Sociologia Ruralis* 52(4), 453–469. Available at: http://dx.doi.org/10.1111/j.1467-9523.2012.00572.x (accessed 4 March 2017).

Nordlund, A. and Garvill, J. (2002) Value structures behind proenvironmental behavior. *Environmental Behavior* 34(6), 740–756. Available at: http://dx.doi.org/10.1177/001391602237244 (accessed 4 March 2017).

North, D.C. (1990) *Institutions, Institutional Change and Economic Performance.* Cambridge University Press, Cambridge, UK.

Petridou, E. and Glaveli, N. (2008) Rural women entrepreneurship within co-operatives: training support. *Gender in Management: An International Journal* 23(4), 262–277.

Pini, B. (2008) *Masculinities and Management in Agricultural Organizations Worldwide.* Ashgate Publishing Ltd, Farnham, UK.

Price, L. (2012) The emergence of rural support organisations in the UK and Canada: providing support for patrilineal family farming. *Sociologia Ruralis* 52(3), 353–376. Available at: http://dx.doi.org/10.1111/j.1467-9523.2012.00568.x (accessed 4 March 2017).

Rokeach, M. (1968) *Beliefs, Attitudes and Values: A Theory of Organization and Change.* Jossey-Bass, San Francisco, California.

Rönningen, K., Renwick, A. and Burton, R.J.F. (2012) Western European approaches to and interpretations of multifunctional agriculture – and some implications of a possible neo-productivist turn. In: Almås, R. and Campbell, H. (eds) *Rethinking Agricultural Policy Regimes: Food Security, Climate Change and the Future Resilience of Global Agriculture.* Emerald Books, Biggleswade, UK, pp. 73–97.

Sattler Weber, S.S. (2007) Saving St. James: a case study of farmwomen entrepreneurs. *Agriculture and Human Values* 24(4), 425–434. Available at: http://dx.doi.org/10.1007/s10460-007-9091-z (accessed 4 March 2017).

Schwartz, S.H. (1992) Universals in the content and structure of values: theoretical advances and empirical tests in 20 countries. *Advances in Experimental Social Psychology* 25, 1–65.

Schwartz, S.H. (1994) Are there universal aspects in the structure and contents of human values? *Journal of Social Issues* 50(4), 19–45. Available at: http://dx.doi.org/10.1111/j.1540-4560.1994.tb01196.x (accessed 4 March 2017).

Schwartz, S.H. (1999) A theory of cultural values and some implications for work. *Applied Psychology* 48(1), 23–47. Available at: http://dx.doi.org/10.1111/j.1464-0597.1999.tb00047.x (accessed 4 March 2017).

Seuneke, P. and Bock, B.B. (2015) Exploring the roles of women in the development of multifunctional entrepreneurship on family farms: an entrepreneurial learning approach. *NJAS – Wageningen*

Journal of Life Sciences 74, 41–50. Available at: http://dx.doi.org/10.1016/j.njas.2015.07.001 (accessed 4 March 2017).

Shortall, S. (2008) Are rural development programmes socially inclusive? Social inclusion, civic engagement, participation, and social capital: exploring the differences. *Journal of Rural Studies* 24(4), 450–457. Available at: http://dx.doi.org/10.1016/j.jrurstud.2008.01.001 (accessed 4 March 2017).

Silvasti, T. (2003) Bending borders of gendered labour divisions on farms: the case of Finland. *Sociologia Ruralis* 43(2), 154–166. Available at: http://dx.doi.org/10.1111/1467-9523.00236 (accessed 4 March 2017).

Statistics Sweden (2015) *Agricultural Statistics 2015 Including Food Statistics – Tables.* Official Statistics of Sweden, Swedish Board of Agriculture, Örebro, Sweden (in Swedish).

Stern, P.C., Dietz, T., Kalof, L. and Guagnano, G.A. (1995) Values, beliefs, and proenvironmental action: attitudes formation toward emergent attitude objects. *Journal of Applied Social Psychology* 25(18), 1611–1636.

Swedish Board of Agriculture's Official Statistics of Sweden (2015) Full-time farming in Sweden 2013. Statistical report JO 65 SM 1401. Available at: http://www.jordbruksverket.se/swedishboardofagriculture/statistics/statsec/holdingsandholders.4.68dc110a12390c69dde8000521.html (accessed 9 February 2017).

van der Ploeg, J.D., Long, A. and Banks, J. (2002) *Living Countrysides. Rural Development Processes in Europe: The State of the Art.* Elsevier, Doetinchem, The Netherlands.

Warren-Smith, I. and Jackson, C. (2004) Women creating wealth through rural enterprise. *International Journal of Entrepreneurial Behavior & Research* 10(6), 369–383.

10 Women Farmers and Agricultural Extension/Education in Slovenia and Greece

M. Černič Istenič[1,*] and C. Charatsari[2]

[1]University of Ljubljana, Ljubljana, Slovenia; [2]Aristotle University of Thessaloniki, Thessaloniki, Greece

Introduction

Education is thought to be one of the fundamental factors in the development of every society and individual. Access to education is a fundamental human right and is essential for the exercise of all other human rights that enable people to understand themselves and the world, thus empowering them to participate in public and political life (UN, 1948; UNESCO, 1960; McCowan, 2011). Moreover, an adequate education is a prerequisite for accomplishment in any profession, which should apply to both genders. However, in reality, an adequate education does not necessarily reap equal outcomes (Loury, 1997; Robeyns, 2006; Bobbitt-Zeher, 2007; Rodríguez Pose and Tselios, 2009).

The importance of education for the empowerment of women in their professional and private lives has been broadly addressed by scholars. The scientific literature encompasses a wide variety of fields and topics such as health, social inclusion, quality of life, macro-economic development and national political stability. However, research attention on the role of education (in particular the contribution of adult education) in the personal and professional development of farm women remains scarce. Women's involvement in agriculture and agribusiness is increasingly recognized. Nevertheless, scholars have mainly focused on women's position in decision making in family farms or their participation in farm labour and off-farm employment (Haugen, 1998; O'Hara, 1998; Brandth, 2002a,b; Little, 2002; Bock, 2006). Conversely, the contribution of formal and informal educational attainment to the empowerment of women in agriculture has received less attention (Seevers and Foster, 2003; Behrman *et al.*, 2004; Meinzen-Dick *et al.*, 2011).

More importantly, the study of factors that propel women to participate in agricultural extension, education or training programmes has attracted a limited number of research projects while comparative studies that could generate integrated insights (e.g. the national influence) and contribute to a better understanding of this phenomenon remain wanting. The analysis presented in this chapter tries to address this gap by answering two questions. First, what are the motivating forces that drive women farmers to participate in agricultural extension/education programmes (AEEPs)? Second, what gives rise to these forces?

In our work, which employs the framework of self-determination theory (Ryan

* Corresponding author e-mail: majda.cernic.istenic@bf.uni-lj.si

and Deci, 2000b), we refer to motivation as the intrinsically or extrinsically regulated process that gives rise to an action performed in order to meet a goal. In that sense, motivation leads to the development of goals, thus directing human behaviour (Deci and Ryan, 2000). In other words, human psychological needs, when unmet, are translated into motives that affect a person's desires and choices (Sheldon and Gunz, 2009). However, the same need may produce different motives and can lead to the pursuit of different goals. Since the satisfaction of these basic needs is fundamental for human thriving, individuals seek different kinds of experiences to satisfy unfulfilled needs (Prentice *et al.*, 2014). Hence, participation in AEEPs could serve as a vehicle to help farmers meet their goals.

In seeking to answer our questions, we conducted a comparative study of Greece and Slovenia, two predominantly rural countries situated on the Balkan Peninsula. They differ in terms of the size of their territory, natural resources, climate conditions and their historic, social, political and economic evolutions; however, to some degree, they are similar in terms of their agrarian social structures. For instance, according to the 2010 Agricultural Census data, 99% of all farms in Slovenia and Greece were family farms (96.9% in the EU-28). In both countries, small family farms prevail, with an average area of less than 10 ha (the average size of an agricultural holding in the EU-28 is 16.1 ha) (European Union, 2015). However, in Slovenia, family farms cultivate more than 90% of the total utilized agricultural area, whereas, in Greece, the corresponding share is 75% (in the whole EU-28, it is 67.4%) (European Union, 2015). Moreover, in Greece, one-third of all farm managers are aged 65 years or above, while, in Slovenia, the equivalent share is a few percentage points lower (the EU-28 average is 31%) (European Union, 2015). Regarding the gender structure of the workforce, the 2010 Agricultural Census data showed that women represent 45.8% of the family farm workforce in Slovenia and 39.6% in Greece (the corresponding share for the EU-27 is 42.3%) (European Commission, 2013). While the share of female farm holders in Greece is 30.3% and 26.3% in Slovenia, in the entire EU-27 it is 28.7% (European Commission, 2012). These characteristics are also mirrored in the share of agriculture in the gross domestic product. In 2012, this figure represented 1.3% of Slovenia's gross domestic product and 2.5% of Greece's (in the EU-27, it was 1.2%) (European Union, 2012). In our study, we sought to highlight the embeddedness of women farmers' willingness to participate in AEEPs in both countries. We assumed that the differences/similarities pertaining to socio-economic settings and agrarian social structures, with the overall social significance of agriculture, impact on the motivations of farm women to participate in AEEPs in either country.

The chapter is partitioned as follows. The next section reviews earlier works on women's participation in AEEPs by providing some sociological explanations of farm women's marginalization in this domain. It also clarifies the specific social contexts of the selected countries in terms of their historical pathways that have led to the particular position of women in agricultural education. Then the methodological basis of our research is outlined. The subsequent two sections present the findings from the analysis of the survey data and discuss them by drawing on the results of previous studies. Finally, the limitations of the research are considered and the main conclusions derived from this study are summarized.

Background and Context

Women farmers' access to educational programmes

One of the more profound challenges faced by farmers in the EU is the need to be/remain competitive in the world food market. Importantly, in the contemporary situation of increased global competitiveness in the agricultural industry, knowledge and its transfer to farmers is becoming a key factor in their success. The EU's new

Rural Development Programme 2014–2020 emphasizes the need for knowledge and innovation transfer in agriculture, forestry and rural development by promoting the first among six priority areas of the programme.

However, the successful implementation of such policies is case sensitive and depends on the characteristics of each target group. Several research findings (e.g. Brown and Bewsell, 2011; Bartoli *et al.*, 2012; Černič Istenič *et al.*, 2012; Lioutas and Charatsari, 2012; De Rosa *et al.*, 2014) indicate that farmers are not homogeneous end users of agricultural knowledge; they have different needs for certain types of content and communication of knowledge depending on their production activities, farm size, development orientation, level of education, age and, of course, gender. Nevertheless, agricultural policy makers and extension planners have paid insufficient attention to the rural concept of gender role stereotypes and its impacts on the diffusion of agricultural knowledge among women farmers.

As a consequence, in many cases, agricultural extension/education systems operate in a way that puts women farmers at a relative disadvantage. This is underpinned by a number of evidences referring to all levels of education worldwide. Indeed, a survey on extension services in 115 countries, carried out in 1989 by the Food and Agriculture Organization of the United Nations (FAO), showed that women, regardless of their significant contribution to food production, received only 2–10% of all extension contacts and only 5% of extension resources worldwide (Marilee *et al.*, 1997). Other studies also indicate that the diffusion of agricultural knowledge to women farmers remains limited (Bock, 1994; Shortall, 1996; Brandth, 2002a; Safilios-Rothschild, 2006; Davran *et al.*, 2007; Rad *et al.*, 2011). The FAO report *The State of Food and Agriculture 2010–2011* also states that in agriculture the world over, women have lower education levels and are less likely to have access to extension services when compared with their male counterparts (FAO, 2011). In addition, according to data from the United Nations Educational, Scientific and Cultural Organization, women have also been consistently and dramatically under-represented in higher education programmes in agriculture (Bradley, 2000). Further, women on family farms are formally educated mainly for non-agricultural occupations and are significantly reliant on the practical experiences obtained from their family members (Ilak Peršurić and Žutinić, 2007). Data from the 2005 Farm Structure Survey for the EU-27 (European Commission, 2012) showed that 23% of men in the EU have some agricultural training (full and basic) compared with only 12% of women.

These facts reveal that the mega-trends for women in education – for example, the female tertiary education attainment rate in the EU-28 in 2012 was already close to 39.9% (34.2% in Greece and 49.6% in Slovenia) while the comparable share for men was 31.5% (27.6% in Greece and 29.5% in Slovenia) (European Commission, 2015) – are not represented in basic agricultural training at the EU or national level. More recent data for Slovenia (Agricultural Census 2010; B. Kutin Slatnar, Ljubljana, 2015, personal communication) show some gender-unbalanced proportions: 8.4% of male farmers and 5.1% of female farmers have undergone full training in agriculture while basic training – mostly provided by the AEEPs through extension services – was received by 20.6% of men and 7.4% of women (Černič Istenič, 2011). Although in Greece there is a lack of official data on the rates of gender participation in agricultural extension, education or training activities, recent survey data indicate that men are the main clients of public and private extension initiatives. For example, Charatsari (2011) found that the proportion of male farmers who participate in AEEPs is significantly higher than the corresponding percentage of female farmers, especially in the case of conventional farming. Lioutas and Charatsari (2012), in their segmentation of Greek farmers according to the use of extension services, discovered that most of the women in their sample were classified in the cluster with the minimum involvement in extension activities.

Tractors for men and jams for women?

Studies from both developing (Jafry and Sulaiman, 2013; Ragasa *et al.*, 2013) and developed countries (Brasier *et al.*, 2009) have demonstrated that the limited access of women farmers to educational programmes is significantly related to the lack of attention within extension services to the needs of women farmers. The reasons for this shortage lie partly in the nature of agriculture. Due to the patrilineal line of farm inheritance, which considers men as the primary farmers and the primary decision makers, men are seen as those who should obtain full as well as basic agricultural knowledge (Collett and Gale, 2009). Since agriculture is still considered a male occupation, women remain largely invisible as farmers (Radel, 2011; Jefferson and Mahendran, 2012). This invisibility and marginalization of women in agricultural education and training is the result of the gendered division of labour within a specifically bounded space and contents of operation, as well as of the culturally constructed notions of femininity and masculinity around farming (Trauger, 2004; Trauger *et al.*, 2008). Therefore, because men are considered farm heads and women only as farmers' wives who do not occupy outdoor spaces and corresponding activities, the designers of AEEPs (mostly men) do not see farm women as 'authentic' farmers (Teather, 1994; Barbercheck *et al.*, 2009; Trauger *et al.*, 2010; Charatsari *et al.*, 2013b).

Hence, farm women gain agricultural knowledge, which goes beyond household chores, mainly through social sources such as spouses (Okwu and Umoru, 2009), relatives (Devi *et al.*, 2012), friends (Yusuf *et al.*, 2013) and other members of the farming communities (Rezvanfar *et al.*, 2007; Agwu and Adeniran, 2010). Conversely, the tenacious focus on commercial agricultural production – which embodies a narrow conception of work that refers only to commercial contributions and does not recognize other work that contributes to farm sustainability as meaningful – further complicates the conceptualization of the work performed by women on farms (Shortall, 1996).

Consequently, the male-focused orientation of AEEPs and the productivist-focused direction of these programmes contribute to women's marginalization from knowledge exchange (Butt *et al.*, 2010; Trauger *et al.*, 2010; Lioutas *et al.*, 2011; Rad *et al.*, 2011). This could explain why – despite women farmers reporting a remarkably positive attitude towards agricultural education/training and stressing a higher willingness to participate in such programmes compared with their male counterparts (Charatsari *et al.*, 2011a) – women remain an underserved population group from agricultural extension/education services (Shortall, 1996; Alston, 1998; Liepins and Schick, 1998; Kiernan, 2005, 2007; Trauger *et al.*, 2008; Barbercheck *et al.*, 2009; Brasier *et al.*, 2009). In addition, AEEP designers often do not see the implicit gender barriers to women's participation and, instead, claim that agricultural training is open to everyone (Shortall, 1999; Trauger *et al.*, 2008).

However, women farmers are discouraged from participating in agricultural extension/education activities not only because of the male-targeted orientation of the agricultural education programmes but also because of factors relating to the programmes themselves, as well as self-perception factors such as their fear of failure and their fear of being unwelcome (Charatsari *et al.*, 2013b). In this vein, AEEPs that exclusively address women farmers appear to have the potential to encourage their participation (Sachs, 1983; Shortall, 1990) and tend to be popular among them (Safilios-Rothschild, 2006; Shortall, 2010). Nevertheless, the appropriateness of such programmes has not been well specified. In the literature, the relative merits and demerits of women-only programmes have been debated without a clear conclusion (Malcolm, 1992; McGivney, 1993). Feminists believe that women-only programmes represent a marginal provision, reinforce stereotypes, do not question social structures and achieve limited, if any, social change. Conversely, counter positions maintain that women-only programmes are positive since women feel more comfortable in this context. Therefore, the question of how to make AEEPs more available, suitable and attractive to women farmers throughout their lifespan is difficult to answer since the explanation cannot be given by only taking into account the characteristics of the AEEPs offered. The answer also needs to encompass, as indicated above, social value systems and the socially and culturally constructed roles of women in farming that include expectations about what skills and knowledge is needed for women farmers.

You do not need to be educated to work on a farm

In this study, we compare Greece and Slovenia, both members of the EU, but each with peculiar socio-historical backgrounds that presumably lead to different meanings in relation to education and the agricultural occupation of male and female farmers. To verify this assumption, we first draw attention to the values and expectations towards educational achievements for both genders in agriculture which have evolved in the two countries over time.

The aforementioned poor representation of women farmers in (full as well as basic) agricultural education in Slovenia is the result of particular historical circumstances that positioned the great majority of the farm population (not just women) at the margins of educational processes. Although after the Second World War, Slovenia (as part of the former Socialist Federal Republic of Yugoslavia) experienced great progress in the inclusion of the entire population at all levels of the educational system – one of the benefits of the socialist regime was free education for all – the rural and particularly the farm population of both genders achieved lower educational results. As some 1990s' analyses of this phenomenon demonstrated (Hribernik, 1993, 1994), after having completed the compulsory eight years of schooling, Slovenia's farm children dropped out of further education processes or chose vocational secondary schools more often than children from other socio-professional groups. There are many reasons explaining these children's poorer educational attainment at higher levels, which consequently led to the very slow educational development of the farm population, a situation that still holds true today.[1] First, the low socio-economic status of the farm population[2] propelled a massive outflow of the young generation from farming, especially the female segment and women who had obtained a formal non-agricultural qualification (Hribernik, 1994).

Moreover, this process was influenced by other factors such as the socio-cultural context of the farm population; for example, their rejection of school as a social institution. From its very beginning, dating back to the 18th century, the school institution was not embraced by farm communities because it took away child labour, an important labour force on which farm production depended at that time. Accordingly, it was believed that the young generation could fully procure all the knowledge needed for the future within the circle of the farm family (Petak, 1976; Schmidt, 1988). This attitude was sustained long into the middle of the 20th century. As research conducted in the 1970s (Krajnc, 1977) revealed, during the time of socialism, educational achievements were still relatively poorly valued by the farm population; during this period, education was considered an investment 'of no value' by 26% of farmers, 'of a moderate value' by 69% and 'of high value' by only 5% of the respondents.

As recent research results show (Knežević Hočevar and Černič Istenič, 2010), such attitudes hardly exist today; both men and women farmers of the younger generation, particularly those living on farms that are subsidized by rural development programme aids, are catching up with the level of education of other professional groups. However, the interest of women farmers in attaining full agricultural education or acquiring additional knowledge depends significantly (despite the general empowerment of women in education in Slovenian society) on their position on a farm. Those with a less important role (e.g. assistant workers) express considerably weaker interests in additional education/training than farm women owners/managers (Černič Istenič, 2006).

In Greece, the situation is more or less similar, despite the differences in the socio-cultural and historical background of the country. After the Second World War and the subsequent Civil War (1946–1949), farm children had very few opportunities to participate in education. Girls, in particular, had limited access to formal education before the 1980s while their entry into post-secondary education remained restricted until the 1990s (Charatsari, 2014) for a variety of reasons, including the economic insecurity of farm families, the early age of marriage of farm women, the exclusion of rural women from paid work and the limited control farm women had over their life choices before the 1990s (Charatsari and Papadaki-Klavdianou, 2017). This entry into education coincided with – and possibly affected the development of – a negative attitude on the part of women towards farm employment (Gidarakou, 1999), which, in combination with the structural changes (organizational, regulatory, technological and economic) in the agricultural sector of the country, urged women to give up agriculture and to seek alternative career paths.

In addition, farmers' involvement in agricultural education or training in Greece is extremely low since the current framework of the official system of agricultural extension/education and training is ineffective and because of the prevailing perception that farming is an occupation based only on practical experience. According to the Organisation for Economic Co-operation and Development (OECD, 2015), while at the level of the EU-28 about 6% of farmers receive agricultural education and/or training, in Greece the corresponding percentage is ten times lower.

In general, research findings and official data suggest that historically in both countries, education in agriculture was not valued as an important precondition of farm occupation for both men and women. This is primarily because for the great majority of farm populations in both countries, agriculture was not a competitive (profit making) activity; it was a livelihood. Therefore, relying on available records of socio-historical developments, the question of what motivates Slovenian and Greek women farmers to participate in agricultural extension/education activities remains unexplained.

Methodology

Since one of the aims of the study was to examine the attitude of women farmers towards the characteristics of AEEPs, a structured questionnaire was designed to determine the relative importance of factors that may influence this attitude. Five-point Likert-type scales were used to measure all items introduced in the questionnaire. The scales concerned the content of AEEPs (40 items), the educational methodology (21 items) and educators' characteristics (14 items). In order to estimate the reliabilities of the scales, we used Cronbach's α. In all cases, α values were found to be above the recommended baseline of 0.60 (Nunnally, 1978, p. 245) but below the proposed maximum value of 0.90 (Lehman *et al.*, 2005, p. 145). However, given that Cronbach's α presupposes the existence of an absolute correlation and equal variances between the scores of the subjects (Sočan, 2000) and is affected by the sample size (Liu *et al.*, 2010), we also calculated Guttman's λ_2. The values of λ_2 were greater than 0.82 in all cases.

In addition, a scale consisting of 12 items relating to the reasons why farmers participate in agricultural education activities was incorporated in the survey instrument. These items referred to individuals' expectation of gaining various benefits from their participation in educational programmes and represented what Lioutas and Charatsari (2012) call 'benefit-oriented motivation'. Four of the statements referred to motives relating to the viability of the farm enterprise (increase of economic returns, broadening of employment options, maintenance of job status, potential access to financial subsidies); three concerned the human tendency to learn (increased knowledge of natural environment, information on innovations, information on modern technology); one was a leisure motive (use of spare time); one was associated with the need for relatedness (widening of social environment); one referenced the curiosity motive (satisfaction of curiosity); while the remaining two concerned the expectation of a better quality of life and the relevance of AEEPs to individuals' interests. The participants were asked to rate their agreement with each statement on a scale from 1 ('strongly disagree') to 5 ('strongly agree'). Guttman's λ_2 for the scale was 0.79 while the value of Cronbach's α was 0.76.

Two samples of farmers, one from each country, were chosen. Data from Slovenia were collected from across the country whereas Northern Greece was selected as the Greek study area. Summary statistics were used to provide an overview of the data, while inferential tests (Mann–Whitney U test, Pearson χ^2 and stepwise linear regression analyses) were employed to answer the research questions.

Research Findings

Participants

A total of 896 farmers participated in the study. The sample from Slovenia included 511 farmers, among whom 134 (26.2%) were women. The mean age of the farmers in the Slovenian sample was 45.97 years (standard deviation = 11.162 years). About one-fifth of the participants had obtained only a primary education whereas only 8.2% had obtained a tertiary education. More than half (53.8%) of the holdings in this sample were livestock farms. In 14.8% of the cases, the farms were dedicated to arable crops while 16.4% of the farmers cultivated several crops in small amounts. The mean proportion of income from agriculture was 62.2% (standard deviation = 30.991%). Most of the farms in the Slovenian sample used only conventional production practices; however, 35.5% of the farmers used integrated crop management (ICM) cultivation practices while 11.5% of the holdings were organic farms.

The sample from Greece consisted of 385 farmers, of whom approximately one-quarter (24.7%) were women. The age of the participants in the Greek sample had a bell-shaped distribution, with the maximum being between 41 and 50 years. Only 11.7% of the participants had a higher education, whereas 24.9% had attained only a primary education. The annual income from farming varied between €9000 and €15,000. Among the cultivations were orchards, vegetables, arable crops, olive oils and vineyards. Most of the Greek participants incorporated sustainable farming methods in their farming (33.0% ICM and 33.2% organic farming).

No differences between men and women regarding the demographic variables or the characteristics of their farms were found in either of the samples. Therefore, any difference between men and women in the analysis that follows can be attributed to gender.

Women farmers and agricultural education

Among the Slovenian farmers, 92.7% had attended at least one AEEP. The analysis revealed that the percentage was equally distributed among men and women farmers as 92.5% of women had participated in AEEPs. On the contrary, only 38.9% of the women farmers in Greece had participated in programmes of agricultural extension/education whereas the respective percentage for male farmers was 42.7%. However, the higher percentage observed in Slovenia can be attributed to the mandatory nature of a considerable part of agricultural education courses since, among the Slovenian women who have attended AEEPs, only 21% have voluntarily participated in such programmes. The corresponding percentage for male farmers was 15.4%, indicating a higher – but not statistically significant (χ^2 = 10.238; P = 0.115) – willingness to voluntarily attend AEEPs for Slovenian farm women. In the Greek data set, when willingness to participate in AEEPs was operationalized as an ordinal variable, the analysis revealed a significantly higher willingness among female farmers (U = 10,934.000; P = 0.002).

In order to investigate the influence of AEEPs' characteristics on women's participation in agricultural extension, education and training activities, we used two linear regression analysis models, one for each sample of women. Three groups of variables were used as predictors. In the first group, variables relating to educators' characteristics were included. The second group consisted of variables regarding the content of AEEPs. In the third group, variables relating to AEEPs' methods were classified. We used these three scales as predictors because they referred to the three basic AEEP elements which influenced farmers' willingness to participate (Charatsari *et al.*, 2011b).

In the Slovenian sample (Table 10.1), the analysis showed that the model could not yield adequate predictions ($P>0.05$). As a result, Slovenian women's participation in AEEPs could not be explained using AEEPs' characteristics. In the sample of Greek women, the analysis confirmed that, among the predictors, only the educators' characteristics could explain the variance of the response variable ($P=0.002$). As Table 10.2 highlights, educators' characteristics explained only a small proportion of the variance of the target variable (adjusted $R^2=0.213$) while the other two predictors did not contribute to the model.

Since both models indicate that women farmers' participation in AEEPs cannot be fully attributed to the AEEP elements, we examined the influence of a set of motives on women's participation in agricultural education. The analysis revealed that in both countries, women's participation in AEEPs is driven by their willingness to broaden their employment options, to be informed about innovations and to gain access to financial subsidies (Table 10.3). In addition, the relevance of AEEPs to women's interests is an important driver of their participation in agricultural extension/education activities. Moreover, Greek

Table 10.1. Change statistics of predictors introduced in the linear regression analysis model for the Slovenian sample (women).

					Change statistics				
Model	*R*	R^2	Adjusted R^2	SE	R^2 change	*F* change	df1	df2	Sig. *F* change
1	0.416[a]	0.173	0.066	0.253	0.173	1.617	14	108	0.086
2	0.721[b]	0.520	0.139	0.243	0.347	1.227	40	68	0.226
3	0.799[c]	0.638	0.060	0.253	0.118	0.731	21	47	0.780

SE, standard error; df, degree of freedom; sig., significance; AEEP
[a]Predictors: (constant), educators' characteristics.
[b]Predictors: (constant), educators' characteristics, AEEP's content.
[c]Predictors: (constant), educators' characteristics, AEEP's content, AEEP's methodology.

Table 10.2. Change statistics of predictors introduced in the linear regression analysis model for the Greek sample (women).

					Change statistics				
Model	*R*	R^2	Adjusted R^2	SE	R^2 change	*F* change	df1	df2	Sig. *F* change
1	0.575[a]	0.331	0.213	0.435	0.331	2.822	14	80	0.002
2	0.810[b]	0.656	0.191	0.441	0.325	0.945	40	40	0.570
3	0.933[c]	0.871	0.364	0.391	0.216	1.516	21	19	0.183

SE, standard error; df, degree of freedom; sig., significance; AEEP
[a]Predictors: (constant), educators' characteristics.
[b]Predictors: (constant), educators' characteristics, AEEP's content.
[c]Predictors: (constant), educators' characteristics, AEEP's content, AEEP's methodology.

women farmers are influenced by their desire to increase their economic returns as well as their willingness to improve their quality of life.

Interestingly, our analysis uncovered some notable similarities regarding the differences in the rates of motivation between men and women farmers in the two countries. As Table 10.4 illustrates, in both cases, women tended to correlate their participation in extension/education activities with the AEEPs' relevance to their interests. In addition, women's perceptions that participation in AEEPs would improve their

Table 10.3. Statistical relationships between examined motives and women's participation in agricultural extension/education programmes (Greece and Slovenia).

	Slovenian women		Greek women	
Motive	Mann–Whitney *U*	*P*	Mann–Whitney *U*	*P*
Broadening of employment options	329.500	0.032*	801.000	0.027*
Improvement of quality of life	501.000	0.612	784.000	0.012*
Increase of economic returns	392.000	0.118	673.000	0.001*
Increased knowledge of natural environment	488.500	0.527	895.000	0.147
Information on innovations	305.500	0.013*	826.500	0.048*
Information on modern technology	351.500	0.044*	1069.000	0.974
Maintenance of job status	422.500	0.214	1028.000	0.713
Potential access to financial subsidies	329.000	0.032*	793.000	0.021*
Relevance to farmer's interests	310.000	0.014*	789.500	0.012*
Satisfaction of curiosity	532.000	0.839	1072.500	0.996
Use of spare time	541.500	0.911	979.500	0.232
Widening of social environment	512.500	0.698	1030.000	0.720

*Significant at the 0.05 level.

Table 10.4. Comparison of the intensity of the examined motives between male and female farmers (Greece and Slovenia).

	Slovenian sample		Greek sample	
Motive	P^a	Comparison of mean scores[b]	P	Comparison of mean scores
Broadening of employment options	0.010	W>M	0.015	W>M
Improvement of quality of life	0.007	W>M	0.000	W>M
Increase of economic returns	0.430	–	0.263	–
Increased knowledge of natural environment	0.000	W>M	0.020	W>M
Information on innovations	0.639	–	0.020	M>W
Information on modern technology	0.949	–	0.000	M>W
Maintenance of job status	0.491	–	0.807	–
Potential access to financial subsidies	0.572	–	0.532	–
Relevance to farmer's interests	0.000	W>M	0.018	W>M
Satisfaction of curiosity	0.005	W>M	0.012	M>W
Use of spare time	0.512	–	0.599	–
Widening of social environment	0.514	–	0.188	–

[a]Mann–Whitney *U* test.
[b]M, men; W, women.

quality of life and expand their employment prospects were higher than those of their male counterparts ($P<0.05$). Another interesting finding is that the desire to learn more about the natural environment was more important to female farmers ($P<0.05$).

What about men?

As mentioned above, although women's participation in AEEPs could not be explained using the characteristics of the educational process, the analysis confirmed that for male farmers, participation was affected by all the three examined groups of variables; that is, the content and methods of AEEPs as well as educators' characteristics (Table 10.5). Following the same technique as that used for the sample of women farmers, we found that the scale concerning educators' characteristics was the most important predictor of men's participation in programmes of extension/education (R^2 change = 0.267) whereas the content (R^2 change = 0.128) and methodology (R^2 change = 0.076) of the educational programmes also contributed to the model.

Discussion

Dependency and invisibility

A common antecedent of Greek and Slovenian women's participation in AEEPs is their willingness to gain access to financial subsidies. To explain this finding, one needs to take into account the significant contribution of agricultural subsidies on the viability of farms. Particularly in the case of small-sized farm enterprises, the provision of subsidies by the EU represents an important economic input (Bojnec and Latruffe, 2013). Nevertheless, an alternative interpretation of this result can be proposed by looking at the issue of (farm) women's economic dependency on men. Despite the fact that around the world several rural policy initiatives have been launched to promote female farmers' economic autonomy and boost their entrepreneurial ambitions (see, e.g., Costa, 2005; Dhungana and Kusakabe, 2010; Paris *et al.*, 2011; Wrigley-Asante, 2012; Torri and Martinez, 2013), gender-equal access to economic resources remains questionable (Quisumbing and Pandolfelli, 2010; Lawal, 2011). This asymmetrical access – which is a reality even for rural women who gain a personal income from off-farm sources (Charatsari, 2013, p. 45) – sustains gender stratification within the family and society (Sorensen and McLanahan, 1987; Gupta, 2007) and contributes to the invisibility of women's roles within their field of work. This interpretation could also partially account for women's willingness to broaden their employment options through the acquisition of additional knowledge and skills. As Aassve *et al.* (2014) note, such a broadening can reduce gender segregation within the family.

According to Hosseini *et al.* (2009), agricultural education is also an important

Table 10.5. Change statistics of predictors introduced in the linear regression analysis model for male farmers (Greece and Slovenia).

					Change statistics				
Model	*R*	R^2	Adjusted R^2	SE	R^2 change	*F* change	df1	df2	Sig. *F* change
1	0.517[a]	0.267	0.250	0.388	0.267	16.220	14	624	0.000
2	0.629[b]	0.395	0.339	0.364	0.128	3.101	40	584	0.000
3	0.686[c]	0.471	0.400	0.347	0.076	3.827	21	563	0.000

SE, standard error; df, degree of freedom; sig., significance; AEEP
[a]Predictors: (constant), educators' characteristics.
[b]Predictors: (constant), educators' characteristics, AEEP's content.
[c]Predictors: (constant), educators' characteristics, AEEP's content, AEEP's methodology.

input that could facilitate women farmers' entrepreneurial success. This could explain why in both countries women farmers believe more than their male counterparts that participating in AEEPs could enhance their professional prospects. In this respect, our results corroborate well with some recent findings (Charatsari *et al.*, 2013a) showing that women farmers' expectation of increasing their operational competencies raises their willingness to attend programmes of agricultural extension/education. Furthermore, our results accord with those of Lioutas *et al.* (2011) that women farmers' dissatisfaction with their (low) social status as farmers and as women also strengthens their willingness to overcome the label of 'invisible' or 'second class' farmer and drives them into the agricultural education classroom.

Moreover, as our findings indicate, among the motives that impel women to participate in AEEPs, the relevance of educational content to their needs is included. Thus, our results support those of Lioutas *et al.* (2010) who discovered that the lack of correspondence between AEEPs' content and the interests of various groups of farmers is the primary factor restricting farmers' willingness to attend extension/education activities. Likewise, as demonstrated in the aforementioned discussion, this discrepancy could be attributed to the male-targeted orientation of agricultural education services. Although there is no available evidence from Slovenia to support this contention, research findings from Greece concur that programmes targeting women continue to be focused on 'feminine issues' (such as the household economy or cooking) whereas topics relating to farm management or technology target male farmers (Safiliou and Papadopoulos, 2004, p. 290; Charatsari *et al.*, 2013b).

Are agricultural innovations a man's affair?

Two seemingly contradictory findings emerged from our analysis. At the one end of the spectrum, the comparison between the male and female participants revealed that the willingness to be informed about agricultural innovations was higher among male farmers in the Greek sample (see Table 10.4). However, at the other end of the spectrum, when we focused our analysis on female farmers, we discovered that women from both countries were driven to participate in agricultural education activities by their willingness to learn about agricultural innovations.

The significant association found between women's participation in AEEPs and their desire to gain knowledge about innovations can be explained as a consequence of the way in which agricultural innovations are promoted to male and female farmers. The relevant literature clearly indicates that innovation diffusion is highly influenced by the position of the individual within the social network (Valente, 1996; Matuschke and Qaim, 2009; Maertens and Barrett, 2013). Consequently, from this finding, it could be anticipated that due to their disadvantaged position within the rural social network, women in the two countries strive for innovation. There are two possible impediments that hold back women's access to information on innovations. The first is an absence of women's agricultural organizations and/or networks, which have been proved to facilitate the diffusion of innovations to female farmers (Ogunlela and Mukhtar, 2009; Barbercheck *et al.*, 2012). The second concerns the absence of strong social capital – which is one of the main drivers in the diffusion of innovations (Bantilan and Padmaja, 2008; Tumbo *et al.*, 2013) – towards women in rural communities (Valdivia and Gilles, 2001).

However, women's access to extension services remains restricted, hampering the stream of information to them (Doss and Morris, 2000; Akpabio *et al.*, 2012), while services aimed at the diffusion of innovations continue to promote innovations based on a dichotomist discrimination that labels women as the 'deuteragonists' in agriculture (Charatsari *et al.*, 2013b) and men as 'farm heads' and, consequently, the main clients of these services. Hence, in most cases, the promotion of innovations is gender biased. In other cases, innovations are divided into two groups: those referred to as technological inputs or organizational changes, which are promoted to male farmers; and 'soft innovations', which concern extra-agricultural activities that target women. However, since our study cannot confirm this argument, more research is needed to explain why farm women's willingness to learn about innovations leads them to seek knowledge through participation in AEEPs.

Conclusions

This study investigated the motives that drive women farmers from two rural countries (Greece and Slovenia) to participate in AEEPs and attempted to explain the sources of these motivations. Although an understanding of women's motivation could advance the promotion of AEEPs to the rural female population, the issue remains largely under-investigated. In this work, rather than conducting a single-country study, we performed a comparative analysis in order to yield a better picture of the factors that propel women farmers to participate in agricultural education activities carried out by extension services. This research strategy permitted us to examine the issue under study within two different socio-cultural and agricultural contexts.

As our analysis shows, women's participation in AEEPs cannot simply be attributed to the characteristics of the existent agricultural extension/education schemes in the two countries. The motives that drive women farmers to attend AEEPs are remarkably similar in the two samples, despite the cultural, economic, social and geomorphological differences, and are most probably due to some similarities in the agrarian social structures between the two study areas. This indicates a relatively common attitude among farm women from the two elected populations towards agricultural education.

A corroboration of the survey results with the extant literature on rural gender studies revealed that these motives are derived from the particular power structures and gender norms that are hidden in the organization of rural societies. Rural societies are hierarchical spaces (which still holds true in both countries under study) where women enjoy fewer privileges than men. The gender stratification within farm families and society and the well-defined issue of women's invisibility in farming (Fortmann, 1979; Sachs, 1983; Kelly and Shortall, 2002; Riley, 2009) continue to resonate in rural gender discourse. Our findings indicate that women's invisibility appears to be the dominant driver that forces them to seek knowledge through participation in AEEPs. The analysis suggests that there are different goal pursuits that activate and energize women farmers and that have the potential to direct them into the agricultural education classroom. This conclusion seems to support the hypothesis – suggested in the literature – that the secondary and dependent role attributed to women farmers is the common denominator behind these pursuits.

Interestingly, the conception of farm women as 'assistant farmers' prevails not only at the family or community level but also, as Shortall (1996) notes, in the way in which agricultural knowledge is diffused. Although vocational education programmes, in general, are usually dominated by gendered subjectivities (Lappalainen *et al.*, 2013), in the case of agricultural knowledge transfer, gender biases are more frequent and sharp because of the peculiarity of agriculture and the socially constructed gender stereotypes it generates and reproduces. The modus operandi of extension services that are aimed at diffusing agricultural information and knowledge is based on traditional assumptions about the masculine nature of farming and the peripheral roles of women in agricultural practice, which was also confirmed in the analysis of both countries. Hence, agricultural extension/education services consider male farmers as the major end users of agricultural knowledge or, as Trauger *et al.* (2010) phrase it, as their 'authentically farming audience'. Still rare are practices whereby agricultural educators consider farm enterprises as family-based businesses and organize knowledge transfer for groups consisting of husband and wife couples (Owen and Williams, 2012).

What this study indicates is that women farmers' participation in AEEPs is the output of their willingness to gain access to economic resources and specific knowledge and to broaden the options of their farm enterprises by accessing spaces of agricultural innovation. Despite the fact that the results of this work are open to some alternative interpretations, the basic conclusion of our study is that the designers of AEEPs should place greater emphasis on farm women's real needs and desires. However, some issues remain obscure and call for further investigation. The first concerns the effect of the mechanisms that thwart the translation of women's motivation into action; that is, into participation in agricultural extension/education activities. The second has to do with the limited attention paid in this study to some forms of intrinsic motivation, such as the kind of motivation that relates to the individual's tendency to learn and assimilate (Ryan and Deci, 2000a). Future studies need to take into consideration these under-investigated issues and include other countries or groups of countries in order to provide a more holistic picture of the linkage between women's participation in AEEPs and drivers relating to their pursuit of success and autonomy.

Notes

[1]For example, according to the Agricultural Census 2010 data, 40% of women farmers and 27% of men farmers have a primary school education (B. Kutin Slatnar, Ljubljana, 2015, personal communication).

[2]During the socialist era, the social reputation of farmers was very low because of political and ideological reasons; they were considered a vanishing social stratum (connected with the bourgeoisie) that would gradually be replaced by the working class.

References

Aassve, A., Fuochi, G. and Mencarini, L. (2014) Desperate housework relative resources, time availability, economic dependency, and gender ideology across Europe. *Journal of Family Issues* 35(8), 1000–1022. Available at: http://dx.doi.org/10.1177/0192513X14522248 (accessed 4 March 2017).

Agwu, A.E. and Adeniran, A.A. (2010) Sources of agricultural information used by arable crop farmers in Isale Osun farm settlement, Osogbo local government area of Osun State. *Journal of Agricultural Extension* 13(1), 24–34 Available at: http://dx.doi.org/10.4314/jae.v13i1.53872 (accessed 4 March 2017).

Akpabio, I.A., Etim, N.A.A. and Okon, S. (2012) Perceptions of constraints affecting adoption of women-in-agriculture programme technologies in the Niger Delta, Nigeria. *International Journal of Agricultural Management & Development* 2(3), 215–222.

Alston, M. (1998) Farm women and their work: why is it not recognised? *Journal of Sociology* 34(1), 23–34. Available at: http://dx.doi.org/10.1177/144078339803400103 (accessed 4 March 2017).

Bantilan, M.C.S. and Padmaja, R. (2008) Empowerment through social capital build-up: gender dimensions in technology uptake. *Experimental Agriculture* 44(1), 61–80. Available at: http://dx.doi.org/10.1017/S0014479707005947 (accessed 4 March 2017).

Barbercheck, M., Brasier, K., Kiernan, N.E., Sachs, C., Trauger, A., Findeis, J., Stone, A. and Moist, L. (2009) Meeting the extension needs of women farmers: a perspective from Pennsylvania. *Journal of Extension* 47(3). Available at: http://www.joe.org/joe/2009june/a8.php (accessed 30 May 2013).

Barbercheck, M., Brasier, K., Kiernan, N.E., Sachs, C. and Trauger, A. (2012) Use of conservation practices by women farmers in the Northeastern United States. *Renewable Agriculture and Food Systems* 29(1), 65–82. Available at: http://dx.doi.org/10.1017/S1742170512000348 (accessed 4 March 2017).

Bartoli, L., De Rosa, M. and La Rocca, G. (2012) The impact of agricultural extension services: an empirical test through AKAP models. Paper presented at *10th European IFSA Symposium*, 1–4

July, Aarhus, Denmark. Available at: http://ifsa.boku.ac.at/cms/fileadmin/Proceeding2012/IFSA2012_WS1.1_Bartoli.pdf (accessed 30 May 2013).

Behrman, J.R., Alderman, H. and Hoddinott, J. (2004) Hunger and malnutrition. In: Lomborg, B. (ed.) *Global Crises, Global Solutions*. Cambridge University Press, Cambridge, UK, pp. 363–420.

Bobbitt-Zeher, D. (2007) The gender income gap and the role of education. *Sociology of Education* 80(1), 1–22. Available at: http://dx.doi.org/10.1177/003804070708000101 (accessed 4 March 2017).

Bock, B.B. (1994) Women and the future of Umbrian family farms. In: van der Burg, M. and Endeveld, M. (eds) *Women in Family Farms: Gender Research, EC Policies and New Perspectives*. Wageningen University, Wageningen, The Netherlands, pp. 83–89.

Bock, B.B. (2006) Rurality and gender identity: an overview. In: Bock, B.B. and Shortall, S. (eds) *Rural Gender Relations: Issues and Case Studies*. CAB International, Wallingford, UK, pp. 279–287.

Bojnec, Š. and Latruffe, L. (2013) Farm size, agricultural subsidies and farm performance in Slovenia. *Land Use Policy* 32, 207–217. Available at: http://dx.doi.org/10.1016/j.landusepol.2012.09.016 (accessed 4 March 2017).

Bradley, K. (2000) The incorporation of women into higher education: paradoxical outcomes? *Sociology of Education* 73(1), 1–18. Available at: http://dx.doi.org/10.2307/2673196 (accessed 4 March 2017).

Brandth, B. (2002a) Gender identity in European family farming: a literature review. *Sociologia Ruralis* 22(3), 227–244. Available at: http://dx.doi.org/10.1111/1467-9523.00210 (accessed 4 March 2017).

Brandth, B. (2002b) On the relationship between feminism and farm women. *Agriculture and Human Values* 19(2), 107–117. Available at: http://dx.doi.org/10.1023/A:1016011527245 (accessed 4 March 2017).

Brasier, K., Barbercheck, M., Kiernan, N.E., Sachs, C., Schwartzberg, A. and Trauger, A. (2009) Extension educators' perception of the educational needs of women farmers in Pennsylvania. *Journal of Extension* 47(3). Available at: http://www.joe.org/joe/2009june/pdf/JOE_v47_3a9.pdf (accessed 30 May 2013).

Brown, M. and Bewsell, D. (2011) Using a market segmentation approach to better target agricultural extension programs – aligning learner needs with learning programs. *Journal of Extension* 48(5). Available at: http://www.joe.org/joe/2010october/a6.php (accessed 28 December 2011).

Butt, T.M., Hassan, Z.Y., Mehmood, K. and Muhammad, S. (2010) Role of rural women in agricultural development and their constraints. *Journal of Agriculture & Social Sciences* 6(3), 53–56. Available at: http://www.fspublishers.org/published_papers/1471_..pdf (accessed 6 July 2017).

Černič Istenič, M. (2006) Farm women in Slovenia. In: Bock, B.B. and Shortall, S. (eds) *Rural Gender Relations: Issues and Case Studies*. CAB International, Wallingford, UK, pp. 63–96.

Černič Istenič, M. (2011) Poticaji i prepreke za cjeloživotno učenje seoskih žena u Sloveniji (Incentives and barriers for lifelong learning among rural women in Slovenia). In: Đugum, J., Sinković, K. and Sečić, B. (eds) *Uloga cjeloživotnog obrazovanja u zapošljavanju seoskih žena u prehrambenom sektoru* (*The Role of Lifelong Learning in the Employment of Rural Women in Food Sector*). Ministarstvo poljoprivrede, ribarstva i ruralnog razvoja, Zagreb, pp. 18–29.

Černič Istenič, M., Knežević Hočevar, D., Udovč, A. and Kapun, S. (2012) *Izzivi in potrebe v prenosu znanja v kmetijsko prakso v Sloveniji* (*Challenges and Needs of Agricultural Knowledge Transfer in Slovenia*). Biotechnical Faculty University Ljubljana, Ljubljana.

Charatsari, C. (2011) Identification of educational needs of organic producers, ICM operators and conventional farmers in Central Macedonia. PhD thesis, Aristotle University of Thessaloniki, Thessaloniki, Greece.

Charatsari, C. (2013) Woman's position in rural society from 1950 onwards: a study based on women's autobiographical memories. Post-doctoral dissertation, Aristotle University of Thessaloniki, Thessaloniki, Greece.

Charatsari, C. (2014) Is this a man's world? Woman in the farm family of Thessaly, Greece from the 1950s onwards. *Gender Issues* 31(3–4), 238–266. Available at: http://dx.doi.org/10.1007/s12147-014-9125-y (accessed 4 March 2017).

Charatsari, C. and Papadaki-Klavdianou, A. (2017) First be a woman? Rural development, social change and women farmers' lives in Thessaly – Greece. *Journal of Gender Studies* 26(2), 164–183. Published online 15 October 2015. Available at: http://dx.doi.org/10.1080/09589236.2015.1095080 (accessed 10 February 2017).

Charatsari, C., Papadaki-Klavdianou, A., Michailidis, A. and Partalidou, M. (2011a) Making the invisible farmers visible: strengthening women's participation in agricultural education programs. Paper presented at *24th European Society for Rural Sociology Congress: Inequality and Diversity in European Rural Areas*, 22–25 August, Chania, Greece.

Charatsari, C., Papadaki-Klavdianou, A. and Michailidis. A. (2011b) Farmers as consumers of agricultural education services: willingness to pay and spend time. *Journal of Agricultural Education and Extension* 17(3), 253–266. Available at: http://dx.doi.org/10.1080/1389224X.2011.559078 (accessed 4 March 2017).

Charatsari, C., Papadaki-Klavdianou, A., Michailidis, A. and Partalidou, M. (2013a) Great expectations? Antecedents of women farmers' willingness to participate in agricultural education programmes. *Outlook on Agriculture* 42(3), 193–199. Available at: http://dx.doi.org/10.5367/oa.2013.0134 (accessed 4 March 2017).

Charatsari, C., Černič Istenič, M. and Lioutas, E.D. (2013) 'I'd like to participate, but . . .': women farmers' scepticism towards agricultural extension/education programmes. *Development in Practice* 23(4), 511–525. Available at: http://dx.doi.org/10.1080/09614524.2013.790345 (accessed 4 March 2017).

Collett, K. and Gale, C. (2009) *Training for Rural Development: Agricultural and Enterprise Skills for Women Smallholders.* City & Guilds Centre for Skills Development, London. Available at: http://apskills.ilo.org/resources/training-for-rural-development-agricultural-and-enterprise-skills-for-women-smallholders (accessed 11 February 2017).

Costa, J.A. (2005) Empowerment and exploitation: gendered production and consumption in rural Greece. *Consumption, Markets & Culture* 8(3), 313–323. Available at: http://dx.doi.org/10.1080/10253860500160353 (accessed 4 March 2017).

Davran, M.K., Darcan, N. and Budak, D.B. (2007) Gender role in the small ruminant sector in mountain area of Turkey. *Journal of Applied Animal Research* 31(1), 93–95. Available at: http://dx.doi.org/10.1080/09712119.2007.9706638 (accessed 6 March 2017).

Deci, E.L. and Ryan, R.M. (2000) The 'what' and 'why' of goal pursuits: human needs and the self-determination of behavior. *Psychological Inquiry* 11(4), 227–268. Available at: http://dx.doi.org/10.1207/S15327965PLI1104_01 (accessed 6 March 2017).

De Rosa, M., Bartoli, L. and La Rocca, G. (2014) Testing extension services through AKAP models. *Journal of Agricultural Education and Extension* 20(5), 513–535. Available at: http://dx.doi.org/10.1080/1389224X.2013.872044 (accessed 6 March 2017).

Devi, U., Varma, S.K., Kaur, B., Singh, D. and Kumar, A. (2012) Technological communication sources utilization pattern of Haryana women farmers for cotton cultivation. *Journal of Cotton Research and Development* 26(2), 281–285.

Dhungana, B.M. and Kusakabe, K. (2010) The role of self-help groups in empowering disabled women: a case study in Kathmandu Valley, Nepal. *Development in Practice* 20(7), 855–865. Available at: http://dx.doi.org/10.1080/09614524.2010.508244 (accessed 6 March 2017).

Doss, C.R. and Morris, M.L. (2000) How does gender affect the adoption of agricultural innovations? The case of improved maize technology in Ghana. *Agricultural Economics* 25(1), 27–39. Available at: http://dx.doi.org/10.1111/j.1574-0862.2001.tb00233.x (accessed 6 March 2017).

European Commission (2012) Women in EU agriculture and rural areas: hard work, low profile. EU Agricultural Economic Brief No. 7. Available at: http://ec.europa.eu/agriculture/rural-area-economics/briefs/pdf/07_en.pdf (accessed 15 June 2015).

European Commission (2013) How many people work in agriculture in the European Union? EU Agricultural Economics Brief No. 8. Available at: http://ec.europa.eu/agriculture/rural-area-economics/briefs/pdf/08_en.pdf (accessed 15 May 2016).

European Commission (2015) Europe 2020 Target: Tertiary Education Attainment. Available at: http://ec.europa.eu/europe2020/pdf/themes/2015/tertiary_education_attainment_20151126.pdf (accessed 14 February 2017).

European Union (2012) Agriculture in the European Union. Statistical and Economic Information Report 2012. Available at: http://ec.europa.eu/agriculture/statistics/agricultural/2012/pdf/full-report_en.pdf (accessed 14 May 2016).

European Union (2015) Agriculture, Forestry and Fishery Statistics 2014. Available at: http://ec.europa.eu/eurostat/en/web/products-statistical-books/-/KS-FK-14-001 (accessed 14 May 2016).

FAO (2011) *The State of Food and Agriculture 2010–2011. Women in Agriculture: Closing the Gender Gap for Development.* Food and Agriculture Organization of the United Nations, Rome. Available at: http://www.fao.org/docrep/013/i2050e/i2050e.pdf (accessed 10 June 2013).

Fortmann, L. (1979) The plight of the invisible farmer: the effect of national agricultural policy of women in Africa. In: Dauber, R. and Cain, M.L. (eds) *Women and Technological Change in Developing Countries.* Westview Press, Boulder, Colorado, pp. 205–214.

Gidarakou, I. (1999) Young women's attitudes towards agriculture and women's new roles in the Greek countryside: a first approach. *Journal of Rural Studies* 15(2), 147–158. Available at: http://dx.doi.org/10.1016/S0743-0167(98)00054-0 (accessed 6 March 2017).

Gupta, S. (2007) Autonomy, dependence, or display? The relationship between married women's earnings and housework. *Journal of Marriage and Family* 69(2), 399–417. Available at: http://dx.doi.org/10.1111/j.1741-3737.2007.00373.x (accessed 6 March 2017).

Haugen, M.S. (1998) The gendering of farming. The case of Norway. *European Journal of Women's Studies* 5(2), 133–153. Available at: http://dx.doi.org/10.1177/135050689800500202 (accessed 6 March 2017).

Hosseini, S.J., Mirdamadi, S.M., Nejad, H.H. and Reza, G. (2009) Extension and education factors influencing the success of entrepreneurship among rural women in northern Iran. *Research Journal of Biological Sciences* 4(9), 967–973.

Hribernik, F. (1993) Vpliv vrednotenja izobraževanja in vrednotenja poklicev na šolanje kmečke mladine (Impact assessment of education and occupations on farmers' youth education). *Sodobno kmetijstvo* 26(6), 251–258.

Hribernik, F. (1994) Education and educational achievements of Slovenian population in 1991. *Research Reports of the Biotechnical Faculty University of Ljubljana* 63, 241–255.

Ilak Peršurić, A.S. and Žutinić, Đ. (2007) Education for family farms and gender aspects in Croatia. *Agriculturae Conspectus Scientificus* 73(1), 43–49.

Jafry, T. and Sulaiman, V.R. (2013) Gender-sensitive approaches to extension programme design. *Journal of Agricultural Education and Extension* 19(5), 469–485. Available at: http://dx.doi.org/10.1080/1389224X.2013.817345 (accessed 6 March 2017).

Jefferson, T. and Mahendran, A. (2012) Calculating the paid off-farm work contributions of women in Australian agricultural and rural communities. *International Review of Business Research Papers* 8(5), 191–204.

Kelly, R. and Shortall, S. (2002) 'Farmers' wives': women who are off-farm breadwinners and the implications for on-farm gender relations. *Journal of Sociology* 38(4), 327–343. Available at: http://dx.doi.org/10.1177/144078302128756714 (accessed 6 March 2017).

Kiernan, N.E. (2005) *Women in Production Agriculture: A Hidden Audience in your County? Tipsheet #77.* Penn State Cooperative Extension, University Park, Pennsylvania. Available at: http://comm.eval.org/HigherLogic/System/DownloadDocumentFile.ashx?DocumentFileKey=6f8bb6fc-4861-4c86-8439-03f5414e8157&forceDialog=1 (accessed 14 February 2017).

Kiernan, N.E. (2007) *A Needs Assessment Strategy for Women in Agriculture. Pennsylvania Example #23.* Penn State Cooperative Extension, University Park, Pennsylvania. Available at: http://ucanr.edu/sites/CEprogramevaluation/files/141183.pdf (accessed 14 February 2017).

Knežević Hočevar, D. and Černič Istenič, M. (2010) *Dom in delo na kmetijah: raziskava odnosov med generacijami in spoloma (Home and Work on Farms: The Study of Generations and Gender Relationships).* Založba ZRC, ZRC SAZU, Ljubljana.

Krajnc, A. (1977) *Izobraževanje, Naša Družbena Vrednota (Education, Our Social Value).* Delavska Enotnost, Ljubljana.

Lappalainen, S., Mietola, R. and Lahelma, E. (2013) Gendered divisions on classed routes to vocational education. *Gender and Education* 25(2), 189–205. Available at: http://dx.doi.org/10.1080/09540253.2012.740445 (accessed 6 March 2017).

Lawal, A.O. (2011) Women's benefits from agricultural technologies: evidence from poultry production among Nigerian fisherfolk. *Development in Practice* 21(3), 371–378. Available at: http://dx.doi.org/10.1080/09614524.2011.558059 (accessed 6 March 2017).

Lehman, A., O'Rourke, N., Hatcher, L. and Stepanski, E.J. (2005) *JMP for Basic Univariate and Multivariate Statistics: A Step-by-Step Guide.* SAS Institute Inc., Cary, North Carolina.

Liepins, R. and Schick, R. (1998) Gender and education: towards a framework for a critical analysis of agricultural training. *Sociologia Ruralis* 38(3), 286–302. Available at: http://dx.doi.org/10.1111/1467-9523.00079 (accessed 6 March 2017).

Lioutas, E.D. and Charatsari, C. (2012) Who is the customer of public agricultural extension/education services? *International Journal of Rural Management* 7(1–2), 83–102. Available at: http://dx.doi.org/10.1177/0973005212459829 (accessed 6 March 2017).

Lioutas, E.D., Tzimitra-Kalogianni, I. and Charatsari, C. (2010) Small ruminant producers' training needs and factors discouraging participation in agricultural education/training programs. *Livestock Research for Rural Development* 22(7). Available at: http://www.lrrd.org/lrrd22/7/liou22126.htm (accessed 26 July 2010).

Lioutas, E.D., Charatsari, C., Tzimitra-Kalogianni, I. and Papadaki-Klavdianou, A. (2011) Women's status anxiety and their willingness to participate in agricultural education programs. *Livestock Research for Rural Development* 23(11). Available at: http://www.lrrd.org/lrrd23/11/liou23234.htm (accessed 15 December 2011).

Little, J. (2002) *Gender and Rural Geography: Identity, Sexuality and Power in the Countryside.* Prentice-Hall, Harlow, UK.

Liu, Y., Wu, A.D. and Zumbo, B.D. (2010) The impact of outliers on Cronbach's coefficient alpha estimate of reliability: ordinal/rating scale item responses. *Educational and Psychological Measurement* 70(1), 5–21. Available at: http://dx.doi.org/10.1177/0013164409344548 (accessed 6 March 2017).

Loury, L.D. (1997) The gender earnings gap among college-educated workers. *Industrial and Labor Relations Review* 50(4), 580–593. Available at: http://dx.doi.org/10.1177/001979399705000402 (accessed 6 March 2017).

Maertens, A., and Barrett, C.B. (2013) Measuring social networks' effects on agricultural technology adoption. *American Journal of Agricultural Economics* 95(2), 353–359. Available at: http://dx.doi.org/10.1093/ajae/aas049 (accessed 6 March 2017).

Malcolm, J. (1992) The culture of difference: women's education re-examined. In: Miller, N. and West, L. (eds) *Changing Culture and Adult Learning.* SCUTREA, Boston, Massachusetts, pp. 52–55.

Marilee, K., Lindley, W.I., Van Crowder, L. and Doron, N. (1997) *Higher Agricultural Education and Opportunities in Rural Development for Women – An Overview and Summary of Five Case-Studies.* Food and Agriculture Organization of the United Nations, Rome. Available at: http://www.fao.org/docrep/w6038e/w6038e00.htm#TopOfPage (accessed 29 May 2013).

Matuschke, I. and Qaim, M. (2009) The impact of social networks on hybrid seed adoption in India. *Agricultural Economics* 40(5), 493–505. Available at: http://dx.doi.org/10.1111/j.1574-0862.2009.00393.x (accessed 6 March 2017).

McCowan, T. (2011) Human rights, capabilities and the normative basis of 'Education for All'. *Theory and Research in Education* 9(3), 283–298. Available at: http://dx.doi.org/10.1177/1477878511419566 (accessed 6 March 2017).

McGivney, V. (1993) Participation and non-participation: a review of the literature. In: Edwards, R., Sieminski, S. and Zeldin, D. (ed.) *Adult Learners, Education and Training.* Routledge, London, pp. 11–30.

Meinzen-Dick, R.S., Quisumbing, A., Behrman, J., Biermayr-Jenzano, P., Wilde, V., Noordeloos, M., Ragasa, C. and Beintema, N. (2011) *Engendering Agricultural Research, Development, and Extension. IFPRI Research Monograph.* International Food Policy Research Institute, Washington, DC. Available at: http://dx.doi.org/10.2499/9780896291904 (accessed 6 March 2017).

Nunnally, J.C. (1978) *Psychometric Theory*, 2nd edn. McGraw-Hill, New York.

OECD (2015) *Fostering Green Growth in Agriculture: The Role of Training, Advisory Services and Extension Initiatives. OECD Green Growth Studies.* OECD Publishing, Paris. Available at: http://dx.doi.org/10.1787/9789264232198-en (accessed 6 March 2017).

Ogunlela, Y.I. and Mukhtar, A.A. (2009) Gender issues in agriculture and rural development in Nigeria: the role of women. *Humanity & Social Sciences Journal* 4(1), 19–30.

O'Hara, P. (1998) *Partners in Production? Women, Farm and Family in Ireland.* Berghahn Books, Oxford, UK.

Okwu, O.J. and Umoru, B.I. (2009) A study of women farmers' agricultural information needs and accessibility: a case study of Apa Local Government Area of Benue State, Nigeria. *African Journal of Agricultural Research* 4(12), 1404–1409.

Owen, W. and Williams, E. (2012) The utilisation of groups for innovation and knowledge transfer. *Studies in Agricultural Economics* 114, 99–105. Available at: http://dx.doi.org/10.7896/j.1218 (accessed 6 March 2017).

Paris, T., Diaz, C. and Hossain, I. (2011) Participatory evaluation of a rice flour mill by poor rural women: lessons from Bangladesh. *Gender, Technology and Development* 15(2), 275–299. Available at: http://dx.doi.org/10.1177/097185241101500205 (accessed 6 March 2017).

Petak, A. (1976) Obrazovanje i profesionalizacija poljeprivrednog zanimanja te socialistička transformacija sela (Education and professionalization of agricultural occupations and socialist transformation of villages). *Revija za sociologiju* 6(2–3), 67–78.

Prentice, M., Halusic, M. and Sheldon, K.M. (2014) Integrating theories of psychological needs-as-requirements and psychological needs-as-motives: a two process model. *Social and Personality Psychology Compass* 8(2), 73–85. Available at: http://dx.doi.org/10.1111/spc3.12088 (accessed 6 March 2017).

Quisumbing, A.R. and Pandolfelli, L. (2010) Promising approaches to address the needs of poor female farmers: resources, constraints, and interventions. *World Development* 38(4), 581–592. Available at: http://dx.doi.org/10.1016/j.worlddev.2009.10.006 (accessed 6 March 2017).

Rad, S.T., Boz, İ., Polatöz, S. and Ateş, H.Ç. (2011) Women's literacy and extension education in rural eastern Mediterranean Turkey. *African Journal of Agricultural Research* 6(12), 2807–2819. Available at: http://www.academicjournals.org/article/article1380908304_Rad%20et%20al.pdf

Radel, C. (2011) Becoming farmers: opening spaces for women's resource control in Calakmul, Mexico. *Latin American Research Review* 46(2), 29–54. Available at: http://dx.doi.org/10.1353/lar.2011.0023 (accessed 6 March 2017).

Ragasa, C., Berhane, G., Tadesse, F. and Taffesse, A.S. (2013) Gender differences in access to extension services and agricultural productivity. *Journal of Agricultural Education and Extension* 19(5), 437–468.

Rezvanfar, A., Moradnezhai, H. and Vahedi, M. (2007) Information needs of farm women related to dairy farming and home management in Ilam State of Iran. *Livestock Research for Rural Development* 19(8). Available at: http://www.lrrd.org/lrrd19/8/rezv19113.htm (accessed 20 December 2007).

Riley, M. (2009) Bringing the 'invisible farmer' into sharper focus: gender relations and agricultural practices in the Peak District (UK). *Gender, Place & Culture* 16(6), 665–682. Available at: http://dx.doi.org/10.1080/09663690903279138 (accessed 6 March 2017).

Robeyns, I. (2006) Three models of education: rights, capabilities and human capital. *Theory and Research in Education* 4(1), 69–84. Available at: http://dx.doi.org/10.1177/1477878506060683 (accessed 6 March 2017).

Rodríguez-Pose, A. and Tselios, P. (2009) Education and income inequality in the regions of the European Union. *Journal of Regional Science* 49(3), 411–437. Available at: http://dx.doi.org/10.1111/j.1467-9787.2008.00602.x (accessed 6 March 2017).

Ryan, R.M. and Deci, E.L. (2000a) Intrinsic and extrinsic motivations: classic definitions and new directions. *Contemporary Educational Psychology* 25(1), 54–67. Available at: http://dx.doi.org/10.1006/ceps.1999.1020 (accessed 6 March 2017).

Ryan, R.M. and Deci, E.L. (2000b) Self-determination theory and the facilitation of intrinsic motivation, social development, and well-being. *American Psychologist* 55(1), 68–78. Available at: http://dx.doi.org/10.1037//0003-066X.55.1.68 (accessed 6 March 2017).

Sachs, C.E. (1983) *The Invisible Farmers: Women in Agricultural Production*. Rowman and Allanheld, Totowa, New Jersey.

Safilios-Rothschild, C. (2006) CAP regulations and farm household relations. In: Bock, B.B. and Shortall, S. (eds) *Rural Gender Relations: Issues and Case Studies*. CAB International, Wallingford, UK, pp. 260–278.

Safiliou, C. and Papadopoulos, A.G. (2004) *Small-Scale Farmers in Greece: Who Will Survive and How*? National Centre for Social Sciences/Gutenberg, Athens (in Greek).

Schmidt, V. (1988) *Zgodovina* Šolstva *in Pedagogike na Slovenskem* (*The History of Education and Pedagogy in Slovenian*). Delavska Enotnost, Ljubljana.

Seevers B. and Foster, B.B. (2003) Women in agricultural and extension education: a minority report. *NACTA Journal* 47(1), 32–37.

Sheldon, K.M. and Gunz, A. (2009) Psychological needs as basic motives, not just experiential requirements. *Journal of Personality* 77(5), 1467–1492. Available at: http://dx.doi.org/10.1111/j.1467-6494.2009.00589.x (accessed 6 March 2017).

Shortall, S. (1996) Training to be farmers or wives? Agricultural training for women in Northern Ireland. *Sociologia Ruralis* 36(3), 269–285. Available at: http://dx.doi.org/10.1111/j.1467-9523.1996.tb00022.x (accessed 6 March 2017).

Shortall, S. (1999) *Gender and Power – Women and Farming*. Macmillan, London.

Shortall, S. (2010) *Women Working on the Farm: How to Promote their Contribution to the Development of Agriculture and Rural Areas in Europe. IP/B/AGRI/IC/2010_046.* European Parliament, Brussels.

Sočan, G. (2000) Assessment of reliability when test items are not essentially tau-equivalent. In: Ferligoj, A. and Mrvar, A. (eds) *Developments in Survey Methodology.* FDV, Ljubljana, pp. 23–35.

Sorensen, A. and McLanahan, S. (1987) Married women's economic dependency, 1940–1980. *American Journal of Sociology* 93(3), 659–687. Available at: http://dx.doi.org/10.1086/228792 (accessed 6 March 2017).

Teather, E. (1994) Contesting rurality: country women's social and political networks. In: Whatmore, S., Marsden, T. and Lowe, P. (eds) *Gender and Rurality.* David Fulton Publishers, London, pp. 31–50.

Torri, M.C. and Martinez, A. (2013) Gender empowerment and equality in rural India: are women's community-based enterprises the way forward? *Journal of International Women's Studies* 12(1), 157–176.

Trauger, A. (2004) 'Because they can do the work': women farmers in sustainable agriculture in Pennsylvania, USA. *Gender, Place & Culture* 11(2), 289–307. Available at: http://dx.doi.org/10.1080/0966369042000218491 (accessed 6 March 2017).

Trauger, A., Sachs, C., Barbercheck, M., Kiernan, N.E., Brasier, K. and Findeis, J. (2008) Agricultural education: gender identity and knowledge exchange. *Journal of Rural Studies* 24(4), 432–439. Available at: http://dx.doi.org/10.1016/j.jrurstud.2008.03.007 (accessed 6 March 2017).

Trauger, A., Sachs, C., Barbercheck, M., Kiernan, N.E., Brasier, K. and Schwartzberg, A. (2010) The object of extension: agricultural education and authentic farmers in Pennsylvania. *Sociologia Ruralis* 50(2), 85–103. Available at: http://dx.doi.org/10.1111/j.1467-9523.2010.00507.x (accessed 6 March 2017).

Tumbo, S.D., Mutabazi, K.D., Masuki, K.F.G., Rwehumbiza, F.B., Mahoo, H.F., Nindi, S.J. and Mowo, J.G. (2013) Social capital and diffusion of water system innovations in the Makanya watershed, Tanzania. *Journal of Socio-Economics* 43(C), 24–36. Available at: http://dx.doi.org/10.1016/j.socec.2013.01.001 (accessed 6 March 2017).

UN (1948) The Universal Declaration of Human Rights. Available at: http://www.un.org/en/documents/udhr/index.shtml (accessed 30 May 2013).

UNESCO (1960) Convention against Discrimination in Education. Adopted by the General Conference at its eleventh session, Paris, 14 December 1960. Available at: http://unesdoc.unesco.org/images/0013/001325/132598e.pdf (accessed 30 May 2013).

Valdivia, C. and Gilles, J. (2001) Gender and resource management: Households and groups, strategies and transitions. *Agriculture and Human Values* 18(1), 5–9. Available at: http://dx.doi.org/10.1023/A:1007608717996 (accessed 6 March 2017).

Valente, T.W. (1996) Social network thresholds in the diffusion of innovations. *Social Networks* 18(1), 69–89. Available at: http://dx.doi.org/10.1016/0378-8733(95)00256-1 (accessed 6 March 2017).

Wrigley-Asante, C. (2012) Out of the dark but not out of the cage: women's empowerment and gender relations in the Dangme West district of Ghana. *Gender, Place & Culture* 19(3), 344–363. Available at: http://dx.doi.org/10.1080/0966369X.2011.572435 (accessed 6 March 2017).

Yusuf, S.F.G., Masika, P. and Ighodaro, D.I. (2013) Agricultural information needs of rural women farmers in Nkonkobe Municipality: the extension challenge. *Journal of Agricultural Science* 5(5), 107–114. Available at: http://dx.doi.org/10.5539/jas.v5n5p107 (accessed 6 March 2017).

11 The Agency Paradox: The Impact of Gender(ed) Frameworks on Irish Farm Youth

A. Cassidy*

NUI, Galway, Republic of Ireland

Introduction

A constant sense of impending crisis has hung over Irish family farming for much of the past 50 years with concerns especially raised around the impact and lure of educational and employment opportunities in the wider social environment. Fathers feared their sons and heirs would be lured away from the farm, never to return, with all the consequences this would have for the transfer of the land into the next generation of the family. Similarly, daughters migrated in such numbers to urban areas that the biological and cultural reproduction of the population was threatened (McNabb, 1964). Yet while there are obvious threats to its continuity, such as decreasing numbers of young farmers and fluctuating incomes, the cultural model of family farming based on the norm of impartible, patrilineal succession continues to hold sway. This is demonstrated by the low level of land sales, with only 0.01% of agricultural land released on to the market in 2011 (Savills, 2011) and only 13% of farms owned by women (Mulhall and Bogue, 2013). The broad question underpinning this chapter is how this norm continues to operate in light of the Irish societal shift towards urban professional lifestyles, equality of opportunities for women and increasing emphasis on choice for young men. In particular, it focuses on the experiences of a cohort of Irish farm youth who migrated to an urban area to attend university. Most of these individuals were highly unlikely to return to the farming industry or take ownership of the landholding in the future. Their pathway towards adulthood is framed by a need to balance a normative and cultural background embedded in the precepts of family farming with potentially contradictory norms that help them build towards a professional career path and the associated lifestyle.

The answer to this question lies, at least in part, in an exploration of the concept of the agency paradox, how this is manifested and its implications for farm youth, particularly those who do not intend to become full-time farmers. Underpinning this agency paradox is the fact that young people are able to switch between contradictory discourse frameworks depending on the context. In thinking of themselves within the farming domain, they filter their positions through the lens of farming norms and operate within the expectations associated with their gender and status in this sphere. Similarly, in thinking of themselves from the

* Corresponding author e-mail: anne.cassidy@nuigalway.ie

point of view of their non-farm positions, they shift towards acting and think in a manner appropriate to this setting. What this means is that a farmer's daughter can, on the one hand, presume she has no rights to be given a share of the farm and, on the other, believe that she should have equal opportunities to advance professionally and to be as independent as possible. For farmer's sons what emerges is that 'off the farm' they believe they have choice and control over their lives and make decisions based on this. However, in relation to the farming domain they often have little choice but to take over the farm or at the very least are the ones who have the most responsibility, among their siblings, to carry around this.

The development of these dual normative systems largely seems to be connected to how their parents simultaneously socialized their offspring into two paradoxical, heavily gendered paradigms. For sons this is based on individualization and the typical dominant 'farmer' position, and for daughters a type of feminism and a suborned position in the succession hierarchy. Here the term feminism should be taken to mean the demand and push for equality of opportunity and independence for women (Winkler, 2010). While on the surface it appears that these young people are exerting individual agency through, for instance, making the decision to attend university, in fact they operate within quite bounded structures. As is argued in this chapter, it is their positioning within contradictory and, yet somehow complementary, norms that enable the dual concerns of helping to establish a family's offspring on the path towards adulthood and retaining ownership of the land to be met.

The chapter first situates the cohort within the wider literature around farming youth and then discusses the theory of individualization. As this chapter focuses largely on agency and how it is acted out in an intricate web of the potentially competing needs of the farm and creating personal pathways towards adulthood, particular attention is paid to Ulrich Beck's theory of individualization. Notwithstanding the fact that there has been some criticism of the individualization concept around, for instance, its overestimation of an individual's ability to reflexively construct their own identities (Furlong and Cartmel, 1997; Evans, 2007), this is still an influential theoretical contribution to the concept of structure-agency as it is examined here. The chapter then moves on to a brief description of the research context. The main focus of the findings section is the agency paradox, how this is manifested, and its implications for young people of both genders who grew up in farming families but who do not intend to become full-time farmers.

Background and Context

Irish family farming

Irish family farming continues to be grounded in cultural norms based on male succession and strong working relationships with the farm, with daughters and non-succeeding sons expected to support this strategy. Nevertheless, in recent decades the number of young men entering the Irish farming profession has dropped dramatically, with only 6.2% of all farm holders under the age of 35 years in 2010 (CSO, 2012). Traditional farming identities were based on the notion of strong male breadwinners, a gendered division of labour and clear delineations of the family's offspring's future pathways. Ní Laoire (2005) argues that although the broad contours of this model remain in place, the social and economic challenges faced by the farming community have led to significant changes. Ireland's economy has moved from a rural, agriculture base to one dependent on urban employment, while at the same time there has been a drop in farm incomes to below the national wage. For example, the average farm income in 2015 stood at €24,244 (Irish Farmers' Association, 2015) while average industrial wages stood at €43,101 (CSO, 2015). In addition, as in other patriarchal rural societies, daughters are encouraged if not compelled to build a life elsewhere and have little chance of taking over the mantle of the 'farmer' or being given an equitable share of the land. The continued focus on patrilineal, impartible inheritance is underpinned by the Irish legal position that farms can be passed on to one heir without needing to make provision for other offspring (Shortall, 2015). Within this context the choice of staying or migrating is often a source of tension for potential successors to the farm, who are caught between a desire to fulfil their family duty and retain their place within a familiar masculinity and the fact that this is a sector coming under increasing internal and external pressure (Ní Laoire, 2005).

From the early years of the 20th century the Irish farming community recognized that the strategy of encouraging differentiated educational paths could ensure not only the basic survival of their non-successor offspring but also their prosperity. The provision of educational opportunities arose from a sense of duty and pride that their children could do well in life, while also helping to reinforce the family's social status (Arensberg and Kimball, 2001). Hannan and Commins (1992) argued that the farming community has long regarded themselves as a separate class of high status. They have made use of educational facilities in a way that, for instance, the working class has not, to take advantage of new employment opportunities in an increasingly industrialized environment. By the 1960s the education farmers' daughters received, especially in more prosperous families, was equal to that of children in urban settings. This acknowledged their need to establish themselves away from the local community and was also a means of compensating them for their exclusion from future ownership of the farm (McNabb, 1964). This trend has continued to the present day, with above-average completion of secondary-level education for females from farm backgrounds (94.8% of farmers' daughters finish school compared with a national average of 89.7%). In addition, farmers' sons have also recorded significant increases in participation in recent decades (Byrne *et al.*, 2009). In the Irish farming family context education continues to be an important element in strategies for survival and acts as a means of aiding the progression of offspring who have not been identified as the 'farmer' (O'Hara, 1998). Thus, young people's pursuit of third-level education is often an important goal both for them and for their parents.

Farm youth in academic research

Farm youth who migrate are usually not included in research studies in the field of agricultural sociology or other related areas of study. Furthermore, most research that has been carried out around young people's engagement with the farming sphere tends to focus on their working relationships with it and the dynamics and impact of this (for example, Elder and Conger, 2000; Wallace *et al.*, 1994; Riley, 2009). Similarly, significant attention has been paid to the notion of succession and land ownership (see, for example, Silvasti, 2003; Bjørkhaug and Wiborg, 2010; Brandth and Overrein, 2013) with particular focus on gendered access to power (O'Hara, 1998; Kelly and Shortall, 2002; Price and Evans, 2006; Price, 2010). However, one piece of research relevant to this cohort, carried out by Shucksmith (2004), demonstrated that while individualized choices can be a feature of life in rural areas they are often predicated on access to family resources, employment and educational opportunities. These asymmetrical chances can impact greatly on the kind of transitions young people make and their relationship with where they grow up.

One of the most obvious gaps in the literature is the relationship the family as a whole has with succession and in particular those offspring whose transition to adulthood is predicated on involvement in further education. Even less attention is paid to female offspring and their position in the farming family vis-à-vis relationships with the homeplace. For instance, in one paper on the relationship between fathering practices, time spent on the farm by children and succession, the term 'children' or 'sons' is used throughout but almost no reference is made to daughters (Brandth and Overrein,

2013). Although this possibly reflects the aims of the study, it nevertheless implicitly collapses the idea of children and sons into one category and overlooks the need to explicitly include the female perspective. In some ways the focus on the (male) successor is understandable, since in being given the farm the individual takes on not just the ownership of the holding but also the values underpinning this status (Daugstad, 2010 cited in Bjørkhaug and Wiborg, 2010). As farm upbringings often stress the handing on of carefully gendered and prescribed roles, implied within this transfer is a passing on of the meanings and performances associated with these. Thus, the transmission of the meaning attached to the farm and the dynamics associated with this become a vital element of the succession process (Bjørkhaug and Wiborg, 2010). It is through this passing of values from one generation to the next that the family farming economy and culture can be secured. However, since non-successors play a crucial role in the family's inheritance strategy, if only through not disputing how the estate is to be divided, their 'place' in the strategy should be considered (Cassidy and McGrath, 2014).

The cultural norm of retaining the landholding in the family has long been seen as vitally important to the continuity of this culture and way of life (Silvasti, 2003; Brandth and Overrein, 2013). However, an in-depth Norwegian study argued that among the younger generations of farmers the urge for succession appears to be less important, with children's rights to make their own choices now viewed as more significant (Brandth and Overrein, 2013). This is partially premised on the idea that unlike their older counterparts, the younger generation considers lifestyle options and their own personal circumstances in their decision to take on the farm or not, and have the possibility of exerting more agency around this (Villa, 1999; Bjørkhaug and Wiborg, 2010). However, this notion of the apparently diluted sense of attachment to the farm might not be as straightforward as it first appears. There is often a kind of double-edged dynamic at work as although the need for rational choices about the future is emphasized, there is still an emotional subtext present. This is centred on the desire to maintain an intact, functioning farm into the next generation (Villa, 1999). Likewise, of three motives advanced in one Scandinavian study for young people agreeing to take on the farm, one was a yearning for the holding to be kept in the family (Andgard *et al.*, 2009 cited in Bjørkhaug and Wiborg, 2010). Grounding this is an attachment to the idea of retaining the farm in the family derived from a sense of obligation to previous generations and a wish to honour the commitment and sacrifices associated with this. Furthermore, the longer the family's history has been enmeshed with that of the farm, the more likely they are to feel a sense of duty to preserve it (Vedeld *et al.*, 2003 cited in Bjørkhaug and Wiborg, 2010). It has also been found that this kind of belonging and the powerful sentiments bound up in the farm prevent the sale of the property even where it appears to be unviable to keep it on (Flemsæter, 2009 cited in Bjørkhaug and Wiborg, 2010). Since the family identity is so closely entangled with the farm, the prospect of losing the land represents more than simply a denial of access to an economically important resource but also the means of expressing who they are and their affiliation to a broader and older ideal (Ní Laoire, 2005).

The notion of choice as a fundamental right of the individual is regarded as being of growing potency in the literature (Villa, 1999; Brandth and Overrein, 2013). This does not mean that the handing on of the farm will soon cease or that family farming will be obliterated from the social landscape, but rather that it will be modified and adapted to meet the changing needs of modern society (Silvasti, 2003). Rural research often focuses on a somewhat black-and-white combination of structure and agency but what this can overlook is how farm youth, for example in Ireland, continue to take account of traditional norms in relation to the farming world and yet adjust to modern social expectations and goals in their off-farm lives. What is also missing from the research focus is the role those who will not succeed to the farm have in the process of transferring land ownership from one generation to the next (Cassidy and McGrath, 2014) and in particular the paradoxical dynamics that underpin this combining of 'old' and 'new' values among this cohort.

Analytical Framework: The Contribution of Individualization Theory

Individualization theory is an important concept to incorporate into this work because of its focus on the level of input that young people have in the decisions they make about their path towards adulthood and the unique narrative shape of their life (Roberts, 2009). In part it attempts to bridge the gap between structure and agency and to highlight the interplay between them (Rudd, 1997). Within this much has been made of the supposed shift from traditional to late modern society with its attendant impact on how lives are shaped and created. Within traditional society, the individual actor was a relatively insignificant unit; decisions were made at a collective level and to wander from accepted norms handed down by community elders and leaders was considered foolish and deviant (Beck and Beck-Gernsheim, 2002). However, as historical patterns of behaviour and lifestyles began to be undermined by industry's voracious demands for a flexible, mobile workforce, new styles of living were needed. In order to survive in the late modern era, the individual actor needs to place themselves at the centre of their world and conduct their lives based on this premise. Beck and Beck-Gernsheim (2002) argue that ties to spatially based communities and stubborn adherence to a particular way of living create a danger of being dismissed as irrelevant or old-fashioned. This suggests that the duty to oneself and to carefully crafting one's own life, with society as an almost extraneous backdrop, has gradually taken precedence over operating as a minor figure in a larger framework within a wider social configuration. Underpinning the concept of individualization is the notion of the 'do-it-yourself (DIY) biography' whereby an individual constructs their own life based on a series of transient choices with often unknown consequences (Beck and Beck-Gernsheim, 2002). This is strongly linked to the fragmentation of social institutions based on the family, distinct social classes and so on (Beck and Beck-Gernsheim, 2009). It raises questions around whether or not the subjective expectations of actors who actively attempt to create their own identity and life path are compatible with the objective conditions they live in. Actors are ostensibly encouraged by society towards the belief that it is possible to be whoever they want to be, irrespective of their social background, their access to resources or their gender (Beck, 1992). However, in reality, individualized agency is not uniformly available to each actor. For example, there are structural influences such as rural–urban disparities of access to services and material wealth which potentially influence the resources available that govern the kind of choices open to an individual (Beck and Beck-Gernsheim, 2002).

The supposed shift towards a more individualized life path is underpinned by the idea that habitual elements of life are increasingly obscured by a haze of questioning and re-tailoring of the taken for granted into something new and unfamiliar. Despite the presence of structural barriers, as part of this, relationships with family and community are supposedly assimilated into the reflexive process so that they are no longer grounded in an automatic sense of obligation but rather a selective and less proximal set of ties. However, evidence from rural sociology indicates that, for instance, devices such as the differentiation of labour inputs enable the status of farmer to remain a male preserve (see, for example, Wallace *et al.*, 1994; Brandth, 1995; Shortall, 1999; Kelly and Shortall, 2002; Silvasti, 2003; Riley, 2009). This demonstrates that at least in the farming community structurally imposed boundaries can still influence the nature of actors' transitions to adulthood.

However, it should be noted that individualization does not suggest that social background is no longer of any import since families with appropriate types of capital continue to help establish their offspring on a life path. This support should be viewed as assisting them in making their own unique transition to adulthood (Roberts, 2009). Perhaps instead, as Shucksmith (2004) argues, it would be better to consider individualization as unevenly accessible and dependent on factors such as place, class and gender. This replaces the universality of Beck's conceptualization with a more nuanced typology, which can take into account varying individual circumstances.

Methodology

The empirical data informing this chapter is based on doctoral research on the relationship between Irish farm youth attending university and the farm or homeplace, as it is often referred to here, that they grew up on. The cohort consisted of 30 young people between 18 and 33 years old, none of whom intended to return to the farm in a full-time capacity. However, a small number of them are likely to be left the landholding, with two or three also contemplating returning to farming in a part-time capacity at some point in the future. The group consisted of 15 females and 15 males, and came from a wide variety of geographical regions and a mixture of farm types. All had self-identified in the recruitment process as having grown up on a family farm in Ireland. Due to concerns about the sensitive nature of the topic of landholding size in Irish family farming culture, the acreage of each farm was unknown. It should be noted that the findings contained here represent the views of most of the participants but the quotes are taken from a smaller number of participants who were very vocal about this issue during the data collection process.

Research Findings

Contradictory gendered norms of behaviour

Gendered agentic frameworks and how they are manifested emerged in the research as one of the most significant influences on the relationship young people have with the farm they grew up on and their subsequent links to it into adulthood. For this research cohort, distinct norms around gender-appropriate roles and behaviours continue to underpin their relationship with the homeplace. These have a profound influence on the position that both sons and daughters occupy within the on-farm gender paradigm by privileging male authority and succession rights and subordinating women into an auxiliary, 'inferior' role. On the other hand, in relation to off-farm positions the roles are often reversed, with greater support often shown for girls' engagement with education. In many cases, male participants were expected to balance an intense involvement with the working life of the farm with their pursuit of an education. By contrast girls were often allowed, if not encouraged, to detach from the farm as they got older and instead focus on their education. In this cohort, sons had more of a dual role to perform; of working on the farm while simultaneously studying for exams that would enable them to attend university. Girls were positioned as helpers, a role with fewer hours of commitment attached, thus enabling them to concentrate to a greater degree on their school work. This mirrors the contrasting relationships these young people are likely to have with the farm in the future, whereby sons are more likely than their sisters to pursue a professional career path while facing the possibility of returning to the farm. By contrast daughters play a supportive role in the succession process and concentrate almost exclusively on building a long-term future away from the farm.

Nevertheless, it should be noted that while cultural norms have a strong impact on attitudes towards male and female positions on and off the farm, they are open to interpretation by the individual and their family. This appears to depend on the context of their relationship with the homeplace and the farm's needs at a particular time. For instance, one woman mentioned that she had not been encouraged to work outside on the farm when she was younger because it was regarded as 'inappropriate' work for a girl. However, her older sister had laboured side by side with her brothers on the farm when she was the same age. The main difference between how gendered work relationships were constructed appears to have been that more labour was required when her sister was growing up because the farm was more heavily stocked and her two brothers were too young to take on all the work, a situation that had changed by the time the participant had grown up. This shift highlights how cultural norms can be opportunely embraced and appropriated to defend and imagine the development of a family member's relationship with the farm depending on the circumstances. However, regardless of where on the gendered spectrum they and their families sat, most participants were socialized into accepting a relatively fixed, differentiated concept of agency and positioning within their family and their movement into adulthood.

Clear and important distinctions existed within most of the cohort's families about expected behaviours and relationships with the farm. This was linked to gendered roles where boys were expected to go outside and 'work', meaning they intensively participated in the ongoing operation of the farm in their spare time. Girls did household chores and took on a 'helper' position on the farm; that is, were regarded as auxiliary sources of labour. Particular perceptions were created around this positioning so that girls who helped were assumed to be relatively removed from the day-to-day life of the farm in a way male 'workers' were not allowed to be. Some female participants presented this detaching as a choice they actively and strategically made, but most appeared to be marginalized by virtue of decisions made by their parents during their upbringing such as whether they were encouraged to drive machinery or not. Either way, even where they might have had a love for the land and an interest in farming, the idea of becoming a farmer never emerged as a viable option for their future transition to adulthood. By contrast in most families there was an expectation that boys had to take an active role on the farm. Even where they might be as uninterested as their female siblings, it was still presumed by their parents that they would be deeply involved. This was the case for James, who spoke of his sister being permitted to disengage from the farm as she got older and yet, while he hated most farm work, he had no choice but to continue with these tasks. Interestingly, this differentiation did not seem to be a source of resentment among participants, regardless of whether they liked working or not. This was Ben's attitude:

> She's [his sister] the youngest, she's a girl, she gets away with not doing a lot but who cares really? She's happy with it, we're happy with it yeah, it's ok we got a good thing going.

Paradoxical agency for women

Within the research there was clear evidence of the agency paradox in how a family's offspring were raised and the nature of the messages parents communicated to them about their futures, whatever the nature of their role on the farm. For daughters significant emphasis was placed on a feminist-imbued notion of independence and equality. Almost all of the female participants spoke of parents encouraging them towards achieving both financial and personal independence through the vehicle of education. In fact, for most, no alternative path had been countenanced or allowed for in their movement towards adulthood. This was emphasized through parental actions, such as facilitating and encouraging their daughters' engagement with school, and language, through frequent verbal reinforcement of the need to be autonomous and self-sufficient in later life. This was highlighted by one participant called Jane:

> My mam always harpered on about it, 'you need an education; get an education'. Go you know make something, be independent is my mam's motto you know, get yourself a good job that was always her thing. Mam'd say 'don't be like us, don't be dependent on kind of one source [of money] you know'.

> She'd always be saying, 'go and work and like to have money'.

Another participant, Shauna, touched on this as well: 'My father has always said to me, "have your own money, never be dependent on any man for money".'

This is in stark and paradoxical contrast to the kind of role and positioning they were framed within in their on-farm life, where they were rarely encouraged towards independence through, for example, being taught core farming skills such as machinery work. This was further reinforced through their positioning as non-successors with a temporary status on the holding. Unlike sons, as is discussed later, the concept of choice was not mentioned by parents in relation to their daughters – it seemed to be the case that this was a taken-for-granted underpinning feature of their lives and, therefore, did not need to be overtly emphasized.

Throughout the interviews, female participants reiterated and supported the norm of male succession. This acceptance by most female respondents of the 'natural' order of a son or sons inheriting their family's landholding appears to be grounded in a psychological disconnect between their upbringing on the farm and the possibility of a long-term, proprietorial relationship with the land. Even in circumstances where there appeared to be little chance of a financial settlement in lieu of being left a portion of the farm, female participants referenced, without any tone of disquiet, the prospect of their brothers assuming ownership of it. One participant called Bridget referred more than once to her parents' material poverty, so it is doubtful whether they would be able to gift her a substantial sum of money in their will. Despite her fragile financial position at the time the research was carried out, this did not lead her to conclude that she should be given some of the farm. The casual gendered presumptions around male succession to the farm are highlighted in Shauna's anecdote about one branch of her extended family. In that family, although the son has built a life quite far away from the homeplace, he will be given the land rather than his sister who is settled nearby to the farm:

> The girl Sabina is living [near home] and we all know that Alex [the son] is going to get the land cos he's the boy. Alex was the farmer, but Sabina never stood foot on the farm so that's why – and cos he's a boy I just presume Alex will get the farm and Sabina will get money. Now I don't know, sure I don't know how she feels about that, but I don't think she will want it anyway.

Even where it is not possible to adhere to this cultural norm, for example in families with no sons, female participants still struggled with their positioning in the succession process. Aisling, who has no brothers, spoke with a degree of wistfulness about this subject; if she had a brother he would have to deal with the tricky dilemma of the farm's future and not her. She claimed this is not because she views farming as unsuitable for women but rather that it is not 'natural' for them to have to take responsibility for this issue. Yet at the same time Aisling was an ambitious budding young professional who was building a career for herself in quite a demanding field. Here are her views on the subject:

> I suppose I wish I had a brother that was interested in farming or just a brother. I know that's very maybe sexist or something but do you know like any of my friends that are from a farm they have brothers who deal with this kind of thing.

Female attitudes to the paradox of their positioning

None of the female participants appeared to identify any conflict between, on the one hand, believing that they did not have a legitimate right to take over the farm, as underpinned by their attendant inferiority in the farming hierarchy, and on the other hand, being pushed and encouraged to see themselves as equal to anyone in the life they create off the farm. They seemed to be able to move into and out of this contradictory position depending on the contextual circumstances and the subject matter at hand. For instance, Aisling could be both a confident young professional with ambitious plans for the future and a doubting potential heir to the farm lamenting the absence of a brother who would assume responsibility for the farm's future. The lack of tension or even awareness about the paradoxical roles they had been socialized into on and off the farm was further apparent in Shauna's interview. In one part of her interview she made the following comment:

> My uncle . . . he'd be slagging [mocking], 'oh we need somebody to help us draw [bring in] the bales. Shauna come out here, and we'll show you how to drive a tractor', and daddy'd be kind of sniggering and laughing under his breath . . . I remember saying it to him, 'sure here I'll drive a tractor, can't be that much to it'. [Her father replied] 'You will not – you'll not drive a tractor – you're not getting up on that tractor, no way!'

From this quote her limited position in the working life of the farm and the strictly bounded expectations her father had of her role on it, are clearly demonstrated. He showed obvious reluctance to teach her essential skills or to even countenance expanding her involvement in the operation of the landholding. Nevertheless, she later discussed how encouraging her father was of her, being given a car as a present, which it itself is a practical and symbolic demonstration of his support for her independence:

> My father said, 'she's going to college, I want her in a good car', so he was all for me driving and now he loves me having a car and he thinks it's great that I've this car.

Despite her father's strongly differentiated attitude, she did not seem to recognize the contradiction between his wish for her to be independent off the homeplace and the simultaneous limitations he placed on her active involvement on the farm and the consequences this would have for her future relationship with it.

The paradox for male participants

In addition to being embedded into farming norms with its emphasis on working on the farm in a kind of apprentice role with an eye to possible future ownership, most male participants' movement into adulthood was underpinned by the concept of choice. This is a key feature of the individualization discourse within which young people are perceived to be the agents of their own destiny. Like their female counterparts, their parents encouraged them to use education as a means of building their life off the farm. However, where independence strongly featured in women's socialization and choice was not mentioned, it was the opposite for male participants. By contrast to the emphasis placed on the independence education could give to daughters, for sons it was predicated on its capacity to provide them with a greater range of choices in the future. For instance, Myles remarked that his parents always said: 'Get a good education and you can do what you want basically.' Another participant, George, also referred to his father's attitude about this:

> He liked the fact that I was going to college and getting some piece of paper, some bit of a degree so I could have a job outside of [the farm] if I was going to do part-time farming.

This duality ignores a paradoxical conflict between their socialization into prevailing farming norms with the responsibilized boundaries this imposes around potential future ownership of the farm and simultaneously being pushed to take advantage of the freedoms and opportunities attending university can bring. Furthermore, in some situations even where another successor had been identified in the family, male participants who had no wish to become farmers positioned themselves as back-up heirs. Others hinted that it was their parents who had the power to decide whether they could pull away from the label of heir. Both of these patterns denote a lack of choice about their future pathways. In addition, male participants spoke of a desire to travel and develop careers, and yet were actively tied into particular work and ownership roles on the farm which could, one day, contradict and temper these aspirations through requiring them to take ownership of the farm.

Attitudes to the paradox among male participants

Like their female counterparts, men were able to switch between these two elements of their identity as they discussed their life on and off the farm without appearing to notice any discrepancies. Neither did they seem to recognize the paradoxical agentic position they occupied in relation to their on- and off-farm lives. This was demonstrated in Donal's interview where he spoke about being asked at a very young age whether he wanted to become a farmer:

> I think I was maybe five or six, I can remember being asked [by his parents] did I want to go farming and I was like no and it was like 'grand that's fine so'.

However, later in the interview he referred to the fact that if during his transition to adulthood his parents had actually demanded he take over the farm, he would have felt compelled to do so:

> I think so [he would take the farm] if they were really pushing it cos then you kind of know that they really want it and that's just kind of letting them down.

It can be seen from this how fragile the concept of choice was in his life – he was brought up on a strong foundation of choice around his education and yet, ultimately, if his parents had wanted him to return to the farm it would have been very difficult for him to refuse to do so. While in Donal's case the farm is likely to be sold despite his deep regret, for many their sense of obligation arising from this paradoxical imagining of agency prevents the farm being sold. The dilemma around this can lead to an internal struggle for participants. One participant called Andrew tried to detach himself from this debate by claiming that he would sell any land he received. Yet this is not as straightforward as he initially tried to present it in the interview, as can be seen in this quote:

> It's always there, it's in the back of your head, the bit of pride over it [the land] even though probably 70% of you just wants to walk away from it but that one 30% is just hanging right in there, going you can't be doing this now, look around you, they're [his family] so proud of what was there before them and I suppose that's why it wouldn't be so easy just to kick it away, there's, there's so much history there, there's so much pride there.

For some male participants the duty to the farm is a frightening position to be put in, in how it can manipulate and shape their futures, but at the same time there was a certain kind of in-built resignation around their potential role as successors or back-up heirs. Seamus was encouraged to go to university by his parents and wanted to build a life far removed from farming, which he detests. However, he does this in the shadow of his father's expectations that he will return to take over the farm when he is too old to manage it. This is further complicated by his disabled brother's yearning to be a farmer, something which he cannot hope to become without Seamus' help. While theoretically Seamus could refuse to comply, were he to do so there would be little solace to be gained from this, for it would mean turning his back on his duty to his family. These were some of his thoughts on the subject:

> It's not what I want . . . but [my brother] wants to do it, so he needs help. It's just that I find that I'm going to be on this farm whether I like it or not, I'm going to have to be involved in farming. I won't be able to escape it.

In some ways, it seems as if it does not matter whether he takes over the farm or not – he will still be trapped by its clay and ditches and by an everlasting sense of owing something to his stricken, but eager brother. Even if he does not take over the farm one wonders would it be a pyrrhic victory because he would still carry his background with him, including the nagging guilt at having gone against his duty. By contrast his sisters have been able to make their transition to adulthood free from the worry of having to return to protect familial ownership of the farm. Participants like Seamus must grapple with the fact that in the immediate future they are free to experiment and be spatially mobile and yet, ultimately, they must return to the farm. This could also have serious consequences for their ability to build personally meaningful careers in geographic regions with few employment and social opportunities.

Discussion: Implications of the Agency Paradox

Transitions to adulthood in the farming community are shaped by exclusionary and inclusionary mechanisms embedded in upbringings, such as working relationships with the farm, which in turn are shaped in large part by these paradoxical positions young people are framed within. Over time these gendered frameworks bring differentiated obligations and quasi-moral responsibilities, which some are eager to grasp and others anxious to avoid, but yet all are shaped by them in some way. These paradoxical positions whereby sons and daughters are socialized into contrasting statuses in relation to their on- and off-farm lives increase the chances of the farm being maintained within the family. At the same time, they help to maximize young people's opportunities for success in the broader societal environment through encouraging their pursuit of education. The relative incompatibility of the individualized life path (see, for example, Beck, 1994; Beck and Beck-Gernsheim, 2002) and the norms still prevailing in the farming community, where family needs still exert a strong emotional pull, can cause conflict for its young people in how they make their transition to adulthood. On one side, there is mobility, urban living and DIY biographies. On the other side, in relation to the farming paradigm there is an emphasis on tradition, rootedness in the homeplace, identities defined in local and temporal terms, and specific cultural norms. Arising from this is the sense that, while farm youth have options to select from in the transition to adulthood they make, the choices open to them began to be shaped many years previously. It is these often jarring and contradictory constructs and dynamics that young people must reconcile as they move towards adulthood.

Despite the continuity of key social norms, the Irish farming community has reacted to changing socio-economic circumstances through, for instance, integrating engagement with third-level education into its survival strategies. This is also set against a backdrop of changing societal trends based on individualized life paths relatively free from the influence of their parental or familial background. Yet regardless of this supposed normative shift, young people who took part in this research continued to view themselves as part of the collective, as demonstrated for instance by their support for their family's succession plan, which prompts a puzzling question of how and why this can be the case. The answer to this appears to be founded on their upbringing within two paradoxical and contradictory gendered frameworks, with one preparing them for life within the family in the context of the farm and one for life outside this. These frameworks do not appear on the surface to be mutually compatible since the gendered requirements and expectations they are grounded in are quite different. Farmers' daughters are not encouraged to see themselves as equal to their brothers on the farm in terms of their roles and ownership rights and yet, off it, are generally urged to pursue professional careers and be as independent as possible. Similarly, sons are pushed towards education and mobility but are often perceived as potential returnees to the farm and, through this, can remain tightly moored to the homeplace. Findings in this research show that in spite of their contradictory nature, the boundaries of the farming paradigm are usually contextually adhered to even after young people move away to university. While male actors build their life away from the farm they do so in the knowledge that this might be curtailed by a duty to return to the farm. Likewise, female actors confidently pursue academic qualifications with the intention of developing vibrant careers but, nevertheless, continue to support succession strategies that are weighted against them.

While much of the literature around succession focuses only on the heir to the farm (for example, Bjørkhaug and Wiborg,

2010; Brandth and Overrein, 2013), this migrated group plays a crucial role in familial succession strategies. The paradoxical gendered engagements they are embedded within are a highly significant feature of family farming culture in Ireland and are vital for the continuity of viable landholdings into the next generation. Through combining paradigms that on the one hand reinforce the idea of the farm being handed on to one successor and on the other the utilization of further education as a vehicle for progression in wider society, families can maximize their chances of holding on to the land while ensuring that their children are successful, functioning members of Irish society. As regards individualization theory, this work shows that while on the surface young people from farming backgrounds appear to have personal capacity to create their very own DIY biography (Beck, 1992), in fact, this is grounded in deeply influential socialization patterns and norms that create boundaries and shape available choices within which personal options are selected. This would fit within Shucksmith's (2004) notion of asymmetrical options being available to young people from rural backgrounds depending on, for instance, the forms of capital they have access to.

Conclusion

Contrary to proponents of individualized DIY biography creation such as Beck, young people are embedded and influenced by their own social circumstances, family relations, etc., as well as the wider societal milieu that expects personal choices and decisions to be the dominant catalyst for transitions to adulthood. This chapter has focused on how migrated Irish farm youth deal with the agentic paradox that these competing tensions create. While not claiming to be a definitive picture of the relationship between migrated farm youth, the homeplace they grew up on and the agentic gendered paradox entwined within these dimensions, this chapter has explored how these young people can manoeuvre within and between the differing normative contexts of the farm and wider society. It demonstrates that Irish farm youth who migrate operate within two gendered sets of normative paradigms that allow them space to make a successful transition to adulthood while increasing the chances of the farm being retained within the family into the next generation. Without wanting to appear overly deterministic, the pathways farm youth follow, which might seem to be based on individual choice, are often choreographed from childhood through decisions made at an early age around the intensity of their involvement in farming and their engagement with education. This paradigmatic duality helps to increase the possibility that the farm is passed on to one successor who takes it on either in a full- or part-time capacity with the support of their siblings. The inculcation of a paradoxical relationship with the concept of agency among a family's offspring is crucial not only to the development of a successful farm succession strategy, but also to the creation of pathways towards adulthood that allow the pursuit of professional careers. As such, the paradoxical relationship with agency should be recognized as a key element in the fight for Irish family farming culture's survival into the future.

References

Arensberg, C. and Kimball, S. (2001) *Family and Community in Ireland*, 3rd edn. CLASP Press, Ennis, Republic of Ireland.

Beck, U. (1992) *Risk Society: Towards a New Modernity*. SAGE Publications Ltd, London.

Beck, U. (1994) The reinvention of politics: towards a theory of reflexive modernization. In: Beck, U., Giddens, A. and Lash, S. (eds) *Reflexive Modernization: Politics, Tradition and Aesthetics in the Modern Social Order*. Polity Press, Cambridge,UK, pp. 1–52.

Beck, U. and Beck-Gernsheim, E. (2002) *Individualization: Institutionalized Individualism and Its Social and Political Consequences.* SAGE Publications, London and Thousand Oaks, California.

Beck, U. and Beck-Gernsheim, E. (2009) Losing the traditional: individualisation and 'precarious freedoms'. In: Elliott, A. and Du Gay, P. (eds) *Identity in Question.* SAGE Publications Ltd, London, pp. 13–36.

Bjørkhaug, H. and Wiborg, A. (2010) Challenges for Succession in Family Farming: Perspectives and Research Questions. Available at: http://www.nordlandsforskning.no/getfile.php/132783/Dokumenter/Arbeidsnotater/2010/Notat_1007_2010.pdf (accessed 14 February 2017).

Brandth, B. (1995) Rural masculinity in transition: gender images in tractor advertisements. *Journal of Rural Studies* 11(2), 123–133. Available at: http://dx.doi.org/10.1016/0743-0167(95)00007-A (accessed 6 March 2017).

Brandth, B. and Overrein, G. (2013) Resourcing children in a changing rural context: fathering and farm succession in two generations of farmers. *Sociologia Ruralis* 53(1), 95–111. Available at: http://dx.doi.org/10.1111/soru.12003 (accessed 6 March 2017).

Byrne, D., McCoy, S. and Watson, D. (2009) *School Leavers' Survey Report 2007.* ESRI and Department of Education and Science, Dublin.

Cassidy, A. and McGrath B. (2014) The relationship between 'non-successor' farm offspring and the continuity of the Irish family farm. *Sociologia Ruralis* 54(4), 399–416. Available at: http://dx.doi.org/10.1111/soru.12054 (accessed 6 March 2017).

CSO (2012) Agricultural Labour Input 2010. Available at: http://www.cso.ie/en/media/csoie/releasespublications/documents/agriculture/2010/aglabinput_2010.pdf (accessed 10 July 2015).

CSO (2015) Earnings and Labour Cost Quarterly. Available at: http://www.cso.ie/en/releasesandpublications/er/elcq/earningsandlabourcostsq32015finalq42015preliminaryestimates/ (accessed 11 February 2016).

Elder, G. and Conger, R. (2000) *Children of the Land: Adversity and Success in Rural America.* University of Chicago Press, Chicago, Illinois and London.

Evans, K. (2007) Concepts of bounded agency in education, work, and the personal lives of young adults. *International Journal of Psychology* 42(2), 85–93. Available at: http://dx.doi.org/10.1080/00207590600991237 (accessed 6 March 2017).

Furlong, A. and Cartmel, F. (1997) *Young People and Social Change: Individualization and Risk in Late Modernity.* Open University Press, Buckingham, UK and Philadelphia, Pennsylvania.

Hannan, D. and Commins, P. (1992) The significance of small-scale landholders in Ireland's socio-economic transformation. *Proceedings of the British Academy* 79, 79–104.

Irish Farmers' Association (2015) Farm Income Review 2015. Available at: http://www.ifa.ie/wp-content/uploads/2016/02/Farm-income-review-2015-Final.pdf (accessed 15 February 2016).

Kelly, R. and Shortall, S. (2002) 'Farmers' wives': women who are off-farm breadwinners and the implications for on-farm gender relations. *Journal of Sociology* 38(4), 327–343. Available at: http://dx.doi.org/10.1177/144078302128756714 (accessed 6 March 2017).

McNabb, P. (1964) Social structure. In: Newman, J. (ed.) *The Limerick Rural Survey: 1958–1964.* Muintir na Tíre Rural Publications, Tipperary, Republic of Ireland, pp. 193–247.

Mulhall, L. and Bogue, P. (2013) National Rural Network Report on 'Women in Agriculture'. Available at: http://www.nrn.ie/wp-content/uploads/2013/12/Final-Case-Study-Women-in-Agriculture1.pdf (accessed 24 October 2014).

Ní Laoire, C. (2005) 'You're not a man at all!': masculinity, responsibility, and staying on the land in contemporary Ireland. *Irish Journal of Sociology* 14(2), 94–114.

O'Hara, P. (1998) *Partners in Production? Women, Farm and Family in Ireland.* Berghahn Books, New York and Oxford, UK.

Price, L. (2010) 'Doing it with men': including men in feminist research practice in order to reveal patriarchy in UK Family Farming. *Gender, Place & Culture* 17(1), 1–22. Available at: http://dx.doi.org/10.1080/09663690903522438 (accessed 6 March 2017).

Price, L. and Evans, N. (2006) From 'as good as gold' to 'gold diggers': farming women and the survival of British family farming. *Sociologia Ruralis* 46(4), 280–298. Available at: http://dx.doi.org/10.1111/j.1467-9523.2006.00418.x (accessed 6 March 2017).

Riley, M. (2009) 'The next link in the chain': children, agri-cultural practices and the family farm. *Children's Geographies* 7(3), 245–260. Available at: http://dx.doi.org/10.1080/14733280903024407 (accessed 6 March 2017).

Roberts, K. (2009) *Youth in Transition: Eastern Europe and the West.* Palgrave Macmillan, Basingstoke, UK.

Rudd, P. (1997) From socialisation to postmodernity: a review of theoretical perspectives on the school-to-work transition. *Journal of Education and Work* 10(3), 257–279. Available at: http://dx.doi.org/10.1080/1363908970100304 (accessed 6 March 2017).

Savills (2011) Irish Farmland Market. Available at: http://www.savills.ie/documents/farmland-update-lr.pdf (accessed 16 June 2014).

Shortall, S. (1999) *Women and Farming: Property and Power.* Macmillan Press, Basingstoke, UK.

Shortall, S. (2015) Gender mainstreaming and the Common Agricultural Policy. *Gender, Place & Culture* 22(5), 717–730. Available at: http://dx.doi.org/10.1080/0966369X.2014.939147 (accessed 6 March 2017).

Shucksmith, M. (2004) Young people and social exclusion in rural areas. *Sociologia Ruralis* 44(1), 43–59. Available at: http://dx.doi.org/10.1111/j.1467-9523.2004.00261.x (accessed 6 March 2017).

Silvasti, T. (2003) Bending borders of gendered labour division on farms: the case of Finland. *Sociologia Ruralis* 43(2), 154–166. Available at: http://dx.doi.org/10.1111/1467-9523.00236 (accessed 6 March 2017).

Villa, M. (1999) Born to be farmers? Changing expectations in Norwegian farmers' life courses. *Sociologia Ruralis* 39(3), 328–342. Available at: http://dx.doi.org/10.1111/1467-9523.00111 (accessed 6 March 2017).

Wallace, C., Dunkerley, D., Cheal, B. and Warren, M. (1994) Young people and the division of labour in farming families. *Sociological Review* 42(3), 501–530. Available at: http://dx.doi.org/10.1111/j.1467-954X.1994.tb00099.x (accessed 6 March 2017).

Winkler, C. (2010) Feminist sociological theory. In: Crothers, C. (ed.) *Encyclopedia of Life Support Systems Historical Developments and Theoretical Approaches in Sociology.* Available at: http://www.eolss.net/Sample-Chapters/C04/E6-99A-21.pdf (accessed 14 July 2016).

12 Rurality and Gender Identity

S. Shortall*
Newcastle University, Newcastle upon Tyne, UK

Introduction

When Bettina Bock wrote the introduction to the identity section of *Rural Gender Relations* in 2006, she noted that it was in the second half of the 1990s that researchers in Western countries really began to look at the construction of gender identities in the rural setting. This is an area that has developed significantly since then. It concentrated initially on the identity of farm women and men. Studies then broadened in scope to consider how the rural might intersect with other factors and shape gender. She also noted that the focus on identity was a predominantly Western one and it was of less interest in the developing world. Ten years on, this remains the case. She commented that with modernization and globalization the economic position and social status of traditional rural professions weaken. More and more farmers, fishers and foresters have difficulties remaining as the primary breadwinner. Marshall (2001), Ni Laoire (2001), Bennet (2004) and Shortall (2006) demonstrate how losing this basic pillar of masculine identity affects men's well-being negatively and also, indirectly, affects the well-being of women. Masculinities and the changing identities of men on farms and other primary industries have been a major area of research over the past 10 years and it is a question considered by three of the chapters (Chapters 15–17) in this section of the current volume.

A brief overview of gender and identity is offered here. The interested reader should refer to a much more complete overview provided in the recently published *Routledge International Handbook of Rural Studies*, Section VIII – 'Gender in rural society' (Shortall, 2016a; Shucksmith and Brown, 2016). All of the contributions in Section VIII of that publication deal with gender and identity (Brandth and Haugen, 2016; Charatsari and Černič Istenič, 2016; Little, 2016; McVay, 2016; Thuesen, 2016) and Shortall's (2016b) contribution specifically deals with the question of gender and identity formation in rural studies.

Gender Identities in Agriculture and Work On and Off the Farm

Initially, research was concerned with understanding how the occupation of farming interacted with the farm family. It was not overtly stated that it was concerned with gender identities and how they were formed, performed and recreated, but it is easy to extract when rereading the classics. In the main, this research was trying to make visible the world of women within the family farm (Gasson, 1980; Bartlett, 1983;

* Corresponding author e-mail: sally.shortall@newcastle.ac.uk

Sachs, 1983; Whatmore, 1991; Shortall, 1992, 1999; Alston, 1995, 1998; O'Hara, 1998; Overbeek *et al.*, 1998; Brandth, 2002; Pini, 2002; Silvasti, 2003; Černič Istenič, 2006). Research considered how men's identity as farmers is tied to their land ownership. Their role identity as farmer comes from owning the means of production. They occupy the occupational position of 'farmer' and they are seen to do the productive agricultural work. On the one hand, gendered identities around farming and farm work have remained stubbornly in place. On the other hand, the increased on-farm and off-farm employment of farm women has meant a considerable change in their work status and identities. Interestingly, both Haugen and Brandth (Chapter 14, this volume) and Riley and Sangster (Chapter 15, this volume) underline just how stubborn gendered identities on farms remain.

Berit Brandth was a pioneer of rural research looking at the construction of masculinities in agriculture. Brandth (1995) analysed the use of imagery in tractor advertisements. She demonstrated how men's mastery of big and heavy machinery serves to underline and symbolize a hegemonic masculinity that is defined by physical strength, control and domination. Women's absence in tractor advertisements supports the symbolic intimacy between men, machinery and management. At the same time it negates women's relation with machinery and management, and marginalizes women's presence in the farm. Such definitions of masculinity and femininity reconfirm the unequal division of labour and power between farming men and women. Similar results were found in studies analysing the construction of agricultural gender identities in the USA, Australia, New Zealand, France and the UK (Liepins, 2000; Morris and Evans, 2001; Saugeres, 2002; Finney *et al.*, 2004). Research on other traditionally male occupations, such as forestry (Brandth and Haugen, 1998, 2000; Reed, 2002), fishing (Marshall, 2001; Grzetic, 2004) and the military (Woodward, 1998), finds that gender identities are constructed in a similar way. Their occupational identities are defined as masculine, based in hard manual labour, physical strength and the capability and courage to endure the dirty, heavy and dangerous work that goes along with surviving and eventually taming and civilizing rural 'wilderness' and nature. The negative impacts that can ensue when this identity is threatened have been a significant feature of recent work on gender and identity in rural studies (Alston and Kent, 2008; Alston, 2012; Brandth and Haugen, 2005, 2010, 2011; Bryant and Pini, 2011; Shortall, 2014).

Women are increasingly working off the farm and in farm diversification activities (see the previous section in this volume on agriculture). With this increased economic independence, scholars expected new identities to emerge in a way that has not happened (Brandth, 2002). As Riley and Sangster (Chapter 15, this volume) discuss, women are often the primary breadwinner, or at least contribute significantly to the survival of the farm (Moss *et al.*, 2000; Kelly and Shortall, 2002; Shortall, 2014). Kelly and Shortall (2002) argue that farm women's increased resources from off-farm work do not contribute to significant renegotiation of domestic responsibilities and gender role expectations. They say that much of the literature on income and gender identities presumes people behave as maximizing individuals. Farm households, however, require analysis at the level of the household to explore what off-farm employment by women means for gender role expectations and the division of labour within the farm family. The farm household behaves as a collective and tries to ensure the well-being of family members by verifying key identities (Wheelock and Oughton, 1996; Bock, 2004). Shortall (2014) has argued that despite their elevated economic status as breadwinner, women on farms continue to do gender identities such that they reinforce men's work identity as a farmer, as the decision maker, and in this way reinforce his masculinity. Again, these processes are clear in Chapters 14 and 15 in this section.

Space, Roles and Rural Gender Identity

Space is gendered. This of course has a cultural and temporal context; some cultures have more gendered space than others and, in general, the rigidities of gendered space change over time. Matysiak (Chapter 17, this volume) and Ni Dhonaill (Chapter 16, this volume) consider gendered space and how it is linked to gendered identities. Like McVay (Chapter 13, this volume), they also look at how the roles men and women assume and take on are shaped by, and further shape, gender identity. Recent research has shown that rural space continues to be used to signify and maintain distinctive gender identities; the men in Riley and Sangster (Chapter 15, this volume) carefully construct their masculine activity in relation to outdoor masculine work. The outdoors is coded as masculine, while indoor activities are coded as feminine (Campbell and Bell, 2000; Pini, 2004; Little and Panelli, 2003; Campbell *et al.*, 2006; Brandth and Haugen, 2010, 2014; Little, 2014). In the outdoors, men undertake hard, physical and sometimes dangerous work such as handling heavy machinery, being foresters and dealing with chemicals. Men have appropriated agricultural technology to underline their identity as farmers (Brandth and Haugen, 1998; Saugeres, 2002; Pini, 2005). Specific activities such as hunting and mining are seen as outdoor male activities (Campbell and Bell, 2000). Riley and Sangster (Chapter 15, this volume) show how men reconstruct their gender activities to maintain their masculine identity. Women's indoor work is predominantly seen as domestic or as socially reproductive work. It is seen as sustaining the household (Whatmore, 1991; Trauger *et al.*, 2010).

Shortall (2016b) notes that while much of the literature acknowledges and discusses the iterative and interactive nature of identity formation, empirical research still has a tendency to focus on the identity formation processes for one group rather than the interactions between different identity groups and how the identities of one influence the identities of the others. It is really interesting to see Haugen and Brandth (Chapter 14, this volume) discuss divorce with men and women and analyse the question in this way, albeit the men and women interviewed were not divorced from each other. We now turn to the chapters in this section.

The Chapters

In Chapter 13, Lori McVay examines the identity of women leaders in Northern Ireland. She focuses on their identity as 'rural', as 'women' and as 'leader', and how they navigate the intersection of these identities. While research on women leaders in the main tends to focus on the barriers women encounter, McVay instead looks at the enablers, and considers the choices and life events that shaped these women's identities. She interviewed leaders in non-government organizations, business, the church and agricultural organizations. Like Barbara Pini, she argues that we need to avoid flattening women's experience and creating a female norm in terms of leadership. It is for this reason she is interested in the intersection of rural, woman and leader. She considers how 'leader' may conflict with the identity as 'woman' and how women, in situations where masculine behaviour is expected, may perceive themselves as inauthentic. She found that rural was a very strong component of identity, and gender was also very strong. But women were less comfortable with the leader identity. McVay's research found that starting in women-only organizations was identified as very important by the women interviewed, who said that it was in these environments they learned the confidence to take other leadership roles. Women rejected a feminist identity, which is particularly interesting given that McVay did not directly ask about feminism. The women saw it as an identity that is militant and promoting unacceptable behaviour. Women tended not to identify as leaders even though others identified them as such. They used words like 'cajole', 'encourager', 'getting on with it' to describe their leadership roles. Women also emphasized their relationships with their 'followers' in organizations and politics, reinforcing her conclusion that the women she interviewed integrate their rural and gender identities with the leader identity in a way that they see as nurturing.

Chapter 14 by Marit Haugen and Berit Brandth explores the construction of gendered farming identities by considering divorce. Since people construct their identity in relation to marriage, divorce requires a new sense of identity. Farm divorce is particularly problematic because it is deeply rooted in a marital economy and historical patterns of inheritance and culture, which tend to privilege men. If assets are divided following a divorce, it has the potential to break up the farm unit and threaten its viability, and they argue this can lead to moral and material pressures on women. They are interested in how gender as an identity can be disrupted, constituted and reconstituted in the context of divorce. They present some of the frames of identity that the literature has identified as important to men and women on farms, such as the 'good farmer', hard-working and successful, and detail how this can have a negative impact when threatened. They also discuss how women's gender equality frame has changed in line with wider societal equality for women. They present four in-depth case studies that were part of a larger study of relational arrangements in Norwegian farming. Two case studies are men, two are women. In each case the women instigated the break-up. The women opposed the farm wife identity within which they were framed and which limited their economic independence and freedom to make individual decisions. They did not claim the assets they were entitled to following divorce. The men related to the good farmer identity and were quite gender equal in their views. Both experienced mental health problems following the break-up. Haugen and Brandth note that despite the gender equality frame, there is a persistence of the patriarchal frame of identity: male privilege and female subordination. Considering identity formation through studying moments when they are shattered is a really interesting approach to researching identity.

In Chapter 15, Mark Riley and Heather Sangster are interested in whether dominant hegemonic masculinities are changing for farm men. How do they negotiate not being the breadwinner, or physical manual work being replaced by machinery? They demonstrate the temporal nature of farm masculinities and they explore the contextual, fluid and interpersonal ways masculinities are performed and identities formed. They contend that there are multiple masculinities. Farming men are often positioned as the breadwinner, with heterosexuality presumed as the norm given the emphasis on producing a son and heir to the farm. Riley and Sangster are particularly interested in the intergenerational transfer and reconfiguration of male farming identities. They review literature that shows the different scales at which masculinities may be defined and reworked: local, regional and global. The importance of space and place in the development of masculine identities is underlined. They also argue that masculinities can be reconfigured over time. Their sample for the chapter is drawn from a larger study in the Peak District of England between 2001–2003 and 2014–2015. Here they focus on eighteen farms where three generations are farming concurrently. They found a number of components of the masculine identity to be particularly important. Patrilineal inheritance and 'keeping the name' on the land remain strong. Alongside this, changes have taken place with spouses working off the farm and the younger men doing more housework and childcare, but they provide a fascinating analysis of how this can be presented in a way that does not threaten the masculine identity on the farm. They also provide an interesting analysis of how masculine identities are protected intergenerationally. They conclude that being the provider and breadwinner are key masculine identities across generations, as well as a continued commitment to intergenerational land transfer. It is clear too that identities are temporal, redefined and reconfigured over time.

In Chapter 16, Caoimhe Ni Dhonaill looks at narratives of masculinity in groups trying to address social exclusion. She compares two different initiatives to tackle social exclusion: one focused on older people in rural areas, men and women; and one specifically aimed at men in urban areas. Previous research shows that men are more at risk of experiencing loneliness and isolation than women. The chapter assumes that social connections are beneficial for people. She very usefully distinguishes between social isolation and loneliness; social isolation refers to objectively measured low levels of social contacts, while loneliness refers to a subjective emotional feeling of having few social contacts. The two are not synonymous. Her research found that it is not the prevalence of contact but the quality of contact that is most import. Given that tackling loneliness in older people is a priority for many organizations, Ni Dhonaill advocates close attention to the quality of relationships fostered in these groups. Men are more vulnerable to loneliness because of masculinity narratives that suggest men should be strong and avoid discussing emotions. The masculine emphasis on the work role means that after retirement, men have lost colleagues and interaction as well as their breadwinning identity. She discusses how services and lower population density may mean older people are more likely to experience social isolation in rural areas. She compares two case studies. One is in a rural area and targeted men and women through a luncheon club and arts and crafts activities. The other is in an urban area and is a Men's Shed, which is part of the global movement to provide a space for men to come together, usually after losing a job or retirement, to tackle social isolation. In Men's Sheds, men do typically masculine activities such as do-it-yourself, wood carving, metal work and use heavy tools. Ni Dhonaill presents a fascinating analysis of the processes of interaction between men and women that reinforce masculine stereotypes and impede men's participation in the mixed group. While the Men's Sheds also serve to reinforce masculine stereotypes, they do so in a positive way. They provide a post-work masculine space that allows men to reconfirm their masculine identity and from the comfort of this safe space they are able to discuss emotional issues. She concludes that men-only groups are a positive initiative and may be a more successful means of addressing social inclusion than gender-mixed groups.

Ilona Matysiak examines the role of village representatives in Poland in Chapter 17. People in this role traditionally liaised between residents and higher levels of authority. She is interested in village representatives because traditionally this was a male domain and it is becoming an increasingly female role, currently at 36% women. Her key question is whether this represents a move away from male dominance in politics or whether it actually represents the opposite. She undertook qualitative research in ten municipalities and supplemented this with secondary data. Matysiak notes that the literature underlines the general under-representation of women in Western Europe and this is also true for rural women. Research has shown that the recent rural structures that developed to implement the EU Leader programme, the Local Action Groups, also have an under-representation of women. Women often participate in local activities, but this is seen as altruistic and does not translate into positions of power. She charts the socio-political history of Poland and outlines the key issues for rural areas. Following the collapse of communism and entry into the EU, she argues that rural areas are spaces that embody long-standing legacies of tradition, alongside processes of dynamic social, economic and cultural change. She is interested in what the intersection of these two processes means for gender roles and perceptions of women's participation in politics. She cautions that there is significant regional variance in Poland and some areas are much more conservative than others, and she demonstrates that these values have a significant impact on gender roles. She offers a fascinating analysis of how the gendered composition of the role of village representative has changed and why more women have become involved. She concludes that it is partly linked to the devaluation of this role, partly to the feminization of the activities of this role and partly to the inevitable processes of socio-cultural change.

Conclusion

To conclude, this section (Chapters 13–17) nicely develops our understanding of rural gender identity. A number of common themes emerge across all of the chapters. They all underline the importance of intersectionality; identities are many and varied and they intersect for all of us, and for men and women in rural areas, on farms, in leadership roles and in divorce. All of the chapters demonstrate how identity is constructed through social interaction and relationships and involves processes of 'doing'. In this way, they also discuss how gender identity is temporal and dependent on time and place.

References

Alston, M. (1995) *Women on the Land: The Hidden Heart of Australia.* University of NSW Press, Sydney, Australia.

Alston, M. (1998) Farm women and their work: why is it not recognised? *Journal of Sociology* 34(1), 23–34. Available at: http://dx.doi.org/10.1177/144078339803400103 (accessed 6 March 2017).

Alston, M. (2012) Rural male suicide in Australia. *Social Science & Medicine* 74(4), 515–522. Available at: http://dx.doi.org/10.1016/j.socscimed.2010.04.036 (accessed 6 March 2017).

Alston, M. and Kent, J. (2008) The big dry: the link between rural masculinities and poor health outcomes for farming men. *Journal of Sociology* 44(2), 133–147. Available at: http://dx.doi.org/10.1177/1440783308089166 (accessed 6 March 2017).

Bartlett, P. (1983) *American Dreams, Rural Realities: Family Farms in Crisis.* North Carolina Press, Chapel Hill, North Carolina.

Bennet, K. (2004) A time for change? Patriarchy, the former coalfields and family farming. *Sociologia Ruralis* 44(2), 147–166. Available at: http://dx.doi.org/10.1111/j.1467-9523.2004.00268.x (accessed 6 March 2017).

Bock, B. (2004) Fitting in and multi-tasking: Dutch farm women's strategies in rural entrepreneurship. *Sociologia Ruralis* 44(3), 245–260. Available at: http://dx.doi.org/10.1111/j.1467-9523.2004.00274.x (accessed 6 March 2017).

Brandth, B. (1995) Rural masculinity in transition: gender images in tractor advertisements. *Journal of Rural Studies* 11(2), 123–133. Available at: http://dx.doi.org/10.1016/0743-0167(95)00007-A (accessed 6 March 2017).

Brandth, B. (2002) Gender identity in European family farming: a literature review. *Sociologia Ruralis* 42(3), 181–197. Available at: http://dx.doi.org/10.1111/1467-9523.00210 (accessed 6 March 2017).

Brandth, B. and Haugen, M. (1998) Breaking into a masculinist discourse. Women and farm forestry. *Sociologia Ruralis* 38(3), 427–442. Available at: http://dx.doi.org/10.1111/1467-9523.00087 (accessed 6 March 2017).

Brandth, B. and Haugen, M. (2000) From lumberjack to business manager: masculinity in the Norwegian forestry press. *Journal of Rural Studies* 16(3), 343–355. Available at: http://dx.doi.org/10.1016/S0743-0167(00)00002-4 (accessed 6 March 2017).

Brandth, B. and Haugen, M. (2005) Doing rural masculinity – from logging to outfield tourism. *Journal of Gender Studies* 14(1), 13–22. Available at: http://dx.doi.org/10.1080/0958923042000331452 (accessed 6 March 2017).

Brandth, B. and Haugen, M. (2010) Doing farm tourism: the intertwining practices of gender and work. *Signs* 35(2), 425–446. Available at: http://dx.doi.org/10.1086/605480 (accessed 6 March 2017).

Brandth, B. and Haugen, M. (2011) Farm diversification into tourism – implications for social identity? *Journal of Rural Studies* 27(1), 35–44. Available at: http://dx.doi.org/10.1016/j.jrurstud.2010.09.002 (accessed 6 March 2017).

Brandth, B. and Haugen, M. (2016) Rural masculinity. In: Shucksmith, M. and Brown, D. (eds) *The Routledge International Handbook of Rural Studies.* Routledge/Falmer, London and New York, pp. 412–424.

Bryant, L. and Pini, B. (2011) *Gender and Rurality*. Routledge, Oxford, UK and New York.

Campbell, H. and Bell, M. (2000) The question of masculinities. *Rural Sociology* 65(4), 532–546. Available at: http://dx.doi.org/10.1111/j.1549-0831.2000.tb00042.x (accessed 6 March 2017).

Campbell, H., Bell, M.M. and Finney, M. (eds) (2006) *Country Boys: Masculinity and Rural Life*. Pennsylvania State University Press, University Park, Pennsylvania.

Černič Istenič, M. (2006) Farm women in Slovenia. In: Bock, B.B. and Shortall, S. (eds) *Rural Gender Relations: Issues and Case Studies*. CAB International, Wallingford, UK, pp. 63–96.

Charatsari, C. and Černič Istenič, M. (2016) Gender, farming and rural social research: a relationship in flux. In: Shucksmith, M. and Brown, D.L. (eds) *The Routledge International Handbook of Rural Studies*. Routledge/Falmer, London and New York, pp. 389–399.

Finney, M.H., Campbell, H. and Bell, M.M. (2004) Rural masculinities: towards a new research agenda. Paper presented at the *XI World Congress of Rural Sociology*, Trondheim, Norway, 25–30 July 2004.

Gasson, R. (1980) Roles of farm women in England. *Sociologia Ruralis* 20(3), 165–180.

Grzetic, B. (2004) *Women Fishes these Days*. Fernwood Publications, Halifax, Nova Scotia, Canada.

Kelly, R. and Shortall, S. (2002) 'Farmers' wives': women who are off-farm breadwinners and the implications for on-farm gender relations. *Journal of Sociology* 38(4), 327–343. Available at: http://dx.doi.org/10.1177/144078302128756714 (accessed 6 March 2017).

Liepins, R. (2000) Making men: the construction and representation of agriculture-based masculinities in Australia and New Zealand. *Rural Sociology* 65(4), 605–620. Available at: http://dx.doi.org/10.1111/j.1549-0831.2000.tb00046.x (accessed 6 March 2017).

Little, J. (2014) The development of feminist perspectives in rural gender studies. In: Pini, B., Brandth, B. and Little, L. (eds) *Feminisms and Ruralities*. Lexington Books, London, pp. 107–119.

Little, J. (2016) Gender and entrepreneurship. In: Shucksmith, M. and Brown, D. (eds) *The Routledge International Handbook of Rural Studies*. Routledge/Falmer, London and New York, pp. 357–367.

Little, J. and Panelli, R. (2003) Gender research in rural geography. *Gender, Place & Culture* 10(3), 281–289. Available at: http://dx.doi.org/10.1080/0966369032000114046 (accessed 6 March 2017).

Marshall, J. (2001) Connectivity and restructuring: identity and gender relations in a fishing community. *Gender, Place & Culture* 8(4), 391–401. Available at: http://dx.doi.org/10.1080/09663690120111627 (accessed 6 March 2017).

McVay, L. (2016) Gender and leadership. In: Shucksmith, M. and Brown, D. (eds) *The Routledge International Handbook of Rural Studies*. Routledge/Falmer, London and New York, pp. 367–379.

Morris, C. and Evans, N. (2001) 'Cheese makers are always women': gendered representations of farm life in the agricultural press. *Gender, Place & Culture* 8(4), 375–390. Available at: http://dx.doi.org/10.1080/09663690120111618 (accessed 6 March 2017).

Moss, J.E., Jack, C.J., Wallace, M. and McErlean, S.A. (2000) Securing the future of small family farms: the off-farm solution. Paper presented at *European Rural Policy at the Crossroads*, Aberdeen, Scotland, 29 June–1 July 2000.

Ni Laoire, C. (2001) A matter of life and death? Men, masculinities and staying 'behind' in rural Ireland. *Sociologia Ruralis* 41(2), 220–236. Available at: http://dx.doi.org/10.1111/1467-9523.00179 (accessed 6 March 2017).

O'Hara, P. (1998) *Partners in Production? Women, Farm and Family in Ireland*. Berghahn Books, New York.

Overbeek, G., Efstratoglou, S., Haugen, M.S. and Saraceno, E. (1998) *Labour Situation and Strategies of Farm Women in Diversified Rural Areas of Europe*. Office for Official Publications of the European Communities, Luxembourg.

Pini, B. (2002) Constraints to women's involvement in agricultural leadership. *Women in Management Review* 17(6), 276–284. Available at: http://dx.doi.org/10.1108/09649420210441923 (accessed 6 March 2017).

Pini, B. (2004) Managerial masculinities in the Australian sugar industry. *Rural Society* 14(1), 22–35. Available at: http://dx.doi.org/10.5172/rsj.351.14.1.22 (accessed 6 March 2017).

Pini, B. (2005) Farm women: driving tractors and negotiating gender. *International Journal of Sociology and Food* 13(1), 1–12.

Reed, M. (2002) Marginality and gender at work in forestry communities of British Columbia, Canada. *Journal of Rural Studies* 19(3), 373–389. Available at: http://dx.doi.org/10.1016/S0743-0167(03)00021-4 (accessed 6 March 2017).

Sachs, C.E. (1983) *The Invisible Farmers: Women in Agricultural Production.* Rowman and Allanheld, Totowa, New Jersey.

Saugeres, L. (2002) The cultural representation of the farming landscape: masculinity, power and nature. *Journal of Rural Studies* 18(4), 373–384. Available at: http://dx.doi.org/10.1016/S0743-0167(02)00010-4 (accessed 6 March 2017).

Shortall, S. (1992) Power analysis and farm wives – an empirical study of the power relationships affecting women on Irish farms. *Sociologia Ruralis* 32(4), 431–452. Available at: http://dx.doi.org/10.1111/j.1467-9523.1992.tb00941.x (accessed 6 March 2017).

Shortall, S. (1999) *Gender and Power – Women and Farming.* Macmillan Press, London.

Shortall, S. (2006) Economic status and gender roles. In: Bock, B. and Shortall, S. (eds) *Rural Gender Relations: Issues and Case Studies.* CAB International, Wallingford, UK, pp. 340–355.

Shortall, S. (2014) Farming, identity and well-being: managing changing gender roles within Western European farm families. *Anthropological Notebooks* 20(3), 67–81.

Shortall, S. (2016a) Changing configurations of gender and rural society. In: Shucksmith, M. and Brown, D. (eds) *The Routledge International Handbook of Rural Studies.* Routledge/Falmer, London and New York, pp. 347–357.

Shortall, S. (2016b) Gender and identity formation. In: Shucksmith, M. and Brown, D. (eds) *The Routledge International Handbook of Rural Studies.* Routledge/Falmer, London and New York, pp. 400–412.

Shucksmith, M. and D. Brown. (eds) (2016) *The Routledge International Handbook of Rural Studies.* Routledge/Falmer, London and New York.

Silvasti, T. (2003) Bending borders of gendered labour division on farms: the case of Finland. *Sociologia Ruralis* 43(2), 154–166. Available at: http://dx.doi.org/10.1111/1467-9523.00236 (accessed 6 March 2017).

Thuesen, A.A. (2016) Gender and rural governance. In: Shucksmith, M. and Brown, D. (eds) *The Routledge International Handbook of Rural Studies.* Routledge/Falmer, London and New York, pp. 379–388.

Trauger, A., Sachs, C., Barbercheck, M., Kiernan, N., Brasier, K. and Schwartzberg, A. (2010) The object of extension: agricultural education and authenic farmers in Pennsylvania. *Sociologia Ruralis* 50(2), 85–103. Available at: http://dx.doi.org/10.1111/j.1467-9523.2010.00507.x (accessed 6 March 2017).

Whatmore, S. (1991) *Farming Women: Gender, Work and Family Enterprise.* Macmillan, London.

Wheelock, J. and Oughton, E. (1996) The household as a focus for research. *Journal of Economic Issues* 30(1), 143–159.

Woodward, R. (1998) It's a man's life! Soldiers, masculinity and the countryside. *Gender, Place & Culture* 5(3), 277–300. Available at: http://dx.doi.org/10.1080/09663699825214 (accessed 6 March 2017).

13 Rural Women Leaders: Identity Formation in Rural Northern Ireland

L. McVay*

Spring Arbor University, Spring Arbor, Michigan, USA

Introduction

The last two decades have seen a marked increase in interest around women's access to leadership positions, as nations and organizations acknowledge the importance of gender parity for sustainability and global improvements in health and socio-economic conditions (UN Women, 2015). The European Parliament, through its Women's Rights and Gender Equality Committee, is pressing for gender mainstreaming in all areas, with particular emphasis on both women's expanding role in decision-making processes and leadership training for women (Mlinar, 2016). Within these frameworks, the question of how women attain leadership positions has focused primarily on examining the many obstacles that women – and rural women in particular – face in accessing positions of power. As an alternative to these approaches, research presented in this chapter undertook to examine positive factors in rural women leaders' development processes: the choices and happenings that shaped their identities as rural, as women, as leaders and as people whose lived experience required navigating the intersection of those identities.

Background and Context: Why Northern Ireland?

Northern Ireland is rich in vibrant rural communities, with 36% of its people living in rural areas (Northern Ireland Assembly, 2012). Having come through the prolonged violence of 'The Troubles',[1] rural women in Northern Ireland face many of the same challenges as their peers across Europe: a persistent pay gap, under-representation in decision-making processes and offices, the dominance of men in land ownership, poor access to training opportunities and work, and the perpetuation of restrictive cultural assumptions regarding gender in the rural idyll (Northern Ireland Assembly, 2012). As women, they face different barriers from their male counterparts, including limited opportunities to return to employment for women whose childcare activities have been prioritized over out-of-home work (Department for Employment and Learning Northern Ireland, 2013). Research shows the persistence of gendered careers advice being linked to the concentration of women in a limited number of Northern Ireland's low-paying job sectors (Northern Ireland Women's European Platform, 2013), and employers appear to lag behind in fully

* Corresponding author e-mail: lori.ann.mcvay@gmail.com

recognizing the skills of women despite women's parity in educational attainment (Department for Employment and Learning Northern Ireland, 2013).

To challenge these and similar entrenched structures and to make gender inequality visible, women around the globe have formed partnerships and networks with other women (Bjørnå, 2012). Women-only spaces are a valuable resource for discussions around shared struggles with gender inequality, encouraging an exchange of information free from concern regarding men's perception and/or judgement (Pini, 2005; Trauger *et al.*, 2008; Debebe, 2011). They further serve as places in which these women's multiple identities and roles can be recognized and valued (Brasier *et al.*, 2014). Although women-only programmes have been criticized for lacking both interactions with people of all genders and the opportunity to network with well-connected men, proponents believe that the opportunity to function as part of a majority group – even on a temporary basis – is a valuable developmental experience that few women have in the workplace (Ely *et al.*, 2011).

In Northern Ireland's rural communities, women-only groups are doing important community development work and have received (limited) funding to continue to do so (Northern Ireland Assembly, 2012). The growing rural population is well served by this dynamic network of rural women's groups, as they work actively to improve the position of women in local, regional and national contexts. Based on their awareness and experience of women's daily lives, the groups have identified leadership development as one means of addressing the needs of Northern Ireland's rural women (Northern Ireland Department of Agriculture and Rural Development, 2006).

Rural women-only groups served as a starting point for conducting the research presented here. Additionally, each of the sectors listed in Macaulay and Laverty's (2007) *Baseline Study of Rural Women's Infrastructure in Northern Ireland* provided excellent representatives of rural women in leadership. These included women serving in non-governmental organizations, politics, business, churches and faith-based organizations, and agricultural organizations.

Methodology

As women's life experiences were the primary source of data for this study, feminism's recognition of women's experiences as a 'legitimate form and source of knowledge' (Pini, 2003) made feminist methods a logical choice. This opened an avenue for exploring not only individual women's lives but other lives that had also informed their experiences. Valuing individual life experiences in this way created the possibility of extrapolating insights to broader social settings (Rosenberg and Howard, 2008). Following Pini's (2005) example of research with rural women, semi-structured interviews loosely following the women's life courses served as the primary research method. These were combined with participant observation/field notes and reflexivity for triangulation purposes.

Analysis of the interviews was undertaken via the 'listening guide' as expounded by Gilligan *et al.* (2003) in 'On the listening guide: a voice-centered relational method'. This method incorporated listening for the entire range of factors participants voiced as significant in their leadership development. Because the 'listening guide' is a voice-centred relational method, each of the interviews was transcribed in the 'naturalistic' mode (Oliver *et al.*, 2005), including verbal starts and stops and vocalized emotional responses such as laughter and sighs. Gilligan *et al.*'s (2003) method of analysis required multiple 'listenings' (p. 159) to the transcripts in order to identify the various voices present in individuals' transcripts and to hear the relationships among those voices through the recognition of their points of harmony and dissonance. Careful listening to the manner in which participants articulated their experiences gave shape to the ways they identified as rural, as women and as leader.

Analytical Framework

Intersectional perspectives

Holvino's writing on intersectionality highlights the necessity of avoiding a 'universal experience or category' of women so that women's unique constellation of experiences may be evaluated in ways that provide insight into how organizational norms can be challenged (2010, p. 264). Pini (2005) shares Holvino's concern and cautions against the limiting activity of conflating *all* women with stereotypes about women. For example, much literature around gender in organizations not only focuses on the male norm, but also flattens the experiences of women and creates a female norm, which excludes differences in areas other than gender (Morley, 2013). In the words of Alvesson *et al.* (2008, p. 19), 'even when people refer to a seemingly shared "we", they imbue this depersonalized collective with diverse and personalized meanings'. As will be shown in the results below, the multifaceted meanings given to the 'rural' and 'woman' and 'leader' identities by the women in this study confirmed Alvesson *et al.*'s findings.

Identities

Identities are constructed through a variety of individual life experiences and situations (Trauger *et al.*, 2008). In particular, relationships are formative influences on self-identity and contribute to changes in how we perceive ourselves and are perceived by others, including which of our identities are acceptable (Billing, 2011). How we make meaning of these 'role relationships' and our choices about appropriate behaviours are constrained by social structures (Brenner *et al.*, 2014, p. 231), and we interpret the appropriateness of others' behaviours using points of reference created by how they attribute meaning to their identities (Carter, 2014). Citing the work of James (1890), Carter (2014, pp. 248–249) notes:

> Identity theory posits that there are as many different selves as there are different positions one holds in the social structure . . . These identities determine how an actor behaves when *alone*, while playing a *role*, or when attached to a *group* . . . All three types of identity can operate simultaneously.

Bock (2006) presented identities as contextually specific. As such, the self-understanding we hold is structured by interactions with 'larger cultural and historical formations' that provide language and norms within which we develop identities (Alvesson *et al.*, 2008, p. 11). From this self-understanding, identities may be seen as more or less salient (likely to be acted upon) and more or less prominent (valuable in relation to other identities); however, salience and prominence are not directly linked (Brenner *et al.*, 2014). In a similar vein, Trauger *et al.* (2008) asserted that identities are prioritized situationally and relationally and are unlikely to remain fixed. The centrality of an identity may also differ between individuals (Dutton *et al.*, 2010).

Rural identity

Saugeres (2002) revealed the depth of impact gender identity holds for rural women when she noted that community acceptance is determined by their ability to carefully conform to gendered expectations. In this way, rural women's gender identities, as with gender identities more broadly, may be seen as resulting from their culture's ideals of what men and women are like (Ridgeway and Smith-Lovin, 1999). The social construction of rural women's gender identity has been linked to inequality (Little and Panelli, 2003), as they have traditionally been framed in less visible roles of support (O'Toole and Macgarvey, 2003; Pini, 2005). And, in spite of massive changes in employment status and the viability of family farms, Shortall (2014) noted that rural women's gender identities – although changing (Shortall, 2002) – have shifted much more slowly than their work and family roles (see also Brandth, 2002; Silvasti, 2003).

This may in part be influenced by the centrality of women's rural identity (Billing, 2011; Brasier *et al.*, 2014) and linked to the necessity of social acceptance for seeing oneself as included in a particular social category (Karelaia and Guillén, 2014). It may also be connected to men's close ties to farming traditions and the reluctance of women to unsettle gender relations and create additional stress for their farmer husbands or partners (Alston and Whittenbury, 2013). Little (2007) has shown that this reluctance to change is also applied to the acceptance of non-heterosexual sexualities, which may be perceived as threatening intergenerational transmission of traditional values and morals. Rural women's sense of self may be intimately connected to their context (Heather *et al.*, 2005), and in the context of Northern Ireland, particularly conservative views on marriage (Breitenbach and Galligan, 2004) can serve as reinforcements for each of these factors.

Importantly, not all rural women are pleased to re-enact established gender relations (Silvasti, 2003) and, as post-productivist changes continue to shape rural areas, the influence of traditional frameworks is being opened to questioning (Sireni, 2015). Women whose difference from traditional rural gender identities defies community expectations are adding to the diversity of rural femininity (Haugen and Brandth, 2015) and demonstrating the adaptability of rural spaces (Wilbur, 2014). As Lear (2013, p. 378) observes, rural communities can be a source of opportunity for 'learning and personal development'. Pini and Leach (2011) note that global changes in rural areas are affecting not only the rural economy but also the mythos of rural spaces in ways that are both destabilizing and full of potential. For example, in Italy, back-to-the-land women with professional backgrounds are discovering that rather than leaving behind their experience and skills, farming allows them to put these assets to use in new ways (Wilbur, 2014).[2]

Gender identity

Both self-claimed identities and identities placed upon us by broader structures are shaped via interactions with others (Brenner *et al.*, 2014). Gender identity and status perceptions are no exceptions (Ridgeway and Smith-Lovin, 1999), being produced and reproduced by processes embedded in everyday life (Acker, 2012) and gaining salience in the process (Carter, 2014). Unlike the limited salience of other identities, gender is central and salient unless the person changes her or his sex category (West and Zimmerman, 1987). Gender is a 'primary category for framing social relations' that is culturally imbedded and highly relevant in creating a sense of difference between people (not just women) (Sheridan *et al.*, 2011) who interact regularly (Ridgeway, 2009, p. 148) and in shaping behaviour across various milieus (Ridgeway and Smith-Lovin, 1999). Feminist writers have pointed out the power of the family and, in particular, parental relationships to begin shaping gender identities in childhood (e.g. Chodorow, 1978), although identities are further shaped by interactions and social relations throughout the lifespan (Saugeres, 2002).

Gender has also been framed as having less mobility than other identities and therefore being less likely to be excluded from self-identity than others (e.g. 'leader') (Karelaia and Guillén, 2014). It is important to note, however, that gender should not be conceived of as a fixed or arbitrary identity. Gender identities may be shifting and multiple (Saugeres, 2002), and biology is an insufficient, although important, marker (Little, 2002). Little's work revealed that there is no monolithic gender identity in rural areas; but as Bock (2006) notes, the construction of gender identities may also be hidden. While gender retains salience in the face of fluctuations in other identities' parameters, gender-atypical behaviour may be redefined as well, resulting in social stigma based on the new parameters – particularly for rural women (Sireni, 2015).

With rural gender relations being identified as ideologies reproduced in everyday, patriarchal interactions (Saugeres, 2002) and contributing to (or being the source of) inequality (Little and Panelli, 2003), gender relations are a valuable resource for uncovering how and why masculine and feminine gender identities exist as they do in rural spaces (Shortall, 2002). Interactions privileging men more often than women combine with cultural beliefs about gender to produce patterns which reinforce structures of power and increase the salience of gender identities for individuals and communities (Ridgeway and Smith-Lovin, 1999), resulting in a strong rural gender binary (Alston and Whittenbury, 2013). This combination of power structure and identity salience results in forms of information exchange in which rural men make assumptions about what rural women need to know and when (Trauger *et al.*, 2008) and positions men and women to approach discussions around gender identities from widely different understandings in spite of women's increased economic activity (Alston and Whittenbury, 2013).

Leader identity

Cultural expectations also shape the internalization and performance of the leader identity for rural women (Smith and Reed, 2010). Pini (2005) purports that rural women's development of a leader identity may parallel men's, although this identity construction may be more difficult for rural than urban women as intersections with the social expectations for their other identities may produce dissonance (Sheridan *et al.*, 2011; Morley, 2013). The quandary faced by leadership scholars in trying to develop a universal (or at least widely accepted) definition of 'leadership' compounds this dissonance (McVay, 2013).

Recent research argues that the gender binary (Baxter, 2014) and the male norm (Billing, 2011) may no longer be as useful for evaluating leader identity as they were once considered. It has also been theorized that internalized beliefs about one's capacity to lead may have more impact on leader selection than gender (Coder and Spiller, 2013). Nevertheless, women's route to internalizing the leader identity may pose more challenges than men's, given the continued conflict over women's authority (Ely and Rhode, 2010 in Ely *et al.*, 2011) and the persistent reinforcement of the perceived leadership–masculinity connection and leadership–femininity disconnect (Pini, 2005; Sinclair, 2005) that has resulted in men being viewed as more 'agentically competent and worthy of status' (Ridgeway, 2009, p. 151).[3] Even when attempting to manage their gender by neutralizing it, women leaders will still be identified and treated as gendered female (Lewis, 2006).

For women entering leadership in situations where masculine behaviours have traditionally been valued and continue to be expected (if not articulated), this often leads to a perception of themselves as inauthentic (Ibarra *et al.*, 2010). In organizations where women are poorly represented, women's gender identity may be seen as less than positive, creating fundamental barriers to women's reconciliation of the leader identity with their sense of self (Karelaia and Guillén, 2014), as well as complicating peers' ability to see them as leaders (Toh and Leonardelli, 2012) and resulting in a lower likelihood that men will follow them (Edwards *et al.*, 2009). Because of the formal relationship between leaders and followers, women may also be framed as unable to simultaneously comply with 'prototypical' leader identities and maintain gender accountability (West and Zimmerman, 1987; DeRue and Ashford, 2010). Pini's (2005, p. 80) study showed that women in agricultural organizations were not only highlighted as different from men (and thereby different from dominant/masculine leader identities), but were expected to 'speak for and about all women' – a difficult situation, especially given the women's self-identification as different from other women. Once the leader identity is internalized, however, it motivates behaviour in line with the identity (Carter, 2014). The leader identity may change as the leader develops skills and competencies, and this lack of linearity is considered an important component of leadership development and the motivation to lead (Chan and Drasgow, 2001; Richardson and Loubier, 2008).

Research Findings

The final step of Gilligan *et al.*'s (2003) 'listening guide' is to revisit the primary research question and ask what has been 'learned about this question through this process' (p. 168). In this study, it became apparent that listening to the women's verbalizations of themselves as rural, as women and as leaders was primary to any discussion of their leadership development. Participants spoke of their 'rural' and 'woman' identities in less formal language than they did when discussing leadership and how they saw themselves filling (and/or not filling) the role of leader. There may be several reasons for this, two of which are: the types of questions asked regarding their definition of leadership (which many experts characterize as nebulous, at best); and their confidence in self-identifying as 'rural' and 'women' (which were well defined in their narratives). Nevertheless, clear articulations of all three identities emerged as the women spoke about their life experiences.

Rural identity

Lived experience may override the lack of a definitive concept of rurality (Mahon, 2007) in developing a rural identity. Although articulating differing conceptions of 'rural', each of the women in this study had a deeply internalized rural identity, reflected in its centrality to their narratives (McVay, 2013). In line with Bock's (2006) claims that identities are contextually specific, many of the participants' narratives conveyed a powerful sense of place-based identity. Barbara's words were typical of the women in this study:

> I will always say that the fact that I was brought up in a rural community, that I've chosen to live there and bring *my* children up in a rural community, it's very much part of my cultural identity. And the *fabric* of my culture and identity. It's another layer.

Resonant with Cassidy and McGrath's (2015) work, even women who were no longer physically connected with their family farm or had moved from rural areas several years before still held an emotional connection to farming and/or rural areas and strongly identified as rural women. Cassidy and McGrath framed this emotional connection as a desire for security and retention of rural insider status. This was true of Eva, who had not lived in a rural area for many years. In her words: 'But, if you're asking me, I'm a peasant woman. And proud of it. And I love rural Ireland.'

Gender identity

In this study, participants had internalized the leader identity to varying degrees – some with an attitude of minimization and others much more boldly. Yet few, if any, of them had 'assimilated' into the masculine norms of leadership (Billing, 2011). Katherine's injunction – 'Well. Don't try to be a man . . . I refused to be turned into a man' – described what many of the women felt. As with Shortall's (2014) findings, these women emphasized their identities as women and as rural over the (traditionally masculine) leader identity, and gave credibility to Karelaia and Guillén's (2014) assertion that positive gender identity can give women the freedom to lead in their own ways.

Although the incorporation of multiple identities may have positive effects for women, Carter (2014) problematizes the ways in which gender becomes stabilized in the context of various interactions and thus reproduces the social norms within which it is negotiated. Pini (2005) showed that women's positions as '*women* leaders in agriculture' (p. 74) required their gender to be both highlighted and subjugated, making gender highly salient as women negotiated a context-specific leader/woman dichotomy and illustrating Brenner *et al.*'s (2014) observation that the salience of an identity may not necessarily equate to an individual's desired performance of that identity.

For many women in this study, organizations served as a framework from which to examine and critique gender roles (McVay, 2013). The negative impacts of gendered organizations may include obstructions to the identity-formation process of women leaders (Ely *et al.*, 2011). Lewis (2006) encountered these attitudes in studying organizations, but questioned whether gendered organizations might be a construct created by those analysing the data. Billing (2011) surmised that younger women in less hierarchical organizations may experience fewer pressures to conform to attempts at gendering.[4] Acker (2012, p. 215) counters these arguments with the concept of 'gendered substructures', which she defines as:

> often-invisible processes in the ordinary lives of organizations in which gendered assumptions about women and men, femininity and masculinity, are embedded and reproduced, and gender inequalities perpetuated . . . These processes and practices are supported by organizational cultures and reproduced in interactions on the job, shaped in part by the gendered self-images of participants.

Bryant and Garnham (2014) point out that women may not be able to see the substructures or norms that act as barriers within organizations and yet articulate their existence.

It must also be cautioned that the gendered nature of organizations should not be seen as proof of inherent differences between genders, nor as the result *solely* of male domination (capitalism's demand for prioritizing work over family life is one of the many other factors) (Billing, 2011). Rather, gendered organizations are socially produced through the interactions of multiple, complicated identities that have themselves been impacted by gender's salience (Ridgeway and Smith-Lovin, 1999; Bryant and Garnham, 2014). Within this framework, some participants in the study asserted that the organizations in which they were involved did not discriminate against women in any way (McVay, 2013):

> some would like to think that rural women are hard done by or whatever . . . I don't feel that in *any* way . . . this organization, and to be honest, this is my personal opinion, do not differentiate between men and women . . . I mean, there's just no mention of male/female . . . there's no discrimination or positive discrimination. . . . I mean, this organization is by *far* a male-dominated organization, but, they respect anyone . . . on their own merit.
>
> (Alice; emphasis hers)

However, others articulated ways in which organizations were most definitely gendered:

> I remember when I was in the [organization], there was women and men in it. And when I'd say, 'Right. We need . . .' if it was a particular type of job, the women would always go for it. And then if it was to do a radio interview or to speak publicly or do the introductions or whatever, suddenly they wouldn't go for it. So what I had to do was prepare those women in advance. And say, 'Right. Today. At staff meeting. I am going to be looking for "x" to do media interview and somebody else to do the introductions. I'm going to be asking you to do it. Do not say no.' [laughs] . . . Sometimes, people had difficulties. The people who always had their hands up had difficulties with the change. But very quickly they got used to that, and then we had a much more dynamic, fair form of leadership.
>
> (Helen)

Experiences in women-only groups and schools demonstrate a positive facet of how gender can function within organizations (Acker, 2012; see also Fitzsimmons *et al.*, 2014). Margaret, like many of the women in this study, was involved in one of these groups and had found it to be a significant source of leadership opportunity and practical experience – named by participants as one of the most necessary components in rural women's leadership development (McVay, 2013):

> I had been in [a rural youth organization] . . . it's not a natural step to move forward from that to [rural women's organization]

> . . . But it was because my mother-in-law said, 'Come on.' And I went along as a twenty-two or twenty-three-year-old . . . I have never, ever regretted it. It was certainly one of the most significant steps I've ever taken in my life. Because when I went there, I became involved.

Participants reacted strongly against being identified as a feminist, including several who espoused feminist views and expressed concern for the position of women in Northern Ireland (McVay, 2013).[5] Dutton *et al.* (2010) attribute this type of distancing to a person's need to maintain their positive reputation when faced with the salience of their membership in a devalued group. Many of the women seemed to see feminism as promoting unacceptable behaviour for rural women, behaviour that would be 'egoist and harmful to others' and give them both a 'bad reputation and a troubled conscience' (Haugen and Brandth, 2015, p. 234).[6] Sara was a good example of this. Like many of the other participants, she took quite an adamant stance against being identified as a feminist, even while critiquing church structures which barred women from leadership and recognizing the pervasiveness of male-dominated organizations in rural areas:

> And I would say as well, I've had exposure to feminists, you know . . . it just gives me images of militants and going out and burning your bra and stuff . . . I was reading something [that] came in the post to me from [women's group] . . . And it said they come at it from a feminist perspective. And all of a sudden I was going [makes shuddering noise]. You know, because I don't see myself as a feminist. I see myself as a family woman, as a mother. As a worker.

Sara and others who shared her rejection of the feminist identity, while positively framing roles that feminists have problematized as rife for oppression, affirmed their non-participation (in feminism) as a 'valid and legitimate choice, made from a place of power' (Shortall, 2008, p. 452).

One common marker of the identity 'woman' among participants (both those who identified with and rejected the label 'feminist') was the capacity to express care for others in some active form, rather than any purely biological construct (Little and Austin, 1996). In telling how she worked for change, Niamh – the one woman in this study who identified strongly as feminist – described what she saw as the root of women's concern around developing a feminist identity:

> Women have suffered so long from internalization, and I see it every single day. They have internalized everything from body image to . . . their lack of ability. We have designed a course . . . for the community activists, and have looked at feminism, misogyny, and the light bulbs are going off all over the place. And you know that it may cause marriages to break up, families to break down . . . And people are worried about that and the changes that that consciousness-raising causes. But [it's] not only training, it's mentoring . . . it's for women to be able to talk about that.

It was at the point of actively demonstrating care that these rural women were most readily able to recognize their role as leaders.

Leader identity

Cultural ideologies about leadership are formative to interactions by which the title of 'leader' is claimed and granted (Ely *et al.*, 2011), as an individual's beliefs about the roles of leaders and followers have usually been formed before involvement with an organization (DeRue and Ashford, 2010) and impact women's willingness to be seen as a leader (Toh and Leonardelli, 2012). Women in this study were no exception. While they appeared to be quite comfortable and clear in identifying themselves as 'rural' and 'woman', they articulated a spectrum of leadership definitions and a high degree of uncertainty around whether they qualified as 'real' leaders. This is illustrated by Olivia and Patricia, who were interviewed together and acknowledged the leadership position of their posts, but did not consider themselves to be demonstrating true leadership skills:

> Patricia: 'I suppose maybe, too, in our current roles – Olivia you can agree or

> disagree with me, that's fine . . . [laughs] – there is a certain element of leadership because we're trying to work with groups to bring groups forward. And, very often, you have to try to show them things that could maybe help them, progress and move forward . . . So I don't know if it's leadership so much as cajoling and . . . [laughs] and giving them a shove in the right direction and, a bit of help when the need it. Rather than leadership.'
>
> Olivia: 'Mm hmm. 'Cause the term I was going to use alongside of that was sort of like an encourager.'

By broadening the 'set of possible meanings' associated with leadership in their everyday practices (Brenner *et al.*, 2014, p. 232), many participants' motivation to lead was increased (Chan and Drasgow, 2001). This, in turn, increased the salience of their leader identity (Carter, 2014). Cara described this process:

> I think for a lot of women, they don't see themselves as leaders but they clearly are within their respective groupings and whatever they're doing . . . I would give them the advice not to think of leadership in a boxed way . . . to think of leadership in the widest possible way . . . I mean if I said to a group of young women from a rural area, 'You're all leaders,' they'd go, 'Aye right. Dead on.' [laughs] But most of them probably are . . . whether it be in school or in their church or in their community or whatever. So, it's getting people to think about leadership in a different way.

Cara's words highlight what several of the women experienced: not recognizing *themselves* as leaders, but being identified by *others* as having the potential to lead. For example, Maureen's internalization of the leader identity began early in life, when she was asked to mentor a younger class in school. That affirmation of her potential to lead at a time when she did not recognize it in herself helped reframe what she saw simply as showing responsibility: 'the teachers . . . chose people who were going to do something. Like, rather than people who, just, didn't want to do anything.'

For all of the women, leadership (which was closely linked in their minds – if not equivalent to – responsibility) appeared in their narratives as 'getting on with it', 'it just needed to be done' and being 'willing to do something' in response to a perceived need (McVay, 2013, p. 35). Maureen summed up this intertwined sense of leadership and responsibility by saying, 'Well if I don't do this . . . nobody else is going to attempt it' (McVay, 2013, p. 35).

Families and other childhood milieus fostered such responsibility and served as the inspiration for an internal drive to work hard and to achieve, along with a strong desire to positively impact others. This echoes Little and Austin's (1996) connection between rural women's gender identities and sense of responsibility to fulfil expectations of caregiving that sustains family and the rural community. It also illustrates DeRue and Ashford's (2010) assertion that the leader identity may be claimed both as a means of creating desired outcomes and of discovering whether leadership fits well within the person's constellation of identities. In the words of Doreen:

> I used go down to [the women's group] after I put the children to bed because we had no computer. So I would sit 'til two and three in the morning doing my assignments . . . [And now, I tell the women] You can do it 'cause I did. And you *can* do it . . . I've just so much compassion for these women.

An ethic of care: rural/woman/leader

Although research often defines rural leaders in terms of their approach or title (Skerratt, 2011), respondents in this study defined leadership as an interactive and supportive connection with followers (Ladkin, 2010). Tracy reflected on this in her interview:

> Have to say . . . when I got involved in politics it wasn't to be elected. My whole reason for becoming involved in politics and for doing what I did was to help . . .. And you know, whatever the party asked of me, I was prepared to do . . . I was there to do my bit.

And, in line with Pini's (2005) work, the importance of people and relationships dominated participants' conceptualization of leadership and was seen as a valuable asset. Katherine's narrative was quite typical:

> To me, I was certainly the senior person, but they were all very important to me. And I mean, my line was, I couldn't have been successful alone in what we achieved. You know, all those other people were important. So therefore you had to work with them. And I was happy to take advice from any of them . . . And I think treating people, treating everybody with respect, regardless of what grade they're at, I think it's important for a leader to be able to do that.

Dutton *et al.* (2010) have theorized that the framing of an identity as consistent with good character (and thereby 'virtuous') results in the identity being considered positive and creates avenues through which to help others. In this way, the active demonstrating of care allowed participants in this study to reduce identity conflict (Karelaia and Guillén, 2014) by reconciling their leadership identity with the socially expected role performance of their identities as rural women, and thereby maintain their 'feminine dignity' (Haugen and Brandth, 2015, p. 227). They were able to combine both the affective component of motivation to lead (leading for the pleasure of leading) and the social-normative component (leading from a sense of responsibility) (Karelaia and Guillén, 2014).

As is obvious in the earlier quote by Niamh, her concern for the women with whom she worked was reflective of a broader ethic of care present in most of the narratives. Rural women have been portrayed within the context of communities that normalize women's self-care as the last on a long list of priorities (Saugeres, 2002; Silvasti, 2003; Heather *et al.*, 2005). 'The Troubles' exacerbated this situation in Northern Ireland, with women taking up multiple roles of responsibility while husbands and fathers were incarcerated (Hinds, 2011). Just as the farm women in Trauger *et al.*'s study were challenging definitions of success to reflect the tenets of 'care, responsibility to the public, and connection to the farm' (2010, p. 53), the women in this study operated within a similar value system, integrating their 'rural' and 'woman' identities into their personal leadership styles in ways that were nurturing (McVay, 2013).

Conclusions

> You know, I think it's important to know the values and the home that you want to have for your family and then you work to get it . . . so that you get a work–life balance. But also to [whispers] enjoy yourself.
>
> (Joan)

Each of the three most study-relevant identities – rural, woman and leader – held different meanings that intersected in fascinating and important ways in participants' development as leaders (McVay, 2013). Using the 'listening guide' to view participants' leadership identity development through the feminist lens of intersectionality allowed the complexities of gender, rurality and leadership to be examined in relation to one another (Moorosi, 2013). While the 'feelings, values and behaviour' associated with identities may be positive as well as conflicting (Alvesson *et al.*, 2008 p. 6) or even threatening to each other (Karelaia and Guillén, 2014), and although each of these identities may have its own way of making meaning (Sireni, 2015), ways of speaking create spaces in which multiple identities may be combined in myriad ways – some positive and some negative – depending upon the social situation (Bargiela-Chiappini *et al.*, 2007 in Baxter, 2014). Participants' self-presentation and narrative construction revealed how they made meaning from the complex events of their lives (Billing, 2011) and clearly demonstrated their navigations of 'complex yet compatible identity structures' as they integrated and operated from multiple identities in positive ways that kept those identities from being subsumed into one another (Dutton *et al.*, 2010, p. 279).

For each of the women, becoming a leader had been both a personal and a social journey, in which they incorporated multiple, interconnected identities. A strong argument could be made that the key to their success as leaders lay in their ways of understanding and utilizing their experience of these identities. For nations and organizations interested in gender parity and the inclusion of rural women, this presents a clear case for deeper qualitative research and the recognition that much can be learned from the examination of individuals' leader identity as it intersects with other identities. Fostering future rural women leaders is not only a matter of providing training and opportunity, but of deeply valuing their experiences as potential leaders, as rural and as women.

Notes

[1]'Following thirty years of political violence . . . there are few families in Northern Ireland's small population . . . that have not been affected by the conflict' (Muldoon *et al.*, 2015).

[2]There may be differing consequences for gender performance, however, dependent on how women form their rural identity: whether by birth or through in-migration (Haugen and Brandth, 2015).

[3]In fact, Morley (2013, p. 126) goes so far as to portray leadership roles as 'virility tests'.

[4]It is important to note here that attempts are still being made to gender the younger women. As Acker (2012) states, 'gross' sexism may have decreased, but it can reappear in subtler forms.

[5]This was especially interesting in light of the fact that none of the interview questions directly addressed feminism.

[6]West and Zimmerman (1987, p. 136) termed this the 'risk of gender assessment', and emphasize that it is individuals and not institutions who will be held accountable for failing to do gender appropriately.

References

Acker, J. (2012) Gendered organizations and intersectionality: problems and possibilities. *Equality, Diversity and Inclusion: An International Journal* 31(3), 214–224. Available at: http://dx.doi.org/10.1108/02610151211209072 (accessed 6 March 2017).

Alston, M. and Whittenbury, K. (2013) Does climatic crisis in Australia's food bowl create a basis for change in agricultural gender relations? *Agriculture and Human Values* 30(1), 115–128. Available at: http://dx.doi.org/10.1007/s10460-012-9382-x (accessed 6 March 2017).

Alvesson, M., Lee Ashcraft, K. and Thomas, R. (2008) Identity matters: reflections on the construction of identity scholarship in organization studies. *Organization* 15(1), 5–28. Available at: http://dx.doi.org/10.1177/1350508407084426 (accessed 6 March 2017).

Baxter, J. (2014) 'If you had only listened carefully . . .': the discursive construction of emerging leadership in a UK all-women management team. *Discourse & Communication* 0(1), 23–39. Available at: http://dx.doi.org/10.1177/1750481313503224 (accessed 6 March 2017).

Billing, Y.D. (2011) Are women in management victims of the phantom of the male norm? *Gender, Work & Organization* 18(3), 298–317. Available at: http://dx.doi.org/10.1111/j.1468-0432.2010.00546.x (accessed 6 March 2017).

Bjørnå, H. (2012) Gender balance and institutions in local government – examples from rural Norway. *Lex Localis – Journal of Local Self-Government* 10(2), 129–152. Available at: http://pub.lex-localis.info/index.php/LexLocalis/article/view/207 (accessed 16 January 2016).

Bock, B.B. (2006) Rurality and gender identity: an overview. In: Bock, B.B. and Shortall, S. (eds) *Rural Gender Relations: Issues and Case Studies.* CAB International, Wallingford, UK, pp. 279–287.

Brandth, B. (2002) Gender identity in European family farming: a literature review. *Sociologia Ruralis* 42(3), 181–200. Available at: http://dx.doi.org/10.1111/1467-9523.00210 (accessed 6 March 2017).

Brasier, K.J., Sachs, C.E., Kiernan, N.E., Trauger, A. and Barbercheck, M.E. (2014) Capturing the multiple and shifting identities of farm women in the Northeastern United States. *Rural Sociology* 79(3), 283–309. Available at: http://dx.doi.org/10.1111/ruso.12040 (accessed 6 March 2017).

Breitenbach, E. and Galligan, Y. (2004) *Gender Equality Indicators for Northern Ireland: A Discussion Document.* Office of the First Minister and Deputy First Minister, Belfast.

Brenner, P.S., Serpe, R.T. and Stryker, S. (2014) The causal ordering of prominence and salience in identity theory: an empirical examination. *Social Psychology Quarterly* 77(3), 231–252. Available at: http://dx.doi.org/10.1177/0190272513518337 (accessed 6 March 2017).

Bryant, L. and Garnham, B. (2014) The embodiment of women in wine: gender inequality and gendered inscriptions of the working body in a corporate wine organization. *Gender, Work & Organization* 21(5), 411–426. Available at: http://dx.doi.org/10.1111/gwao.12045 (accessed 6 March 2017).

Carter, M.J. (2014) Gender socialization and identity theory. *Social Sciences* 3(2), 242–263. Available at: http://dx.doi.org/10.3390/socsci3020242 (accessed 6 March 2017).

Cassidy, A. and McGrath, B. (2015) Farm, place and identity construction among Irish farm youth who migrate. *Journal of Rural Studies* 37, 20–28. Available at: http://dx.doi.org/10.1016/j.jrurstud.2014.11.006 (accessed 6 March 2017).

Chan, K.Y. and Drasgow, F. (2001) Toward a theory of individual differences and leadership: understanding the motivation to lead. *Journal of Applied Psychology* 86(3), 481–498. Available at: http://dx.doi.org/10.1037/0021-9010.86.3.481 (accessed 6 March 2017).

Chodorow, N. (1978) *The Reproduction of Mothering.* University of California Press, Berkeley, California.

Coder, L. and Spiller, M.S. (2013) Leadership education and gender roles: think manager, think '?'. *Academy of Educational Leadership Journal* 17(3), 21–51. Available at: http://www.alliedacademies.org/articles/aeljvol17no32013.pdf (accessed 6 March 2017).

Debebe, G. (2011) Creating a safe environment for women's leadership transformation. *Journal of Management Education* 35(5), 679–712. Available at: http://dx.doi.org/10.1177/1052562910397501 (accessed 6 March 2017).

Department for Employment and Learning Northern Ireland (2013) *DEL Oral Statement: Gender Issues Relating to Employment, Learning and Skills.* DELNI, Belfast.

DeRue, D.S. and Ashford, S.J. (2010) Who will lead and who will follow? A social process of leadership identity construction in organizations. *Academy of Management Review* 35(4), 627–647.

Dutton, J.E., Roberts, L.M. and Bednar, J. (2010) Pathways for positive identity construction at work: four types of positive identity and the building of social resources. *Academy of Management Review* 35(2), 265–293.

Edwards, B., Dillard, M. and Juska, A. (2009) Gender and leadership in the Lithuanian rural community movement. *Transitions* 47(2), 15–28.

Ely, R.J., Ibarra, H. and Kolb, D.M. (2011) Taking gender into account: theory and design for women's leadership development programs. *Academy of Management Learning & Education* 10(3), 474–493.

Fitzsimmons, T.W., Callan, V.J. and Paulsen, N. (2014) Gender disparity in the C-suite: Do male and female CEOs differ in how they reached the top? *The Leadership Quarterly* 25(2), 245–266. Available at: http://dx.doi.org/10.1016/j.leaqua.2013.08.005 (accessed 6 March 2017).

Gilligan, C., Spencer, R., Weinberg, M.K. and Bertsch, T. (2003) On the listening guide: a voice-centered relational model. In: Camic, P.M., Rhodes, J.E. and Yardley, L. (eds) *Qualitative Research in Psychology: Expanding Perspectives in Methodology and Design.* American Psychological Association Press, Washington, DC, pp. 157–172.

Haugen, M.S. and Brandth, B. (2015) When farm couples break up: gendered moralities, gossip and the fear of stigmatisation in rural communities. *Sociologia Ruralis* 55(2), 227–242. Available at: http://dx.doi.org/10.1111/soru.12065 (accessed 6 March 2017).

Heather, B., Skillen, L., Young, J. and Vladicka, T. (2005) Women's gendered identities and the restructuring of rural Alberta. *Sociologia Ruralis* 45(1–2), 86–97. Available at: http://dx.doi.org/10.1111/j.1467-9523.2005.00292.x (accessed 6 March 2017).

Hinds, B. (2011) *The Northern Ireland Economy: Women on the Edge? A Comprehensive Analysis of the Impacts of the Financial Crisis.* Women's Resource and Development Agency, Belfast.

Holvino, E. (2010) Intersections: the simultaneity of race, gender and class in organization studies. *Gender, Work & Organization* 17(3), 248–277. Available at: http://dx.doi.org/10.1111/j.1468-0432.2008.00400.x (accessed 6 March 2017).

Ibarra, H., Snook, S. and Guillen Ramo, L. (2010) Identity-based leader development. In: Nohria, N. and Khurana, R. (eds) *Handbook of Leadership Theory and Practice*. Harvard Business School Publishing, Boston, Massachusetts, pp. 657–678.

Karelaia, N. and Guillén, L. (2014) Me, a woman and a leader: positive social identity and identity conflict. *Organizational Behavior and Human Decision Processes* 125(2), 204–219. Available at: http://dx.doi.org/10.1016/j.obhdp.2014.08.002 (accessed 6 March 2017).

Ladkin, D. (2010) *Rethinking Leadership: A New Look at Old Leadership Questions*. Edward Elgar Publishing, Cheltenham, UK and Northampton, Massachusetts.

Lear, G. (2013) Ready for action and civic engagement: resilient Third Age women learners in rural Australia. *Australian Journal of Adult Learning* 53(3), 375–393.

Lewis, P. (2006) The quest for invisibility: female entrepreneurs and the masculine norm of entrepreneurship. *Gender, Work & Organization* 13(5), 453–469. Available at: http://dx.doi.org/10.1111/j.1468-0432.2006.00317.x (accessed 6 March 2017).

Little, J. (2002) Rural geography: rural gender identity and the performance of masculinity and femininity in the countryside. *Progress in Human Geography* 26(5), 665–670. Available at: http://dx.doi.org/10.1191/0309132502ph394pr (accessed 6 March 2017).

Little, J. (2007) Constructing nature in the performance of rural heterosexualities. *Environment and Planning D: Society and Space* 25(5), 851–866. Available at: http://dx.doi.org/10.1068/d2605 (accessed 6 March 2017).

Little, J. and Austin, P. (1996) Women and the rural idyll. *Journal of Rural Studies* 12(2), 101–111. Available at: http://dx.doi.org/10.1016/0743-0167(96)00004-6 (accessed 6 March 2017).

Little, J. and Panelli, R. (2003) Gender research in rural geography. *Gender, Place & Culture* 10(3), 281–289. Available at: http://dx.doi.org/10.1080/0966369032000114046 (accessed 6 March 2017).

Macaulay, T. and Laverty, B. (2007) *Baseline Study of Rural Women's Infrastructure in Northern Ireland*. Rural Community Network and the Rural Women's Sectoral Programme Consortium, Cookstown, Northern Ireland.

Mahon, M. (2007) New populations; shifting expectations: the changing experience of Irish rural space and place. *Journal of Rural Studies* 23(3), 345–356. Available at: http://dx.doi.org/10.1016/j.jrurstud.2007.01.006 (accessed 6 March 2017).

McVay, L.A. (2013) *Rural Women in Leadership: Positive Factors in Leadership Development*. CAB International, Wallingford, UK.

Mlinar, A. (2016) *Report on the Work of Gender Mainstreaming in the European Parliament*. Committee on Women's Rights and Gender Equality, Brussels.

Moorosi, P. (2013) Constructing a leader's identity through a leadership development programme: an intersectional analysis. *Educational Management Administration & Leadership* 42(6), 792–807. Available at: http://dx.doi.org/10.1177/1741143213494888 (accessed 6 March 2017).

Morley, L. (2013) The rules of the game: women and the leaderist turn in higher education. *Gender and Education* 25(1), 116–131. Available at: http://dx.doi.org/10.1080/09540253.2012.740888 (accessed 6 March 2017).

Muldoon, O., Schmid, K., Downes, C., Kremer, J. and Trew, K. (2015) *The Legacy of the Troubles: Experience of the Troubles, Mental Health and Social Attitudes*. The Queen's University of Belfast, Belfast.

Northern Ireland Assembly (2012) *Research and Information Briefing Paper: Empowerment of Rural Women in Northern Ireland – Impacts on Poverty, Development and Current Changes*. Northern Ireland Assembly, Belfast.

Northern Ireland Department of Agriculture and Rural Development (2006) *A Gender Equality Strategy 2006–2016*. DARD, Belfast.

Northern Ireland Women's European Platform (2013) *Submission to the Committee on the Convention for the Elimination of all forms of Discrimination against Women (CEDAW), Seventh Periodic Report of the United Kingdom of Great Britain and Northern Ireland: Shadow Report Northern Ireland Women's European Platform*. Northern Ireland Women's European Platform, Belfast.

Oliver, D.G., Serovich, J.M. and Mason, T.L. (2005) Constraints and opportunities with interview transcription: towards reflection in qualitative research. *Social Forces* 84(2), 1273–1289. Available at: http://dx.doi.org/10.1353/sof.2006.0023 (accessed 6 March 2017).

O'Toole, K. and Macgarvey, A. (2003) Rural women and local economic development in south-west Victoria. *Journal of Rural Studies* 19(2), 173–186. Available at: http://dx.doi.org/10.1016/S0743-0167(02)00072-4 (accessed 6 March 2017).

Pini, B. (2003) Feminist methodology and rural research: reflections on a study of an Australian agricultural organisation. *Sociologia Ruralis* 43(4), 418–433. Available at: http://dx.doi.org/10.1046/j.1467-9523.2003.00253.x (accessed 6 March 2017).

Pini, B. (2005) The third sex: women leaders in Australian agriculture. *Gender, Work & Organization* 12(1), 73–88. Available at: http://dx.doi.org/10.1111/j.1468-0432.2005.00263.x (accessed 6 March 2017).

Pini, B. and Leach, B. (eds) (2011) *Reshaping Gender and Class in Rural Spaces.* Ashgate, Aldershot, UK.

Richardson, A. and Loubier, C. (2008) Intersectionality and leadership. *International Journal of Leadership Studies* 3(2), 142–161. Available at: http://www.regent.edu/acad/global/publications/ijls/new/vol3iss2/IJLS_V3Is2_Richardson_Loubier.pdf (accessed 6 March 2017).

Ridgeway, C.L. (2009) Framed before we know it: how gender shapes social relations. *Gender & Society* 23(2), 145–160. Available at: http://dx.doi.org/10.1177/0891243208330313 (accessed 6 March 2017).

Ridgeway, C.L. and Smith-Lovin, L. (1999) The gender system and interaction. *Annual Review of Sociology* 25, 191–216.

Rosenberg, K.E. and Howard, J.A. (2008) Finding feminist sociology: a review essay. *Signs* 33(3), 675–696. Available at: http://dx.doi.org/10.1086/523711 (accessed 6 March 2017).

Saugeres, L. (2002) Of tractors and men: masculinity, technology and power in a French farming community. *Sociologia Ruralis* 42(2), 143–159. Available at: http://dx.doi.org/10.1111/1467-9523.00207 (accessed 6 March 2017).

Sheridan, A., McKenzie, F.H. and Still, L. (2011) Making visible the 'space of betweenness': understanding women's limited access to leadership in regional Australia. *Gender, Place & Culture* 18(6), 732–748. Available at: http://dx.doi.org/10.1080/0966369X.2011.617909 (accessed 6 March 2017).

Shortall, S. (2002) Gendered agricultural and rural restructuring: a case study of Northern Ireland. *Sociologia Ruralis* 42(2), 160–175. Available at: http://dx.doi.org/10.1111/1467-9523.00208 (accessed 6 March 2017).

Shortall, S. (2008) Are rural development programmes socially inclusive? Social inclusion, civic engagement, participation, and social capital: exploring the differences. *Journal of Rural Studies* 24(4), 450–457. Available at: http://dx.doi.org/10.1016/j.jrurstud.2008.01.001 (accessed 6 March 2017).

Shortall, S. (2014) Farming, identity and well-being: managing changing gender roles within Western European farm families. *Anthropological Notebooks* 20(3), 67–81.

Silvasti, T. (2003) Bending borders of gendered labour division on farms: the case of Finland. *Sociologia Ruralis* 43(2), 154–166. Available at: http://dx.doi.org/10.1111/1467-9523.00236 (accessed 6 March 2017).

Sinclair, A. (2005) *Doing Leadership Differently: Gender, Power and Sexuality in a Changing Business Culture.* Melbourne University Publishing, Carlton South, Australia.

Sireni, M. (2015) Reinventing rural femininities in the post-productivist Finnish countryside. *European Countryside* 7(1), 42–56. Available at: http://dx.doi.org/10.1515/euco-2015-0003 (accessed 6 March 2017).

Skerratt, S. (2011) A critical analysis of rural community leadership: towards systematised understanding and dialogue across leadership domains. *The Journal of Contemporary Issues in Business and Government* 17(1), 87–107.

Smith, D.G. and Reed, K. (2009) Appalachian women leaders: Products of culture and life events. *Journal of Leadership & Organizational Studies* 17(1), 87–99. Available at: http://dx.doi.org/10.1177/1548051809347107 (accessed 6 March 2017).

Toh, S.M. and Leonardelli, G.J. (2012) Cultural constraints on the emergence of women as leaders. *Journal of World Business* 47(4), 604–611. Available at: http://dx.doi.org/10.1016/j.jwb.2012.01.013 (accessed 6 March 2017).

Trauger, A., Sachs, C., Barbercheck, M., Kiernan, N.E., Brasier, K. and Findeis, J. (2008) Agricultural education: gender identity and knowledge exchange. *Journal of Rural Studies* 24(4), 32–439. Available at: http://dx.doi.org/10.1016/j.jrurstud.2008.03.007 (accessed 6 March 2017).

Trauger, A., Sachs, C., Barbercheck, M., Brasier, K. and Kiernan, N.E. (2010) 'Our market is our community': women farmers and civic agriculture in Pennsylvania, USA. *Agriculture and Human*

Values 27(1), 43–55. Available at: http://dx.doi.org/10.1007/s10460-008-9190-5 (accessed 6 March 2017).
UN Women (2015) *Progress of the World's Women 2015–2016 Summary: Transforming Economies, Realizing Rights.* UN Women, New York.
West, C. and Zimmerman, D.H. (1987) Doing gender. *Gender & Society* 1(2), 125–151. Available at: http://dx.doi.org/10.1177/0891243287001002002 (accessed 6 March 2017).
Wilbur, A. (2014) Back-to-the-house? Gender, domesticity and (dis)empowerment among back-to-the-land migrants in Northern Italy. *Journal of Rural Studies* 35, 1–10. Available at: http://dx.doi.org/10.1016/j.jrurstud.2014.03.005 (accessed 6 March 2017).

14 Gender Identities and Divorce among Farmers in Norway

M.S. Haugen[1],* and B. Brandth[2]

[1]*Centre for Rural Research, Trondheim, Norway;* [2]*NTNU, Trondheim, Norway*

Introduction

This chapter takes break-up among farmers as a starting point for studying gender identities. Since people construct their identities in relation to marriage, divorce requires a 'new sense of identity' (Wallerstein and Blakeslee, 1990; Giddens, 1991, p. 11). In using divorce to explore gender the chapter follows Morgan (1992), who suggests that masculinity may be studied in situations where it becomes problematic. Extending this approach to femininity as well, the chapter regards family break-up as a strategic point for studying challenges to gender identity.

Family break-up is a common phenomenon in most Western countries. In Norway the divorce rate has roughly doubled since the beginning of the 1970s (Statistics Norway, 2015). Although farmers are less likely to divorce than the population as a whole, their divorce rate is rising (Follo and Haugen, 2010). With a few exceptions (Price and Evans, 2006; Haugen and Brandth, 2015; Haugen *et al.*, 2015), farm divorce has received scant attention in social research, particularly as it relates to consequences for gender identity.

Farm couples' divorces are especially fraught with problems because family farming represents a way of life that is deeply rooted in a marital economy and historic patterns of inheritance and culture (Price and Evans, 2006; Jacobs, 2010). Heterosexual relationships, which provide the basis for the continuation of the family farm into the next generation, are an important organizing principle of agricultural societies. According to Price and Evans (2006), family farm survival has been dependent on women's compliance with a patriarchal ideology that demands they be 'good as gold'. Individual aspirations, status and social security benefits must be sacrificed for the sake of farm continuity (Contzen and Forney, 2017). This compliance may be challenged when male farmers marry women from non-farm backgrounds who are thought not to understand the way of life. If women rebel against the self-sacrifice that is expected of them or are perceived as deviating from whatever the family farm demands, they are seen as threatening its future (Price and Evans, 2006). In contrast, in their study of divorce among farm couples in Norway, Haugen *et al.* (2015) found that women demand less than they might be legally entitled to in order not to jeopardize the farm. To the extent that divorce represents an individualization process for women, it does not seem to displace their care for the family and their commitment to farm survival. Farms may not be viable if ex-wives

* Corresponding author e-mail: marit.haugen@bygdeforskning.no

withdraw money as well as labour. This leads to moral as well as material pressures on women, as it touches the heart of family farming (Jacobs, 2010).

This chapter explores what family dissolution among farmers can tell about stability and change in gender identities. Of particular interest are the processes by which divorce exposes salient, sometimes conflicting gender identities.

Analytical Framework: Changing Narratives of Gender Identity in Farming?

In the analysis, gender is understood as a social identity that can be disrupted, constituted and reconstituted in the context of divorce. According to Bruner (1990), people attribute meanings to themselves and their practices through narratives that culture makes available for them. As long as they act in accordance with tradition or accepted norms, this process proceeds rather unreflectively (Bruner, 1990). But when they step outside the normal range of action, as is the case when farm couples break up, they must interpret these new actions in relation to recognizable narratives. Goffman's (1974) rather similar concept 'frame' was developed in order to analyse how people make sense of practices and events. Actions are constituted and given meaning by particular frames (or narratives) and when the frame changes, so too do the situational actions of the participants. Different frames and practices are complexly intertwined and a particular practice may be traced back to several frames (or narratives). By applying this perspective, it is possible to observe the invisible norms that guide everyday practices and to see how relationships with others are connected to the ways people understand themselves. Using the concept of 'frames', one theoretically conceives identities as relational, variable and fluid for men and women farmers.

A much described cultural frame in family farming concerns the theme of male domination/female subordination. Within this framing it is the male farmer who owns the farm, participates in farming organizations and is subject to taxation. Women's position is distinctly different from that of men, as they undertake the domestic work in the farm household and childcare. Entering farming by marrying a farmer, they usually do not have ownership to the land, the farm or the farm home. Considering the interconnections of landownership, gender and power, Shortall (1999) has shown that established farming principles provide women with much less power than men because women do not enjoy equal access to and control over property. They do, however, contribute significantly to work on the farm, often invisibly and cast in the role of assistant.

Thus, traditional narratives of gender identities in farming have often been regarded as unambiguous. Farm women's identities are described as being constructed around wifehood and practices of reproduction and caring in order to ensure the continuation of the farm, local traditions and community life (Brandth, 2002; Little, 2002). Daily work revolves around enhancing other people's well-being and meeting their emotional needs. The material and moral dimensions of women's family responsibilities are interwoven. This configuration of work and family contradicts contemporary ideals of gender-equal intimate partnerships (Bryant and Pini, 2011).

One dominant narrative of agricultural masculinity is the ideal of the 'good farmer' (Burton, 2004; Burton and Wilson, 2006). This identity has been related to production-oriented goals, symbolized by 'crop yields, tidy farms/fields and good stockmanship (Riley, 2016, p. 96). Good farmers are also assessed by their experience-based competence, which is epitomized by inheriting the farm from their father (Saugeres, 2002). Failure to fulfil the moral norms of hard work and good farming practices may result in a loss of self-respect and status in the farming community. Researchers have been puzzled by the persistence of this ideal, given the scope of recent changes in agriculture. Heggem (2014a) explains the stability of the good farmer identity through male farmers' fears of losing the sense of self, meaning and socio-cultural rewards it confers amidst ongoing socio-economic transformation.

At the same time, the idealized masculine identity 'good farmer' carries negative consequences. The threat of status loss or a felt inability to act in line with the conventional narrative of farm masculinity may result in depression (Valkonen and Hänninen, 2012). Several studies have found that the hegemonic ideal undermines men's mental health (Parr *et al.*, 2004) and silences their emotional distress (Coen *et al.*, 2013). In other words, the power of masculinity in its conventional or 'monologic' form (Peter *et al.*, 2000) tends to downgrade men who do not fulfil its norms. Moreover, as the dominant narrative of masculine success also values family life, partnership and continuity on the land (Bartlett, 2006, p. 48), a family break-up may be deeply felt as an additional blow to farmers' masculine identity (Haugen and Brandth, 2015).

Focusing on the dynamic and situated character of gender, research has increasingly demonstrated variable ways of being women and men in agriculture, adding complexity and nuance to gender narratives within a rural and agricultural context. It has, for instance, raised doubts about the advantages of male privilege (Alston and Kent, 2008; Bryant and Garnham, 2015) and demonstrated farm women's increased status and more gender-equal positions as social change has tended to put women in the forefront (Riley, 2009; Brandth and Haugen, 2010; Brasier *et al.*, 2014; Sireni, 2015). Since gender identity is relationally constructed, masculinities are often influenced by changes in women's practices and positions and vice versa.

Over the years, Norwegian society has changed considerably in ways that are reflected in the practices of rural families (Brandth and Overrein, 2013; Brandth, 2016). One of the most marked shifts concerns the gender division of labour. The traditional differentiation between male and female tasks and responsibilities is regarded as unsustainable. The ideal of the mother-carer and the father-provider has practically disappeared and been replaced by a dual earner/dual carer model (Ellingsæter and Leira, 2006). This development may have affected farming and contributed to a narrative of gender equality in farm families, although rural societies and farm communities are sometimes seen as lagging behind in regard to gender equality (Heggem, 2014b).

Farm women's and men's narratives of the divorce situation demonstrate that tradition is being confronted by transformative forces as agricultural gender norms come up against gender equality ideals. A family break-up is interesting in this respect as it unsettles matters and requires the construction of new subject positions. When women and men work out their positions for themselves, they utilize cultural narratives, speaking and acting within the frames that are available to them. This chapter explores the frames within which divorced farmers locate themselves. Do they interpret their actions within 'the family farm discourse' of male domination/female subordination and complementary models of femininity and masculinity in farming? To what extent do they locate themselves within narratives of gender equality? How do these frames relate to each other? These questions are analysed by exploring four case histories, the accounts of family break-up given by two women and two men.

Methodology

As part of a larger study of relational arrangements in Norwegian farming, 25 farm women and men were interviewed in 2009–2011. The main objective of the study was to identify how farm couples organized their relationships and handled marriage and cohabitation agreements, including how they experienced the dissolution of intimate partnerships. Interviewees were located in a variety of ways, through agricultural extension services, professional networks and a newspaper notice. All interviews were with individuals. Most of them were conducted in their homes, tape-recorded and then fully transcribed before being analysed.

Seven women and four men in the sample had experienced a divorce or break-up. In this chapter, four cases – two divorced women and two divorced men – have been chosen for analysis. These cases are not selected as ideal types, nor can they be said to be fully representative with regard to background characteristics other than women's and men's different entry points into farming. In Norway, despite the legal shift that makes the right of farm succession gender neutral, most women come to the farm through marriage while men more often enter farming by succeeding their parents (Storstad and Rønning, 2014). The main reason for selection of cases is that these interviewees gave the most ample information about their situation before the divorce and offered rich details and facts concerning gender identities. Thus, the sample is not saturated, but the selected cases illustrate some of the pressures on couples' relationships in contemporary farming and they enable 'hidden' stories (Scott, 1990) about masculinity and femininity and their framings to be revealed. Gender was not an explicit subject in the interviews, so the interviewees' gendered narratives are a more accurate expression of their subjective identities than their answers to direct, conventionally framed questions about gender relations would have been.

One advantage with a small number of cases is that case studies produce context-dependent knowledge and can follow a story and present it more in detail than a study cutting across many cases (Flyvbjerg, 2006). Using rich descriptions, insight is gained into the processual aspects of the relations between the spouses and how gender identities play out during their marriage and after their break-up. Each case represents a unique story in which the divorce indicates how gender identities are being tested, but each case may also represent processes that are more widely recognized. More generally, in qualitative research the generalizations that can be made are about the nature of a process and not about a group or population (Gobo, 2006). The aim here is to produce knowledge about how gender identities may be recast in a context of divorce among family farmers in Norway.

Because only one of the spouses was interviewed, it is not possible to compare the stories of former wives with those of their former husbands to determine the extent of convergence in their understandings of the situation. In all the cases analysed here, as in the majority of break-ups among farm couples in the larger sample, the women initiated the break-up. Thus, women's stories might be interpreted as justifying why they wanted to leave and men's stories as justifying why they did not understand what led to the break-up.

Research Findings: Four Cases of Divorce and Identity Change

Never accepted as an equal partner in production

Petra met her husband Peter at an agricultural college. When they married, she moved to his farm with what she described as 'self-confidence, agricultural education, and a genuine interest in farming'. It was a relatively small, mixed farm with a dairy operation, and according to Petra it was an awful lot of work. Petra's plan was to work on the farm together with her spouse: 'I was not going to work in the local shop or something like that; I wanted to have my workplace on the farm.'

Petra had many ideas about how to create her own job on the farm, but she soon realized that conservative frames regarding women's position in farm families put constraints on her possibilities. She described these frames in the following way: 'When you come to a farm [marry a farmer] you just have to act in accordance with the rules and code of conduct which applies, especially when you have your in-laws living on the farm.' She claimed that, despite her education in agriculture, she was never accepted as an equal partner in the farm business by her husband or his parents. She participated in farm work, but had very little decision-making power as Peter preferred to discuss farm matters with his father, not with her. Petra had expected to share authority: 'It was we who should farm together, not him and his father.'

Deeply disappointed, she felt 'trapped' in a traditional farmer's wife position being responsible for domestic work and care work, in addition to assisting on the farm. Petra described her busy working days and emphasized how little recognition she received:

> You have kids, and they need help with schoolwork, you get them to bed, and you do the dishes and tidy up and make sure that he [her husband] can take a rest before he goes to the barn. You do a lot to facilitate him carrying out his job. But this does not count [either for him or for his parents]. . . . It is a bit 'Taliban', this view of women. It's a matter of feminism. We have still a long way to go. And the women themselves are keeping other women down.

Reflecting on the gap between women's subordinated position in farming and gender equality in society, Petra argued that her rural community is 'lagging behind' urban areas. 'I was interested in a husband to farm together with and who cared for his children and his family.' Instead, her husband preferred to participate in politics and leisure activities rather than being with his wife and children. Leisure time was a luxury she never could ask for: 'It would have been nice to do something else once a while, but they [men] have a greater need, which you have to consider.' Petra had ideas about how to diversify the farm by starting a farm-based enterprise in which she could play a more active part, but while she saw the opportunities 'all those around me saw only the limitations. . . . I believe that he didn't like the idea; he just wanted to do the milking as fast as possible, and then do his own things.'

After more than 15 years of marriage, she reached a point where she could not suppress her own interests any longer. Petra made a New Year's promise to herself: 'If it's not my turn soon on this farm, if there isn't room for me soon to do what is right for me, then I simply have nothing here to do.' When she told her husband, his reaction was 'to look around for someone else'. The knowledge that he was seeing another woman strengthened her decision to divorce. 'It is simply a nightmare, a farm divorce is like hell', Petra asserted, giving examples from other women's experiences as well as her own. She attributed farm women's inferior position to their not having shared ownership of the farm property. Petra encountered the attitude that 'she cannot claim anything from this farm' because it was passed down in her husband's family, and they refused to make her a partner.

During the break-up process, she had to find the strength to organize a new life for herself and the children:

> I understood that in order to stay strong for my kids and myself, I had to do it my own way. I had to keep calm and do it my way. Moreover, I was very determined that I was going to buy my own house. Because I would never ever be in the same situation again, that I didn't have control. Having children made me strong; I had something to fight for, not only for myself, but for my kids.

A year after the divorce, Petra bought an old house in the neighbourhood so that the children, who lived with her, could continue to have frequent contact with their father and spend time on the farm. She received 'transitional benefit' and took further education in order to find a new job. Petra managed to retain her identity as a caring and considerate mother. She was determined that the children should not suffer because of the divorce. She concluded that she had become independent, but not at the cost of the sustainability of the farm: 'I could have claimed more [in the divorce settlement], but then it would have ended with a compulsory sale of the farm, and that's not in anyone's interest.' Her choice to divorce was a step towards liberation and a new identity, although she sacrificed her financial interest in the farm to do so. Looking at the divorce in retrospect, she concluded that, although the process was difficult, 'you need to be a kind of a businesswoman, you need to be practical, to know what is needed to manage fairly well.'

Farm wife's emancipation

Astrid married a farmer at the age of 20. Her husband Arne, who was 10 years older, had just taken over the farm from his parents. She had already completed a course in home economics and one year of commercial training. She planned to continue her education after marriage and birth of her first child, 'but it was never time to do that'. Astrid wanted an off-farm job, but neither her husband nor his parents supported that idea. Her mother-in-law emphasized that she should be a farm wife and assist with the farm work, as the farm was relatively large-scale and had long traditions. Astrid explained: 'I felt very soon that I got a different life from what I had expected, than I had imagined.' She sacrificed her own individual aspirations and adapted to life on the farm. Her father-in-law tried to teach her how to do various farm tasks, such as managing the dairy,

> but I knew that if I learnt it all, it would be me and him [father-in-law] doing the dairy, as I eventually learnt that he [her husband] was more interested in participating in activities outside the farm than in staying home and farming. I decided 'I am not going to learn it all', because then it would be me who had to do it all.

Although Astrid participated in much of the farm work, she felt like an outsider. Her husband 'went to his parents to discuss farm business matters instead of me. I was told that it was no use to discuss these things with me, as I didn't have any knowledge of the issues.' In this way Astrid was framed as an assistant in a subordinate position in relation to her husband.

Unlike many farm women in the 1980s, Astrid was, however, registered with a taxable income based on her hours of farm work. As most pensions and social insurance rights depend on taxable income, acknowledging farm women's occupational position within the family farm business was an important issue for increasing gender equality in agriculture. But Astrid did not have actual economic independence, as the farm income was transferred to the farm's trading account which her husband administered. Her husband insisted that as the farm operator, he needed full overview and disposal of the accounts.

After many years of raising three children and participating in heavy farm work, Astrid developed serious health problems and could no longer do the farm work that was expected of her. She applied for a retraining course and got an off-farm job, but Arne retained access to her bank account and used her salary to subsidize the farm operation and pay for running the household. Astrid was annoyed, but accepted the situation, being constantly told that they had to reduce private expenditure. When Astrid decided to get a divorce after 25 years of marriage, she felt that she broke with the gendered expectations of a farm woman. She described the frame in the following way: 'When you are married on a farm, you should stay there. You should consider the farm, first and foremost, and there you should stay.'

Although Astrid had been aware that their relationship was ailing and they were drifting apart, her husband was surprised when she moved out:

> I do not think he realized that I really left him. And I didn't believe it would be permanent either, or think that I would manage. He could also change, I hoped. The children, at least the youngest, hoped it was only temporary and that I would come home again. They asked me to move back home because the house was dirty and cold, and no meals were prepared.

Only one of the children still lived at home, and he stayed with his father until Astrid had rented and furnished an apartment.

Like Petra, Astrid did not demand the amount of money she was entitled to in the divorce settlement, as her former husband would not have been able to pay her without selling the farm. 'I did try to find a solution where he could continue to operate the farm, but I knew I had to get something, otherwise I was down and out.' Buying her own home was important to her, and by the time of the interview she had started building a new house. After the divorce, both financially and in practical terms, Astrid demonstrated a new identity as an emancipated woman.

Summing up the women's stories

Petra and Astrid were both positioned as a farmer's wife, which is located within the dominant frame of family farming and characterized by tightly constricting boundaries. Both women experienced a lack of respect and recognition within their marriage. They wanted to be active partners in farm operations and in investments and other financial decisions, but their husbands preferred to discuss business matters with their parents. Petra wanted to become an entrepreneur by starting her own business on the farm, and Astrid wanted to have an off-farm job. Both, however, complied with the cultural expectation that they stay on the farm and assist with farm work in addition to doing the domestic work and raising the children. They were never recognized as equal partners on the farm, nor were their husbands involved in domestic work.

Both Petra and Astrid opposed the farm wife identity that they were framed within but struggled to live in accordance with it for many years. They were in favour of gender equality and expressed an identity in accordance with a gender equality frame, but as the farm had no room for it they became more alienated from and resistant to the farmer's wife identity. After the break-up both Petra and Astrid became economically independent, making decisions and taking responsibility for their own future. Experiencing a lack of gender equality within the marriage, they demonstrated the ability to establish a life of their own after the divorce.

Petra and Astrid did not, however, transgress the moral obligations of motherhood and care for others, including their ex-husband and the family farm. They continued to locate themselves within a narrative of care by balancing their own individual interests with the interests of others.

'Good farmer' identity in crises

When talking about their lives, the men in the study described themselves as hard-working farmers concerned with providing for their families. Geir, who had been married to Gunhild for 10 years, lived in a small community where farming and forestry had been major industries but today most residents work in a nearby town. He operated a large dairy farm that he had inherited from his family, while Gunhild was employed off the farm and did not participate in farm work. Geir expanded the farm operation by buying two additional farms and leasing neighbouring land. Describing his work, he said:

> I never saw work as a problem. Nothing stopped me, never! When I changed to loose housing I rebuilt the barn myself; in half a year's time, expanding it by 200 square metres. Dad and a relative helped out. And, before building started, I cut the timber in my own forest and sawed the boards myself . . . And when I bought the second farm and started working on it, I did eight additional hours in the barn.

Before their two children were born he even had an off-farm job, which he gave up because he had to do the farm work in the evenings. Geir recounted that his time-consuming work created tensions in his marriage:

> She was annoyed at me because I was too much away [working in the fields] in the evenings. But then I felt I didn't do enough on the farm. It was when the children slept or something that I could do some work, right. But I never managed . . . So, I quit my employment, I did, to become a full-time farmer while the children were small.

He attributed his 10- to 12-hour workday to the nature of farming: 'This is just the way it is, and I don't think she [his wife] was ever able to adapt to it, to having to make the farm the highest priority all the time.'

Particularly in the spring and during the harvest season, when he worked 18 hours a day, nothing else could be on his agenda. As one of the women we interviewed explained, 'Setting aside time for family and children has not been common in the countryside.' It was not part of the good farmer narrative.

Nevertheless, Geir described himself as a good father and modern husband who wanted to contribute to the indoor work. He admitted it was hard to do that alongside all his farm work, so most of the housework fell to his wife. Geir stated that he had never wanted a 'farmer's wife' who was 'number two' on the farm. Rather, he envisioned a wife who could sometimes join in the work and be a partner in decisions concerning the farm. In return, he would involve himself in the children and family matters. He described his ideal: 'She was to work off the farm, and I was going to be home with our child. That's what I wanted, and I looked forward to it! . . . I was going to be a home-working dad – and a farmer.' In addition to childcare, he cooked dinner and baked bread. Reflecting on the divorce, he said somewhat ironically:

> She was annoyed because I didn't hit her, I wasn't a bad dad – there was nothing wrong with me – so she couldn't blame the divorce on such things. Really, I think it was . . . well, she is from an urban area. She never adapted to life on a farm, simply put!

Following their divorce, Geir experienced significant distress and personal problems. He suffered from insomnia and could no longer manage all the work on the farm. At the same time, his parents were taken ill and eventually moved to a nursing home. During this difficult period, Geir told no one about his problems. He explained: 'It started with me not managing to take care of the animals. I skipped milking sometimes, the cleaning was inadequate, and some of the calves died.' Other animals had to be slaughtered. The veterinarian was the first person to notice that he was having problems and urged him to see a doctor. Geir consulted a physician but, unfortunately, she did not understand the seriousness of his condition. 'She made a mistake, but I am to blame a bit myself, too, because I am not the type of guy who tells a lot about myself and what is wrong. It does not show on my outside.'

Geir and his wife decided to share custody of the children, aged eight and nine, who alternated between their parents' homes every two weeks. In the divorce settlement his wife got the car and the monetary equivalent of what she had contributed to their common loan over the years. Geir kept the farm and continued to operate it, but he quit dairying, concentrated on the production of meat and grain, and reduced his plans to modernize the operation. He said: 'I am trying to fight my way back into a position where I am counted on again. I was in the forefront of farming before this unexpected blow, and now I am at the very back.' At the time of the interview he was not very optimistic that he would be able to advance very much and be a 'good farmer' again because he lacked both energy and money. With his wife and parents gone from the farm, his children were all that remained of his family, which made him locate himself within a 'good father' narrative. He longed for the weeks when the children were living with him, and he prioritized his time accordingly.

Identity trouble for 'partners in production'

Harald and his wife, Hilde, married when they were in their late 20s and they got divorced after 20 years of marriage. They operated a farm that his parents had initially leased from the church. Fourteen years after they had taken over the operation, they bought the farm together. In contrast to Geir's situation, Harald's wife was both a co-owner and an active, full-time farmer. This arrangement allowed for diversified production, including dairy cows, sheep and pigs. In addition, Harald had an extra job in the winter clearing snow off the roads, and Hilde was responsible for all the indoor work and the care of their four children. They renovated the buildings and modernized the equipment: 'We bought the farm in 1998, and since then we have installed a muck tank, and then I rebuilt the old silo, and then a new milking parlour, and all this required loans.' Harald was active in the farmers' union, politics, church work, the dairy cooperative and the swine producer cooperative. He represented the farm in producer organizations and in public.

Harald praised his wife for being a good farmer, and she received a prize for her work with the pigs. He also commended his wife for her domestic work: 'I value indoor work just as much [as outdoor]! When we used hired labour it was extra work for her . . . On a farm it is a piece of work to cook, you know.' These statements indicate that he relates to a gender equality frame as well as a frame of complementary work as he himself never took part in the housework.

Harald and Hilde shared loans and income equally between them. Sharing income equally was important to Hilde, and Harald hinted that she was a 'redstocking' – that is, a feminist. They were each responsible for separate farm operations; she was primarily responsible for the sheep and the pigs, while he was primarily responsible for the dairy cows. Yet Harald appreciated the extent to which they worked side by side. Thinking back, he said:

> I was fine when we lived together. I felt terrific! We worked together, and that was nice. She learned how to drive . . . I taught her to drive the tractor, and she joined in the haymaking. We hired a babysitter for her to be able to join . . . and she drove the tractor. And we brought coffee and waffles, and sat there in the hay – very nice. These are good memories from the time we worked together.

After sending their four children off to school, she would often come to the barn to help him milk. She also shared his interest in hunting and joined his hunting party, which Harald had very mixed feelings about:

> I felt that it became . . . I didn't refuse it, but it is clear . . . we discussed it once: what if I joined the women's club [sewing circle] you go to every month? That would be strange, wouldn't it? Likewise, perhaps we men have a need to be men by ourselves once in a while and go on men's trips and such.

During the last years of their marriage, Hilde stopped coming to work in the barn in the mornings and instead took a part-time job off the farm. In retrospect Harald suspected this shift was part of a plan to divorce him. To Harald, her wanting a divorce after 20 years of marriage 'came out of the blue'. He never thought he would get divorced and he had no good explanation for why it happened. The divorce triggered mental health problems. He tried to stay tough and carry on, but he lost motivation:

> I became depressed . . . I forced myself to get up . . . forced, forced! I went to see the doctor; he described antidepressants and talk therapy. My brothers and sisters became very worried. They feared that I was suicidal. I didn't say I was, but this was what they feared . . . And I did think a lot about it; [I] had the plans ready.

Harald could not manage all the work by himself; lambs and piglets died, and his emotional problems reduced his capacity to take care of the farm. 'It came to nothing because I couldn't get up at night and watch the lambing. The veterinarian said: "Harald, you have to quit this, because you can't take care of it." . . . So I quit the pigs and the sheep, and it was a relief.' The loss of his wife's labour, coupled with the loss of his own working capacity due to his depression, made it necessary to reduce the farm operation.

The eldest child had moved from the parental home before the divorce, the second child stayed with Harald, and the two youngest moved with their mother but often came to stay with their father. After his divorce, Harald's children became very important to him.

Summing up the men's stories

Despite the differences in their situations, Geir and Harald both relate to two main frames: 'good farmer' and 'gender-equal'/ 'women-friendly' models of masculinity.

There are also indications of a third frame in their testimonies, the 'male domination/female subordination' frame: the imperatives of the farm economy continue to privilege men and produce gender inequality on farms despite a general egalitarian setting.

We see the good farmer narrative in Geir's explanation of his divorce: his wife could not adjust to or understand why he worked so much, as he lived up to the good farmer morality. After the divorce both men developed emotional problems, which were expressed in lack of energy and difficulties in managing the farm work. They sought to re-establish their failed identity as good farmers, but both had to make substantial adjustments. Although they kept the farm and continued to operate it, they had to reduce production substantially and bring further investments to a halt.

The ways Geir and Harald located themselves within a gender equality frame differed to some degree. Both wanted to present themselves as modern men. Geir stressed an identity as an actively participating father of small children who shared childcare and housework, but he unrealistically planned to care for the children at home during the day and at the same time be a full-time, hard-working farmer. In practice, he was not able to fulfil both of his ideals and the 'good farmer' took precedence over the 'modern father'. Harald expressed gender-equal attitudes by praising his former wife's work, which fulfilled the ideal of a good farmer, and by accepting the fairness of her claim for an equal share in the divorce settlement because they had been equal partners. He followed her feminist orientation to some extent, but he was ambivalent, particularly as she identified with men's areas of activity and tended to be a better farmer than he was. Harald did not frame himself as a participant in domestic and care work. He assumed authority in the public realm, while she was responsible for the private.

Both men framed themselves as good fathers and husbands trying their best to provide for their families. They expressed sorrow and grief at having been divorced, and they were especially keen to keep a strong relationship with their children. The identity they re-established after dissolution of their marriage emphasized fatherhood.

Conclusion

This chapter has used Goffman's concept of 'frames' as an analytical tool to explore how divorce challenges gender identities in family farming. With rising divorce rates, identity narratives were expected to become somewhat altered. The four cases presented indicate the existence of several frames that is made culturally available for men and women farmers. Particularly the women's testimonies show persistence of patriarchy: male privilege and female subordination. The division of labour remains unequal. Women receive little recognition for their labour on farms, and property ownership often remains unequal with wives accessing farms through marriage. Husbands appear able to dominate financial decision making and control. Women's demands for independence, autonomy or simply equality were met with incomprehension, interpreted as infringing femininity and threatening the farm family. One reason for this may be that the farm's viability rests in part on the wife's labour.

Comparing women and men, the chapter finds that this traditional narrative of family farming means trouble for both sets of gender identities, but in different ways. The women oppose being located within the family farm-wife narrative. They object especially to the lack of recognition as equal partners and dual earners and the lack of opportunity to follow their own interests and aspirations. While they never question being framed as 'good mothers', they criticize the men's lack of involvement in household and care work. Women's decision to not take all due to them financially upon divorce indicates the importance placed on their maintenance of 'good' relationships (see Haugen *et al.*, 2015). It also reflects the material situation that the farms are not viable if not part of a family economy.

Men's stories show that their identities are located within the hegemonic frame of 'good farmers' and the men succeeded as 'good farmers' when they were married. The stories of the two men show that they worked hard, expanding and modernizing the farm. They thrived with a wife and children, living up to the ideology of family farming. Their identity troubles start when they no longer can fulfil this ideal. The women are most vulnerable in the marriage, while the men are vulnerable after the divorce when their failure to live up to the good farmer frame results in mental health problems. After the divorce the women gain economic independence through employment and being able to control their own lives strengthens their identity as modern, emancipated women. Men, on the other hand, lose women's services and taken-for-granted contribution to family life on which their own identity as good farmers depend. Both women and men depart from the family farm narrative: the women voluntarily and the men because they have to.

The gender equality frame is important for the women's new identities after the divorce, especially in terms of being economically independent. The narrative of women-as-carers remains salient for them, however. The men seek to re-establish their identity as good farmers, but in a modified way. After the divorce their identities as fathers assume greater importance. This may be interpreted as clinging on to the farm family frame, but it may also demonstrate a certain location within a gender equality narrative. Taken together, these four cases indicate that the patriarchal family farm narrative and gender equality narrative are both active within the contemporary culture of agriculture, but that they may be difficult to reconcile in terms of identity.

Acknowledgements

This article is part of the study 'Love on the farm: a risky business?' funded by the Research Council of Norway and the Agricultural Agreement Research Fund. The authors would like to thank Gro Follo for her participation in the project and Susie Jacobs for her constructive comments to an earlier version of this chapter.

References

Alston, M. and Kent, J. (2008) The big dry: the link between rural masculinities and poor health outcomes for farming men. *Journal of Sociology* 44(2), 133–147. Available at: http://dx.doi.org/10.1177/1440783308089166 (accessed 6 March 2017).

Bartlett, P.F. (2006) Three visions of masculine success on American farms. In: Campbell, H., Bell, M.M. and Finney, M. (eds) *Country Boys: Masculinity and Rural Life*. Pennsylvania State University Press, University Park, Pennsylvania, pp. 47–66.

Brandth, B. (2002) Gender identity in European family farming: a literature review. *Sociologia Ruralis* 42(3), 181–200. Available at: http://dx.doi.org/10.1111/1467-9523.00210 (accessed 6 March 2017).

Brandth, B. (2016) Rural masculinities and fathering practices. *Gender, Place & Culture* 23(3), 435–450. Available at: http://dx.doi.org/10.1080/0966369X.2015.1013454 (accessed 6 March 2017).

Brandth, B. and Haugen, M.S. (2010) Doing farm tourism – the intertwining practices of gender and work. *Signs* 35(2), 425–446. Available at: http://dx.doi.org/10.1086/605480 (accessed 6 March 2017).

Brandth, B. and Overrein, G. (2013) Resourcing children in a changing rural context: fathering and farm succession in two generations of farmers. *Sociologia Ruralis* 53(1), 95–111. Available at: http://dx.doi.org/10.1111/soru.12003 (accessed 6 March 2017).

Brasier, K.J., Sachs, C.E., Kiernan, N.E., Trauger, A. and Barbercheck, M.E. (2014) Capturing the multiple and shifting identities of farm women in the Northeastern United States. *Rural Sociology* 79(3), 283–309. Available at: http://dx.doi.org/10.1111/ruso.12040 (accessed 6 March 2017).

Bruner, J.S. (1990) *Acts of Meaning.* Harvard University Press, Cambridge, Massachusetts.

Bryant, L. and Garnham, B. (2015) The fallen hero: masculinity, shame and farmer suicide in Australia. *Gender, Place & Culture* 22(1), 67–82. Available at: http://dx.doi.org/10.1080/0966369X.2013.855628 (accessed 6 March 2017).

Bryant, L. and Pini, B. (2011) *Gender and Rurality.* Routledge, London.

Burton, R.J.F. (2004) Seeing through the 'good farmers' eyes: towards developing an understanding of the social symbolic value of 'productivist' behavior. *Sociologia Ruralis* 44(2), 195–215. Available at: http://dx.doi.org/10.1111/j.1467-9523.2004.00270.x (accessed 6 March 2017).

Burton, R.J.F. and Wilson, G. (2006) Injecting social psychology theory into conceptualisations of agricultural agency: towards a 'post-productivist' farmer self-identity. *Journal of Rural Studies* 22(1), 95–115. Available at: http://dx.doi.org/10.1016/j.jrurstud.2005.07.004 (accessed 6 March 2017).

Coen, S.E., Oliffe, J.L., Johnson, J.L. and Kelly, M.T. (2013) Looking for Mr. PG: masculinities and men's depression in a northern resource-based Canadian community. *Health & Place* 21, 94–101. Available at: http://dx.doi.org/10.1016/j.healthplace.2013.01.011 (accessed 6 March 2017).

Contzen, S. and Forney, J. (2017) Family farming and gendered division of labour on the move: a typology of farming-family configurations. *Agriculture and Human Values* 34(1), 27–40. Available at: http://dx.doi.org/10.1007/s10460-016-9687-2 (accessed 6 March 2017).

Ellingsæter, A.L. and Leira, A. (2006) *Politicizing Parenthood in Scandinavia.* Policy Press, Bristol, UK.

Flyvbjerg, B. (2006) Five misunderstandings about case-study research. *Qualitative Inquiry* 12(2), 219–245. Available at: http://dx.doi.org/10.1177/1077800405284363 (accessed 6 March 2017).

Follo, G. and Haugen, M.S. (2010) *Sivil status i landbruket. Ekteskap, samboerskap og skilsmisser. Report 3/2010.* Centre for Rural Research, Trondheim, Norway.

Giddens, A. (1991) *Modernity and Self-Identity. Self and Society in the Late Modern Age.* Polity Press, Cambridge, UK.

Gobo, G. (2006) Sampling, representativeness and generalizability. In: Seale, C., Gobo, G., Gubrium, J.F. and Silverman, D. (eds) *Qualitative Research Practice.* SAGE Publications Ltd, London, pp. 405–426.

Goffman, E. (1974) *Frame Analysis: An Essay on the Organization of Experience.* Harvard University Press, Cambridge, Massachusetts.

Haugen, M.S. and Brandth, B. (2015) When farm couples break up: gendered moralities, gossip, and the fear of stigmatisation in rural communities. *Sociologia Ruralis* 55(2), 227–242. Available at: http://dx.doi.org/10.1111/soru.12065 (accessed 6 March 2017).

Haugen, M.S., Brandth, B. and Follo, G. (2015) Farm, family, and myself: farm women dealing with family break-up. *Gender, Place & Culture* 22(1), 37–49. Available at: http://dx.doi.org/10.1080/0966369X.2013.855708 (accessed 6 March 2017).

Heggem, R. (2014a) Diversification and re-feminisation of Norwegian farm properties. *Sociologia Ruralis* 54(4), 399–415. Available at: http://dx.doi.org/10.1111/soru.12044 (accessed 6 March 2017).

Heggem, R. (2014b) Exclusion and inclusion of women in Norwegian agriculture: exploring different outcomes of the 'tractor gene'. *Journal of Rural Studies* 34, 263–271. Available at: http://dx.doi.org/10.1016/j.jrurstud.2014.03.002 (accessed 6 March 2017).

Jacobs, S. (2010) *Gender and Agrarian Reforms.* Routledge, New York and London.

Little, J. (2002) Rural geography: rural gender identity and the performance of masculinity and femininity in the countryside. *Progress in Human Geography* 26(5), 665–670. Available at: http://dx.doi.org/10.1191/0309132502ph394pr (accessed 6 March 2017).

Morgan, D.H.J. (1992) *Discovering Men.* Routledge, London.

Parr, H., Philo, C. and Burns, N. (2004) Social geographies of rural mental health: experiencing inclusions and exclusions. *Transactions of the Institute of British Geographers* 29(4), 401–419. Available at: http://dx.doi.org/10.1111/j.0020-2754.2004.00138.x (accessed 6 March 2017).

Peter, G., Bell, M., Jarnagin, S. and Bauer, D. (2000) Coming back across the fence: masculinity and the transition to sustainable agriculture. *Rural Sociology* 65(2), 215–234. Available at: http://dx.doi.org/10.1111/j.1549-0831.2000.tb00026.x (accessed 6 March 2017).

Price, L. and Evans, N. (2006) From 'as good as gold' to 'gold diggers': farming women and the survival of British family farming. *Sociologia Ruralis* 46(4), 280–298. Available at: http://dx.doi.org/10.1111/j.1467-9523.2006.00418.x (accessed 6 March 2017).

Riley, M. (2009) Bringing the 'invisible farmer' into sharper focus: gender relations and agricultural practices in the Peak District (UK). *Gender, Place & Culture* 16(6), 665–682. Available at: http://dx.doi.org/10.1080/09663690903279138 (accessed 6 March 2017).

Riley, M. (2016) Still being the 'good farmer': (non-)retirement and the preservation of farming identities in older age. *Sociologia Ruralis* 56(1), 96–115. Available at: http://dx.doi.org/10.1111/soru.12063 (accessed 6 March 2017).

Saugeres, L. (2002) The cultural representation of the farming landscape: masculinity, power and nature. *Journal of Rural Studies* 18(4), 143–159. Available at: http://dx.doi.org/10.1016/S0743-0167(02)00010-4 (accessed 6 March 2017).

Scott, J.C. (1990) *Domination and the Arts of Resistance: Hidden Transcripts.* Yale University Press, New Haven, Connecticut.

Shortall, S. (1999) *Women and Farming: Property and Power.* Macmillan Press, London.

Sireni, M. (2015) Reinventing rural femininities in the post-productivist Finnish countryside. *European Countryside* 7(1), 42–56. Available at: http://dx.doi.org/10.1515/euco-2015-0003 (accessed 6 March 2017).

Statistics Norway (2015) Marriages and Divorces, 2015. Available at: https://www.ssb.no/en/ekteskap/ (accessed 15 February 2017).

Storstad, O. and Rønning, L. (2014) *Trender i norsk landbruk 2014. Report 6/14.* Centre for Rural Research, Trondheim, Norway.

Valkonen, J. and Hänninen, V. (2012) Narratives of masculinity and depression. *Men and Masculinities* 16(2), 160–180. Available at: http://dx.doi.org/10.1177/1097184X12464377 (accessed 6 March 2017).

Wallerstein, J. and Blakeslee, S. (1990) *Second Chances: Men, Women and Children a Decade after Divorce.* Ticknor & Fields, New York.

15 Merging Masculinities: Exploring Intersecting Masculine Identities on Family Farms

M. Riley* and H. Sangster

University of Liverpool, Liverpool, UK

Introduction

Following on from the feminist-inspired studies that have considered the role(s) and gender(ed) identities of women on farms, work on farming men and masculinities has recently started to proliferate (Brandth and Haugen, 2005a; Brandth, 2006; Riley, 2016a).[1] Such work has noted the continued prevalence of hegemonic masculinities (after Connell, 1995), which celebrate the importance of farm men as 'stoic', hard-working, hard-bodied and lacking in emotion, or at least a willingness to articulate this emotion (Bryant and Garnham, 2015). Despite the seemingly continued entrenchment of these hegemonic masculinities there is now a growing discussion of the multiple contexts in which these may be subject to change, contestation and reworking. At one level, there has been a consideration of how structural changes and economic downturns have challenged farming men's position as 'breadwinner' (Filteau, 2015) and how the necessity to diversify away from a sole focus on agricultural production has been accompanied by different gender performances (Brandth and Haugen, 2010). Allied to this, others have explored how changes in agricultural technologies have negated much of the need for hard toil and physically demanding work, with authors such as Brandth and Haugen (2005b) and Riley (2009a) examining how the operation of farm machinery has become reworked as an embodied performance of 'men's work' and men (re)positioned not just as 'workers' but as farm 'managers' (Pini, 2008). Alongside this, recent work has begun to unpick the temporal dynamicity of farming masculinities, reflecting on how changing societal norms impact (or not) farming men across different generations (Brandth, 2016) and across the farming life course (Riley, 2011, 2012, 2016a).

This chapter takes forward the call within the masculinities literature to develop a more relational approach to masculinity (see Hopkins and Noble, 2009) and to give fuller recognition of the contextual, fluid and interpersonal ways in which masculinities are achieved and performed. Specifically, the chapter draws on a consideration of family farms in the Peak District (UK) to consider the multiple ways that masculinities intersect between different generations of farming men on the same farm. The chapter thus aims to expand existing understandings of how masculinities are (re)configured in specific contexts.

* Corresponding author e-mail: mark.riley@liverpool.ac.uk

Analytical Framework: Conceptualizing Farming Masculinities

(Re)considering hegemonic masculinity

Connell's (1995) ideas of hegemonic masculinity have provided a key touchstone in theorizing on gender and are useful for approaching the discussion of farming men. Important within this discussion of hegemonic masculinity is the recognition that there may be *multiple* masculinities in different geographical and temporal contexts (Kimmel, 1995). The performance of masculinity, this research suggests, is developed in particular social and cultural contexts which shape how individuals 'do gender' (after West and Zimmerman, 1987). As Brandth (2016) argues, far from being pre-given, there is an iterative engagement with the structural conditions in particular contexts in which individuals consider the gender discourses, practices and performances available to them. As feminist scholarship has highlighted, gendered subject positions are seen to result from dominant discourses and gender 'is understood as a discursive process by which people and practices become understood as "masculine" or "feminine"' (Bryant and Garnham, 2015, p. 70). Research on masculinities within farming has noted the importance of patrilineal inheritance and the associated patriarchal relations in which 'keeping the name on the land' becomes a central aspect of farming subjectivities (Riley, 2010a). Farming men are thus commonly positioned as 'breadwinner',[2] with heterosexuality – and the expectation that marriage will be followed by the birth of a farming (grand)son – becoming normalized within the system (Little, 2003). Central, therefore, to this chapter is giving attention to relation(s) between generations, exploring both how specific gender norms might be reproduced through the often referred to processes of socialization (e.g. Riley, 2009b) or, perhaps, challenged and reworked over time and across generations.

Connell and Messerschmidt (2005) draw the distinction between the term 'hegemonic masculinity' – which places emphasis on hegemony as a political mechanism that results in the structural dominance of men over women – and masculinities which are 'socially dominant' within particular contexts. As Beasley (2008) suggests, their conflation has been one of the 'slippages' within the discussion of masculinity and they, instead, note that socially dominant masculinities are those which are celebrated, powerful and often commonplace in a given context. While it is recognized in this reformulation that hierarchies may exist between different masculinities, not *all* socially dominant masculinities are necessarily hegemonic because, as Filteau (2014, p. 397) argues: 'men who legitimate an unequal gender order may only represent a minority of men who are not the most socially celebrated in a local, regional, or global context'. That is, 'dominant forms of masculinity are not analogous to hegemonic forms of masculinity, because the former may not always legitimate men's power over women' (Messerschmidt, 2010, p. 23). Important to this discussion is the observation that non-hegemonic masculinities may be dominant in a particular context and, more importantly, these may serve to challenge the previously dominant forms of hegemonic masculinity (Connell and Messerschmidt, 2005). Thus, as Filteau (2014) notes, drawing distinction between dominant non-hegemonic masculinity and hegemonic masculinity is crucial to understanding gender relations and gender dominance within a particular context. Implicit within this is a hierarchical relationship in which Messerschmidt (2012) identified four distinct masculinities and femininities: dominant, dominating, subordinate and equality. 'Dominating' masculinities refer to those which 'involve commanding and controlling specific interactions and exercising power and control over people and events – "calling the shots" and "running the show"' (Messerschmidt, 2012, p. 38). 'Subordinate' masculinities, by contrast, refer to those 'situationally constructed as less than or aberrant and deviant to hegemonic masculinity [and] dominant/dominating masculinities' (Messerschmidt, 2012, p. 39).

Messerschmidt goes on to suggest that, depending on specific context, this subordination may be conceptualized, among many others, in relation to class, race, age and bodily display. A secondary element of subordination may occur, Messerschmidt (2012, p. 39) argues, when there is an 'incongruence within the sex–gender–heterosexuality structure', such as when individuals are perceived as male but perform practices defined as feminine, such as being physically weak, displaying attraction to the same sex, practising celibacy or being shy. 'Equality' is the authentication of an equal relationship between men and women and, Messerschmidt (2012, p. 40) argues, this does not 'assume a normal and natural relationship to sex and sexuality and usually they are not constructed as naturally complementary'. Crucial to these processes are the issues of spatiality and the role of scale, and Connell and Messerschmidt (2005) point to three levels of analysis:

> (1) *Local*: constructed in the arenas of face-to-face interaction of families, organizations, and immediate communities, as typically found in ethnographic and life-history research; (2) *Regional*: constructed at the level of the culture or the nation-state, as typically found in discursive, political, and demographic research; and (3) *Global*: constructed in transnational arenas such as world politics and transnational business and media, as studied in the emerging research on masculinities and globalization.

Sherman (2009, 2011) offers a useful explication of this framework in her study of a town in California during a period of economic decline. Here she observed how broad-scale unemployment led to a disjuncture as global and regional-level masculinity – associated with forestry – was no longer available to men at the local level. As a result, men in this area began to construct new forms of non-hegemonic, but locally dominant, masculinity. Changes were seen in the rules of what is considered legitimate masculine practice in which there was a refocusing of 'masculine identity away from breadwinning and segregated spheres . . . [toward] sharing power, child care, and household chores with women' (Sherman 2009, p. 161). Sherman (2009) contrasts this more 'flexible' masculinity with the 'rigid' masculinity which emphasizes distinct male and female spheres and limited access, for women, to labour and money. This form of flexible masculinity, it was observed, was able to gain a socially dominant position at the local scale – this, despite being subordinate to the hegemonic masculinities which still predominate among forest workers at the regional and global scale.

Agricultural masculinities

In relation to agriculture, authors such as Filteau (2015, p. 6) have noted that structural changes – which might include technological innovations and changing economic climates – alter the conditions under which farming men accomplish 'being a man'. Research in this vein has shown that due to these structural changes, men see men undertaking tasks and performing roles that were previously seen as those for women and hence are, at the local level, feminized (Peterson, 2005). As Filteau (2015, p. 7) argues, however, such assertions fail to take account of 'men's ability to define masculinity at the local level in ways that contrast from definitions of masculinity at regional and global levels due to situational constraints'. Here, inspiration can be drawn from that work which considered how, in periods of change such as those associated with economic decline, men (re) construct masculinities at the local level. Examples of this have included farmers' ability to reconstruct masculinity in line with being business 'managers', rather than being tied to ideas of physical labour and hard work (Barlett and Conger, 2004), and deriving masculinity from other activities such as farm tourism (Brandth and Haugen, 2010). As Filteau (2015, p. 9) argues, what we see is a 'new nonhegemonic masculinity that ascends to a dominant position at the local level'. Important to note, here, is that although these local forms' performances may diverge from traditional forms of hegemonic masculinity, they may reinforce gender inequalities through the exclusion of women from these locally specific masculinized arenas (Ni Laoire, 2002).

Taken together, these examples are important for framing the key questions of this chapter in several ways. First, they illustrate the different scales at which masculinities may be defined and, importantly, reworked. Although there are common general references made to farming masculinities, these are often done without spatial or scalar reference. What this has done, for our previous understandings of farming masculinity, has led to a somewhat generalized treatment of masculinity and often very singular and particular definitions used. Second, the importance of space and place comes to the fore. While geographers have made substantial inroads into the discussion of place-based masculinities, there is important ground still to be covered on the importance of localized types of masculinity and particularly those associated specific contexts of farming and the farm. Third, and interrelated, the literature shows that there may be disparity between more regional and global forms of masculinity related to specific occupations such as agriculture and that the localized forms of socially dominant masculinity do not always reproduce, or even align with, those at other scales. Fourth, they illustrate that masculinities can be reconfigured over time. Although many of these studies do not look at change from the perspective of the individual, they show that there is potential for men to take up different gender subject positions over different time horizons. Such observations are important for the intergenerational approach taken in the current research – allowing a recognition that different farming masculinities may be taken on and reworked across the farming life course, and also allowing conceptual space to explore how these might be reworked (or not) among different farming generations. Fifth, although more implicit in the aforementioned literature, is that these masculinities are drawn not only in relation to women – as the classic discussions of hegemonic literature suggest (e.g. Connell, 1995) – but also in relation to other men in local, regional and global contexts. As previous scholars of rural gender have noted: 'gender identity is not fixed or given but is made and re-made constantly in response to countless different national and local sets of meanings and practices' (Little, 2006, p. 369).

Methodology: Exploring Gender Relations in Agriculture – A Case Study

The material drawn upon within this chapter comes from a wider study interested in farming culture and changing agricultural practices. The focus of the research was on changing farming practices and the adoption of environmentally sensitive farming practices (see Riley, 2006, 2008, 2016b), but in order to fully understand these issues it was necessary to develop a more holistic, 'farm life history' approach (see Riley, 2010). This approach involved repeat interviews (discussed in Riley and Harvey, 2007) on 64 farms in the Peak District,[3] UK and a period of participant observation on 20 of these farms. The initial study took place in 2001–2003 and these farmers were revisited in 2014 and 2015. The term *farm life history* is used to highlight the focus on not only individual biographies per se but also on the life history of the farm as a nexus where individual and collective histories intersect. Like Brandth (2016), the research took a multi-generational approach – interviewing, where possible, all generations involved in the farming operation. The material drawn on here comes from the most recent, second phase, of this study and is taken from 18 of these farms, where three generations were farming the same farm concurrently. For the purposes of the following discussion, the 'older generation' of farmers are those working with their sons and grandsons and who self-defined as still working full-time, and ranged in age from their early 60s to their late 80s.[4] The 'middle generation' is the one working with their father and son, and the 'younger generation' is those working with their fathers and grandfathers and self-defining as working full-time on the farm (as opposed to still being in education). These family farms were selected in order to explore not only the perspectives of individuals but also how these intersect with other generations, and both collective and individual interviews were conducted with the respective participants.[5] Interviews were transcribed verbatim and each transcript read several times and coded manually. Several overarching themes were identified using this thematic coding and are discussed in the following sections.

Research Findings

Taking things on – younger-generation masculinities

In starting the analysis from the interviews, extracts from across three generations of farmers are illustrative of a dominant emerging theme:

> My son has taken it on, and my grandson is in place to take it on.
>
> (3O)[6]

> I was pleased when [my son] said they wanted to take it on . . . it means that all our hard work will not be lost.
>
> (8M)

> I'm hoping I'll meet someone and we'll have a family to keep the family name on the land.
>
> (2Y)

As would perhaps be expected within the discussion of family farming, patrilineal inheritance is a central element and the above quotes illustrate the crucial importance of previous and future generations in the way that farming men think about their farm and themselves. Multiple examples can be seen in previous research of how such inheritance patterns are central to farming masculinities and, as the quotes of 3O and 8M suggest, maintaining the 'name on the land' becomes an important *raison d'être* of agriculture which, as farmer 2Y shows, feeds into a normalizing of heterosexuality and expectations regarding marriage and having children (cf. Little, 2003). Although, taken together, these opening quotes from three different farming generations suggest a level of continuity and stasis, the interviews with the younger generation of farmers showed that some change did take place. As Brandth (2016) notes, younger farmers are more likely to be influenced by wider cultural understandings of 'doing gender', particularly in relation to childcare and domestic relations. Among those interviewed, this generation contained more examples of wives working off-farm and more instances – relative to their fathers and grandfathers – of men undertaking household tasks and childcare. Important, however, was how these tasks and changing positions were narrated in relation to wider farm operations and trajectories. Two farmers commented on the different elements to this, the first pointing out: 'there's not the money in the job these days [my wife] has to bring in extra income' (13Y) and the second noting 'I have to do a lot more in the house now my wife works as well . . . because we've got dad and granddad here, we need the extra . . . in time she'll be able to give it up and stay at home' (8Y). Taken in isolation, farmer 13Y's quote suggests a softening of the rigid 'inside–outside' dichotomy observed as prevalent on many farms (see Brandth, 2002). When, however, the quotes are viewed in the context of the latter part of farmer 8Y's statement, it can be seen how they relate closely to more traditional forms of hegemonic masculinity. Off-farm employment is most often sought by farm women and thus often means little change to farming men's masculine performances. For farmer 8Y, while he accepted roles which might historically be seen as 'women's work', he was able, to a large extent, to avoid de-masculinization (Sherman, 2009) by making reference to the temporary nature of these arrangements. Here, this was done relationally to other men in the farming family. This showed it was, in part, a result of three generations remaining on the farm and the associated delayed retirement of his grandfather – a theme returned to later in the chapter – that he was temporarily performing this 'less masculine' task. Such observations have a twofold relevance to our wider understandings of farming masculinities. First, they show that the 'local' scale at which new socially acceptable forms of masculinity may be 'situationally constructed' (Messerschmidt, 2012) – which might contrast or challenge those prevalent at regional or (inter)national scales (Connell and Messerschmidt, 2005; Filteau, 2015) – may be as specific as the farm or farm-family scale. In cases such as these, household tasks are something observed only by other men in that specific farm setting, but this becomes acceptable because of how it ties in with their positions on the farm. Second, and interrelated, such evidence shows that performances of farming masculinities which diverge away from more hegemonic masculinity (or what Sherman, 2009 refers to as more 'flexible' masculinities) may be permissible, in this local context, because they are consciously recognized as transient and temporary positions.

For others, their discussion of increasing involvement in domestic tasks was framed in reference to growing participation of farm women in other farm-related tasks. Farmer 16Y suggested: 'my wife has the brains to do all the paperwork so I have to take over her work' and farmer 7Y noted 'we're expanding our herd and I need [my wife's] help feeding all the calves'. Although we might see a reworking of farming masculinities at this micro scale, in which there is a move away from the rigid position of not doing housework or domestic tasks, we see that more longstanding forms of farming masculinity remain at work. In both cases, the position of 'manager' was still prevalent, with both tasks undertaken by farm women positioned as subsidiary to their own. Moreover, the words of farmer 7Y show that the priority of expanding the herd and farm size serve to magnify his position as 'breadwinner' and hence compensate for, and over-write, his activities within the domestic space of the farm. In both cases we see the continued importance of hegemonic forms of masculinity in which the farmer, and the male farmer's role within this, takes precedence over that of women on the farm.

An area of change observed in previous research is that associated with the decline in physical labour and the potential receding of hegemonic farming masculinity associated with being strong and hard-bodied (Brandth, 2006). The relational quality, here between generations, was seen in the interview with farmer 1Y:

> I know great-granddad worked himself into an early grave . . . silly things really, out in all weather [. . .] he had pneumonia twice, but wouldn't stop going. Dad has been much the same, he's got rheumatism bad [. . .] I'm not going to kill myself like that . . . you've got to be sensible like that . . . it's a marathon, not a sprint. There is no need for us to be doing that today. We've no mortgage these days.

For this farmer, we see a movement away from stoicism associated with physical, often painful, labour. Instead, attention was given to looking after his body and a more 'sensible' form of masculinity which does not see the body put under the same pressure. While such a position might be viewed as an evolution away from the monolithic focus on hard, physical labour, we see how it becomes intertwined with wider cultures of the family farm. Here, the ability to take this more 'sensible', less physically demanding approach, as farmer 1Y's reference to 'giving us what we have' intimates, is only possible because of the success of his predecessors. The hard work of forebears in this case, which means not having to have a mortgage and also having substantial land and livestock assets, meant an ability to move away from such labours. Although evidence from across the interviews suggested a move from manual to mechanized field activities and the proliferation of 'labour-saving' technologies,[7] it was noted that 'hard work' (4Y) still played a significant role within farmers' self-definition. Again, this was written relationally to fathers and grandfathers, with reference to wanting to show 'I can do a hard day's work like everyone else has' (9Y), and while this might not be the same volume of physically intense work, it was compensated by similar levels of long hours and 'working from the crack of dawn until bedtime' (11Y) and a call to be the 'heroic' figure noted in previous considerations of men (Barlett and Conger, 2004). As farmer 1Y's reference to farm life being 'a marathon not a sprint' suggests, we see a substitution of physically demanding work being replaced by another form of stoicism and arguably hegemonic form of masculinity: that associated with a focus on farming longevity.

The nature of work and the relational masculinities on farms were seen in a more practical sense where reference was made to how three generations work together. As farmer 14Y explained:

> I'm working my backside off to be fair [. . .] dad can't do as much as he did, so I try to rush around doing that bit more . . . I don't want him to think he's not coping, so I take more on. . . . He [grandfather] has had heart-trouble . . . we didn't think dipping was good for him, so I said I'd get the licence for it . . . sort of an excuse really.

These examples put hard work back at the centre of the discussion and lean towards the depiction of heroic masculinities (cf. Bryant and Garnham, 2015). Interestingly, this worked not only at a practical level with younger farmers often compensating, through their increased efforts, for the reduced input of others, but also at an emotional level in which the masculinity among older farmer was buttressed by the actions of their sons and grandsons. The reference to 'not wanting him to think he's not coping' and using 'an excuse' to take over tasks showed how farming masculinities intersect and, through this working in tandem, younger farmers may seek to prop up the working masculinities of other generations.

'Earning the right'– managing the farm and maturing masculinities

What might be characterized as the 'middle' generation of farmers was the one where more relational masculinities were most clearly evident. At once, this group was involved in taking on a lead role in the farm business (where fathers were still present) and working out shared responsibilities with their sons. An important observation emerging from the interviews with this group was how, despite seemingly rigid forms of masculinity, there was a sense of continual revision, at the farm scale, of these masculinities. Three extracts help exemplify this point:

> I can see dad giving me a sideways glance when I say I'm getting the tea on or I'm in the house hanging a curtain rail or something [. . .] but I let him know enough to know that Jane [wife] is not blowing the milk cheque on fake tan and designer handbags.
>
> (18M)

> I don't thrash about like I used to . . . I think I've passed some of that work on to [my son], I think I've earned the right to do these [less manual] jobs.
>
> (12M)

> David [son] is a hell of a worker . . . he's built like the side of a house. I can't pretend to keep up with him, but I think I've got a bit more experience and think about things more . . . don't waste as much time as him.
>
> (6M)

The first extract highlights a similar scenario to that discussed in the previous section, in which their surveillance by other men on (or associated with) the farm shapes the individual's own position. While he reports on what may be seen as a socially less acceptable activity in the context of the farm and his father's eyes – those associated with the domestic sphere – farmer 18M regains a dominant masculine position relationally through reference to his wife. As Price and Evans (2006) note, there are very particular farming narratives on what activities are thought appropriate for farm women – the use of fake tan and the purchase of designer handbags not being one of them, farmer 18M's quote suggests – and the avoidance of divorce and splitting up of farms through divorce settlements is central to underpinning certain farming masculinities. Structurally, and at the regional and national scales, it is the exclusion of women from both farm decision making and farm finances which have been central to reinforcing farming masculinities. At the farm level, we see a reflection of this, but a slightly more flexible masculinity at work. The apparent control and surveillance of his wife's spending habits meant farmer 18M was able to maintain a socially dominant role at the farm scale, even when freely admitting to undertaking domestic tasks.

The references from farmers 12M and 6M show that ageing might be accompanied by a refining of masculinities. Rather than this being an example of older age serving to subordinate masculinities (cf. Messerschmidt, 2012), drawing links to their farming sons allowed them to retain an important position at the micro scale. For farmer 12M, it was not just this current manual labour which shaped his masculine identity, but also his past efforts – or 'thrashing about' – which had 'earned' him the right to do less today. Important for thinking more broadly about masculinity is the recognition that performances may be cumulative and that socially dominant masculinities apparent in particular contexts may be based on previous roles and performances. Important for our wider understanding of masculinity is that the process of 'situationally constructing' (Messerschmidt, 2012) socially dominant masculinities in particular contexts is temporally as well as spatially layered. It is not only current performances, these examples suggest, which serve to frame farming masculinities, but also longer term, past, involvements and actions.

Farmer 6M's reference to his son's capabilities chimes with the observation that conjoining different generations can serve to mask or play down the contribution (or lack of) for any one individual and this preserves that individual's sense of masculinity. He took this further, however, to challenge the dominance of this form of physical masculinity practised by his son, suggesting its limits. Here, reference was made to his own skill of being selective and thoughtful about which activities he undertook. Emphasis was placed instead on his experience and accumulated knowledge of how to undertake his farming activities without expending unnecessary energy. Scenarios similar to this took place on a number of farms and are important to how we think of different geographical contexts of masculinity, and particularly how we treat the 'local'. In previous research, this has focused primarily at the scale of particular geographically bounded communities – such as Filteau's (2015) discussion of an oilfield community in the USA. If farmer 6M's quotes are viewed within this frame, it can be seen that farming neighbours and others at the local scale would see a functioning farm and would not necessarily be party to the distribution of work within it – so the reduced contribution of any one generation would not be visible. Digging below this, however, reveals a finer scale at which masculinities are challenged within the farm. Moreover, it can be seen that different masculine characteristics may jostle for position at the farm scale. Not only are farmers such as farmer 6M able to bring their experience and decisive decision making (see Liepins, 2000) to the fore, they are able to use this in maintaining a dominant position in relation to younger generations on the farm.

'Working together' – maintaining masculinities in older age

The older generations of farmers – those typically over the age of 65 – showed several examples of linked masculinities. Previous work has noted how older age may act as a significant challenge towards masculinity, particularly in relation to the perceived loss of hard-bodied masculinity often associated with agriculture (Alston and Kent, 2008). Important to nearly all of those spoken to was a de-individualizing of the farm and farming activity. Here, references were made to 'we are going great guns . . . really motoring' (5O) and 'we're getting through lots of work [. . .] we do well together' (11O). Rather than offering an alternative conception of farming masculinity, as seen earlier for the middle generation of farmers in relation to their sons, the approach here was to narrate the farm, and farming activity, so that individual contributions were erased or left undefined. The collective 'we' meant that reference to the success of the farm was something that these older farmers could take credit for and which could be seen to reinforce masculine identities. For those who were more explicit about drawing out their specific contributions, this was again framed in relation to other men on the farm:

> They're adding to things . . . building on what we've done. They keep refining the herd that we've built up over many years.
> (17O)

> Sometimes [my grandson] gets a bit ahead of himself and thinks he's the big boss, and I have to remind him where all this comes from.
> (14O)

In both cases, their relative position on the farm was, in part, articulated through their past contribution. Rather than being a simple memory, however, these contributions are made active by their link to the farm's current operation. For farmer 17O, it was reference to the bloodlines and stock which, the tone implicitly suggests, is only possible today because of his previous contribution in the past.[8] Farmer 14O takes this further by revealing the power relations that exist between these. While he is no longer technically in charge of the farm (it had been 'signed over' to his son), he is quick to point out his contribution to the farm that his grandson now operates. As Messerschmidt (2012) argues, social and economic power are central to the reconstruction of masculinity at the local level and the examples here show how this may not be just direct financial control, but also that a 'dominating' masculinity might be born out of the role that they have played in the overall accumulation of this financial status. Unlike the challenge to masculinity which older age transitions – such as retirement – might bring as individuals leave the workplace, these farmers were able to avoid such challenges to masculinity by their continued presence, both physically on (or around) the farm as well as it being embodied in the overall capital (and associated facets such farm livestock) of the farm.

Perhaps most significant among the older generation of farmers was the performance of what might be seen as conventionally less manly (after Connell and Messerschmidt, 2005) activities and also more subsidiary tasks within farming. Common among the older generation was their role in looking after grandchildren and great-grandchildren, and two interesting perspectives were given on this:

> I like looking after [great-grandchildren], you've got to think of the future [. . .] telling them about farming life and the proper way of doing things.
> (13O)

> I never really had the time with our children, and [my son] hasn't now, he's doing so much [. . .] so I try to make sure I can give them that time.
> (15O)

The example of farmer 13O echoes some of the earlier sentiments, that older generations are important to the socialization of farm children (Riley, 2009b). Part of this process, other research has suggested, is the passing on of particular knowledge. For farmer 13O, and several other farmers, their concern was about more 'traditional', longer-standing, forms of knowledge relating to their specific, familial history on the farm, as well as specific techniques such as building dry stone walls or the handling of livestock. Such examples showed the relationship between masculine identities as well as being a masculine performance in itself – that is, demonstrating their own knowledge and understanding. Important within this was the link to their sons (or grandsons), who, both quotes above suggest, did not have the time to dedicate to these tasks. While it was observed earlier that younger farmers may undertake work which enables any decline in contribution of their fathers and grandfathers, we see in the quote of farmer 15O that there may be a reciprocal relationship here in which fathers and grandfathers take on some of the nurturing responsibilities. Any talk of these being less manly tasks was thus written out of the narratives of these farmers as they saw it as part of 'keeping the place running' (3O) and 'preparing them [grandchildren] well for farming' (12O). Both farmer 13O and farmer 15O show that there are different elements of farming masculinities that they wish to develop in younger farming generations: 'Whilst [my son] shows [my grandson] how to operate the machinery, I make sure that he can throw a wall up' (13O). Such examples have relevance for our wider understandings of farming masculinities. In addition to each farming generation subtly redrawing (or reinforcing) what are appropriate masculine behaviours, different generations may feed into what they feel the younger generation needs to 'be a man' – with the case here showing that the middle generation might provide more technical understandings and the older generation providing more traditional knowledge.

Conclusions

This chapter has given attention to the ways that farming masculinities are (re)created relationally among farming men. The observation that socially dominant masculinities are developed in intersection with other men has been well made in previous research, but the consideration of this within the context of the family farm offers several insights. Although we see that there may be generational differences in the wider cultural narratives of how to 'do gender' on which farming men may draw, that of 'breadwinner' and 'provider' remains central across generations. Keeping the family name on the land, and perpetuating the farm business, are central goals and serve to relationally construct, regardless of age or generation, what it is to be a farm man. While there is evidence from other research that patrilineal inheritance structures – so often underpinning the system of hegemonic farming masculinities – are being challenged, the evidence from those farmers considered here suggests that the role of farm women often remains secondary to that of the breadwinning farming man.

The chapter has seen the centrality of place and scale to the (re)constitution of farming masculinities. Developing on from the local–regional–global distinctions (after Connell and Messerschmidt, 2005) often drawn upon in masculinities research, the chapter has seen the importance of the micro scale of the farm. Most notably, it has been seen that when viewed at the local scale – such as neighbouring farming community – a farm containing two or three generations of farming men can be seen as functioning effectively and the masculine status of all those involved may, to these outside observers, remain unchanged or undiminished. When we move to the farm scale, we start to see a more diverse internal dynamic in which there may be discrepancies in contribution. Within this, processes such as de-individualizing (by older farmers) of the farm operations and the conscious compensating (for older farmers by younger farmers) mean that to the outside world, each generation can, again, maintain a coherent and unfractured farming masculinity.

Temporality, too, is an important aspect which this chapter contributes to the broader literature on masculinities. It has been seen that certain farming masculinities may lose their currency over time. So, for example, the idea of highly demanding physical labour, which leads to obvious bodily scars and injuries, is one which younger farmers are keen to move away from. Important to note is that 'hard work' does not cease to become valorized; it merely takes on another manifestation – such as working long hours, or having a longer working life overall. Further, such examples show that these masculinities are (re)worked temporally as well as spatially – with experiences of the previous generations utilized in how younger generations do gender. Alongside this, temporality is also significant, particularly in older generations, in terms of their past contributions. For middle and older generations, in having 'served their time' – by performing hard physical labour for example – they have 'earned' the right to take up alternative or modified masculine subject positions. Older generations of farmers are able, to some extent, to avoid older age leading to subordinate masculinities (cf. Messerschmidt, 2012) through both the memory of their past contribution per se as well as a discursive re-activating of this in the present, such that their contribution, and position as a farming man, remains relatively undisturbed.

Notes

[1]Indeed, Brandth (2016) suggests that studies of farming men may now outnumber those of farming women.

[2]Although see Kelly and Shortall (2002) on the role of women and the challenge to this 'breadwinner' status.

[3]This is an upland region of the UK. Farming here is characterized by a mixture of dairying, sheep and beef farms. For full details on the sample and how it was derived, see Riley (2008).

[4]As Riley (2016a) notes, being in receipt of state pension (65 years old in the UK) cannot be taken as a proxy for being retired, with many farmers continuing to work well beyond the age of 65.

[5]See Riley (2014) for a critical reflection on the use of such joint interview approaches. It should also be recognized that this sample – where different generations of men are working together – is one where traditional patrilineal succession is/has taken place. Future research could usefully explore those cases, albeit likely to be fewer in number, where farming women have taken sole responsibility for the farm.

[6]To maintain their anonymity, farmers are given numbers and a letter (Y = younger; M = middle; O = older) to represent their generational position on respective farms.

[7]Examples of this are numerous and might include lifting equipment (tractor loaders, telehandlers, etc.), sheep-handling equipment and quad bikes, to name just a few.

[8]See Riley (2011) for a fuller discussion of these themes of farmer ageing and livestock.

References

Alston, M. and Kent, J. (2008) The big dry: the link between rural masculinities and poor health outcomes for farming men. *Journal of Sociology* 44(2), 133–147. Available at: http://dx.doi.org/10.1177/1440783308089166 (accessed 6 March 2017).

Barlett, P.F. and Conger, K.J. (2004) Three visions of masculine success on American farms. *Men and Masculinities* 7(2), 205–227. Available at: http://dx.doi.org/10.1177/1097184X03257409 (accessed 6 March 2017).

Beasley, C. (2008) Rethinking hegemonic masculinity in a globalizing world. *Men and Masculinities* 11(1), 86–103. Available at: http://dx.doi.org/10.1177/1097184X08315102 (accessed 6 March 2017).

Brandth, B. (2002) Gender identity in European family farming: a literature review. *Sociologia Ruralis* 42(3), 181–200. Available at: http://dx.doi.org/10.1111/1467-9523.00210 (accessed 6 March 2017).

Brandth, B. (2006) Agricultural body-building: incorporations of gender, body and work. *Journal of Rural Studies* 22(1), 17–27. Available at: http://dx.doi.org/10.1016/j.jrurstud.2005.05.009 (accessed 6 March 2017).

Brandth, B. (2016) Rural masculinities and fathering practices. *Gender, Place & Culture* 23(3), 435–450. Available at: http://dx.doi.org/10.1080/0966369X.2015.1013454 (accessed 6 March 2017).

Brandth, B. and Haugen, M.S. (2005a) Doing rural masculinity – from logging to outfield tourism. *Journal of Gender Studies* 14(1), 13–22. Available at: http://dx.doi.org/10.1080/0958923042000331452 (accessed 6 March 2017).

Brandth, B. and Haugen, M.S. (2005b) Text, body, and tools: changing mediations of rural masculinity. *Men and Masculinities* 8(2), 148–163. Available at: http://dx.doi.org/10.1177/1097184X05277716 (accessed 6 March 2017).

Brandth, B. and Haugen, M.S. (2010) Doing farm tourism: the intertwining practices of gender and work. *Signs* 35(2), 425–446. Available at: http://dx.doi.org/10.1086/605480 (accessed 6 March 2017).

Bryant, L. and Garnham, B. (2015) The fallen hero: masculinity, shame and farmer suicide in Australia. *Gender, Place & Culture* 22(1), 67–82. Available at: http://dx.doi.org/10.1080/0966369X.2013.855628 (accessed 6 March 2017).

Connell, R. (1995) *Masculinities.* Allen and Unwin, Sydney, Australia.

Connell, R.W. and Messerschmidt, J.W. (2005) Hegemonic masculinity: rethinking the concept. *Gender & Society* 19(6), 829–859. Available at: http://dx.doi.org/10.1177/0891243205278639 (accessed 6 March 2017).

Filteau, M.R. (2014) Who are those guys? Constructing the oilfield's *new* dominant masculinity. *Men and Masculinities* 17(4), 396–416. Available at: http://dx.doi.org/10.1177/1097184X14544905 (accessed 6 March 2017).

Filteau, M.R. (2015) A localized masculine crisis: local men's subordination within the Marcellus Shale region's masculine structure. *Rural Sociology* 80(4), 431–455. Available at: http://dx.doi.org/10.1111/ruso.12072 (accessed 6 March 2017).

Hopkins, P. and Noble, G. (2009) Masculinities in place: situated identities, relations and intersectionality. *Social & Cultural Geography* 10(8), 811–819. Available at: http://dx.doi.org/10.1080/14649360903305817 (accessed 6 March 2017).

Kelly, R. and Shortall, S. (2002) 'Farmers' wives': women who are off-farm breadwinners and the implications for on-farm gender relations. *Journal of Sociology* 38(4), 327–343. Available at: http://dx.doi.org/10.1177/144078302128756714 (accessed 6 March 2017).

Kimmel, M. (1995) *Manhood in America.* Free Press, New York.

Liepins, R. (2000) Making men: the construction and representation of agriculture-based masculinities in Australia and New Zealand. *Rural Sociology* 65(4), 605–620. Available at: http://dx.doi.org/10.1111/j.1549-0831.2000.tb00046.x (accessed 6 March 2017).

Little, J. (2003) 'Riding the rural love train': heterosexuality and the rural community. *Sociologia Ruralis* 43(4), 401–417. Available at: http://dx.doi.org/10.1046/j.1467-9523.2003.00252.x (accessed 6 March 2017).

Little, J. (2006) Gender and sexuality in rural communities. In: Cloke, P., Marsden, T. and Mooney, P. (eds) *Handbook of Rural Studies.* SAGE Publications Ltd, London, pp. 365–378.

Messerschmidt, J.W. (2010) *Hegemonic Masculinities and Camouflaged Politics: Unmasking the Bush Dynasty and Its War against Iraq.* London, Routledge.

Messerschmidt, J.W. (2012) *Gender, Heterosexuality, and Youth Violence: The Struggle for Recognition.* Rowman & Littlefield, Lanham, Maryland.

Ni Laoire, C.N. (2002) Young farmers, masculinities and change in rural Ireland. *Irish Geography* 35(1), 16–27. Available at: http://dx.doi.org/10.1080/00750770209555790 (accessed 6 March 2017).

Peterson, S.V. (2005) How (the meaning of) gender matters in political economy. *New Political Economy* 10(4), 499–521. Available at: http://dx.doi.org/10.1080/13563460500344468 (accessed 6 March 2017).

Pini, B. (2008) *Masculinities and Management in Agricultural Organizations Worldwide.* Ashgate, Aldershot, UK.

Price, L. and Evans, N. (2006) From 'as good as gold' to 'gold diggers': farming women and the survival of British family farming. *Sociologia Ruralis* 46(4), 280–298. Available at: http://dx.doi.org/10.1111/j.1467-9523.2006.00418.x (accessed 6 March 2017).

Riley, M. (2006) Reconsidering conceptualizations of farmer conservation behaviour: the case of conserving hay meadows. *Journal of Rural Studies* 22(3), 337–353. Available at: http://dx.doi.org/10.1016/j.jrurstud.2005.10.005 (accessed 6 March 2017).

Riley, M. (2008) Experts in their fields: farmer–expert knowledges and environmentally friendly farming practices. *Environment and Planning A* 40(6), 1277–1293. Available at: http://dx.doi.org/10.1068/a39253 (accessed 6 March 2017).

Riley, M. (2009a) Bringing the 'invisible farmer' into sharper focus: gender relations and agricultural practices in the Peak District (UK). *Gender, Place & Culture* 16(6), 665–682. Available at: http://dx.doi.org/10.1080/09663690903279138 (accessed 6 March 2017).

Riley, M. (2009b) 'The next link in the chain': children, agri-cultural practices and the family farm. *Children's Geographies* 7(3), 245–260. Available at: http://dx.doi.org/10.1080/14733280903024407 (accessed 6 March 2017).

Riley, M. (2010) Emplacing the research encounter: exploring farm life histories. *Qualitative Inquiry* 16(8), 651–662. Available at: http://dx.doi.org/10.1177/1077800410374029 (accessed 6 March 2017).

Riley, M. (2011) 'Letting them go' – agricultural retirement and human–livestock relations. *Geoforum* 42(1), 16–27. Available at: http://dx.doi.org/10.1016/j.geoforum.2010.08.004 (accessed 6 March 2017).

Riley, M. (2012) 'Moving on'? Exploring the geographies of retirement adjustment amongst farming couples. *Social & Cultural Geography* 13(7), 759–781. Available at: http://dx.doi.org/10.1080/14649365.2012.725136 (accessed 6 March 2017).

Riley, M. (2014) Interviewing fathers and sons together: exploring the potential of joint interviews for research on family farms. *Journal of Rural Studies* 36, 237–246. Available at: http://dx.doi.org/10.1016/j.jrurstud.2014.09.003 (accessed 6 March 2017).

Riley, M. (2016a) Still being the 'good farmer': (non-)retirement and the preservation of farming identities in older age. *Sociologia Ruralis* 56(1), 96–115. Available at: http://dx.doi.org/10.1111/soru.12063 (accessed 6 March 2017).

Riley, M. (2016b) How does longer term participation in agri-environment schemes [re]shape farmers' environmental dispositions and identities? *Land Use Policy* 52, 62–67. Available at: http://dx.doi.org/10.1016/j.landusepol.2015.12.010 (accessed 6 March 2017).

Riley, M. and Harvey, D. (2007) Oral histories, farm practice and uncovering meaning in the countryside. *Social & Cultural Geography* 8(3), 391–415. Available at: http://dx.doi.org/10.1080/14649360701488823 (accessed 6 March 2017).

Sherman, J. (2009) *Those Who Work, Those Who Don't: Poverty, Morality, and Family in Rural America*. University of Minnesota Press, Minneapolis, Minnesota.

Sherman, J. (2011) Men without sawmills: job loss and gender identity in rural America. In: Smith, K. and Tickamyer, A. (eds) *Economic Restructuring and Family Well-Being in Rural America*. Pennsylvania State University Press, University Park, Pennsylvania, pp. 82–104.

West, C. and Zimmerman, D.H. (1987) Doing gender. *Gender & Society* 1(2), 125–151. Available at: http://dx.doi.org/10.1177/0891243287001002002 (accessed 6 March 2017).

16 Creating 'Masculine' Spaces for 'Feminine' Emotions – Men and Social Inclusion

C. Ni Dhonaill*
Queen's University Belfast, Belfast, UK

Introduction

This chapter explores narratives of masculinity in the context of groups dedicated to increasing social inclusion. The research discussed in the chapter compares two different types of social activities: one targeting rural older people and the other specifically targeted at older men, focusing on traditionally 'male' activities, in an urban setting. The research questions explored are based on the premise that being socially connected is beneficial for people, in terms of physical, mental and emotional well-being (Cattan *et al.*, 2005), and that older people are at particular risk of loneliness and social isolation (Edelbrock *et al.*, 2001). The chapter explores the male experience of ageing in the context of loneliness, isolation and emotional well-being through the consideration of two groups attempting to reduce isolation in older people. The chapter also explores notions of gender, as men and women experience loneliness, isolation and ageing very differently, with men being significantly more at risk of experiencing loneliness and isolation than women (Tijhuis *et al.*, 1999; Dykstra and de Jong Gierveld, 2004). In comparing a mixed-gender with a single-gender group, this chapter promotes the opinion that single-gender social inclusion projects have significantly more success in reducing isolation than do mixed-gender groups.

The chapter begins with a review of relevant literature, exploring isolation, loneliness and social exclusion in the context of gender and geographical place and space. This is followed by methodology and findings. The chapter concludes with a discussion on the role of segregated gendered groups for male emotional well-being.

Analytical Framework: Social Isolation and Loneliness

This chapter begins with the premise that social connections are beneficial for people in terms of their mental and physical well-being, but that particular areas of society are particularly vulnerable to *social isolation* and *loneliness*. These are distinct yet related concepts, which are often used interchangeably within literature, with social isolation referring to *objectively* measurable low levels of social contacts and loneliness referring to a *subjective* emotional feeling of having few social contacts (Cattan *et al.*,

* Corresponding author e-mail: c.nidhonaill@qub.ac.uk

2005). Generally, loneliness and isolation tend to share comparable prevalence rates ranging between 10 and 30% of the population, and more recently reported as high as 43% (Smith and Hirdes, 2009). The two concepts are not synonymous, however: a person can experience a subjective feeling of loneliness in spite of having high levels of social contact and a person can feel that they have deep emotional connections while having, objectively, few social contacts. Loneliness is seen as 'the evaluation of a discrepancy between the desired and the achieved network of relationships as a negative experience' (de Jong Gierveld *et al.*, 2006, p. 485). Indeed, it was the finding of this research that the *quality* of contact is significantly more important to participants than the simple prevalence of contact. Loneliness is certainly a possible outcome of social isolation; however, it is not a definite outcome. Should an objectively small network provide a person with the personal and emotional support that they feel they need, subjectively, this can be more beneficial than having an objectively large social network. Or, to put this more simply: some relationships matter more than others (Stephens *et al.*, 2011). Much research has consistently shown that older people are particularly vulnerable to isolation and are at risk of loneliness as a consequence (Edelbrock *et al.*, 2001; Warburton and Lui, 2007; Grenade and Boldy, 2008), and therefore groups attempting to reduce isolation for older people are a focus for numerous public bodies. Without considering the quality of relationships fostered in said groups, however, a subsequent reduction in loneliness may not become apparent.

The fact that older people are at significant risk of isolation and loneliness is of increasing importance as the UK faces an expanding ageing population, with one in six of the population being over the age of 65, representing 272,000 people in Northern Ireland (OFMDFM, 2013). This older population tends to have higher proportions of health, visual and mobility issues than their younger counterparts. One in four persons in the 60+ age bracket has mobility issues, and one in five over the age of 75 has some sight loss, increasing to one in two after the age of 90 (RNIB, 2008); all of which are considered as increasing this vulnerability to social isolation. Hearing loss is also an important factor in loneliness in older people as it affects communication by telephone, which has both a physical and a psychological impact on making older people feel isolated (Dugan and Kivett, 1994). Physical health and mobility issues related to becoming older have an impact on the likelihood of being socially isolated, and include sight and hearing problems, arthritis, stroke, osteoporosis and dementia (Victor *et al.*, 2002). Limited physical mobility can impact on ability to travel, which is a risk factor in isolation, particularly when mobility-related health conditions impact on ability to drive or simply to enter and exit cars (Alsnih and Hensher, 2003; Metz, 2003). This is often compounded in rural areas, where lack of access to transport is a major contributor to older people not accessing services (Lucas, 2006). As service uptake reduces services move out of areas, increasing this exclusion and creating a link between socially excluded areas and socially excluded people (Scharf *et al.*, 2002).

Health, loneliness and gender

The relationship between health and loneliness and isolation is two-way; health issues can be a cause of isolation and loneliness, but social isolation and loneliness also have profound effects on both physical and mental health, with people who report as isolated or lonely having poorer heath levels than those who report as less lonely (Stephens *et al.*, 2011; Hawton *et al.*, 2011). Fifty-five per cent of the 'most lonely' were depressed compared with just 4% of the least lonely (Beach and Bamford, 2014). Loneliness also has a linkage with cognitive decline; lonely people are at twice as high a risk of Alzheimer's than their socially engaged counterparts (AGENI, 2014).

The concept that men and women experience ageing differently is certainly not a new theory (Levy, 1988); however, there can be a tendency for older people to be treated as a homogeneous group, often seen as losing their individual identity as they age and being seen only as an 'older person'. This chapter contends that ageing, loneliness and exclusion are experienced in very different ways by men and women, which therefore leads to different pathways to address their isolation. It has been well established that older men as a group are more vulnerable to loneliness than their female counterparts (Tijhuis *et al.*, 1999; Dykstra and de Jong Gierveld, 2004; Dykstra and Fokkema, 2007). Indeed, men report loneliness more frequently than women, whether they live alone or in a shared home (Flood, 2005), emphasizing that loneliness and isolation are not synonymous concepts. Loneliness increases as we age, particularly after the age of 70 (Victor and Yang, 2012), as mobility decreases, working life ends, families move away, spouses become widowed and friends are lost through death.

Moreover, this is compounded by male aversion to discussing emotional issues with peers. Even with the most traumatic emotional events, such as bereavement, older men consider discussing emotional issues as weak or feminine – 'sissy stuff' (Bennett, 2007, p. 347). The loss of a spouse is associated strongly with emotional loneliness; whether through the end of a relationship or bereavement, although bereavement naturally becomes more likely as a person ages (Bisconti *et al.*, 2004). This tends to be a particular source of stress for men, as they are socialized to be non-emotional and are now faced with grieving and loss. Furthermore, through this socialization process, men tend not to have a wide range of confidantes, usually having their wife as their primary and often only source of emotional support (de Jong Gierveld *et al.*, 2006). Men remain at high risk of remaining in this state of loneliness, however, as they tend to have less social support than women. Over the course of their lifetime, men tend to prioritize the working role over social and emotional bonds and friendships. Post-retirement, this can be problematic as, alongside losing the important worker role, they immediately lose significant amounts of social contact through their colleagues. As the older man ages, family moves away and he becomes widowed, this lack of social contact and emotional support built over a lifetime exacerbates the feelings of loneliness brought on by bereavement. Men are also seen as being less naturally inclined to join social groups than their female counterparts, with women seen as being more likely to want to 'join in' (Marhankova, 2014). This has led to an increase in targeting of men in groups focused on reducing social isolation.

Ageing and isolation in a rural setting

Older people in rural areas are particularly affected by a number of issues. They have a tendency to be poorer than their urban counterparts, although issues for people living in rural areas are usually more complex than being based purely on a lack of resources, however. The Rural Community Network (2004) 'emphasise that it is necessary to take into consideration deprivation of resources, access to services, isolation and quality of life rather than only highlighting income based indicators' (p. 1). Importantly, people in rural areas may find themselves to be socially excluded. According to Philip and Shucksmith (2003), 'Social exclusion is a multi-dimensional, dynamic concept which emphasises the processes of change through which individuals or groups are excluded from the mainstream of society and their life-chances reduced' (p. 461).

Older people are physically further away from services, making older people living in rural areas 'disadvantaged simply because of where they live' (Walsh and O'Shea, 2008). Moreover, the lack of an appropriate transport system will further disadvantage many older people living in rural areas. The restricted public transport in Northern Ireland has been criticized by older rural people in Northern Ireland as 'a major barrier to access' (Rural Community Network, 2004).

Due to the dispersed nature of rural dwelling, along with having a lack of access to services due to issues around transport, older rural people are placed at a significant risk of social isolation and related loneliness. They tend to have a small social circle due to the geographical distance to neighbours and family. These factors cumulatively leave older people in rural areas vulnerable to loneliness, particularly post-widowhood, as their primary social companion is lost (Dugan and Kivett, 1994).

In summary, the key points from the literature relevant to this chapter are that social inclusion is beneficial to people and that certain groups are particularly vulnerable to loneliness and exclusion, including people in rural areas and men. Moreover, and compounding the difficulties of loneliness and exclusion, men tend to have a smaller social support network than women and often have very few people to lean on for emotional support; a small circle which tends to decrease as people age. This chapter explores these themes, while identifying pathways to reduce isolation for men.

Methodology

This section discusses the methodology used in the research upon which this chapter draws, to compare different types of social activities addressing the needs of two specific groups of older people. The two groups were investigated as part of two different projects: the work on the rural group of men and women was part of the author's PhD research, which was funded by the Department of Education and Learning; and the urban male-only group was a part of a research project focusing on the specific needs of older men, funded by the Public Health Agency and commissioned by the Men's Working Group, a collaboration of several community and voluntary organizations, established by Age Partnership Belfast. Both projects took a participant observation approach, followed by semi-structured narrative interviews, to data collection. All interviews were conducted by the author, either at the participant's home or in the meeting place of the relevant organization, with the choice of venue being decided by the participant to maximize their comfort with the research. All interviews were taped and transcribed by the author. Although two different research projects were involved, the similar data collection style and research questions involved in each allowed for comparison between the two.

The two major differences between the participant profiles of these groups is that one was based in a rural setting, the other urban; and one focused on the needs of rural older people, male and female, and the second focused on the needs of urban older men. With the exception of these, the participant profiles were fairly similar. Both groups involved quite a diverse subsection of the community in terms of socio-economic and religious background, both having a dedicated cross-community ethos. There was no 'typical' member in terms of educational background, with some members having no basic qualifications and others being highly qualified. There was a subsequent lack of typical job types; some jobs in earlier life were high-skilled, some low. Members had highly diverse levels of health in these groups, some having quite serious health conditions, and the groups attempted to accommodate this as much as possible.

The first group was based in Cookstown, Northern Ireland and targeted older people in the surrounding rural areas. The aim of this group was based on the notion that rural older people are at high risk of being socially excluded, and that contact with other older people would have a positive impact on their lives. The two specific areas of the project this chapter focuses on are: (i) the Luncheon Club, in which participants paid a small fee to come to a community centre and eat a meal together, in order to foster a sense of friendship and community; and (ii) an arts and crafts activity, which ran after the Luncheon Club on a Wednesday. This was attended primarily by women, as was the Luncheon Club, and two reasons were offered for this. First, women were considered more likely to be socially involved and want to attend social groups; and second, women would be more 'naturally' interested in arts and crafts. In both groups, the small number of male attendees tended to sit together and had a lower level of participation in the arts and crafts offered, generally using the time to talk to each other.

The second project this chapter addresses is an urban 'Men's Shed' project. Men's Sheds are an increasingly popular support and inclusion project for older men in attempting to combat isolation and depression (Ballinger *et al.*, 2009). Originating in Australia in the late 1990s, Men's Sheds provide a male-only space for older men to come together and work on do-it-yourself (DIY) type projects, thereby learning new skills and keeping active and in company (Brown *et al.*, 2007). This group specifically targeted men, as the organizers aimed to address men's social isolation. The Men's Shed group uses DIY crafts as a tool to bring isolated and lonely men together to again form friendships and social connections. This chapter contends that creating a male-only space, in which men take part in typically 'masculine' activities, fosters a healthy space for the participants not just to form these social connections, but also to address some of the deeper emotional and psychological issues that the non-gender-specific group cannot meet as successfully.

The following section addresses each group in turn, considering the push and pull factors at play in participants becoming involved in the two groups. The push/pull factors are crucial, as they illuminate issues around gender roles in older age. For the purpose of this chapter, *push* factors in men's motivation for attendance relate to wanting or feeling the need to be out of the house, and are not always related to the activities provided; the feeling that there is no specific place for them was most common, and usually connected to gendered notions of the domestic sphere as being female. In contrast, *pull* factors refer to men's motivations to attend related to being drawn to the group, as it fulfils a need or desire through the activities of the group itself. This is followed by a discussion regarding how successful each group was in tackling emotional loneliness among the participating men. This chapter aims to show that in older age, men are at particular risk of emotional loneliness due to a lifetime of gender stereotyping, as older generations tend to have quite entrenched notions of gender roles in the home (Inglehart, 2008). In terms of the rural group considered, it is important to note that these entrenched traditional gender roles also maintain influence in rural Ireland (Shortall, 2002). This is compounded with a lack of 'male space'; in the home, through a lack of work and in social settings. It is shown that by creating this 'male space', the Men's Shed group was significantly more successful than its rural counterpart at addressing male loneliness.

Research Findings

Cookstown Arts and Crafts

This group was formed on the premise that social isolation is a negative state in which to live. It further argued that older rural people are at particularly high risk of being isolated and lonely, for a number of reasons, in keeping with findings from the literature: older people lose spouses, family moves away, and they may find that deterioration in their physical health, mobility, hearing, sight, etc. becomes preventive in continuing social bonds. Moreover, rural older people have particular risks of isolation, due to the dispersed nature of rural populations. Simple social contacts, such as meeting a person in a shop, post office or bank, are made difficult due to the above-mentioned physical decline associated with older age. This group planned to tackle these issues through providing a space to bring people together and providing transport, to address the issues of transport and travel, in the context of physical mobility issues and population dispersal. This group did arts and crafts activities, including glass painting, flower arranging and some baking projects, as a means of providing some entertainment for the group; however, its basic premise was that in tackling social isolation, bringing people together was the most important means of addressing this aim.

There were two distinct groups of men who attended the rural group in terms of social isolation – widowers and those who were still married – and each attended for different reasons.

Widowers typically came to the group due to pull factors, primarily due to the provision of lunch by the Luncheon Club which took place every day between Monday and Friday and was followed by the Arts and Crafts group on Wednesdays. This group of people had usually heard about the project through word of mouth, typically through other men who were widowed, or through family who wanted to ensure that their fathers were getting at least one square meal per day. This appears to be due to the stereotypically gendered nature of households among older rural people. During their lifetime, this group of men did not tend to take part in work in the domestic sphere, with this being the job of the wife. After the loss of said spouse, men found themselves without some of the more basic knowledge of how to provide for themselves in terms of simple cookery (Bennett *et al.*, 2003). This appeared to be quite a demasculinizing realization, as one participant commented he had not realized he had been 'looked after' his entire life by his wife and now was being 'looked after' by a community group. The men maintained a sense of self and pride when discussing this by distancing themselves from those who had food provided by the related Meals on Wheels service as they, at least, still had the physical ability to leave their homes.

Those who were still married typically attended the group due to push factors, again related to stereotypical gender roles in the household. Men described the household as being the 'woman's space' and a typical sentiment coming from the group was that they were constantly 'getting underfoot' in the home. In some cases, their wives were looking after grandchildren and, again, the men felt that they were constantly in the way of the women's ability to care. There was a very clear gendered split here: the men in this situation were attending this group not from the desire or need to be in attendance, but from the feeling that they were 'in the way' at home. This will have come from a lifetime of leaving the home to go to work, creating the idea that the male sphere is outside the home, and the female, inside (Mowl *et al.*, 2000).

The Arts and Crafts group, however, while providing a physical space for men to be outside the home and no longer 'underfoot', was not successful in creating a male space for the participants. The men therefore still did not feel like they truly belonged there, any more than they did in the home. The gap left by retirement, and the male space and identity created during the working life, was not being filled by the activities offered by this project. The types of activities on offer at the Arts and Crafts group were considered to be typically female activities. Essentially, the group replaced one female sphere for another. It was evident that the men involved did not particularly enjoy taking part in the activities, especially those deemed very feminine, such as flower arranging. In this instance, the men collected the material and left it in front of them, not actually taking part in the activity and using the time to chat among themselves. When men did take part in activities deemed overly feminine, the group tended to attempt to correct this, through gentle mocking. On one occasion, a participant picked a picture of a butterfly to paint and was mocked by the rest of the group for wanting to take part in 'girly' things. While the men did attend the Arts and Crafts group, they were very careful to distance themselves from anything that could potentially be deemed 'girly'. This was observed by the organizers and female participants, but there appeared to be no impetus to tailor the activities to suit the male attendees. The general attitude was that men generally don't want to get involved in social and group activities, and that women were 'just naturally better at joining in'. Comments such as 'sure, they just prefer sitting and talking among themselves anyway' were made repeatedly by female members of the group, essentially blaming male lack of participation on a stereotypically male attitude as opposed to considering that the choice of activity on offer may have been related to non-participation. This was not the attitude of the men, who discussed among themselves the types of activities they may have enjoyed. The idea of a walking club would be discussed, then quickly dissipate, as someone would remind the group that not all male participants would be physically able for that. These conversations were held internally at the small male table and not generally discussed with the wider group, perhaps because of the overwhelmingly female nature of the group as a whole; again, symptomatic of the men not wanting to interject into a female space.

It is important to note also that while transport was provided to enhance social inclusion and well-being, on the day of the Arts and Crafts session, this may have had the unintended side-effect of trapping the men in an activity that they neither enjoyed nor would have elected to take part in otherwise. Those who came for the daily lunch did enjoy the food and certainly appeared to enjoy the social contact with other men that provided here. They discussed the essential role of the transport provided by the project in collecting them from their homes, in order to be able to come and spend time with their friends. On the day that Arts and Crafts took place however, the transport had a different interpretation. Rather than being a positive means of enhancing social contact, it became a means of trapping participants at the activity. Comments around having to stay and wait for the bus were repeated by male participants; this was in contrast to their attitude on other days and, indeed, to the attitude of the women on all days of how positive it was that they had a mode of transport to be able to take part in activities.

In the cases of both widowers and spouses, the group did provide a means of social contact for the participants; however, in not tailoring the work to the needs of men, it did not have the effect of providing male space outside the home, often leaving the participants feeling peripheral to the overall experience of group participation, creating a small subgroup which did not take part in most group activities.

Men's Shed

The Men's Shed project has some areas of similarity to the rural Arts and Crafts group. Again, it is based on the assumption that social isolation is generally a negative lived experience and that particular groups are at high risk of being socially isolated. In this case, its focus was on men, who, it was noted, had often lost many of their most crucial identifying roles, particularly that of worker. Like the rural group, the men were from a mixed background in regard to marital status, with some widowed, some still married and some lifetime bachelors. Due to this, participants did acknowledge having some of the same push factors in feeling that the domestic space was the woman's sphere and wanting to give their wives space in their home. Very similar sentiments were discussed in both groups, with a typical sentiment expressed being:

> Well, I think one of the things about men, certainly from talking to the chaps, is they've been out all their life in work . . . if you're lucky enough to be employed and then all of a sudden that stops, and you're in the wife's way, the partner's way and . . . because she's not used to you being in the house so you all of a sudden have to find some other things to do to get out of . . . because that's her space.
>
> (M, Men's Shed)

Here the dual difficulty for many men is clearly expressed: the male sphere is outside the home, and when work ends, they take up space in the feminine sphere, and are essentially just in the way. The Men's Shed, like the rural group above, provides an outlet for this particular push factor in giving participants a physical space to remove themselves from the home. Unlike the rural group, however, this project also creates a positive, male space for the participants. They are able to feel both like they have given their partners the space at home that they desire, while also having a place that is truly theirs to attend, so moving them out of the female domestic space and into a male, working space. This has significantly more positive impacts for the participants than the mixed-gender project.

The pull factors involved in this group were quite different from those in the rural project, as the Men's Shed project provided a post-work masculine space that allowed for a more positive environment for the participants and, as a consequence, created a space in which they were able to fully address emotional issues that otherwise they may have deemed too female. In some ways, the activities on offer were not entirely dissimilar to those on offer in the rural group, in that they were based on arts and crafts projects. However, the arts and crafts involved were more tailored to a male group, with them more closely related to physical DIY-style projects. Rather than painting and flower arranging, projects on offer here involved wood carving, metal work and pyrography, all of which involved the use of heavy tools and some sense of physical prowess. This was an essential and crucial difference between the two groups. There was no consideration that taking part in any of the activities could be considered overly feminine and the men encouraged each other to attempt all aspects of the projects, rather than the gentle mocking seen when participants attempted to involve themselves in the rural setting. This allowed the men a sense of pride in their completed projects, often taking them home as presents for family members, about which they happily boasted:

> B: 'You're buzzing, as I say, at the end of the day, I brought my wee wood carving home one day and her and the daughter were fighting over it and we have a wee shelf and she said "put it up there", so they all want a wood carving.'
>
> I: 'That real sense of pride, that you've. . .'
>
> B: 'That's it! You're getting that . . . you've brought it home.'
>
> (B and Interviewer, Men's Shed)

The act of creation and manufacture was of crucial importance to the men, as it allowed them the sense of pride and usefulness that employment had brought during their working life. Many of the men described being able to use skills they had accumulated over a lifetime of work in activities provided by the Men's Shed. Primarily, this was related to physical tasks; members who had been bricklayers or in any type of crafts work had directly transferable skills, and often used these to guide and support other members of the group. However, even when employment skills were not directly transferable, participants discussed the ways in which their working life experience enhanced their experience in the Men's Shed or how they could use their experience to contribute to the project. One member, for example, had been a project manager, and discussed using his skills at managing teams of workers to help create small working teams for larger projects in the Men's Shed. Another had worked in information technologies, and discussed creating simple Excel files to monitor attendance and organize future planning. It was very evident that continuing to relate to their working identity was of significant importance to the men. This finding is supported in the literature about ageing through ideas around continuity theory, in which people attempt to maintain aspects of their working life activities post-retirement, in order to maintain a sense of identity (Atchley, 1989). This is particularly important for men, who tend to place more importance on their working identity than their female counterparts (Forret *et al.*, 2010).

While it is clear that both groups were successful in combating social isolation, by quantifiably increasing the attendees' number of social contacts, the urban Men's Shed group was significantly more successful in reducing emotional loneliness than the rural Arts and Crafts group. Men were freer in discussing emotions and issues of personal importance. This they related directly to the masculine nature of the work involved in the project. There was an acknowledgement that women were significantly more open in their ability to talk about subjects that are difficult and sensitive, and due to this, talking about these emotional issues, in and of itself, was seen as a feminine activity. In providing this masculine space, however, the Men's Shed was successful in creating a space in which men could talk about traditionally 'feminine' topics, such as emotions, health and well-being, without fear of being deemed a 'sissy'. The physical space, and the masculine nature of the tools being used in the Men's Shed, could be used as physical props to emphasize masculinity, while taking part in a feminine activity:

> And another thing about it is, you talk to people easier when you're doing something, especially men. It's okay when you're whacking away [imitates hammering] and can chat to people. Women can sit like this [gestures at us sitting over a table] and talk about all sort of things but men, would maybe talk about football [laughs]. That's it.
> (J, Men's Shed)

Conclusion

Comparing these two social groups has illuminated the needs for older men in the context of isolation and loneliness. The rural Arts and Crafts group rightly identify the increased risk of isolation for older people in rural areas. The dispersed nature of rural dwelling, along with decreased mobility associated with ageing, creates a high risk of isolation in older people, supporting the findings of Walsh and O'Shea (2008), who noted the spatial dimension in isolation in a rural context. The group's approach to countering this was simple: remove the issue of mobility and transport through providing a door-to-door pick-up service and combat isolation by bringing older people together, which addressed transport needs identified by Heehan (2006). This approach, however, does not consider the more complex needs of older men, specifically when it comes to dealing with emotional bonding (de Jong Gierveld *et al.*, 2006; Bennett, 2007). The entrenched gender norms evident in older rural people became preventive in allowing men to fully take part in the activities involved, due to the feminine nature of the activities offered. The consequence of this was that the male participants in this group did not actually participate and instead sat separately from the main group.

In contrast, the urban Men's Shed group was successful in both reducing isolation and increasing emotional social bonds through creating a masculine space for men that allowed them to discuss feminine emotions. Creating a space which did not denigrate their masculinity in any way, but rather fostered and celebrated masculinity, in fact allowed a space for the feminine, because masculinity could not be called into question in any way. Being able to identify strongly with masculinity in the Men's Shed was crucial for the participants as they were able to use this to fulfil an emotional need that otherwise would have been unavailable due to their understanding of gender roles. This research emphasizes the importance of masculine spaces for older men as a major pathway to reducing isolation and loneliness, leading to benefits for emotional and mental well-being.

Acknowledgement

The author would like to acknowledge and thank Drs Paula Devine, Lorna Montgomery and Janet Anand, who invited her to take part in the Men's Shed's research as part of their team.

References

Atchley, R.C. (1989) A continuity theory of normal aging. *The Gerontologist* 29(2), 183–190.

Alsnih, R. and Hensher, D.A. (2003) The mobility and accessibility expectations of seniors in an aging population. *Transportation Research Part A: Policy and Practice* 37(10), 903–916. Available at: http://dx.doi.org/10.1016/S0965-8564(03)00073-9 (accessed 6 March 2017).

Ballinger, M.L., Talbot, L.A. and Verrinder, G.K. (2009) More than a place to do woodwork: a case study of a community-based Men's Shed. *Journal of Men's Health* 6(1), 20–27. Available at: http://dx.doi.org/10.1016/j.jomh.2008.09.006 (accessed 6 March 2017).

Beach, B. and Bamford, S.M. (2014) *Isolation: The Emerging Crisis for Older Men: A Report Exploring Experiences of Social Isolation and Loneliness among Older Men in England.* Independent Age and International Longevity Center–UK, London.

Bennett, K.M. (2007) 'No sissy stuff': towards a theory of masculinity and emotional expression in older widowed men. *Journal of Aging Studies* 21(4), 347–356. Available at: http://dx.doi.org/10.1016/j.jaging.2007.05.002 (accessed 6 March 2017).

Bennett, K.M., Hughes, G.M. and Smith, P.T. (2003) 'I think a woman can take it': widowed men's views and experiences of gender differences in bereavement. *Ageing International* 28(4), 408–424. Available at: http://dx.doi.org/10.1007/s12126-003-1012-x (accessed 6 March 2017).

Bisconti, T.L., Bergeman, C.S. and Boker, S.M. (2004) Emotional well-being in recently bereaved widows: a dynamical systems approach. *The Journals of Gerontology Series B: Psychological Sciences and Social Sciences* 59(4), 158–167. Available at: http://dx.doi.org/10.1093/geronb/59.4.P158 (accessed 6 March 2017).

Brown, M., Foley, A., Harvey, J. and Gleeson, L. (2007) *Men's Sheds in Australia: Learning through Community Contexts.* National Centre for Vocational Education Research, Adelaide, Australia.

Cattan, M., White, M., Bond, J. and Learmouth, A. (2005) Preventing social isolation and loneliness among older people: a systematic review of health promotion interventions. *Ageing & Society* 25(1), 41–67. Available at: http://dx.doi.org/10.1017/S0144686X04002594 (accessed 6 March 2017).

de Jong Gierveld, J., van Tilburg, T. and Dykstra, P.A. (2006) Loneliness and social isolation. In: Perlman, D. and Vangelisti, A. (eds) *The Cambridge Handbook of Personal Relationships.* Cambridge University Press, Cambridge, UK, pp. 485–500.

Dugan, E. and Kivett, V.R. (1994) The importance of emotional and social isolation to loneliness among very old rural adults. *The Gerontologist* 34(3), 340–346. Available at: http://dx.doi.org/10.1093/geront/34.3.340 (accessed 6 March 2017).

Dykstra, P.A. and de Jong Gierveld, J. (2004) Gender and marital-history differences in emotional and social loneliness among Dutch older adults. *Canadian Journal on Aging* 23(2), 141–155. Available at: http://dx.doi.org/10.1353/cja.2004.0018 (accessed 6 March 2017).

Dykstra, P.A. and Fokkema, T. (2007) Social and emotional loneliness among divorced and married men and women: comparing the deficit and cognitive perspectives. *Basic and Applied Social Psychology* 29(1), 1–12. Available at: http://dx.doi.org/10.1080/01973530701330843 (accessed 6 March 2017).

Edelbrock, D., Buys, L.R., Waite, L.M., Grayson, D.A., Broe, G.A. and Creasey, H. (2001) Characteristics of social support in a community-living sample of older people: the Sydney Older Persons Study. *Australasian Journal on Ageing* 20(4), 173–178. Available at: http://dx.doi.org/10.1111/j.1741-6612.2001.tb00382.x (accessed 6 March 2017).

Flood, M. (2005) *Mapping Loneliness in Australia.* Australia Institute, Canberra.

Forret, M.L., Sullivan, S.E. and Mainiero, L.A. (2010) Gender role differences in reactions to unemployment: exploring psychological mobility and boundaryless careers. *Journal of Organizational Behavior* 31(5), 647–666. Available at: http://dx.doi.org/10.1002/job.703 (accessed 6 March 2017).

Grenade, L. and Boldy, D. (2008) Social isolation and loneliness among older people: issues and future challenges in community and residential settings. *Australian Health Review* 32(3), 468–478. Available at: http://dx.doi.org/10.1071/AH080468 (accessed 6 March 2017).

Hawton, A., Green, C., Dickens, A.P., Richards, S.H., Taylor, R.S., Edwards, R., Greaves, C.J. and Campbell, J.L. (2011) The impact of social isolation on the health status and health-related quality of life of older people. *Quality of Life Research* 20(1), 57–67. Available at: http://dx.doi.org/10.1007/s11136-010-9717-2 (accessed 6 March 2017).

Heehan, D (2006) The factors influencing access to health and social care in the farming communities of County Down, Northern Ireland. *Ageing & Society* 26(3), 373–391. Available at: http://dx.doi.org/10.1017/S0144686X06004697 (accessed 6 March 2017).

Inglehart, R.F. (2008) Changing values among western publics from 1970 to 2006. *West European Politics* 31(1–2), 130–146. Available at: http://dx.doi.org/10.1080/01402380701834747 (accessed 6 March 2017).

Levy, J.A. (1988) Intersections of gender and aging. *The Sociological Quarterly* 29(4), 479–486. Available at: http://dx.doi.org/10.1111/j.1533-8525.1988.tb01429.x (accessed 6 March 2017).

Lucas, K. (2006) Providing transport for social inclusion within a framework for environmental justice in the UK. *Transportation Research Part A: Policy and Practice* 40(10), 801–809. Available at: http://dx.doi.org/10.1016/j.tra.2005.12.005 (accessed 6 March 2017).

Marhankova, J.H. (2014) 'Women are just more active' – gender as a determining factor in involvement in senior centres. *Ageing & Society* 34(9), 1482–1504. Available at: http://dx.doi.org/10.1017/S0144686X13000275 (accessed 6 March 2017).

Metz, D. (2003) Transport policy for an ageing population. *Transport Reviews* 23(4), 375–386. Available at: http://dx.doi.org/10.1080/0144164032000048573 (accessed 6 March 2017).

Mowl, G., Pain, R. and Talbot, C. (2000) The ageing body and the homespace. *Area* 32(2), 189–197. Available at: http://dx.doi.org/10.1111/j.1475-4762.2000.tb00129.x (accessed 6 March 2017).

OFMDFM (2015) *A Profile of Older People in Northern Ireland: Annual Update 2015*. Office of the First and Deputy First Minister, Statistics and Research Branch, Belfast. Available at: https://www.executiveoffice-ni.gov.uk/publications/profile-older-people-northern-ireland-annual-update-2015 (accessed 20 September 2016).

Philip, L.J. and Shucksmith, M. (2003) Conceptualizing social exclusion in rural Britain. *European Planning Studies* 11(4), 461–480. Available at: http://dx.doi.org/10.1080/09654310303646 (accessed 8 March 2017).

RNIB (2008) Key information and statistics. Available at: http://www.rnib.org.uk/knowledge-and-research-hub/key-information-and-statistics (accessed 17 November 2015).

Rural Community Network (2004) *Rural Network News, Spotlight on Rural Poverty*. RCN, Cookstown, Northern Ireland.

Scharf, T., Phillipson, C., Smith, A. and Kingston, P. (2002) Older people in deprived areas: perceptions of the neighbourhood. *Quality in Ageing and Older Adults* 3(2), 11–21. Available at: http://dx.doi.org/10.1108/14717794200200011 (accessed 6 March 2017).

Smith, T.F. and Hirdes, J.P. (2009) Predicting social isolation among geriatric psychiatry patients. *International Psychogeriatrics* 21(1), 50–59. Available at: http://dx.doi.org/10.1017/S1041610208007850 (accessed 6 March 2017).

Shortall, S. (2002) Gendered agricultural and rural restructuring: a case study of Northern Ireland. *Sociologia Ruralis* 42(2), 160–175. Available at: http://dx.doi.org/10.1111/1467-9523.00208 (accessed 6 March 2017).

Stephens, C., Alpass, F., Towers, A. and Stevenson, B. (2011) The effects of types of social networks, perceived social support, and loneliness on the health of older people: accounting for the social context. *Journal of Aging and Health* 23(6), 887–911. Available at: http://dx.doi.org/10.1177/0898264311400189 (accessed 6 March 2017).

Tijhuis, M.A., de Jong-Gierveld, J., Feskens, E.J. and Kromhout, D. (1999) Changes in and factors related to loneliness in older men. The Zutphen Elderly Study. *Age and Ageing* 28(5), 491–495. Available at: http://dx.doi.org/10.1093/ageing/28.5.491 (accessed 6 March 2017).

Victor, C.R. and Yang, K. (2012) The prevalence of loneliness among adults: a case study of the United Kingdom. *The Journal of Psychology* 146(1–2), 85–104. Available at: http://dx.doi.org/10.1080/00223980.2011.613875 (accessed 6 March 2017).

Victor, C.R., Scambler, S.J., Shah, S., Cook, D.G., Harris, T., Rink, E. and de Wilde, S. (2002) Has loneliness amongst older people increased? An investigation into variations between cohorts. *Ageing & Society* 22(5), 585–597. Available at: http://dx.doi.org/10.1017/S0144686X02008784 (accessed 6 March 2017).

Walsh, K. and O'Shea, E. (2008) Responding to rural social care needs: older people empowering themselves, others and their community. *Health & Place* 14(4), 795–805. Available at: http://dx.doi.org/10.1016/j.healthplace.2007.12.006 (accessed 6 March 2017).

Warburton, J. and Lui, C.W. (2007) Social isolation and loneliness in older people: a literature review. Available at: http://espace.library.uq.edu.au/view/UQ:160390 (accessed 30 January 2016).

17 Gender Desegregation among Village Representatives in Poland: Towards Breaking the Male Domination in Local Politics?

I. Matysiak*

Maria Grzegorzewska University, Warsaw, Poland

Introduction

The aim of this chapter is to analyse the causes and consequences of the process of gender resegregation in the role of the village representative (*sołtys*), the institution of local government in Poland that is closest to rural residents.[1] 'Village representative' is a traditional liaison between village residents and higher levels of local authorities. Historically speaking, this role was a male domain, as men were the main public actors in communities. However, in the last few decades the number of rural women serving as village representatives has increased significantly. In 1958, women amounted to only 0.8% of village representatives in Poland, whereas in 2014 this percentage was 35.9% on the national level and exceeded 40% in some provinces.[2]

Searching for answers, the chapter also deals with a fundamental underlying question. Is this change a step towards breaking the male dominance in the sphere of local politics, or is it the opposite – does it preserve the weaker position of women compared with men? The analysis presented in this chapter allows us to develop a better understanding of the transformation of gender roles under the conditions found in rural areas, which are peripheral in relation to Western countries. The predominance of traditional gendered norms and values seems to be particularly strong in these places. However, at the same time, social changes often take place spontaneously, in a manner that is not motivated or required by official pro-equality programmes or public policies (Bock, 2014). The analysis conducted in this chapter also contributes greatly to the debate on political representation of women in general.

This analysis is based mostly on the outcomes of a qualitative study conducted in ten municipalities located in different regions of Poland, and supplemented with secondary data. First, a brief overview of the literature on rural women and politics is presented. Next, a short description of the main changes in Polish rural areas pre- and post-1989 is provided, as well as details about the function of the village representative. The subsequent section presents the theoretical framework: the process of gender desegregation and its relationship to the institution of the village representative. After that, the research problem and methodology are discussed. The final parts of the

* Corresponding author e-mail: ilona.matysiak@gmail.com

chapter are devoted to presenting and discussing the results of the analysis.

Background and Context

Rural women and political power

In the Western European literature, underrepresentation of rural women in politics has particularly been underlined. In the case of rural areas, male domination in politics in general overlaps with male domination in the fields of agriculture, forestry and rural development, which are perceived as 'male fields of expertise' (Bock, 2010). As a result, rural women are often excluded from decision-making processes regarding policies targeted at the development of rural areas. In the case of EU Member States, the situation was not made any better by implementation of new mechanisms and structures for the governing of rural areas, which were supposed to operate from the bottom up and assumed participation of various social actors on the local level (Bock and Derkzen, 2008). For instance, the available data show that women are rarely represented equally in Local Action Groups operating in the frame of the LEADER programme (Bock, 2010). Gender mainstreaming regarding the Common Agricultural Policy and rural development exists only on paper, or is limited to random projects addressed to specific groups of women, and has not led to systemic improvement of the situation of female rural residents or increases in their participation in political decision-making processes (Bock, 2014; Shortall, 2014).

The historically and structurally conditioned male domination in politics is maintained by various social and cultural mechanisms (cf. Bourdieu, 2002). For example, research conducted in the Netherlands indicates the exclusion of women from political decision-making processes by use of specific language practices underlining the importance of technical and scientific language and depreciating language and knowledge based on experience (Bock and Derkzen, 2008). Research on women in local government structures in Australia indicated open sabotaging of female mayors by local male politicians: 'these [types of sabotage] include minimising women's power as mayor, excluding women from networks, knowledge and information, denigrating women and sexualising women' (Pini, 2008, p. 300). As shown in Little's (1997) work, due to the persistence of traditional gender roles and division into the private and public sphere, in Britain women often fall into the trap of performing informal social activities on behalf of their villages. Altruism is attributed to them as a 'natural' concern and thus they are expected to engage in various voluntary works for free. Involvement of this kind can be a source of a sense of belonging and of some influence in their communities. However, voluntary work is perceived as 'non-professional', 'apolitical' and less important than paid work (Little, 1997). Because of the hard-to-grasp mechanisms of male dominance (Bourdieu, 2002), the social and economic resources of women in rural areas still rarely translate to political power. Oedl-Wieser's (2008) research in Austria has shown that the growing number of women among farm managers does not mean they are becoming better represented in the structures connecting farmers and institutions that co-create agricultural and rural development policies.

This chapter contributes to the debate on the mechanisms that shape male dominance in local politics in rural areas, explaining how they change and the reasons for these changes in the Polish context.

Changes of rural areas in Poland

In Poland, a relatively high share of the population resides in the rural areas of the country: 39.2% in 2011 (Central Statistical Office, 2012, p. 47). The Polish agriculture sector is characterized by a great number of small family farms. In 2010, the average area of a farm was 9.6 ha (Poczta, 2012, p. 84). In contrast with many Western European countries, the Polish countryside is covered in a dense network of rural settlements. Another special feature of rural areas in Poland is their spatial diversification. Regional differences pertain to many aspects of life, including the size and productivity of the farms, degree of urbanization, demographic structure, access to communication infrastructure, voting preferences, economic performance, types of local organizations and even types of social capital (Nowak *et al.*, 2000; Zarycki, 2002; Bartkowski, 2003; Halamska, 2007; Herbst and Swianiewicz, 2008). This diversification is due to the historical background of how the regions were shaped or consolidated in the partition period (1795–1918). Separate regions of Poland were under the control of three partitioning powers (Russia, Prussia and the Austro-Hungarian Empire), which each implemented different state, social and cultural policies. The effects of the partition period then overlapped with later change processes that took place in the interwar period (1918–1939), as well as after the Second World War.

After the Second World War, Poland was included in the Soviet sphere of influence and a communist regime was implemented in the country. Under communist rule, reforms were made which led to the liquidation of large private land estates and the establishment of state farms. However, attempts to collectivize agriculture and create production cooperatives failed, and Poland remained the only country in the Eastern bloc that retained individual farming. In general, the communist policy towards farmers resulted, among other things, in the perseverance of a large number of small family farms and excess workforce in rural areas. It is also worth mentioning that Poland's borders were changed at the end of the Second World War. The loss of its eastern territories, which were incorporated into the Ukrainian, Belarusian and Lithuanian Soviet Republics, was 'compensated' for by the addition of formerly Prussian lands in the west and north. These territorial changes contributed significantly to the spatial diversification of rural and urban areas observed today.

In 1989, a major political and economic transformation was implemented in Poland. The communist regime was replaced with democracy and state-controlled economy – with a free market. The key processes of these systemic changes included development of the dual agricultural model leading to the creation of a group of large-scale production farms on the one hand, and social (no production or small-scale production) farms on the other, while maintaining the generally unfavourable agrarian structure (Bukraba-Rylska, 2011; Rosner, 2011). At the same time, the social structure of the rural population has also been changing. Due to the de-agrarianization of employment, the percentage of rural households that depend on farming only for their livelihoods has dropped below 10% (Wilkin, 2011, p. 117). The educational aspirations of the rural population are growing and the migration flow from the cities to the countryside – in particular, into the suburbs – is increasing. Halamska (2013) indicates that, as a result of all these changes, a rural middle class consisting of white collar workers and specialists is emerging. Also, due to the availability of EU funds, the quality of the rural technical infrastructure (including things like roads and sewage systems) has been improving, as well as the social infrastructure – for example, public utility buildings (schools, community centres, etc.). It should be underlined, however, that the developmental changes in specific communities differ greatly in terms of their intensity, depending on the characteristics of the local context.

Consequently, rural areas in Poland are spaces where legacies of longstanding traditions interact with processes of dynamic social, economic and cultural changes of a global nature. These surely affect, among others, gender roles and perceptions of women's participation in local politics. However, the spatial diversification of Polish rural areas indicates the significance of regional and local factors.

Function of the village representative

The function of the village representative (*sołtys*) is rooted in the processes of settlement under German law in the Medieval Age. In the following periods, the obligations of village representatives and the nature of their role changed repeatedly as political and economic conditions shifted. Today, village representatives are included in the structure of local government, recreated after decades of state centralization under the communist regime (Matysiak, 2014a,b).

Poland is divided into 16 provinces, several hundred districts and almost 2500 municipalities. The latter are the strongest entities in the system in terms of the scope of competences and financial independence. Municipalities located in rural areas (rural and urban–rural municipalities) are further divided into rural sub-municipal auxiliary units (*sołectwo*) operating at the level of a village.[3] According to the concept of decentralization, there are separate executive and legislative bodies at each level (see Table 17.1).

Village representatives are the executive officers of rural sub-municipal auxiliary units. The other institutions of these units include a village assembly (*zebranie wiejskie*) – the legislative body – and a village council (*rada sołecka*) – the advisory body to the village representative. It has to be emphasized that, in terms of the structure of power, rural sub-municipal auxiliary units are dependent both financially and institutionally on municipal authorities; they have neither their own full legal identity nor an independent budget.[4]

Village representatives and members of the village council are elected by the residents of the village. The election involves a secret, direct voting method, which mostly takes place at the village assembly. All residents can present themselves as a candidate for village representative or village council member.[5] The term of office for the village representative is usually 4 years. Village representatives embody the interests of the residents and are expected to manage the daily problems of a given village, such as issues associated with the local infrastructure. They also perform administrative tasks, such as informing the residents about the local authorities' plans and decisions or organizing village assemblies, and they are expected to mobilize the residents for collective initiatives. In fact, the scope of activities of the village representative differs from one municipality to the other, as these activities are not strictly defined by the legal regulations, but rather become determined by the expectations and needs of the residents, the attitudes of particular municipal authorities and the intentions of the village representatives themselves. In general, the impact of the village representative on local decision-making processes consists of 'backstage' activities, or exerting informal

Table 17.1. The organs of local government entities in Poland in 2013. (From Central Statistical Office, 2014, pp. 38–39.)

Entities	Executive bodies	Legislative bodies
16 provinces (*województwa*)	Province executive board (*zarząd województwa*) chaired by province marshal (*marszałek*)	Province regional assembly (*sejmik wojewódzki*)
314 districts (*powiaty*)	District executive board (*zarząd powiatu*) chaired by starost (*starosta*)	District council (*rada powiatu*)
66 cities with district status (*miasta na prawach powiatu*)	City president (*prezydent miasta*)	City council (*rada miasta*)
2479 rural, urban–rural and urban municipalities (*gminy*)	Mayor (*wójt/burmistrz*)	Municipal council (*rada gminy*)
40,583 rural sub-municipal auxiliary units (*sołectwa*)	Village representative (*sołtys*) supported by village council (*rada sołecka*)	Village assembly (*zebranie wiejskie*)

influence over the local government. It has to be noted that village representatives can participate in municipal council meetings, but they do not retain the right to vote. Four times a year, many village representatives collect local taxes from the residents of the village on behalf of the municipal authorities and receive the equivalent of a small percentage of the collected sum as a gratuity for their efforts (Matysiak, 2014a,b).

Analytical Framework: Processes of Occupational Gender Desegregation

Basically, occupational gender segregation is understood as the tendency for women and men to work in different occupations and job positions. Such a tendency is observed in many countries, especially the wealthy and industrialized ones (e.g. Blackburn *et al.*, 2002). The research literature distinguishes between two different forms of occupational gender segregation: horizontal and vertical (Anker, 2001). The former refers to the distribution of men and women across occupations. For example, occupations associated with women in Europe and America in recent decades include nursing, primary-school teaching, hairdressing and other beauty work, whereas men tend to concentrate in occupations such as mining, driving, plumbing and car sales (Hesmondhalgh and Baker, 2015). Vertical segregation, in contrast, refers to the distribution of women and men in the same occupation focusing on a lower proportion of women than men occupying managerial positions (Renzetti and Curran, 2005). Job and workplace segregation are also distinguished as any job, in terms of function and level signals among other things, is characterized by a particular conception of gender (Tienari, 1999).

Occupational gender segregation contributes significantly to inequality on the labour market. Predominantly female jobs and occupations command lower wages compared with those that are predominantly male, even when controlling for required skills and experience (Skuratowicz and Hunter, 2004). Moreover, women are more likely than men to work in occupations which provide few or no benefits, have shorter career ladders and undertake part-time work (Gatta and Roos, 2001; England, 2010). Occupational gender segregation is explained by theories of human capital and rational choice, and feminist and sociological theories of patriarchy. The former focus on women's lesser employability due to, among others, their lower skills and load with household responsibilities, while the latter emphasize the gender discrimination (Blackburn *et al.*, 2002) and the persistence of gender stereotypes (Hesmondhalgh and Baker, 2015). According to the preference theory (Hakim, 2006), women and men prioritize professional or family roles depending mostly on their lifestyle choices.

However, occupational gender resegregation is also observed; that is, changes in the proportion of women and men in given occupations or jobs over time. In fact, these changes have largely taken the form of women moving into male-dominated fields (Gatta and Roos, 2001). For example, in recent decades an increase in the concentration of women in public relations, systems analysis, bartending, advertising and insurance has been noted in many countries (Hesmondhalgh and Baker, 2015). According to the literature, the process of occupational gender desegregation may lead to: (i) genuine integration, which means that women achieve equity with their male counterparts; (ii) ghettoization, which indicates that women and men in the same occupations perform different jobs and/or are employed in different companies; or (iii) resegregation, which refers to the situation when persons of one gender replace those of another in a given trade or profession (Gatta and Roos, 2001; Charles and Grusky, 2004).

According to the well-known queuing theory (Reskin and Roos, 1990; Reskin, 1991), women's entry into male occupations can be explained through the idea of job queues and labour queues. Labour markets function on the basis of queuing of the potential employees and employers: the former aim at getting the best available positions, the latter at hiring the best possible candidates. Gender constitutes one of the most significant lines of identifying employees as more or less desirable in the eyes of employers. Reskin (1991, pp. 174–180) identified the main factors that condition the process of women's entry into so-called male professions. The most powerful one was job deterioration; that is, a decrease in a job's attractiveness due to lowering of salaries and limitation of additional benefits, technological changes and transformation leading to deskilling, or limitation of promotion and development opportunities. The other factors, however less significant, included: quick development of a given trade, resulting in demand for more employees; demand for female employees, enforced by anti-discrimination provisions; and changes of social attitudes towards gender roles.

The main lines of criticism of the queuing theory refer to its focus mainly on one vertical axis, in which men dominate the best jobs (Charles and Grusky, 2004). According to Blackburn and Jarman (2002, p. 291), the queuing theory overestimates male agency and underemphasizes women's ability to resist or support the circumstances in which they have an unequal status. Gatta and Roos (2001) indicated that not only does gender shape the patterns of occupational desegregation, but class and race also play a part in these processes. Women tend to enter into traditionally male occupations at the upper levels of the occupational hierarchy, while men – blacks, Hispanics, immigrants and low-skilled workers – tend to move into traditionally female occupations. Other researchers claim that the structural changes of labour markets (e.g. deindustrialization) (Blackburn *et al.*, 2002), dynamics of local labour markets and cultural factors (Charles and Grusky, 2004), as well as the impact of expansion and contraction of the labour supply (Blackburn and Jarman, 2006), should also be taken into account while explaining occupational gender desegregation.

Despite these legitimate reservations, the queuing theory provides inspiration in terms of explaining gender desegregation occurring in particular occupations or positions, including political ones. Like in the labour market, in the sphere of political activity, the pool of positions that are attractive due to prestige, power, finances or other benefits is limited (e.g. parliament seats, state institution jobs, etc.). Gender segregation in politics is observed, for example, in a much smaller number of women involved in prominent decision-making positions than men (Epstein Fuchs, 1981). Also, female members of municipal councils, more often than men, are members of commissions dealing with problems perceived as being 'female-related', such as education and health care (Dzieniszewska-Naroska, 2004; Fuszara, 2006). However, an in-depth analysis of changes in terms of gender desegregation should not be focused only on the 'men's flight' theme, but must incorporate other factors as well, including ones related to the cultural and local context.

Polish women officially acquired the right to run in various elections in the year 1918. However, women living in the rural areas were not expected to even take part in village assemblies, as the role of representing the voice of a given family was traditionally ascribed to men. The discrepancy between women's formal rights and the societal reality was also an aspect in political institutions at higher levels such as the Polish Parliament (Siemieńska, 1990).

According to the existing literature, women were not really present in the institutions of rural sub-municipal auxiliary units either before the Second World War or shortly after (Jakubczak, 1976). However, the available data indicate a significant increase in the percentage of women among village representatives in Poland in recent decades. In 1958, the share of women performing this function amounted to only 0.8% of all village representatives; however, in 1967, the percentage had already reached 2.8%. A survey of a representative sample of Polish villages, conducted in the mid-1990s, showed that the share of female village representatives was 11.0% at that point (Ostrowski, 1995). Currently, as of the end of 2014, the percentage has increased to 35.9% (Table 17.2).

It is worth noting that the share of women among village representatives is

Table 17.2. The share of women village representatives in Poland for the years 1958, 1967, 2009, 2012 and 2014. (Data from Central Statistical Office, 1968; Central Statistical Office, Local Data Bank (www.stat.gov.pl).)

Year	Village representatives (total number)	Female village representatives (total number)	Share of women among village representatives (%)
1958	40,589	317	0.8
1967	39,822	1,099	2.8
2009	40,254	12,144	30.2
2012	40,402	14,192	35.1
2014	40,365	14,491	35.9

clearly higher than among municipal councillors (25%), district councillors (19%), province regional assembly members (25%), members of the Lower House (24%) and Upper House of Parliament (13%), and also mayors and city presidents (9%).[6] In general, this confirms the well-known observation that women are more likely to be present in the lower positions of political or business structures (Fuszara, 2006).

Interestingly, the process of gender desegregation among village representatives seems to be related to the spatial diversification of Polish rural areas. The percentages of female village representatives are visibly higher mostly in the north-western part of Poland and lower in the south-east (Fig. 17.1).

In the available data, the spatial distribution of the share of women among village representatives seems to be constant. As early as the late 1960s, the highest number of female village representatives was found in the western and northern parts of Poland, and the lowest in the east. Women more often became village representatives in areas that were more developed economically and less agricultural (Central Statistical Office, 1968, p. xvii). These observations also indicate that contextual factors must be taken into account while analysing the determinants underlying the increase in female village representatives.

The aim of this chapter is to address questions about the possible reasons for and consequences of the observed gender desegregation among village representatives in Poland. In addition to factors that condition women's entry into professions dominated by men, it looks at changes in the attractiveness of the village representative role in the context of the benefits that are achievable by the position holders, as well as the associated social prestige and scopes of competences and tasks. This analysis also includes changes in social attitudes with regard to gender roles. The remaining factors listed by Reskin (1991), such as fast development of a given trade, increase in the number of positions and impact of antidiscrimination legislation, are not included, as they are not directly associated with the function of the village representative.[7]

It may be assumed that the growing number of women among village representatives is caused by the low prestige and importance of this office; that women are more often elected as village representatives due to the fact that this position is not attractive for the local men. Alternatively, the mechanisms underlying the increasing share of female village representatives could be related more to changing gender roles and ideas of women's and men's participation in private and public spheres in Polish society generally. In other words, women become village representatives more often due to cultural changes and changing perceptions of the position, which are defined as more adequate for women or men (Inglehart and Norris, 2003).

As discussed previously, in the case of some provinces in Poland, the share of women among village representatives is significantly higher than in other parts of the country. As a result, this chapter attempts to

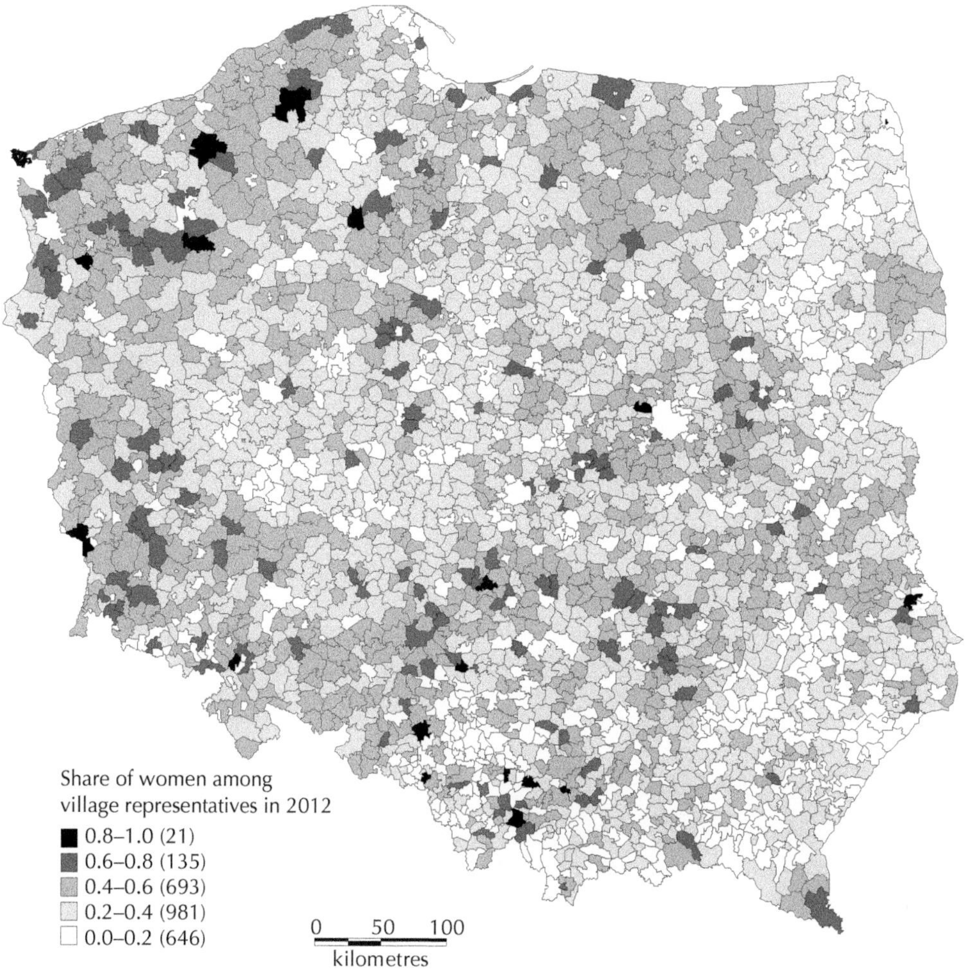

Fig. 17.1. Share of women among village representatives in Poland in 2012. The intervals refer to the share of women among village representatives in a municipality; the numbers in parentheses represent the number of municipalities characterized by a given share of female village representatives. (Data from Central Statistical Office, Local Data Bank (www.stat.gov.pl).)

grasp the impact of context-related factors on women's entry into the position of the village representative. The spatial diversification of Polish rural areas is conditioned by the complicated history of Poland, on the one hand, and recent development processes, on the other. Therefore, the aim is to verify which factors are more significant regarding the gender desegregation of the function of the village representative: the specific history of a given region and its cultural traits, the changes in gender roles, or the actual characteristics of the local context.

As for the consequences of the growing number of female village representatives, the question is whether women perform this function in a different way from men. According to the literature on women in politics, it is assumed that female politicians, due to their particular experiences in both the private and public spheres, 'make a difference' in legislative processes and political leadership by expressing greater concern about childcare, health, education and women's presence in politics (Lovenduski and Norris, 2003). Therefore, it can be hypothesized that female village representatives are more eager to engage in activities in such areas compared with men.

Methodology

The analysis presented is based primarily on qualitative data collected from ten municipalities located in different provinces of Poland. The municipalities were chosen because of their different proportions of women serving as village representatives and municipal councillors. Two municipalities – one 'feminized' and one 'masculinized' – were examined in each of five chosen provinces of Poland (Table 17.3). A 'feminized' municipality was defined as one where the percentage of women among village representatives and municipal councillors was close to 50%, whereas 'masculinized' municipalities were those where the percentage of women serving as village representatives and municipal councillors did not exceed 30%.[8]

Fieldwork was completed between November 2009 and December 2010. In total, 108 individual in-depth interviews were conducted in ten municipalities. The respondents consisted of 51 village representatives and 57 interviewees from local authorities, public institutions and organizations. Among the village representatives interviewed were 28 women and 23 men. They were most often 40–70 years of age and working on their own farms, retired, or receiving disability benefits. The remaining interviewees consisted of 34 women and 23 men: representatives of municipal authorities (mayors, councillors) and leaders of local organizations, as well as local public officers.

The qualitative empirical material is supplemented with select quantitative data from national representative surveys used to describe the overall trends in redefinition of gender roles in Poland. As well as that, Central Statistical Office and National Electoral Commission data from 2005 to 2013, obtained through the web service Local Data Bank[9] and the non-governmental portal My Polis,[10] were included. These data have been used to perform multiple regression analyses aiming to identify the strength and direction of the relationships between the share of women among village representatives and selected contextual factors.

Research Findings

'Attractiveness' of the function of the village representative

Benefits, prestige and competences

According to the interviewees, the village representative's role does not entail any significant material benefits; it is perceived as a voluntary activity: 'It is very important, but it is unprofitable and impossible to make a living out of it' (Z.D.9_f.v.).[11] Annual income from collecting local taxes in villages and for taking part in meetings of the municipality's council is minimal: 'I get 3400 PLN [about €850] per year from the local tax collection. A laughable sum' (M.K.6_f.v). Some municipalities' authorities introduce telephone and petrol cost allowances for their village representatives, but such arrangements were not mentioned in the ten municipalities studied.

Due to the aforementioned strong institutional and financial dependence of rural sub-municipal auxiliary units on municipal authorities that is embedded in the current regulations, the function of the village representative is perceived as being deprived of power and significance. According to the constitutional principle of subsidiarity, the local authorities of each municipality are free to decide what range of responsibilities should be passed to the rural sub-municipal auxiliary units within their administrative territory. However, in fact, often they are not eager to pass them any significant tasks, because they perceive it as a weakening of their own field of power. Individuals performing village representative function cannot formally request anything from the municipality's authority;

Table 17.3. Characteristics of municipalities subject to research, Poland, 2009. (Data from Central Statistical Office, Local Data Bank (www.stat.gov.pl).)

Province	Zachodniopomorskie		Mazowieckie		Lubelskie		Wielkopolskie		Podkarpackie	
Municipality	Dobra Szczecińska	Mielno	Kałuszyn	Wyszogród	Serniki	Ryki	Murowana Goślina	Czerwonak	Lutowiska	Łańcut
Type	Rural	Rural	Urban–rural	Urban–rural	Rural	Urban–rural	Urban–rural	Rural	Rural	Rural
Population	15,569	4,935	6,086	5,939	4,876	20,505	16,248	25,300	2,195	20,849
Share of women among municipal councillors (%)	53.3	33.3	53.3	6.7	40.0	19.0	53.3	28.6	13.3	9.5
Share of women among village representatives (%)	66.7	12.5	75.0	11.8	46.2	23.3	25.0	36.4	66.7	11.1

they can only attempt to influence it in such a way that the needs of their residents are reflected in the commune's budget:

> The village representative can't do much. . . . The ideas have to be presented to the municipality's authorities to convince the councillors to vote for a given investment to be carried out in this year and not in two or three years.
>
> (Z.D.10_f.v)

The village representative's function is perceived as less prestigious than before the system's transformation, which began in 1989. As one interviewee stated: 'In general, village representatives do not have the same esteem in the eyes of the society as they had, let's say, 20 years ago' (P.Ł.2_m.l). During the period of state socialism, village representatives could access limited resources, such as fertilizers, construction materials and farming machines, more easily than ordinary residents of the village: 'In the past . . . it was hard to get anything, under communist rule, and this position [of village representative] allowed us to deal with the[se types of] matters more easily' (M.W.6_m.c). It seems that, at that time, the function of the village representative was more strictly associated with politics in terms of relations to local administration and the ability to benefit from those relations.

Scope of responsibilities

Interviewees pointed to increasing social expectations with regard to the activities of village representatives. First of all, the demands of rural inhabitants are growing: they expect a high standard of local infrastructure and quick solving of reported problems. In particular, the female village representatives indicated being treated by the inhabitants as a 24-hour emergency service for everything. Moreover, village representatives are increasingly expected to act as local leaders, building the social capital of the village and committing people to participate in joint projects for local development: 'there is [a] demand, rather, for an activist, a leader' (L.R.4_m.c). Some interviewees pointed out that it is expected of them to possess the skills to fundraise from external sources, the EU included, in order to implement local initiatives.

At the same time, changes in the scope of obligations and tasks assigned to village representatives can be analysed also in the context of various activities that are regarded as 'male' and 'female'. Activities directly associated with investing in local infrastructure were perceived by the interviewees as more masculine; that is, more appropriate for men than for women. On the contrary, activities related to integration of the residents of the village and cultural initiatives were perceived as more feminine. During the period of socialism, representatives were often responsible for the coordination of so-called 'voluntary social work' (*czyn społeczny*). Many local roads and municipal buildings were constructed by village inhabitants, and local authorities delivered the construction materials. In the opinion of some respondents, a recent satisfaction of the most urgent needs in this regard raises expectations that other types of initiatives will take place. They said that village representatives should focus more on integrating the residents and motivating them to participate in the life of the community:

> The man, who was [the village representative] here, he contributed to construction of the gas supply network in the village, we appreciate this. These were important matters, surely, but they did not integrate the inhabitants . . . there was no such work, like, we'd do something for ourselves or for others.
>
> (Z.D.10_f.v)

This shift in expectations does not mean that all needs concerning local infrastructure are satisfied completely. Today, the inhabitants often engage in some work on behalf of their village (e.g. small repairs, painting or cleaning of public spaces). However, heavy work, such as improvements to the conditions of the local roads or snow removal, are outsourced and paid for by the municipal authorities. Therefore, these heavy projects, which used to be done by village residents, no longer require village representatives to supervise them, which means that male authority is no longer perceived to be necessary to be successful in the role.

Men's interest in the function of the village representative

Many interviewees explicitly stated that there are more women among village representatives now because men are less interested in taking on this role: 'Maybe it is due to the fact that men no longer want this sort of job . . . because it is voluntary activity, not really well paid' (Z.D.8_f.v). The growing expectations of village representatives and lack of benefits mean that many men consider it no longer worth their time to take on that role.

In many interviews, it was indicated that for young and middle-aged men, the priority is professional work which generates income, and not the voluntary work of a village representative: 'Young men, between 30 to 50 years old, travel to work, and women stay home, locally and have time for this [dealing with the village's affairs]' (L.S.5_m.v).

> Men. . . they go to work in the morning, they return home very late, and the ladies, those who get involved, these are usually not working women. They simply have more time in the afternoon, which they can dedicate [to village activities].
>
> (P.L.8_f.v)

In the case of some localities, it was indicated that men looked for jobs further away and they often visit their village only on the weekends: 'So this is a specific village, because there are a lot of men who work outside of the village. A guy doesn't have the time for the village stuff, he is not here' (M.W.8_f.c). While internal migration for income and work commutes were mentioned more frequently, migration abroad was mentioned only sporadically – specifically in the context of seasonal migration for fruit-picking (for men and women). Of course, in Poland, it is possible to indicate regions in which women tend to leave their village more often than men to look for work elsewhere in the country or abroad (Slany, 2008). Nevertheless, it should be noted that the division of paid work, seen as more suitable for men, and voluntary work, considered to be more appropriate for women, continues to be maintained (cf. Little, 1997).

Attitudes towards gender roles

Redefinition of gender roles in Poland – overall trends

According to Siemieńska (2011), Poland, as a country undergoing systemic transformation, constitutes a specific mosaic of traditional and non-traditional systems of values and behaviours. However, as in many other countries, Poland is witnessing a redefinition of traditional gender roles which have assumed that family life is women's responsibility. Increase in the level of education attained by women, and their growing aspirations, motivates them to become more involved in the public sphere, and they are becoming more accepted in such roles (Siemieńska, 2011). These trends are highly present in Poland, despite a concurrent backlash in the form of the growing visibility of conservative public discourse and conservative public policy implementation (Gerber, 2011).

Results of the World Values Survey and European Values Survey indicate that in the years 1980–2005, acceptance for professional work and political involvement of women increased markedly in Poland, as well as acceptance for transformation of the family model, which makes it possible to reconcile work with family duties (Siemieńska, 2011). According to the 2012 World Values Survey results, only 31.4% of Poles (39.4% of men and 28.3% of women) agreed with the statement that 'generally speaking, men are better political leaders than women' (WVS, 2012, unpublished).

Similar changes in social attitudes towards gender roles are also observed among inhabitants of rural areas, although to a lesser extent in comparison with cities. According to a representative 2013 national study, 50% of Polish women preferred a partnership family model, 19% a traditional one and 23% a so-called un-proportional one (husband/partner spends more time on salaried work, while wife/partner in addition to salaried work also takes care of the household duties). Among the rural female population, 43% were in favour of the partnership model, 28% the traditional model and 20% the un-proportional one (Public Opinion Research Center, 2013, p. 5). According to the Polish General Social Survey, in 1992 57.4% of men and 45.0% of women agreed that 'governing the country is best when left to men'. The respective percentages among rural men and women agreeing with that statement were 68.3% and 57.1%. By 2010, only 35.7% of male respondents and 23.0% of female respondents agreed with this statement; in rural populations, however, those percentages remained much higher, with 46.4% of men and 32.2% of women agreeing that it would be best if only men governed the country.[12]

In 1970, just 0.2% of rural women had a university degree and 5.7% of them had completed secondary education. For rural men levels of educational attainment were similar, at 0.6% and 4.9%, respectively (Central Statistical Office, 1981, p. 45).[13] In 2011, 12.1% of rural women had achieved a higher education degree and 27.9% were educated up to the secondary level. Among men, these percentages were 7.7% and 23.1%, respectively (Central Statistical Office, 2012, p. 60).[14] Thus, the low level of education among rural women is no longer an argument against entrusting them with local public functions.

Female village representatives' social resources

According to this qualitative research conducted in ten municipalities, women taking on the role of village representative had a slightly more favourable educational attainment level when compared with their male counterparts. However, it needs to be noted that, irrespective of gender, individuals with higher education were very rare to come by (Table 17.4).

Nearly half of the interviewees, women and men alike, were involved in various kinds of local social activity before taking on the village representative role. Activity within formal structures was more typical among men than women; there was also a trend of gender segregation in the types of activities they engaged in. In the case of female village representatives, the dominant types of activity were those perceived as typically 'feminine'; that is, activities relating to parental committees and councils at local schools, rural women's organizations and informal activities. Men more often were previously engaged in local government structures such as village or municipal councils, or were involved in traditionally male-dominated organizations, such as the voluntary fire service, farming associations and sports clubs. Despite the fact that the majority of the interviewed village representatives were farmers or beneficiaries of old-age or disability pensions, most of the interviewees (both female and male) had some experience in professional work outside agriculture. It also transpired that among those who had little or no experience in social activity prior to taking on

Table 17.4. Level of education of village representatives interviewed according to gender, Poland, 2009–2010.

Education	Women (n=28)	Men (n=23)
PhD	–	1
Higher	2	4
Secondary (including vocational secondary)	19	7
Basic vocational	5	6
Elementary	–	2
No data available	2	3

the village representative role, some of the interviewees – men slightly more often than women – pointed out they had been elected as village representatives mainly thanks to their professional experience, which warranted them 'visibility' in the local community and extensive possibilities of establishing relationships with people.

It needs to be noted that roughly half of village representatives in this study continued family traditions of involvement in local public life. Family role models were most often male figures; women mentioned tended to be involved in a smaller array of activities. Interviewees mentioned their fathers, who were municipal council members, village representatives and members of village councils, involved in the farmers' organizations (*kółka rolnicze*), voluntary fire services (*ochotnicze straże pożarne*), and folk ensembles and theatres; some interviewees also mentioned their grandparents and uncles active in the interwar period or earlier (council members, peasants' movement activists, village representatives). Mothers or mothers-in-law who were mentioned were mostly involved in rural women's organizations (*koła gospodyń wiejskich*); only few persons spoke of female council members, village representatives or librarians.

It can be said then, that contemporary rural women, like men, have various social resources which enable them to take on local public roles. The comparison of their experiences with the previous generation of women depicts the process of an increasing female involvement into various local structures and activities.

Gender roles in different local contexts

This research pointed to clear differences of opinion with regard to gender roles, as well as women's involvement in local public life between the 'feminized' and 'masculinized' municipalities. According to the interviewees, women perform public or decision-making functions mostly in educational and cultural institutions, and they comprise more of the employees in the local public service. Particularly in the 'masculinized' municipalities, local politics were perceived as the domain of men, and social and cultural work was that of women. In some municipalities it was stated that there is an inherent 'glass ceiling' within the local government structures, according to which women take good care of issues at a village level but not when deciding about municipality-scale investments: 'Women are suited better for a village representative than for a municipal councillor. . . . Councillors [deal with] investments, financial matters, the municipality's budget' (P.Ł.3_f.c).

It is interesting that in the 'masculinized' municipalities, traditionally male organizations, such as the voluntary firefighting service, were listed as the most active and visible in the local public sphere. This decreases the chances of women gaining social support and different experiences which could strengthen their position in the local community. In contrast, in the 'feminized' municipalities, the number of active local organizations, including male-dominated ones, was smaller, in general. Thus, it can be expected that the lower density of social capital, especially in terms of 'masculinized' local structures, contributes to greater openness of the local public sphere for women. An illustrative example of this is given by statements from one of the municipalities, in which the most visible local organization was the volunteer firefighting service and the local labour market was based on 'male' professions associated with work in the woods:

> Our [municipality's] council was set up with 13 [men] to two [women], chiefly because there was a firefighting service electoral committee, which is made of men mostly, and the firefighting service and the forest district were involved in the making of councillors.
>
> (P.L.5_m.l)

In the 'feminized' municipalities, the greater share of women among village representatives and municipal councillors was accompanied by more positive and comparison-oriented opinions on gender roles and women performing local public functions. In these 'feminized' municipalities, the interviewees tended to point out changes in the general way of thinking about the equality of men and women: 'It seems to me that social awareness has increased . . . women no longer fit the stereotype of only taking care of the house, children, cleaning, cooking' (L.S.8_f.l). Apart from that, so-called gender-specific traits, such as sensitivity, scrupulousness, serenity and self-composure, as well as a sense of responsibility and discipline, were acknowledged here as particularly useful in performing the functions of the village representative or municipal council member. Moreover, according to the interviewees, women were more sensitive than men regarding social problems, such as issues relating to children in deprived families. Emphasis was also put on the greater readiness of women to cooperate with local organizations and engage in activities on behalf of the common good: 'They are more committed, more willing to cooperate. And men tend to go their own ways. Unless they're party members already . . . and they know there are benefits in it. Women are more focused on cooperation, on giving' (Z.D.6_f.c). In those municipalities, women performing the services of a village representative or municipal councillor were perceived as competent and well prepared for such tasks.

In the 'masculinized' municipalities, the majority of interviewees, including women, tended to point out the persistence of more traditional gender roles: 'With regard to female village representatives, for ages women have not been allowed to participate in power, only hearth and home. A guy was more able to attend to everything, representing outside of home and outside of the village' (M.W.8_f.c). Some of them suggested that there is a strong fear of failure and humiliation which prevents local women from having social commitments, and that a lot of women may be afraid: 'Why should they bad-talk about me?' . . . like "You are not up to it!"' (P.Ł.2_m.l). Interestingly, in the 'masculinized' municipalities, the particular attributes described by interviewees as 'female' ones were, in many cases, the same as previously mentioned. However, they were more often assessed negatively or vaguely by both men and women. Moreover, the declarations of the few women serving as village representatives or municipal council members in the male-dominated municipalities indicated the presence of the phenomenon of tokenism (Kanter, 1977). First of all, these women experienced a sense of uncertainty and perceived themselves as outsiders. For many, the sense of isolation was associated with the classic situation of being excluded from male social networks: 'If there's a group of men, they gather together, after the session, they go to have a drink together or something' (M.W.7_f.v).

These outcomes support the statement that the visible presence and activity of women performing public functions significantly influence the perception of gender roles in a given society and increase the acceptance of women's entry in politics (cf. Siemieńska, 2005).

Influence of contextual factors

As previously discussed, in the case of some provinces in Poland, the share of women among village representatives is significantly higher in comparison with that in other parts of the country. Differences are particularly visible between south-eastern Poland (mainly the region of Galicia[15] and some parts of the former Congress Poland[16]) versus the north-western territories added after the Second World War.

According to the literature, the territories added to Poland after the Second World War are characterized by greater openness to innovations, such as women serving in local government roles. Their primary inhabitants fled during the war or were forced to leave shortly after it ended. Afterwards, these territories were settled by people from different parts of pre-war Poland. In this process, new post-migration local communities were created which, according to the literature, are less 'rooted' and have loosened traditional norms and social control. It results in, on the one hand, a higher level of pathological behaviour but, on the other, greater acceptance for the new and uncommon. In comparison, in the southern and eastern regions of Poland, local communities are characterized by conservatism and preserved continuity of traditions (Gorzelak and Jałowiecki, 1998; Bartkowski, 2003).

The multiple regression analysis was conducted in order to verify the hypothesis that the historical differences between Polish regions, especially the specificity of the territories added to Poland after the Second World War, are related to the process of gender desegregation among village representatives more strongly than other factors related to socio-cultural traits and actual characteristics of the local context. The multiple regression analysis used allows checking in what way a specified rural municipality's characteristics tie in with the share of women among its representatives, after taking into account the influence of other factors on the variable being explained.

Based on public statistical data available on a municipality level (NTS-5), this research analysed the strength and direction of relationships between the dependent variable – the share of women among village representatives in a rural municipality in 2013 – and the independent variables which can be described as:

1. Location in a given historical region (Galicia; the areas of the former Prussian and Russian partitions; the territories added after the Second World War);
2. Cultural traits (types of local organizations: traditional, indicated by voluntary firefighting services; modern, indicated by non-governmental organizations, progressive voting preferences);
3. Gender roles (the share of women among municipal councillors; the share of men among working population);
4. Characteristics related to local development (proportion of households with income from agriculture; the quality of local infrastructure measured by the water-supply system; the wealth of the municipality; migrations).

The analysis confirmed a relatively strong relationship between the dependent variable and the municipality's location within a given historical region – in particular in the former Austrian partition region (Galicia) and in the western and northern territories.[17] The analysis also showed significant, but relatively weaker relationships between the dependent variable and the share of women among municipal councillors, types of social/voluntary organizations and the share of households with agricultural activity income (Table 17.5). The model does not include statistical values or significance levels because the sample is exhaustive; that is, it includes all rural municipalities in Poland.[18]

The analysis showed the importance of the impact of historical factors on the different Polish regions with regard to the process of gender desegregation among village representatives. This process is the most advanced in the territories added to Poland after the Second World War, described in the literature as the areas most willing to accept innovations, and the slowest in the conservative region of Galicia. At the same time, the significance of other contextual factors related to culture and gender roles was also shown. As discussed already in terms of qualitative data, the types of local organizations and the presence of women in local politics are significant as well. These factors seem to be much more relevant for understanding gender desegregation among village representatives than the actual economic and developmental characteristics of rural municipalities.

However, the regression model explains 21% of the variance in the share of women among village representatives. This means that, on the one hand, the structure of available statistical data may not provide an exhaustive picture of the current situation of rural areas in Poland.[19] On the other hand, it can be assumed that taking on the role of village representatives by women is linked with some phenomena or processes present at the lowest local level – rural settlements – for which there are no detailed public statistical data.

Table 17.5. Multiple regression analysis: independent variables and linear model coefficients predicting the share of women among village representatives in a rural municipality in Poland in 2013.

Independent variable	Coefficient
(Intercept)	37.210
Proportion of households with income from agriculture (2010)	–0.125
Share of the municipality's residents using the water-supply system (2012)	0.014
Share of women among municipal councillors (2013)	0.330
Value of EU projects per resident – Human Capital Programme (2013)	0.002
Share of men among the working population (2013)	0.035
Number of residents migrating for work per 1000 residents (2006)[a]	0.024
Number of men opting out from the residents' register per 1000 residents (2013)	–0.070
Historical region – Prussian partition area versus Russian partition area	1.863
Historical region – Austrian partition area (Galicia) versus Russian partition area	–7.638
Historical region – western and northern territory versus Russian partition area	2.691
Non-governmental organizations per 1000 residents (2013)	0.214
Municipality's income per resident (2013)	0.000
Share of votes for D. Tusk in the second round of presidential elections (2005)[b]	0.006
Number of voluntary fire services per 1000 residents (2013)	–2.431

[a]More recent data are not available.
[b]In the 2005 presidential election D. Tusk strongly attracted voters of a progressive hue.

The consequences of women taking on the village representative function

Types of village representatives' initiatives

In the light of the qualitative study's findings, the actions implemented by both men and women as village representatives turned out to be quite similar. Most often they included the improvement of local technical infrastructure (roads, sewage and water-supply systems, gas networks, Internet networks, etc.), the improvement of local social infrastructure and public spaces (rural community centres, playgrounds, sports fields, village centres, caretaking and village aesthetics overall), establishing residents' integration–building initiatives (fairs, special occasion parties, etc.), and coordinating philanthropic and charitable activities (food collections, helping the elderly and disabled). It seems that both female and male village representatives tried to address the most urgent needs in a given locality.

However, an in-depth analysis of the interviews allowed the capture of some interesting differences. Men were more focused on the improvement of the local road infrastructure, sewage systems and so forth, while women tended to more often prioritize the livening up of local social life. Women found the integration and public spaces building initiatives (rural community centres, playgrounds, etc.) to be especially important: 'Last year we did this park, we have benches, you can have a rest there. Even with kids – moms are meeting up and this is so nice' (Z.D.1_f.v). Men also take on such activities, but they tended to list them later, giving more focus in their interviews to solving problems regarding local technical infrastructure. They did not pay as much attention as women to the social or integrative functions of a well-managed and maintained local public space. The female village representatives slightly more often than their male counterparts initiated the setting up of local social or voluntary organizations: 'Caritas' branches[20] (one female), village associations (four females, one male) and a voluntary fire service (one male, one female). Only female village representatives initiated philanthropic and charitable initiatives – visiting the elderly and those living alone during Christmas or Easter, helping people in difficult circumstances – whereas men tended to engage in such actions organized by other local actors. The women were also more patient and open in contacts with residents. More often than men, they pointed out that residents of the village were confiding in them with private problems: 'It is just that this role [of the village representative] is very needed, because sometimes you just have to listen to them, because something happened, and you have prick up one's ears and be a psychologist' (Z.D.9_f.v).

The position of village representative as a 'springboard'

According to statements by both female and male village representatives, performing this function allows them to increase their social capital: broaden their contact networks, gain or strengthen support in the local community. Women and men became public personas, being recognizable by the municipality's residents and authorities. What is significant is that half of the male and female village representatives pointed out that, while already in office, they had also taken on another function within public life or got involved in some activity of a local organization. Most often these were either a municipal councillor position or an involvement in social or voluntary organizations, also as a member of their management boards: 'I managed to [get elected as municipal councillor] because as a [female] village representative I have close and direct contact with people' (P.L.8_f.v). 'Firstly I was a [male] village representative, then a municipal councillor. . . . People noticed my activity' (W.MG.5_m.v). These types of experiences were slightly more often mentioned by men than women, but the differences were small.

The results indicate that performing the functions of a village representative exerted a positive influence on women. Only women described personal development and increasing self-confidence as being benefits of the position:

> Now I am more confident because, initially, it was difficult for me to talk to them [municipal councillors and local officials] at all. It's different when you talk to your family or friends and different when you go to an office to deal with an issue.
> (M.W.11_f.v)

Benefits of this kind were mentioned by women in the context of self-development, self-esteem boosts and a sense of well-being associated with receiving inhabitants' appreciation for their work.

Conclusions

The process of gender desegregation among village representatives is partially explained by the devaluation of this position. The function of village representative is now perceived more as a voluntary activity deprived of any formal power and, therefore, as being more appropriate for women than in the past. However, in practice, it requires a lot of political involvement (the ability to negotiate and gain the support of others). Compared with the period of the socialist rule, the village representative role requires quite a substantial commitment today but no longer brings calculable financial or material benefits. At the same time, the nature of activities undertaken by village representatives has changed in recent decades: the coordination of 'heavy' investment projects, such as local roads or communal buildings constructed by the inhabitants, is less frequent, whereas a greater emphasis has been put on 'soft' activities aimed at integrating the residents and getting them involved in the social and cultural life of the village. Heavy projects are described by the respondents as being more 'masculine', whereas the soft initiatives are perceived as being more 'feminine'. Such interplay between socio-cultural factors regarding the nature of the village representative role and its social perception has resulted in its increasing availability for women. It can be said that the village representative role is becoming less and less 'masculinized'; that is, less associated with characteristics attributed to men, such as political authority, physical strength, 'specialist' technical knowledge and practical skills in construction, or the necessity to 'manage' the male workforce. The analysis conducted herein has also shown that women do not lack social resources, such as formal and informal social involvement, that provide the competencies necessary to perform public functions. However, these resources are clearly marked by gender segregation. The male village representatives who were interviewed more often got involved in the local government structures and the 'male' local organizations, and the female village representatives in structures, organizations and informal social initiatives perceived as important, but distant from the sphere of local politics, and specifically those dealing with perceived 'female' issues. This confirms the strength of the impact of the mechanisms of increasing male dominance, discussed in the literature presented above.

Interestingly, when it comes to the consequences of gender desegregation among village representatives, the results of these analyses suggest that being a woman or a man does not make a significant difference in how this role is performed. However, women more than men seem to emphasize initiatives aimed at increasing social cohesion. It can be said that women bring to the rural local government a particular 'added value', but all in all gender is not a strong factor differentiating the types of implemented activities. It transpires, from the majority of female village representatives in the qualitative study, that the benefits related to holding this office were more likely to take the form of personal development. However, some of them, although slightly less often than men, were able to use their skills and experience to take on relatively more formal and influential roles, especially the role of municipal councillor (cf. Lowndes, 2005, p. 214). Therefore, it is not that 'random' women fill in the gap left by men, who simply prefer to be involved somewhere else. It seems that the role of the village representative has been deprived of formal power, but not necessarily of social prestige. However, this function does not provide prestige automatically – it depends on each of its holders' ambitions and engagement in local affairs whether its potential will be used in terms of pursuing any further career in local politics.

Therefore, it is possible to consider the increase of the number of women among village representatives in Poland as a manifestation of slow but inevitable processes of socio-cultural change. As has been noted by Inglehart and Norris (2003), a prerequisite for real redefinition of gender roles towards equality is a far-reaching cultural change. The great significance of culture is due to the fact that both men and women accept the attitudes, values and beliefs associated with division of gender roles that are dominant in their society. Wherever traditional values prevail, women are limited not only by the society – they limit their chances on their own. In the case of Poland, the reasons for spatial distribution of women among village representatives should be sought also in the specific regional and local cultural traits resulting from historical circumstances. A special case in this context is represented by the western and northern parts of the country, comprised largely of post-migrant communities that have looser social ties and local roots, and which, at the same time, are free from the burden of tradition, known to hinder the acceptance of social behaviours that are not typical in a given society.

Notes

[1]This text either develops or complements the issues discussed in other publications, in particular Matysiak (2014a,b).

[2]Central Statistical Office, Local Data Bank (www.stat.gov.pl).

[3]A rural sub-municipal auxiliary unit usually covers one village, but in some cases it may cover the territory of two small villages or a part of a larger village.

[4]Since 2009, the municipal councils have had the ability to create a village fund (*fundusz sołecki*), used to provide financial support for the rural sub-municipal auxiliary units. However, it should be noted that some municipalities implement their own mechanisms of providing financial support for their rural sub-municipality auxiliary units, independent of the village fund.

[5]All persons who are Polish citizens, aged 18 years or older, permanent residents of the village and having all public rights.

[6]Data from the Central Statistical Office. With regard to the local government's structures, these refer to the 2010–2014 governing period. Gender breakdown tables for the last local elections (2014) were still not available at the time of the writing of this text.

[7]The number of village representatives in Poland has changed only slightly in the last 50 years. In 2011, the 35% quota for the national and local election lists with the recommendation of placing at least one woman among the top three candidates was introduced. The introduction of the electoral quota regulations in Poland definitely contributed to public debates on gender equality. However, the quota regulations do not apply to the rural sub-municipal auxiliary units.

[8]The sampling criteria for municipalities used were determined by the aims of the author's PhD study, funded by the Ministry of Science and Higher Education (N N116 433 237).

[9]http://stat.gov.pl/bdl/app/strona.html?p_name=indeks

[10]http://www.mojapolis.pl/

[11]Labelling of interviews: the first letter indicates the province ('Z'=Zachodniopomorskie, 'M'=Mazowieckie, 'L'=Lubelskie, 'W'=Wielkopolskie, 'P'=Podkarpackie), the second letter(s) refers to the first letter(s) in the name of the municipality where the interview was conducted; the number identifies which of the interviews from that location is being referred to; the letter 'f' or 'm' refers to the respondent's gender, female or male; and the last letter indicates the category of the respondent ('v'=village representative, 'l'=local leader, 'c'=municipal councillor, 'o'=public official or representative of another local institution).

[12]The Polish General Social Survey (PGSS), which began in 1992, is a permanent programme of statutory research at the Institute for Social Studies of the University of Warsaw. The details on the PGSS, the data and surveys can be accessed through the Social Data Archives website (http://www.iss.uw.edu.pl/pgss/).

[13]To compare with the urban Polish population: in 1970, 6.4% of men and 3.1% of women had university degrees, and 20.7% of women and 19.2% of men completed secondary education (Central Statistical Office, 1981).

[14]Again comparing with the urban Polish population, in 2011 in that group there were 23.2% female university degree holders and 37.3% women with secondary education. For men, it was 19.4% and 33.1%, respectively (Central Statistical Office, 2012).

[15]Galicia is a historic and geographic region in Eastern Europe, which straddles the border between present-day Poland and Ukraine. As a result of the partitions of Poland (1772–1795), Galicia was granted to the Habsburg Empress Maria-Theresa.

[16]The Congress Poland was created in 1815 within the territory granted to the Russian Empire in the course of the partitions of Poland (1772–1795).

[17]The municipality's location within the historical Russian partition region is the reference category here.

[18]So-called urban–rural municipalities (where there is a town as well as rural areas within the borders) have been excluded from this analysis, due to limited availability of the urban–rural breakdown of the public statistical data.

[19]The data on the structure of the labour market, employment and household income sources on municipality levels (NTS-5) are incomplete and out of date; this prevented inclusion of such factors into the multiple regression model. Up-to-date data on migration are based on information regarding opting out from the residents' register; it does not reflect the complete scale of mobility, including the circular type.

[20]'Caritas' is a confederation of Catholic charity, development and social service organizations operating in over 200 countries and territories worldwide.

References

Anker, R. (2001) Theories of occupational segregation by sex: an overview. In: Fetherolf Loutfi, M. (ed.) *Women, Gender and Work. What Is Equality and How Do We Get There?* International Labour Office, Geneva, pp. 129–155.

Bartkowski, J. (2003) *Tradycja i polityka. Wpływ tradycji kulturowych polskich regionów na współczesne zachowania społeczne i polityczne*. Wydawnictwo Akademickie „Żak", Warsaw.

Blackburn, R.M. and Jarman, J. (2006) Gendered occupations: exploring the relationship between gender segregation and inequality. *International Sociology* 21(2), 289–315. Available at: http://dx.doi.org/10.1177/0268580906061380 (accessed 27 March 2015).

Blackburn, R.M., Browne, J., Brooks, B. and Jarman, J. (2002) Explaining gender segregation. *British Journal of Sociology* 53(4), 513–536. Available at: http://dx.doi.org/10.1080/0007131022000021461 (accessed 18 May 2016).

Bock, B. (2010) *Personal and Social Development of Women in Rural Areas in Europe.* European Parliament, Brussels.

Bock, B. (2014) Gender mainstreaming and rural development policy: the trivialisation of rural gender issues. *Gender, Place & Culture* 22(5), 731–745. Available at: http://dx.doi.org/10.1080/0966369X.2013.879105 (accessed 26 March 2015).

Bock, B. and Derkzen, P. (2008) Barriers to women's participation in rural policy making. In: Morell, I.A. and Bock, B. (eds) *Gender Regimes, Citizen Participation and Rural Restructuring.* Elsevier, Amsterdam, pp. 263–281.

Bourdieu, P. (2002) *Masculine Domination.* Stanford University Press, Redwood City, California.

Bukraba-Rylska, I. (2011) O polskiej wsi i tych, co „chłopu żywemu nie przepuszczą". In: Bucholc, M., Mandes, S., Szawiel, T. and Wawrzyniak, J. (eds) *Polska po 20 latach wolności.* Wydawnictwa UW, Warsaw, pp. 274–293.

Central Statistical Office (1968) *Sołectwa i sołtysi.* Central Statistical Office, Warsaw.

Central Statistical Office (1981) *Rocznik Statystyczny 1981.* Central Statistical Office, Warsaw.

Central Statistical Office (2012) *Report on the Results. The National Census of Population and Housing 2011.* Central Statistical Office, Warsaw.

Central Statistical Office (2014) *Statistical Yearbook of Voyvodships.* Central Statistical Office, Warsaw.

Charles, M. and Grusky, D.B. (2004) *Occupational Ghettos: The Worldwide Segregation of Women and Men.* Stanford University Press, Stanford, California.

Dzieniszewska-Naroska, K. (2004) *Radny – Sąsiad i polityk. Reprezentacja polityczna na szczeblu lokalnym.* Wydawnictwo „Trio", Warsaw.

England, P. (2010) The gender revolution: uneven and stalled. *Gender & Society* 24(2), 149–166. Available at: http://dx.doi.org/10.1177/0891243210361475 (accessed 30 March 2015).

Epstein Fuchs, C. (1981) Women and elites: a cross-national perspective. In: Epstein Fuchs, C. and Rose, C.L. (eds) *Access to Power: Cross-national Studies of Women and Elites.* Allen & Unwin, London, pp. 3–15.

Fuszara, M. (2006) *Kobiety w polityce.* Wydawnictwo „Trio", Warsaw.

Gatta, M.L. and Roos, P.A. (2001) Rethinking occupational integration. Revised version of a paper presented at the Annual Meeting of the American Sociological Association, Anaheim, California. Available at: http://www.rci.rutgers.edu/~roos/pdfiles/integration1103.pdf (accessed 12 May 2016).

Gerber, A. (2011) Cultural categories of worth and Polish gender policy in the context of EU accession. *Social Politics* 18(4), 490–514. Available at: http://dx.doi.org/10.1093/sp/jxr021 (accessed 30 March 2015).

Gorzelak, J. and Jałowiecki, B. (eds) (1998) *Koniunktura i mobilizacja społeczna w gminach.* Studia Regionalne i Lokalne, Warsaw.

Hakim, C. (2006) Women, careers, and work–life preferences. *British Journal of Guidance & Counselling* 34(3), 279–294. Available at: http://dx.doi.org/10.1080/03069880600769118 (accessed 29 March 2015).

Halamska, M. (2007) Regionalne zróżnicowanie rozwoju wsi. In: Gorzelak, G. (ed.) *Polska regionalna i Lokalna w świetle badań EUROREG-u.* Wydawnictwo Naukowe „Scholar", Warsaw, pp. 70–90.

Halamska, M. (2013) *Wiejska polska na początku XXI wieku. Rozważania o gospodarce i społeczeństwie.* Wydawnictwo Naukowe „Scholar", Warsaw.

Herbst, J. and Swianiewicz, P. (2008) Kapitał społeczny Dolnego Śląska i Małopolski na tle ogólnopolskiego zróżnicowania regionalnego. In: Swianiewicz, P., Herbst, J., Lackowska, M. and Mielczarek, A. (eds) *Szafarze Darów Europejskich. Kapitał Społeczny a Realizacja Polityki Regionalnej w Polskich Województwach.* Wydawnictwo Naukowe „Scholar", Warsaw, pp. 54–124.

Hesmondhalgh, D. and Baker, S. (2015) Sex, gender and work segregation in the cultural industries. *The Sociological Review* 63(Suppl. 1), 23–36. Available at: http://dx.doi.org/10.1111/1467-954X.12238 (accessed 17 May 2016).

Inglehart, R. and Norris, P. (2003) *Rising Tide: Gender Equality and Cultural Change around the World.* Cambridge University Press, Cambridge, UK.

Jakubczak, F. (ed.) (1976) *Być w Środku życia. Wydawnictwo „Książka i Wiedza"*, tom 3, Warsaw.

Kanter, R.M. (1977) Some effects on proportions on group life: skewed sex ratios and responses to token women. *American Journal of Sociology* 82(5), 965–990.

Little, J. (1997) Constructions of rural women's voluntary work. *Gender, Place & Culture* 4(2), 197–210. Available at: http://dx.doi.org/10.1080/09663699725431 (accessed 17 May 2016).

Lovenduski, J. and Norris, P. (2003) Westminster women: the politics of presence. *Political Studies* 51(1), 84–102. Available at: http://dx.doi.org/10.1111/1467-9248.00414 (accessed 25 March 2015).

Lowndes, V. (2005) It's not what you've got, but what you do with it: women, social capital, and political participation. In: O'Neill, B. and Gidengil, E. (eds) *Gender and Social Capital*. Taylor & Francis, New York, pp. 201–226.

Matysiak, I. (2014a) The feminization of governance in rural communities in Poland: the case of village representatives (*soltys*). *Gender, Place & Culture* 22(5), 700–716. Available at: http://dx.doi.org/10.1080/0966369X.2013.879104 (accessed 26 March 2015).

Matysiak, I. (2014b) *Rola sołtysów we współczesnych społecznościach wiejskich. Płeć jako czynnik różnicujący kapitał społeczny*. Wydawnictwo Naukowe „Scholar", Warsaw.

Nowak, A., Kuś, M., Urbaniak, J. and Zarycki, T. (2000) Stimulating the coordination of individual economic decisions. *Physica A: Statistical Mechanics and Its Applications* 287(3–4), 613–630. Available at: http://dx.doi.org/10.1016/S0378-4371(00)00397-6 (accessed 28 March 2015).

Oedl-Wieser, T. (2008) The rural gender regime: the Austrian case. In: Morell, I.A. and Bock, B. (eds) *Gender Regimes, Citizen Participation and Rural Restructuring*. Elsevier, Amsterdam, pp. 283–297.

Ostrowski, L. (1995) *Rada solecka w systemie samorządu terytorialnego*. Instytut Ekonomiki Rolnictwa i Gospodarki Żywnościowej, Warsaw.

Pini, B. (2008) Men, masculinities and the (re)gendering of local government in rural Australia. In: Morell, I.A. and Bock, B. (eds) *Gender Regimes, Citizen Participation and Rural Restructuring*. Elsevier, Amsterdam, pp. 299–315.

Poczta, W. (2012) Przemiany w rolnictwie ze szczególnym uwzględnieniem przemian strukturalnych. In: Wilkin, J. and Nurzyńska, I. (eds) *Polska wieś 2012. Raport o stanie wsi*. FDPA, Warsaw, pp. 65–99.

Public Opinion Research Center (2013) O roli kobiet w rodzinie. Komunikat z badań nr BS/30/2013. Available at: http://www.cbos.pl/SPISKOM.POL/2013/K_030_13.PDF (accessed 30 March 2015).

Renzetti, C.M. and Curran, D.J. (2005) *Kobiety, mężczyźni i społeczeństwo*. Wydawnictwo Naukowe PWN, Warsaw.

Reskin, F.B. (1991) Labor markets as queues: a structural approach to changing occupational sex composition. In: Huber, J. (ed.) *Macro-Micro Linkages in Sociology*. SAGE Publications Inc., Newbury Park, California, pp. 170–192.

Reskin, F.B. and Roos, P.A. (1990) *Job Queues, Gender Queues: Explaining Women's Inroads into Male Occupations*. Temple University Press, Philadelphia, Pennsylvania.

Rosner, A. (2011) Zróżnicowanie przestrzenne obszarów wiejskich. In: Nurzyńska, I. and Drygas, M. (eds) *Rozwój obszarów wiejskich w Polsce. Diagnozy, strategie, koncepcje polityki*. IRWiR PAN, Warsaw, pp. 155–174.

Shortall, S. (2014) Gender mainstreaming and the Common Agricultural Policy. *Gender, Place & Culture* 22(5), 717–730. Available at: http://dx.doi.org/10.1080/0966369X.2014.939147 (accessed 29 March 2015).

Siemieńska, R. (1990) *Płeć, Zawód, Polityka. Kobiety w Życiu Publicznym w Polsce*. Oficyna Wydawnicza „Forum", Poland.

Siemieńska, R. (ed.) (2005) *Płeć, wybory, władza*. Wydawnictwo Naukowe „Scholar", Warsaw.

Siemieńska, R. (2011) Kobiety i mężczyźni – aktorzy 'płynnej rzeczywistości'. Niespójności postaw, zachowań i pozycji – dwie dekady przemian w Polsce i na świecie. In: Bucholc, M., Mandes, S., Szawiel, T. and Wawrzyniak, J. (eds) *Polska po 20 latach wolności*. Wydawnictwa UW, Warsaw, pp. 444–564.

Skuratowicz, E. and Hunter, L.W. (2004) *Where Do Women's Jobs Come From? Job Resegregation in an American Bank. Working Paper*. Wharton School, University of Pennsylvania, Philadelphia, Pennsylvania. Available at: http://fic.wharton.upenn.edu/fic/papers/01/0127.pdf (accessed 12 May 2016).

Slany, K. (ed.) (2008) *Migracje kobiet. Perspektywa wielowymiarowa*. Wydawnictwo Uniwersytetu Jagiellońskiego, Cracow, Poland.

Tienari, J. (1999) The first wave washed up on shore: reform, feminization and gender resegregation. *Gender, Work & Organization* 6(1), 1–19. Available at: http://dx.doi.org/10.1111/1468-0432.00065 (accessed 14 May 2016).

Wilkin, J. (2011) Jak zapewnić rozwój wsi w warunkach zmniejszającej się roli rolnictwa. In: Halamska, M. (ed.) *Wieś jako przedmiot badań naukowych na początku XXI wieku*. Wydawnictwo Naukowe „Scholar", Warsaw, pp. 117–137.

Zarycki, T. (2002) *Region jako kontekst zachowań politycznych*. Wydawnictwo Naukowe „Scholar", Warsaw.

18 Gender and International Development

B.B. Bock* and M. van der Burg
Wageningen University, Wageningen, The Netherlands

Introduction

Gender equality has gained a prominent place on the agenda of international development during the early decades of the 21st century. Rising concerns about food insecurity, in combination with increasing worries about the effects of climate change, other natural disasters and political conflicts, have contributed to this change. Criticism of the impacts of the structural adjustment programmes of the 1990s and a drastic questioning of the efficiency of investments in development have led to a re-examination of many obvious development routines. Attempts to secure pathways to fair and sustainable development induced the formulation and ratification of the United Nations' (UN) Millennium Development Goals (MDGs) in 2000, which were to be reached by 2015. In 2015, the MDGs were followed by a new initiative of the UN: the Sustainable Development Goals (SDGs) of the 2030 Agenda for Sustainable Development, in which sustainability, social justice and gender equality are featured prominently. Moreover, the SDGs emphasize that sustainable development is a target to be reached everywhere in the world and not only in the global South.

Worldwide, funding agencies, top research institutions and governmental as well as non-governmental development agencies have reformulated their missions and reframed their efforts to contribute to these goals. An inherent challenge of the UN goals was to overcome thematic and disciplinary divides and renew more integrated approaches accordingly. This all shaped the context for a renewed interest in gender studies expertise. As a consequence, many longstanding debates separating transformative idealists from hands-on pragmatists seem to be converging into shared commitments according to the slogan 'Promoting gender equality is not only good for women; it is also good for agricultural development' (see, for example, FAO, 2011; Quisumbing *et al.*, 2014; Haddad, 2015).

The following offers a brief overview of these new lines of research and development. First the theoretical development of the field is addressed, after which the research topics that have been considered most relevant in recent research on gender and international development are presented. Finally, the chapters included in this section of the book (Chapters 19–23) are introduced.

* Corresponding author e-mail: bettina.bock@wur.nl

The Theoretical Development of the Field

Intersectionality, femininities and masculinities

The first feminist studies in international development focused on the role and position of women. In time, the focus shifted gradually from women in/and development (WID and WAD) towards gender and development (GAD) (Razavi and Miller, 1995; Bock, 2006; Moser, 2014). In the early decades of the 21st century, it became common to consider the *intersection* of gender with other social dimensions, such as generation/age, class/wealth, race, ethnicity, religion, global or geographical locality, health status and sexuality. This has been increasingly taken up in gender studies research and practices. According to Cho *et al.* (2013), there is no single definition of an intersectionality framework; but, in their introduction to a special issue of *Signs* (2013) on intersectionality, they stress the common cause in the objective 'to spark further inquiry into the dynamics of intersectionality both as an academic frame and as a practical intervention in a world characterized by vast inequalities' (Cho *et al.*, 2013, p. 807).

This new approach to interrelated social complexities also helped to reframe gender equality efforts within other programmes, for instance around poverty eradication. By taking into consideration how gender intersects with, for instance, age or generation, both women and youth can be seen as marginally positioned in relation to a dominant-adult-male-based standard or model. This recognition also pushed us to acknowledge that neither men nor women are homogeneous groups and that masculinity as well as femininity should be addressed in plural. Wanner and Wadham (2015), however, stress that although men and masculinities are increasingly mentioned in policy documents, this rarely goes beyond paying lip service and is seldom integrated into the programmes themselves.

Transnationality and translocality

Acknowledging complexities, diversity and differences around gender equality also deepened the question of whether universal claims – such as human rights – can be upheld and actualized without imposing the values of the global dominant model on 'others'. Since Mohanty's criticism (1986, 2013) that Western feminist values had been imposed on women in the South, a feminist post-colonial perspective on global relations made space to interpret development programmes critically and to encourage the acknowledgement of local and situated circumstances and voices. Combined with the growing attention to globalization and geopolitics, this led to a critical analysis of the nation as a constituent institution, to which Patil (2013) added an analysis of intersections of state and family in which patriarchal values have been anchored, both in colonial and post-colonial contexts.

This acknowledgement allowed centring the focus on initiatives separate from the policies of nation states or international governing bodies such as UN institutions. Seen as a source of empowerment, initiatives from below – particularly those that increasingly transgress national borders – received renewed attention. It is commonly acknowledged that increased use of information and communications technologies, combined with expanding and accelerating migration, has enlarged the potential of transnational movements. Women's transnational solidarity has been newly presented as a sustainable pathway to alternative development patterns (see also Harcourt, Chapter 20, this volume; Bock, Chapter 2, this volume).

Especially in the context of agriculture-based development, the environment and the use of natural resources have been central to debates on sustainable livelihoods and well-being. Gender studies contributed to that debate by foregrounding how access to natural resources differed according to gender, while also intersecting with other dimensions of social difference and inequality. Most academic gender studies scholars now use the feminist political ecology (FPE) approach as an entry point to uncover and unravel such differences and to promote inclusive sustainable development, which aims at both safeguarding environmental and natural resource potential as well as advancing social constellations that allow its sustainable and fair management and use. *Geoforum* (2012) gives a good overview of the points under debate in its special issue, which is well introduced by Elmhirst (2012) and updated by Harcourt (Chapter 20, this volume).

The Thematic Development of Gender Research

Food security and food production have always been important subjects of research in the field of gender in international development, and this has remained true into the 2000s. Whereas research has tended to focus on questions of farm-based agricultural production, we now see research that includes the whole value chain from production to consumption, studies that concentrate on urban food production and consumption, and research into food policy and food governance in relation to food safety and food security among different citizen groups. Increasing emigration from rural to urban areas, for instance, has changed the organization of production in rural areas in terms of production size, crops and production methods. It has also supported the attention to and growth of organic and urban agriculture and the trading of food between rural and urban areas, with women playing an important role in both (see Sachs and Garner, Chapter 19, this volume). Foreign investments and so-called 'land grabbing' promote a shift towards plantation-based production, and with it a shift towards growing wage labour among the rural population, in turn affecting gender relations as well (Sachs and Garner, Chapter 19, this volume; Canning, Chapter 21, this volume). Looking into such questions immediately evidenced the need for integrative approaches and gender-specific information (for example, Centrone *et al.*, Chapter 22, this volume).

The organization of food production at the level of the farm – with attention to the gender-specific division of labour, decision making, time allocation and access to resources such as land, water, seeds and machinery – has been further researched in relation to the above-mentioned ongoing challenges. Margreet Zwarteveen opened the field of research into masculinities in the irrigation sector (Zwarteveen, 2008). The themed issue 'Global Geographies of Gender and Water' of *Gender, Place & Culture* (2009), introduced by Kathleen O'Reilly, Sarah Halvorson, Farhana Sultana and Nina Laurie (O'Reilly *et al.*, 2009), shows other interesting new ways in which this more integrative approach has been taken up in relation to water issues. Another special issue, 'Land, Gender, and Food Security' in *Feminist Economics* (2014), introduced by Cheryl Doss, Gale Summerfield and Dzodzi Tsikata (Doss *et al.*, 2014), expands on land issues and food security. New research regards land titling and inheritance as matters of (inter)national legal regulation (Jacobs, Chapter 23, this volume). Another trend is the growing attention to gendered aspects of the use and impact of new technologies, such as new varieties engineered by biotechnology or genetic modification (Gouse *et al.*, 2016; National Academies of Sciences, Engineering, and Medicine, 2016). Some studies stress the positive effects of these technologies in terms of quality of life, as new varieties require fewer chemicals, thereby reducing negative health effects, or can increase production size or quality; others stress reduced manual labour and engagement in other more gainful activities as positive results (Subramanian *et al.*, 2010). Critiques note that new technologies may be difficult to access and may enhance gender inequalities (Gouse *et al.*, 2016). For organic agriculture and conservation agriculture, the specific gender aspects were outlined to support the research and development agendas of those fields (Farnworth and Hutchings 2009; Farnworth *et al.*, 2016). Technologies can, however, also lead to unintended consequences. There is much more to uncover and integrate concerning gendered impacts on health and bodies, for instance from the chemicals used by labourers or ingested by consumers, or from growing workloads if some but not all tasks are mechanized.

The realization that not only the farm but the whole agrifood system is gendered became evident when looking for ways to enhance food security along all stages of the vertical chain of food production, often phrased as 'the pathway from farm to table'. Research on new partnerships and market relationships included gender issues in agricultural associations and cooperatives, and along the whole chain from seed selection to planting, maintaining, harvesting, postharvest processing, storage and marketing (see, for example, Sachs and Patel-Campillo, 2014; Mayanja *et al.*, 2016; Sachs and Garner, Chapter 19, this volume). This led to studies that unravelled the gender aspects of trade laws and licence agreements and of access to external farm inputs and institutional support extending beyond agriculture-specific services, such as that provided through education/extension, associations and organizations, auctions, storehouses and transport facilities (see also Shortall, Chapter 7, this volume). These new research interests reflect the change in the organization of food production.

Under the slogan of 'closing the gender gap in agriculture' (for example, Quisumbing *et al.*, 2014), new research and development initiatives have aimed at combining the advancement of gender equality, food security and climate resilience through gender mainstreaming. For instance, mainstream research institutes of the Consultative Group for International Agricultural Research (CGIAR) agreed on making gender an overarching theme, developed a firm and integral Consortium Gender Strategy, and established a focal point. CGIAR's strategy entails overarching cross-programme cooperation in specific and integrative research programmes, including all of its centres, while including gender budgeting and monitoring mechanisms throughout its research processes and research programming. Two overarching gender programmes were developed, of which the Women's Empowerment in Agriculture Index (WEAI) has attracted the most interest and has led to many discussions and adjustments. This effort pushed women in farming into the world of Big Data in its attempt to quantify indicators and measure the gradual achievements of empowerment (Alkire *et al.*, 2013; Centrone *et al.*, Chapter 22, this volume). The GENNOVATE research programme addresses how gender norms and agency influence men, women and youth to access, adopt and benefit from innovations in agriculture and natural resource management. It uses qualitative research to address hidden norms within societies, especially in the field of gender and agriculture, that intersect with other dimensions, particularly wealth, generation, ethnicity and religion (Badstue *et al.*, 2014).

The objective of mainstreaming is to understand the relevance of gender relations in order to support and monitor the more efficient production of food, partially by addressing inequality in access to resources (Quisumbing *et al.*, 2014; Russell *et al.*, 2015). However, this objective has also been criticized. One might argue that, on the institutional level, gender empowerment has become instrumental for achieving other goals, even beyond specific programmes that underscore the importance of a transformative dual approach to feeding one another. Feminists have criticized such an approach before and continue to do so, as reflected in a recent publication by Cornwall (2016). She underscores that the transformational approach that characterizes feminist research gets lost when strengthening women's economic position does not coincide with a fundamental change in gender relations and ideologies. Gammage *et al.* (2016) point to the problems arising when mobilizing women into more productive work or entrepreneurship does not converge with a redistribution of household labour and redefinition of norms and values (see also Sachs and Garner, Chapter 19, this volume; Canning, Chapter 21, this volume). They also underline the importance not only of focusing on women but also of considering the whole complex web of social relations in which – in addition to gender – generation and class play an important role. Otherwise, existing social inequalities can be easily reinforced and increased or new inequalities produced.

The Chapters

In Chapter 19, Carolyn Sachs and Elisabeth Garner show that agrifood systems have undergone significant transitions in recent decades. Family farm operations have been reorganizing, large-scale commodity production has been increasing and new models of agriculture have been emerging. To illuminate these trends from a gender perspective, they review new literature in light of shifting masculinizations and feminizations of agriculture in both the global North and South contexts. Such shifts are related to changes in family farms and large-scale commodity agriculture, with a specific focus on labour and on new trends in agriculture, such as new partnerships in organic and urban agriculture. In this way, they show how gender identities shift across scale and geographical contexts in the productive spheres of agriculture while also illuminating the need for more integrative research on the socio-cultural contexts that are obstructing or restricting changes in, for instance, the fair distribution of family income or the redistribution of household, family and care work.

Chapter 20 by Wendy Harcourt addresses glocal networking as a way of moving toward more interactive policy making that respects and values diversity. She illustrates this with case studies in rural India, Tanzania and Italy. In glocal networking, Harcourt connects advocacy, intermediators and action research within the feminist political ecology (FPE) framework. This includes purposely engaging with local actors and having their situations and voices echoed in national and international institutions. She stresses the role of so-called 'interlocutors' in representing local voices at the intersection between the 'sidestreams' of women's knowledge and experiences and the 'mainstream' of gender and sustainable development policy. By labelling local voices and activities as 'sidestreams', Harcourt locates their 'otherness' in juxtaposition with 'mainstreams', allowing her to value them as constructive alternatives rather than as acts of resistance or opposition. Because local actors are influenced by global-scale actions, these intermediators can also help by safeguarding local 'sidestreams' through vertical upscaling and can help bolster their space and sustainability through spreading and interconnecting similar activities horizontally.

Megan Canning explicates in Chapter 21 how plantation systems shape workers' lives and rural communities in a gendered way, based on her case study of the sugarcane industry in Malawi. Interviews with both male and female workers and community members reveal that traditional gender roles are redefined as both men and women are increasingly entering wage work. Although women have been entering only the lowest levels of the workforce, they have increased their bargaining power at home and some have even achieved financial independence from men altogether. Canning also notes the persistence of gender ideologies. At the plantation, women were particularly vulnerable to super-exploitation by male-only supervisors, including sexual intimidation and assault. At home, most women workers lengthened and intensified their total labour time, while only some could hire additional labour at home or rely more on their daughters. Plantation labour was largely associated with the masculine values of physical strength and hardship, which sustained the status of men as the 'breadwinners,' caused them to be hired most often, and enabled them to keep controlling the family income and avoiding household and care labour.

In Chapter 22, Francesca Alice Centrone, Bettina Bock, Angela Mosso and Angela

Calvo address how gender indicators can help optimize research and development project design. Their chapter first reviews the main recent multidimensional gender indices with a focus on the inclusion of indicators doing justice to rural women. The authors stress that these multidimensional indices enrich analysis beyond the use of statistics as such, allowing us to move beyond unilateral interpretations and to have various types of information combined in a single number. However, these indices are usually developed at the national level and therefore often hide important diversities at regional or local levels. This is why the authors present and discuss their experimentation with these same indicators to support practitioners – especially small and local actors – in rural gender projects at the local level in three diverse rural regions in Senegal.

Susie Jacobs explores the impact of agricultural land titling on women in Chapter 23. Jacobs defines land 'titling' as the formalization of land tenure law practices, which gives full or partial property rights to users. It has been presented as aiming at gender equality because it supersedes customary tenure practices. Jacobs observes that the customary law offered relative security to women, particularly with regard to rights of access and food cultivation. Yet in practice, such customs have proven to be fragile and often violated. Jacobs argues that the new approach has dismantled common property regimes advantageous to women's access to land. Jacobs found that current research assesses both joint and individual titling as positive to women's lives but also argues that this recent development is difficult to assess. She also warns of the effect of a 'Trojan horse', causing further privatization and loss of land to increasing state and corporate land acquisition. She calls upon the state, women's rights organizations and public bodies to exercise their key roles in support of these women 'on the ground'.

References

Alkire, S., Meinzen-Dick, R., Peterman, A., Quisumbing, A.R., Seymour, G. and Vaz, A. (2013) The women's empowerment in agriculture index. *World Development* 52, 71–91. Available at: http://dx.doi.org/10.1016/j.worlddev.2013.06.007 (accessed 6 March 2017).

Badstue, L., Kantor, P., Prain, G., Ashby, J. and Petesch, P. (2014) Concept Note: Innovation and Development through Transformation of Gender Norms in Agriculture and Natural Resource Management: A Global Comparative Research Initiative. Available at: https://gender.cgiar.org/wp-content/uploads/2015/09/CN-Global-Study-GenderNormsAgency_modified-Jan-2015.pdf (accessed 23 September 2016).

Bock, B.B. (2006) Introduction: rural gender studies in north and south. In: Bock, B.B. and Shortall, S. (eds) *Rural Gender Relations: Issues and Case Studies*. CAB International, Wallingford, UK, pp. 1–18.

Cho, S., Crenshaw, K.W. and McCall, L. (2013) Toward a field of intersectionality studies: theory, applications, and praxis. *Signs* 38(4), 785–810. Available at: http://dx.doi.org/10.1086/669608 (accessed 6 March 2017).

Cornwall, A. (2016) Women's empowerment: what works? *Journal of International Development* 28(3), 342–359. Available at: http://dx.doi.org/10.1002/jid.3210 (accessed 6 March 2017).

Doss, D., Summerfield, G. and Tsikata, D. (2014) Land, gender, and food security. *Feminist Economics* 20(1), 1–23. Available at: http://dx.doi.org/10.1080/13545701.2014.895021 (accessed 6 March 2017).

Elmhirst, R. (2012) Introducing new feminist political ecologies. *Geoforum* 42(2), 129–132. Available at: http://dx.doi.org/10.1016/j.geoforum.2011.01.006 (accessed 6 March 2017).

FAO (2011) *The State of Food and Agriculture 2010–11. Women in Agriculture – Closing the Gender Gap for Development.* Food and Agriculture Organization of the United Nations, Rome. Available at: http://www.fao.org/docrep/013/i2050e/i2050e.pdf (accessed 18 February 2017).

Farnworth, C. and Hutchings, J. (2009) *Organic Agriculture and Women's Empowerment.* IFOAM, Bonn, Germany.

Farnworth, C.R., Baudron, F., Andersson, J.A., Misiko, M., Badstue, L. and Stirling, C.M. (2016) Gender and conservation agriculture in East and southern Africa: towards a research agenda. *International Journal of Agricultural Sustainability* 14(2), 142–165. Available at: http://dx.doi.org/10.1080/14735903.2015.1065602 (accessed 6 March 2017).

Feminist Economics (2014) Special Issue: Land, Gender, and Food Security. *Feminist Economics* 20(1), 1–248.

Gammage, S., Kabeer, N. and van der Meulen Rodgers, Y. (2016) Voice and agency: where are we now? *Feminist Economics* 22(1), 1–29. Available at: http://dx.doi.org/10.1080/13545701.2015.1101308 (accessed 6 March 2017).

Gender, Place & Culture (2009) Themed Issue: Global Geographies of Gender and Water. *Gender, Place & Culture* 16(4), 365–495.

Geoforum (2012) Themed Issue: New Feminist Political Ecologies. Elmhirst, R. (ed.) *Geoforum* 42(2), 129–183.

Gouse, M., Sengupta, D., Zambrano, P. and Falck Zepeda, J. (2016) Genetically modified maize: less drudgery for her, more maize for him? Evidence from smallholder maize farmers in South Africa. *World Development* 83, 27–38. Available at: http://dx.doi.org/10.1016/j.worlddev.2016.03.008 (accessed 6 March 2017).

Haddad, L. (2015) Equity: not only for idealists. *Development Policy Review* 33(1), 5–13. Available at: http://dx.doi.org/10.1111/dpr.12089 (accessed 6 March 2017).

Mayanja, S., Barone, S., McEwan, M., Thomas, B., Amaya, N., Terrillon, J., Velasco, C., Babini, C., Thiele, G., Prain, G. and Devaux, A. (2016) *Prototype Guide for Integrating Gender into Participatory Market Chain Approach.* International Potato Center, Lima, Peru.

Mohanty, C.T. (1986) Under Western eyes: feminist scholarship and colonial discourses. *Boundary 2* 12(3), 333–358.

Mohanty, C.T. (2013) Transnational feminist crossings: on neoliberalism and radical critique. *Signs* 38(4), 967–991. Available at: http://dx.doi.org/10.13060/12130028.2015.127.2.216 (accessed 6 March 2017).

Moser, C.O.N. (2014) Gender Planning and Development: Revisiting, Deconstructing and Reflecting. DPU60 Working Paper Series: Reflections no. 165/60. Available at: https://www.academia.edu/30466990/Gender_planning_and_development_Revisiting_deconstructing_and_reflecting?auto=download (accessed 21 February 2017).

National Academies of Sciences, Engineering, and Medicine (2016) *Genetically Engineered Crops: Experiences and Prospects.* The National Academies Press, Washington, DC.

O'Reilly, K., Halvorson, S., Sultana, F. and Laurie, N. (2009) Introduction: global perspectives on gender–water geographies. *Gender, Place & Culture* 16(4), 381–385. Available at: http://dx.doi.org/10.1080/09663690903003868 (accessed 6 March 2017).

Patil, V. (2013) From patriarchy to intersectionality: a transnational feminist assessment of how far we've really come. *Signs* 38(4), 847–867. Available at: http://dx.doi.org/10.1086/669560 (accessed 6 March 2017).

Quisumbing, A., Meinzen-Dick, R., Raney, T.L., Croppenstedt, A., Behrman, J.A. and Peterman, A. (2014) Closing the knowledge gap on gender in agriculture. In: Quisumbing, A., Meinzen-Dick, R., Raney, T.L., Croppenstedt, A., Behrman, J.A. and Peterman, A. (eds) *Gender in Agriculture. Closing the Knowledge Gap.* FAO and Springer, pp. 3–31. Available at: http://libcatalog.cimmyt.org/download/general/98958.pdf (accessed 30 September 2016).

Razavi, S. and Miller, C. (1995) *From WID to GAD: Conceptual Shifts in the Women and Development Discourse. UNRISD Occasional Paper No.1 for the Fourth World Conference on Women in Beijing.* United Nations Research Institute for Social Development, Geneva, Switzerland.

Russell, N., Karlsson, K., Ashby, J. and Mascarenhas, M. (2015) *Change in the Making: Progress Reports on CGIAR Gender Research. Issue No. 1: Toward gender-equitable control over productive assets and resources.* CGIAR Gender and Agriculture Research Network, CGIAR Consortium Office and International Center for Tropical Agriculture (CIAT), Cali, Colombia.

Sachs, C. and Patel-Campillo, A. (2014) Feminist food justice: crafting a new vision. *Feminist Studies* 40(2), 396–410. Available at: http://www.jstor.org/stable/10.15767/feministstudies.40.2.396 (accessed 6 March 2017).

Signs (2013) Special Issue: Intersectionality: Theorizing Power, Empowering Theory. Cho, S., Crenshaw, K.W. and McCall, L. (eds) *Signs* 38(4), 785–1060.

Subramanian, A., Kirwan, K., Pink, D. and Qaim, M. (2010) GM crops and gender issues. *Nature Biotechnology* 28(5), 404–406. Available at: http://www.nature.com/nbt/journal/v28/n5/full/nbt0510-404.html (accessed 6 March 2017).

Wanner, T. and Wadham, B. (2015) Men and masculinities in international development: 'menstreaming' gender and development? *Development Policy Review* 33(1), 15–32. Available at: http://dx.doi.org/10.1111/dpr.12090 (accessed 6 March 2017).

Zwarteveen, M. (2008) Men, masculinities and water powers in irrigation. *Water Alternatives* 1(1), 111–130. Available at: http://www.water-alternatives.org/index.php/volume1/v1issue1/19-a-1-1-7/file (accessed 6 March 2017).

19 Gender Transitions in Agriculture and Food Systems

C. Sachs* and E. Garner
The Pennsylvania State University, State College, Pennsylvania, USA

Introduction

Along with the liberalization of international trading regulations, agrifood systems have seen significant transitions in the flow of goods and services. These shifts are shaping agriculture not only on a global scale but also in local contexts. As a result, family farm operations are reorganizing, large-scale commodity production is increasing and new models of agriculture are emerging. Gender transitions in agriculture and food systems are occurring globally with these shifts but they take different forms, including different feminizations of agriculture and shifting masculinities. Despite these shifts in agricultural spaces, care work and subsistence responsibilities still fall heavily on women.

Women's participation and recognition in agriculture are increasing, in particular in the global South where men's migration and employment takes them off the farm, and in the global North where women are entering farming at faster rate than men. The term 'feminization of agriculture' identifies women's increased responsibility for agricultural work, but also is attentive to the barriers to agricultural production that women face. In order to enlighten the emerging trends in gender and agriculture, this chapter looks at the shifting masculinizations and feminizations of agriculture in the following six contexts that span scale and geography: family farms in the global North and global South; large-scale commodity agriculture, focusing on labour, in the global North and global South; and emerging trends in agriculture in the global North and global South. By positioning these trends together we see how gender identities shift across these contexts in the productive spheres of agriculture, while recognizing that care work remains a predominantly feminized space.

This chapter explains shifts in gendered responsibilities, work and identities in the global North, with a focus on the USA, and in the countries of the global South. The following section reviews the literature on food regimes and the transitions in the global agrifood system, the multiple feminizations and masculinizations of agriculture in different spaces, and the continued role of women in care work. Next, the reasons for using the different contexts are provided, before presenting the results and discussion of the analysis of the shifting gender identities on family farms and on large-scale agriculture, focusing on labour, as well as on the new models of agriculture. The chapter concludes with possible directions for more just agrifood systems.

* Corresponding author e-mail: xyl@psu.edu

Background and Context

This literature review frames the relationships between the current transitions in the global agrifood systems and the emerging masculinizations and feminizations of agriculture. It is evident that food system dynamics are (re)shaping gendered identities in the productive spaces of agriculture, while these transitions are not as dynamic in the reproductive, domestic spaces that are still considered feminine. By aligning the concepts of food regimes, shifting gender identities and care work, the chapter analyses agriculture at different scales highlighting the linkages between global food systems, the farm and the family.

Food regimes: transitions in the global agrifood systems

Since the 1990s, the liberalization of trade policies has increased the mobility of food, both within regions and between nations. As a result, scholars and policy makers devote increased attention to the relationships between countries and actors in agrifood markets, specifically through international value chains. This results in a 'dual market' system where export-oriented, high-value products characterize one flow of the system, bringing high returns on crop yields to those who are connected. Local food production and sale characterize the other, low-value, market in this dual system.

This liberalization of policies has resulted in what many refer to as the 'third food regime', in which decreased political regulation and increased consumer influence characterize the demand for and transportation of food (McMichael, 2009b; Pechlaner and Otero, 2010). As Otero *et al.* describe, food regimes are 'grounded in relatively stable (albeit typically uneven or asymmetrical) international trade relations' (2013, p. 270). Given the international context and power differentials, this mobility does not necessarily create mutually beneficial transactions or stable economies (McMichael, 2009a).

Further, the emphasis on high-value, export-oriented food production, coupled with decreases in public spending, has decreased livelihood options in rural areas. The global market offers substantial opportunity and resources for agricultural producers in the global South, but it also comes with limitations and unequal distributions of resources (Weinberger and Lumpkin, 2007). Local markets are less profitable than the markets agro-exporting industries often access. In the global South, this has resulted in increased migration from rural areas, in particular among men (Broughton, 2008), and surplus labour for large-scale agriculture, in particular among women (ILO, 2010). Because of these shifts in agriculture, women's and men's participation in agrifood systems involves shifting gender roles and identities, the result of which varies across different spaces and different scales of agriculture.

The liberalizing of global markets has created a space for food chains to internationalize and has increased the importance of the supermarket within these chains (Busch and Bain, 2004). Due to this process, the role of the consumer, consumer demands and the supermarket have become strong forces in shaping this production–consumption system (Busch and Bain, 2004). The oligopolistic nature of this market has not decreased competition, but rather shifted from 'price competition to nonprice competition where variety, convenience, quality, and year-round supply are as important or even more important than price' (Busch and Bain, 2004, p. 329). As part of the competition surrounding quality, labelling has become an important piece of marketing food products, particularly those from international markets where consumers have increased awareness and concern surrounding labour practices and fair pricing.

Shifting gender identities: feminization and masculinization of agriculture

Women's responsibility for agriculture is increasing, especially in many countries in the global South, as men migrate or seek other forms of employment. This phenomenon, referred to as the 'feminization of agriculture', involves the increasing ratio of women to men participating in agricultural production and the increasing proportion of women who are employed in agriculture. Globally, the best data available suggest that women comprise 43% of the workforce in agriculture (FAO, 2011), but even this number under-represents women's unpaid work on farms. The Food and Agriculture Organization of the United Nations (FAO) did not report similar data for the USA, but women's participation in agriculture in the USA has also increased. The percentage of US farms operated by women almost tripled in a similar time period, from 5% in 1978 to 14% in 2007. As can be seen in Fig. 19.1, the proportion of women in agriculture increased in all regions except Europe from 1980 to 2010, resulting in what many refer to as the 'feminization of agriculture'. De Schutter (2013) argues that the feminization of agriculture is not one thing but actually the result of three very different phenomena that do not necessarily empower women as either farmers or farm workers.

As De Schutter (2013) notes, each form of the feminization of agriculture results in different sets of gender relationships, that are also driven by transitions in global agrifood systems. In the first case, women take over farming when male adults on farms migrate or take alternative employment, leaving women to provide for household food security or the subsistence needs of their families. In the second instance, women take over farming the family land to produce primarily for the market rather than the subsistence needs of their household. The third type of feminization of agriculture involves women's employment as farm workers, typically in large-scale corporate agriculture enterprises.

Adding to de Shutter's categorizations, we argue that there is a fourth type of feminization of agriculture occurring in the USA

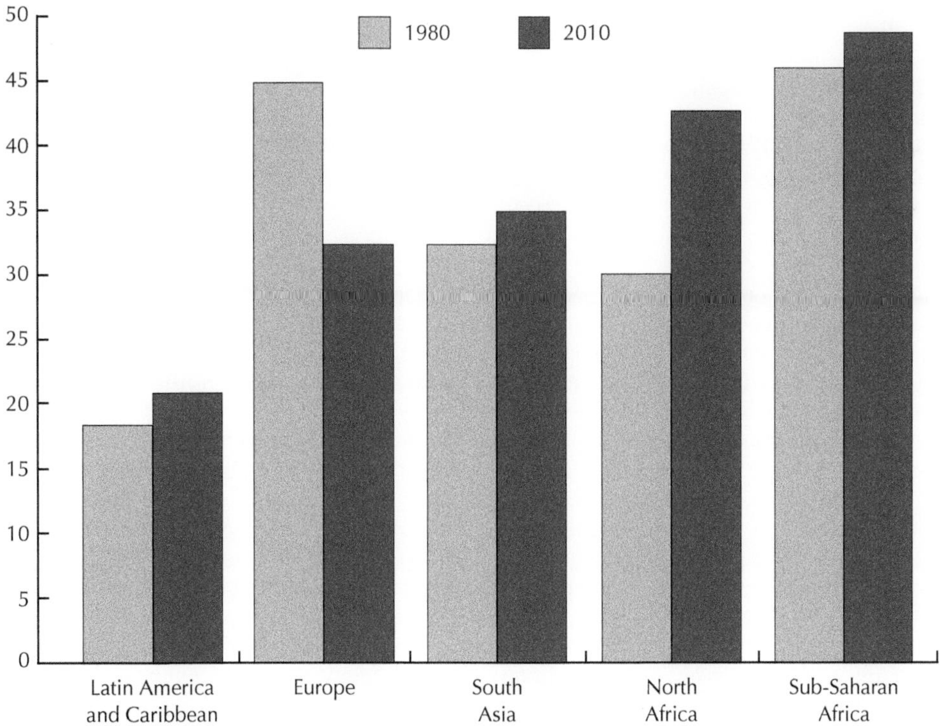

Fig. 19.1. Female share of the agricultural labour force. (From FAO, 2011.)

and Europe which takes a different form from the agricultural transitions that are happening in much of the global South. On smaller-scale farms in the global North, we see a feminization of agriculture with women choosing the role of farmer, moving towards more gender-equitable relationships in agricultural households and communities (Sachs *et al.*, 2016).

As global structures construct femininities and masculinities simultaneously (Connell, 1998), it is also important to consider the transitions of masculinities in this third food regime. Along with rural ideals of femininity, constructions of masculinity also work to distinguish and enforce gender roles that exclude and marginalize women from agricultural production in farming families (Saugeres, 2002b). Traditional conceptions of farming masculine identities include a control over nature (through machinery), physical strength and one's occupation (Brandth, 2002; Little, 2002; Saugeres, 2002a).

Barlett and Conger (2004) identify industrial and agrarian as two well-established masculine ideologies in the rural USA; however, these 'older visions of masculinity are being altered in accord with a new paradigm of systems thinking and sustainability' (Barlett and Conger, 2004, p. 207). These masculine identities often correlate in the definition of success and how power is identified within the different forms of agricultural production; performance and interactions with other men (homosocial) are continuously shaping these identities (Bell, 2000; Peter *et al.*, 2000; Barlett and Conger, 2004). As such, transitions in daily responsibilities, management of farms, geographically located socio-cultural histories and agrifood market structures link directly to shifts in gender constructs in rural contexts. Accordingly, along with the emergence of sustainable agriculture comes a new form of masculinity.

In addition, a fourth masculinization is occurring in relation to large-scale commodity farms. For example, in the USA, large-scale corporate farms are increasingly employing men (primarily from Mexico) as hired farm workers. In many countries in the global South, as agriculture shifts from subsistence, small-scale production to larger-scale production for commercial markets, men are assuming the primary role in decision making. In addition, many programmes of land tenure reforms re-entrench men's ownership of land, a phenomenon which Archambault and Zoomers (2015) refer to as 'the masculinization of rural space'.

Additional shifts in masculinity are occurring in the rural global South; however, further research is needed to illuminate this trend. Changes in labour markets can tie to men's feeling of loss of power within their families and challenge existing masculine identities defined through men's labour and position as breadwinner (Chant, 2000). The emphasis on the identity and responsibilities of men as the primary breadwinner uniquely situates them as responsible for leaving the home to find work either in urban centres or by migrating abroad. Yet, migration for employment is not unique to men. As women also migrate, this can challenge gender identities as well (Preibisch and Encalada Grez, 2010).

Subsistence and care work

Attention to the new scholarship on subsistence and care work bolsters our understanding of shifting masculinities and femininities in agriculture. Recently, development institutions and feminist economics have focused on valuing caring labour (Esquivel, 2014) and confronting women's heavy and unequal responsibility for care labour (Kidder *et al.*, 2014). Food-related work comprises much of the care work performed in households. Women perform the majority of food-related work throughout the world including subsistence production, obtaining firewood, acquiring food, processing food, planning meals, cooking, serving food, cleaning kitchens and washing dishes (Allen and Sachs, 2007).

Esquivel (2014) argues that the meanings of care are contested where a narrower understanding of care focuses only on caring for dependants, work frequently performed by women. Tronto exemplifies the transformative approach defining care as the activities 'that we do to maintain, continue, and repair our "world" so we can live in it as well as possible', including care for people (ourselves, dependants and others), objects and our environment (Tronto, 2012, p. 33). A more transformative approach seeks to lighten women's care burdens and redistribute care work while also challenging the idea that care work is exclusively the domain of females and of less importance than productive or paid work. Redistribution of care labour must go beyond the household level and consider the role of the state and the market in lessening caregiving burdens, especially of poor families (Elson, 2008).

The position of women within the context of domestic responsibilities provides important insights into both the opportunities and vulnerabilities faced by rural women in the North and South. In looking at women in England, Stalford *et al.* (2003) identified that domestic violence was a more acute problem for farming families due to the limited services in rural areas and the social isolation. In the global South, care work is a time burden that acts as a gendered barrier to improving women's agricultural production (FAO, 2011). Further, the gendered responsibility for ensuring the dietary needs of the family also leave women more vulnerable to negative shocks due to the gender gap in rural development as well as the gendered responsibilities of home and food-related care work (De la O Campos and Garner, 2014).

Methodology

Looking at food production at multiple scales is essential for understanding gendered differences (Sachs and Patel-Campillo, 2014) as different compositions and organizations of agriculture correspond to different forms of masculinities (Barlett and Conger, 2004; Peter *et al.*, 2000) and feminizations (De Schutter, 2013; Sachs *et al.*, 2016). In order to analyse the impacts of shifting international food regimes on the composition of agriculture, this chapter uses conceptual framings of feminization and masculinization of agriculture to look at gendered divisions of labour across the following agriculture/rural spaces: family farms; large-scale commodity farming; and emerging challenges to the global industrial food system. The analysis also looks at the different contexts between the global North and global South. By discussing these six contexts in conjunction, the chapter responds to the current gap in analysis that typically positions these shifts and spaces in isolation.

Research Findings

Family farms

Family farms in many regions of the world have historically been characterized and organized by patriarchal relations, with the oldest men owning land, controlling labour of other family members and making decisions about agricultural production (Sachs, 2015). Women have long been involved in growing crops and raising livestock on family-operated farms, but the extent of their labour and their contributions have been undercounted and undervalued. As agriculture changes and gender relations in the non-agricultural realm shift, women are often taking more responsibility on farms and are increasingly recognized as essential to food production. However, women in both the global North and global South continue to face gendered barriers to agricultural participation, often lacking access to key resources for farming including land, credit, inputs, technology and agricultural education.

Family farms: USA and Europe

On family farms in the USA, males still comprise the majority of farmers (70%) who self-identify as principal operator (USDA, 2014). However, women are increasing in agriculture in the USA at a faster pace than are men, where they tend to operate smaller and have lower sales than farms with men principal operators (Hoppe and Korb, 2013). Between 1982 and 2007, the number of women-operated farms in the USA increased from 121,600 to 306,200, while the number of men-operated farms declined (Hoppe and Korb, 2013). This could be for two reasons: more women are identifying as farmers or more women are entering farming (Brasier *et al.*, 2014).

In part because of rising economic pressures, many family farms in the USA have increased their participation in off-farm employment for additional income to support the family and continued farm operations. Both women and men principal operators have increased their off-farm work: between 1982 and 2007, women operators working off the farm increased from 42 to 59%, while men increased from 53 to 65% (Hoppe and Korb, 2013). Farm factors are more likely to influence men's decision to work off the farm, while demographic characteristics of the women and other household members are more likely to influence women in this decision (Caillavet *et al.*, 1994). The presence of young children is more likely to limit women's off-farm participation in 'traditional families'; however, the husband's inheritance of the farm positively correlates to women working off-farm, as her husband is more likely to stay and work the farm (Bharadwaj and Findeis, 2003).

Further contextualizing gendered roles on family farms, women in the USA do the majority of household food-related work (Bowen *et al.*, 2014). While the responsibility for making food-related decisions and preparing food still falls largely to women, women have felt both burdened and empowered by this position (Cairns and Johnston, 2015). Men do participate in food-related work, but to a lesser extent than women. On an average day in 2009, 40% of men did food preparation or clean-up compared with 68% of women (Bureau of Labor Statistics, 2010).

In Europe, family farms make up the overwhelming majority of farming operations, where women constitute on average 41% of the family labour force, are more likely to hold very small holdings and are less likely to inherit the farm (Shortall, 2010). Farm diversification, or 'pluriactivity', and women taking off-farm work characterize the shifts in rural landscapes. Funds from the EU Rural Development Programme have encouraged many farms to diversify their activities and women are active in this diversification (Gorman, 2006; Brandth and Haugen, 2010, 2011). However, women often do small-scale diversification activities that fit their other caring and domestic commitments, thereby reinforcing rather than transforming gender relations on the farm (Brandth and Haugen, 2010).

In working off-farm, many women are the principal breadwinner or contribute significantly to the survival of the farm (Shortall, 2014, 2016). However, women's contributions to the household and farm through off-farm work have done little to change domestic work and gender expectations (Kelly and Shortall, 2002). In more recent work, Shortall (2014) argues that women may have more economic status from their off-farm work, but they perform gender identities that reinforce men's masculine identities as farmers and decision makers. As a result, this transition has two contradicting tendencies: a retraditionalization of gender roles, on one hand, and the improved socio-economic status of women, on the other (Asztalos Morell and Brandth, 2007).

Family farms: global South

For female farmers, the barriers to equal participation in global value chains stem from unequal access to land and capital, credit, information and agricultural education (FAO, 2011), as well as their limited ability to hire labour, purchase inputs and access marketing channels (Razavi, 2009). Land in particular is a crucial resource for food security in rural areas; women tend to manage smaller plots of land than do men, often of inferior quality and with insecure tenure (FAO, 2011). Women also tend to have unequal and inadequate access to financial resources (credit, savings and insurance), not hold bank accounts and lack collateral (FAO, 2011; Fletschner and Kenney, 2014). Differences in access to inputs and participation in farming organizations influence a farmer's ability to upgrade farm operations and maintain market participation (Rubin and Manfre, 2014).

Women may face additional burdens to participate in high-value markets due to time burdens associated with their dual roles in productive and reproductive labour (Jiggins *et al.*, 1998). Housework in the global South can be an extremely time-consuming activity, where collecting water and food production depend on cultural tradition and household maintenance (Lee-Smith and Trujillo, 1992, p. 79). Women's traditional roles in society and the household can exacerbate the gender gap in rural development (De la O Campos and Garner, 2014) as women spend more time in unpaid labour than men (Antonopoulos, 2008).

Access to land is essential for agricultural production and evidence from across the globe shows that large gender inequities exist in land ownership and control (Archambault and Zoomers, 2015). Worldwide, the FAO estimates that 20% of landowners are women (FAO, 2010). Table 19.1 demonstrates the great range in women's land ownership across the global South. While these data reveal gender disparities in land ownership, Doss *et al.* (2015) warn that these data should be interpreted with caution and are sceptical about using a single measure of women's land rights.[1] They warn that ownership is often not clearly defined, joint-title to land may be counted differently in different contexts, and communal land and land owned by tribes, institutions or governments, and not by individuals, are not included in individual-level measurement of land ownership.

Women's rights to land often come from secondary rights to land, in which they rely on others (primarily men) such as husbands, fathers or traditional authorities to gain access to land. In these instances, women are vulnerable to losing their rights to land through widowhood, divorce, male migration or decisions to sell the land. More secure land rights for women bring a number of benefits including access to credit to purchase inputs, improving household bargaining power, and increasing entrepreneurship and household income (Fletschner and Kenny, 2014).

Labour on large-scale farms

Within the third food regime, large-scale farms have come to dominate food production, outcompeting small farms and

Table 19.1. Disparity of women's land ownership in the global South. (Data from FAO's Gender and Land Rights Database.[2])

Region	Country	Percentage of women's land ownership
West, Central and Northern Africa	Mali	3
	Senegal	9
	Lebanon	8
	Saudi Arabia	1
Southern Africa	Botswana	35
	Malawi	32
	Lesotho	31
Asia	Bangladesh	5
	Thailand	27
Latin America	Guatemala	8
	Peru	31

changing rural contexts. As small farms become less viable within this context, larger farms increasingly depend on wage labour in general and migrant labour in particular. This transition is reshaping agricultural labour in rural contexts in the global North and South.

Labour on Northern large-scale commodity farms

The rural context in the USA and Western countries is experiencing shifts as family farms fade in place of large agricultural corporations and production operations. Despite the increased value of agricultural products sold, the exit of family farms and the growth of large-scale commercial agriculture continue to shape the agricultural production landscape in the USA (USDA, 2012). Although there has been an increase in small-scale farms, the disappearing middle has resulted in the decline of medium-sized farms (MacDonald *et al.*, 2013; Guptill and Welsh, 2014). The top 2% of farms in the USA supply 47% of the agricultural products and 55% of all farms account for only 1% of agricultural production (Hoppe *et al.*, 2010). In contrast, only 4.4% of holdings in the EU are not family farms (Shortall, 2010, p. 11). The increasing cost of farm operations and the value of farms, lack of access to government commodity payments and market power, as well as the ageing population of farmers are some of the reasons for farm exit among non-corporate farmers (Guptill and Welsh, 2014).

In the USA, the industrialization of the agrifood system has also transitioned food processing from the once mainly home-run operations to primarily commercial factories (Sachs, 1991). Women make up the majority of the labour 'on the line', directly handling the processing of the food, while men hold jobs as supervisors or drivers. These jobs are labour intensive with high injury rates and also one of the lowest paying occupations for women in the USA, with median earnings less than US$400 per month (US Department of Labor, 2014). In 2005, women comprised 75% of graders and sorters of agricultural products (US Department of Labor and US Bureau of Labor Statistics, 2005). The intersection of race and gender is particularly important as black women disproportionately hold janitorial positions and perform the 'dirty work' in the factories and as processing plants increase recruitment efforts to immigrants from the Americas (Gray, 2014).

Migrant labour

Agrifood market shifts directly influence the structure of labour in farming. While the majority of farm workers in the USA are still family farm workers, both family-operated and corporate farms have increasingly relied upon day- and seasonal-field labour (Martin and Jackson-Smith, 2013). In 2007, the US Department of Agriculture estimated that 22% of US farms hired a worker (USDA-NASS, 2009). In contrast to the global South, agricultural wage workers in the USA and Canada are overwhelmingly men. Clearly ethnicity, citizenship and gender intersect in the preference for workers on corporate farms. In 2012, 82% of hired farm workers in the USA were male. The majority, 68%, of agricultural workers were born in Mexico, 41% were not US citizens, and many were undocumented workers. Average hourly earnings of farm workers are low, at $US10.80 per hour in 2012 (USDA, 2016).

Many attribute the increased reliance on and presence of immigrant labour in commodity farming in the USA and Canada to the North American Free Trade Agreement, implemented in 1994, which sped up rural-to-urban migration in Mexico by both emphasizing industrial exports and decreasing support to rural communities (Broughton, 2008; Pechlaner and Otero, 2010). The emphasis on decreasing trade regulation between Mexico, Canada and the USA increased agricultural commodity competition as cheaper food from the USA entered Mexico and Mexican industrial agriculture expanded (Broughton, 2008). Broughton (2008) argues that a reinvention of masculinities surrounding work, family and place occur through this physical relocation. In addition, masculinity is reconfiguring where workers are often in vulnerable living conditions and states of residential status in the USA.

Migrant labour in Europe is also mostly male, with the majority of short-term/seasonal labour coming from Eastern Europe to work on family farms, as documented in Norway (Rye, 2014; Rye and Andrzejewska, 2010), Greece (Kasimis *et al.*, 2003) and Southern Europe (McAreavey, 2012). These studies also position migrant labour in response to shifting trends in agriculture. Outmigration from small farms in rural areas in the face of increased competition from large-scale agriculture and transnational shifting of goods creates space and need for migrant labour (Labrianidis and Sykas, 2009; Rye and Andrzejewska, 2010; McAreavey, 2012). While the literature interrogates relationships of migrant labour with international trends, little research on migrant agricultural labour in Europe forefronts a gendered or intersectional analysis.

In 2012, while women made up 47% of all US wage and salary workers, they comprised only 18% of all hired farm workers (USDA, 2016). Despite evidence that women are a growing proportion of migrant farm workers in North America, little is known about the unique experiences these women face. By tracing women's migration from Mexico to Canada through state-created programmes as temporary farm workers, Preibisch and Encalada Grez (2010) give visibility to a population on which Canada's family farms are increasingly reliant. The power relations based on race, gender and citizenship shape women's vulnerability as applicants as well as workers and are reinforced through state and employer treatment of these workers.

Additional research demonstrates that women farm workers are at high risk of sexual assault due to the masculinity surrounding migrant farm labour, the environment in which the labour takes place (e.g. the lack of visibility due to shrubs or distance between employees) and the vulnerable position of women immigrants (Waugh, 2010). Women are particularly vulnerable to sexual assault and violence from labour contractors and supervisors upon whom they depend for jobs, housing and transportation (Southern Poverty Law Center, 2010). Further, women face uncertainty regarding their labour rights, in particular their right to report harassment, and the norms of the US workplace (Waugh, 2010). Despite the attempts to regulate the behaviour of female farm workers, migrant workers have found points of advocacy by developing personal relationships and navigating citizenship regulations (Preibisch and Encalada Grez, 2010).

Labour on large-scale commodity farms in the global South

In the global South, the history of colonialism has shaped much of the large-scale, export-oriented agriculture. With capitalism replacing colonialism as the hegemonic market structure, the emphasis is on export-oriented, high-value market chains (McMichael, 2009b). Transnational corporations now control most of the agricultural production, due to the shift of influence from nation-states to institutions to transnational corporations (Busch and Bain, 2004). For the international food market, an increased demand for fresh produce year-round has encouraged a switch from traditional export crops such as bananas, coffee, cotton and sugarcane to non-traditional high-value crops such as fresh fruits and vegetables (Friedmann, 1999), as well as livestock and fisheries such as poultry and shrimp farming.

Women's employment as farm workers

As agricultural production shifts away from traditional products, off-farm employment and wage labour have become essential to supporting family agricultural practices (Smith and Stevens, 1988). Employment also diversifies risk and substitutes for the frequent lack of formal finance opportunities in rural settings (Hernandez *et al.*, 2010). The feminization of agriculture is rampant here; women are often the preferred workers in the harvesting, processing and packaging of these new crops. These products are typically grown for export markets and sold through value chains operated by large international firms to supply fresh produce to developed countries. Table 19.2 shows the percentage of women workers in selected value chains in flower (range: 65–87%), vegetable (70–95%), fruit (45–53%), poultry (90%) and shrimp (36–62%) production and processing in various countries (ILO, 2010).

Women often comprise the majority of workers in these value chains, but have less favourable employment conditions than men. Women are more likely to be employed as casual, temporary workers compared with men, who often have longer-term employment. While these value chains offer possibilities for better-quality and higher-wage jobs than traditional agricultural production, the verdict is still mixed on whether value-chain employment improves women's lives and household well-being (ILO, 2010).

The justification of gender segregation often entails stereotypical descriptions of women's and men's capabilities, resulting in occupational segregation. Women are perceived as possessing dexterity, the ability to work with their hands, and care in handling easily damaged fruits and vegetables. Men, on the other hand, are perceived as more capable in operating machinery and equipment. Thus, men's occupations are defined as more highly skilled and this gender segregation is used to legitimize lower wages for women workers (ILO, 2010).

The difference in men's and women's wages varies by country and crop. In India, women's wages for casual agricultural work are 30% lower than are men's (World Bank, 2007). In Pakistan, women's wages are 50% of men's wages in sugarcane production whereas in Mexico women's wages are 80% of men's wages in flower production (ILO, 2010). Barrientos (2013) analyses how global production networks in agriculture exploit women workers. Global production networks demand increased flexibility from workers, which includes compulsory overtime and the use of casual day labourers, often women, who can be hired on a daily basis depending on the demand. Women's responsibilities for both productive and reproductive work leave them with weak bargaining positions in which they often must accept these low wages. Consequently, they are 'exploited as a low-cost source of value enhancement' (Barrientos, 2013, p. 46).

New models of agriculture in the global North and global South

In response to the liberalization of government regulation and the corporatization of the food system of the third food regime, many producers and consumers seek to

Table 19.2. Percentage of women workers in selected value chains. (From ILO, 2010.)

Value chain	Country	Percentage of women workers	Percentage of women casual workers
Flowers	Kenya	75	65
	Columbia	65	n/a
	Zimbabwe	70–87	42
Vegetables	Peru	70 field 95 processing	65–75
Fruit	South Africa	53	65–75
	Chile	45	55–70
Poultry	Brazil	90	Majority
Shrimp	Bangladesh	36 62 processing	90

create and facilitate alternative food market systems. The push towards alternatives to global industrial production may open new opportunities for women in agriculture and flexibility in masculine farming identities.

New models of agriculture in the global North

In the global North, many women farmers manage or co-manage agricultural operations that produce for organic, sustainable and local markets (Liepins, 1998; DeLind and Ferguson, 1999; Hassanein, 1999; Trauger *et al.*, 2010; Sachs *et al.*, 2016). These types of operations provide a challenge to the global agro-industrial food system, but represent a small although growing proportion of agricultural production. Women comprise 50% of organic farmers in the UK, 33% in Canada and 22% in the USA (Jarosz, 2011). These percentages far exceed women's representation on conventional farms. Women's involvement in more environmentally conscious agriculture such as organic and sustainable agriculture parallels women's greater concern for environmental issues in general (Barbercheck *et al.*, 2012; Sachs *et al.*, 2016). Women consistently express more concern about environmental issues and risks than men across a wide range of issues (Zelezny *et al.*, 2000; Goldsmith *et al.*, 2013; Xiao and McCright, 2015).

The recent surge in the local food movement involves an increasing number of farmers' markets, direct sales of farm products to consumers and restaurants, and community-supported agriculture. Although limited demographic data exist on characteristics of 'local' farmers, a recent study in Washington State found that women outnumbered men as vendors and managers in farmers' markets (Ostrom, 2014). Another study in Washington found that women predominated as farmers in community-supported agriculture (Jarosz, 2011). Jarosz (2011) expresses concern that many may dismiss women's work as merely urban gardeners rather than farmers, but she argues that they have chosen this type of agriculture because it nourishes their families and communities. Research from Michigan further adds that participation in community-oriented, or 'value-added', agriculture does not automatically result in a more empowered position or in a shifting feminine identity (Wright and Annes, 2016).

The growing interest in new agriculture models has also created space for new farming masculinities. Through interviews with sustainable farmers in Iowa, Peter *et al.* (2000) describe the associated farming masculinity as *dialogic masculinity* that involves caring for the land, sharing ideas and knowledge with other farmers, and moving towards more equitable gender relations in their households. Barlett and Conger (2004) contrast this masculinity to the more established ideals of industrial and agrarian agriculture in their comparison of farming communities in Georgia, Illinois and Iowa in the USA. They found that demonstrations of successful masculinity for sustainable farmers emphasize experimentation and use of test plots in contrast to financial status as breadwinner and ability to support their wives as homemakers (Barlett and Conger, 2004). In addition, masculine identities shaped and were complemented by the roles and expectations of the women in the household, regardless of their own expectations or measurements of success.

New models for agriculture in the global South

Challenges to the global industrial food system also come from the Fair Trade movement and the food sovereignty movement in the global South. Fair Trade relies on labelling and certification of commodities to improve the position of small-scale producers while encouraging consumers in the North to focus on ethical consumption. Food sovereignty is a rights-based approach that calls for a redefinition of food and agriculture that corrects for the marginalization of smallholder farmers in the control over production and consumption. Further, there is increasing awareness of women's efforts in very small-scale agriculture, including in urban spaces, that fosters connections between people and food, and to one another (Sachs and Patel-Campillo, 2014). These movements demonstrate a response to the third food regime that calls for more equitable ownership, compensation and regulation of agrifood production, with particular attention to gender.

One of the goals of the Fair Trade movement is to empower women (Lyon *et al.*, 2010; Le Mare, 2012). This is addressed by including standards of non-discrimination in recruitment, inclusion and benefits, as well as requiring adherence to national or collective bargaining agreements (FLO, 2011). Research on fair trade and gender empowerment has found mixed results. Lyon *et al.* (2010) document in South Mexico and Guatemala that women's increased involvement in coffee production and also how the confluence of both fair trade practices and organic certification requirements has increased women's role as farm operators and landowners, as well as improved their access to organizations, property and income. However, women moving into this value chain as prices are low could take up labour-intensive, poorly remunerated work as men turn to other income-earning strategies including migration. Research by Jones *et al.* (2012) on fair trade collectives in Kenya, Tanzania, Uganda, India, Nepal, Nicaragua and Mexico finds that collective learning and pooled resources were among the benefits of participation. However, fair trade collectives do not automatically lift women out of poverty without strong support to improve their access to capital and markets and efforts to improve gender equity and equality.

In the mid-1990s, a coalition of peasants, farm workers and small producers formed La Via Campesina to challenge the global neoliberal food regime (Wittman *et al.*, 2010). At the heart of La Via Campesina is the push for food sovereignty, which is 'the right of nationals and people to control their own food systems, including their own markets, production modes, food cultures and environments' (Wittman *et al.*, 2010, p. 2). Peasant women and women farmers participated in the food sovereignty movement from the outset. Women in the food sovereignty movement insist that their work in production for the market, for the family and in social production be recognized and valued.

Very small-scale, especially urban, women producers' contributions to household and community food security are often overlooked and underestimated (Sachs, 2015). In many cities in the global South, women play key roles in producing vegetables and livestock in kitchen gardens and small plots of land, as the FAO documents in India, the Philippines, Ghana, Kenya and Peru (FAO, 2011). Much of this work in food production goes unrecognized and more efforts by city planners and local government officials could provide land and other resources to support these endeavours (Sachs and Patel-Campillo, 2014).

Discussion

Transitions in gendered roles work differently across the global North and global South and the different scales and operations of agriculture. By looking at these six contexts, we see how the changing global systems of agriculture connect with shifts in women's productive labour on family farms, industrial farming, as well as emerging models of agriculture.

Family farms are making great transitions in both the global North and global South as the character of the third food regime increases competition and challenges the viability of family farms in rural contexts. In response, off-farm employment has come to hold greater importance in supplementing the income of these farm operations. In both contexts, despite women's increased participation and recognition in family farms, the identity of farmer remains masculine and women in both spaces still face barriers to entry and authority in agriculture.

In response to consumer demands and decreased regulation, large-scale commodity farming has become essential in the distribution of food. As a result, the majority of agricultural production in the global North, especially in the USA, occurs on large-scale farms that depend on migrant and seasonal labour. In the global South, increased reliance on female, temporary workers shapes export-oriented agriculture. While men and women both face difficult labour conditions in both contexts, women are disproportionately marginalized and undervalued in their labour in terms of wages, job security and even physical safety. Intersectional approaches further characterize the experiences of women in these value chains, as do discussions of men as 'breadwinners' and their mobility and value as labourers.

Emerging models of agriculture are responding to the decreased regulation in the third food regime and the call for greater accountability in and control over food systems. This rising interest in the global North positions women at the forefront of diversifying farm activities that reconnect farms and their communities. In the global South, Fair Trade and food sovereignty movements are also responding to smallholders' lack of voice in global value chains. In both contexts, agricultural models are contesting the expanding, global nature of agrifood systems, while challenging the traditional models of gender.

Women's labour is moving into spheres that are more public across these contexts as it is gaining greater attention and visibility. While there are noticeable impacts on men's labour and challenges to traditional ideals of masculinity, care work and private/domestic responsibilities still fall largely to women. While this is true to a certain extent within emerging models of agriculture, global responses to the decreased government regulation and increased inequality demonstrate a demand for more equitable agricultural practices.

Conclusion

This chapter demonstrates that the transitions in the global agrifood system result in identifiable feminizations and masculinizations of agriculture, while gendered care responsibilities remain largely unshifted. As current food regimes include increased pressure for farmers to produce for global value chains and corporate organization of agricultural production, we see transitions in gender identities vary across the different contexts of family farms, large-scale commodity farms and emerging models of agriculture in the global North and South. This demonstrates that research must situate men and women within their historical, cultural contexts to account for race, class, age and citizenship. Further, the manifestations of these identities continue to transition as food regimes change and policies seek to address or redress discrimination of women in agriculture.

More studies on shifting masculinities in agriculture, especially in the global South, and their intersections with the feminization of agriculture would provide deeper insights into how households adjust in food regimes, improving policy and development interventions. So far, the emphasis has been on the integration of women into agrifood market systems, as well as the potential for and barriers to women's increased capital gains. In the meantime, women's care labour and responsibilities still go largely undervalued and unrecognized. Masculinizations are also still strongly shaped by the public spaces involving market participation and market goals. Therefore, expanding understandings of shifting gender identities incorporate the relation between transitions in intrahousehold divisions of labour and agrifood systems.

Sachs and Patel Campillo argue that a feminist food justice requires revaluing 'women's social reproduction work with food' (Sachs and Patel-Campillo, 2014, p. 406). The authors envision that involving men and boys more in food preparation and nutritional concerns will reframe traditional gender relations by reinstating the fluidity of gender roles. We must also re-evaluate the hiring practices and labour regulations that place women in vulnerable working conditions because of their assumed 'need' to fulfil social and domestic responsibilities. Scholars, development agencies and agricultural institutions must study and push for gender justice and broader social justice within agrifood systems.

As presented here, the current conversations regarding shifting women's and men's gendered identities and participation in agriculture occur in parallel. While our analysis contributes to the literature on masculinizations and feminizations within the context of shifting agrifood systems, a new framework is needed to unpack the relationships between gendered identities at different scales. This may provide greater insight into how these identities shape and reshape each other, as well as the potential for fluidity and flexibility in their performance. This could also address the gap in previous research that analyses the emergence of new feminizations of productive spaces without simultaneously interrogating the shifts in reproductive, feminine spaces.

Notes

[1]Also, the primary source of FAO data for land ownership comes from national-level agricultural censuses which are usually conducted every 10 years and often date back to the 1990s, so the data are often dated. In addition, agricultural censuses collect information on landholders who make decisions about the operation rather than landowners.

[2]The FAO's user-friendly Gender and Land Rights Database provides country-level statistics from 83 countries on the percentage of men and women landholders, information on women's tenure rights, customary laws and practices, and civil society organizations engaged in gender equitable land tenure reform (http://www.fao.org/gender-landrights-database/country-profiles/en/, accessed 22 February 2017).

References

Allen, P. and Sachs, C. (2007) Women and food chains: the gendered politics of food. *International Journal of Sociology of Agriculture and Food* 15(1), 1–23.

Antonopoulos, R. (2008) *The Unpaid Care Work–Paid Work Connection. Working Paper No. 541*. The Levy Economics Institute of Bard College, Annandale-on-Hudson, New York.

Archambault, C. and Zoomers, A. (2015) *Global Trends in Land Tenure Reforms: Gender Impacts*. Routledge, London.

Asztalos Morell, I. and Brandth, B. (2007) Family and gender in the transformation of the countryside. *Journal of Comparative Family Studies* 38(3), 371–377. Available at: http://www.jstor.org/stable/41604163 (accessed 28 February 2017).

Barbercheck, M.E., Brasier, K.J., Kiernan, N.E., Sachs, C. and Trauger, A. (2012) Use of conservation practices by women farmers in the Northeastern United States. *Renewable Agriculture and Food Systems* 29(1), 65–82. Available at: http://dx.doi.org/10.1017/S1742170512000348 (accessed 22 February 2017).

Barlett, P.F. and Conger, K.J. (2004) Three visions of masculine success on American farms. *Men and Masculinities* 7(2), 205–227. Available at: http://dx.doi.org/10.1177/1097184X03257409 (accessed 22 February 2017).

Barrientos, S. (2013) Corporate purchasing practices in global production networks: a socially contested terrain. *Geoforum* 44, 44–51. Available at: http://dx.doi.org/10.1016/j.geoforum.2012.06.012 (accessed 22 February 2017).

Bell, D. (2000) Farm boys and wild men: rurality, masculinity and homosexuality. *Rural Sociology* 65(4), 547–561. Available at: http://dx.doi.org/10.1111/j.1549-0831.2000.tb00043.x (accessed 22 February 2017).

Bharadwaj, L. and Findeis, J.L. (2003) Off-farm work among farm women: motivations, earnings, and benefit receipt. Paper presented at the *Annual Meeting of the American Agricultural Association*, Montreal, Canada, 27–30 July 2003.

Bowen, S., Elliott, S. and Brenton, J. (2014) The joy of cooking? *Contexts* 13(3), 20–25. Available at: http://dx.doi.org/10.1177/1536504214545755 (accessed 22 February 2017).

Brandth, B. (2002) Gender identity in European family farming: a literature review. *Sociologia Ruralis* 42(3), 181–200. Available at: http://dx.doi.org/10.1111/1467-9523.00210 (accessed 22 February 2017).

Brandth, B. and Haugen, M. (2010) Doing farm tourism: the intertwining practices of gender and work. *Signs* 35(2), 425–446. Available at: http://dx.doi.org/10.1086/605480 (accessed 22 February 2017).

Brandth, B. and Haugen, M. (2011) Farm diversification into tourism – implications for social identity? *Journal of Rural Studies* 27(1), 35–44. Available at: http://dx.doi.org/10.1016/j.jrurstud.2010.09.002 (accessed 22 February 2017).

Brasier, K., Sachs, C.E., Kiernan, N.E., Trauger, A. and Barbercheck, M.E. (2014) Capturing the multiple and shifting identities of farm women in the Northeastern United States. *Rural Sociology* 79(3), 283–309. Available at: http://dx.doi.org/10.1111/ruso.12040 (accessed 22 February 2017).

Broughton, C. (2008) Migration as engendered practice: Mexican men, masculinity, and northward migration. *Gender & Society* 22(5), 568–589. Available at: http://dx.doi.org/10.1177/0891243208321275 (accessed 22 February 2017).

Bureau of Labor Statistics (2010) *American Time Use Survey – 2009 Results*. Bureau of Labor Statistics, US Department of Labor, Washington, DC. Available at: http://www.bls.gov/news.release/archives/atus_06222010.pdf (accessed 15 November 2016).

Busch, L. and Bain, C. (2004) New! Improved? The transformation of the global agrifood system. *Rural Sociology* 69(3), 321–346. Available at: http://onlinelibrary.wiley.com/doi/10.1526/0036011041730527/abstract (accessed 28 February 2017).

Caillavet, F., Herve, G. and Lifran, R. (eds) (1994) *Agricultural Household Modelling and Family Economics*. Elsevier, New York.

Cairns, K. and Johnston, J. (2015) *Food and Femininity*. Bloomsbury, New York.

Chant, S. (2000) Men in crisis? Reflections on masculinities, work and family in north-west Costa Rica. *The European Journal of Development Research* 12(2), 199–218. Available at: http://link.springer.com/article/10.1080/09578810008426772 (accessed 28 February 2017).

Connell, R.W. (1998) Masculinities and globalization. *Men and Masculinities* 1(1), 3–23. Available at: http://dx.doi.org/10.1177/1097184X98001001001 (accessed 22 February 2017).

De la O Campos, A.P. and Garner, E. (2014) *Women's Resilience to Food Price Volatility: A Policy Response*. Food and Agriculture Organization of the United Nations, Rome. Available at: http://www.fao.org/docrep/019/i3617e/i3617e.pdf (accessed 4 February 2016).

De Schutter, O. (2013) The agrarian transition and the 'feminization' of agriculture. Paper presented at *Food Sovereignty: A Critical Dialogue*, Yale University, New Haven, Connecticut, 14–15 September 2013.

DeLind, L.B. and Ferguson, A.E. (1999) Is this a women's movement? The relationship of gender to community-supported agriculture in Michigan. *Human Organization* 58(2), 190–200. Available at: http://dx.doi.org/10.17730/humo.58.2.lpk17625008871x7 (accessed 22 February 2017).

Doss, C., Kovarik, C., Peterman, A., Quisumbing, A. and van den Bold, M. (2015) Gender inequalities in ownership and control of land in Africa: myth and reality. *Agricultural Economics* 46(3), 403–434. Available at: http://dx.doi.org/10.1111/agec.12171 (accessed 22 February 2017).

Elson, D. (2008) The three R's of unpaid work: recognition, reduction and redistribution. Paper presented at the *Expert Group Meeting on Unpaid Work, Economic Development and Human Well-Being*, United Nations Development Programme, New York, 16–17 November 2008.

Esquivel, V. (2014) What is a transformative approach to care, and why do we need it? *Gender & Development* 22(3), 423–439. Available at: http://dx.doi.org/10.1080/13552074.2014.963303 (accessed 22 February 2017).

FAO (2010) *Gender and Land Rights: Understanding Complexities; Adjusting Policies*. Food and Agriculture Organization of the United Nations, Rome. Available at: http://www.fao.org/economic/es-policybriefs/briefs-detail/en/?no_cache=1&uid=40497 (accessed 22 February 2017).

FAO (2011) *The State of Food and Agriculture 2010–11. Women in Agriculture: Closing the Gender Gap for Development*. Food and Agriculture Organization of the United Nations, Rome. Available at: http://www.fao.org/docrep/013/i2050e/i2050e.pdf (accessed 22 February 2017).

Fletschner, D. and Kenney, L. (2014) Rural women's access to financial services: credit, savings, and insurance. In: Quisumbing, A.R., Meinzen-Dick, R., Raney, T.L., Croppenstedt, A., Behrman, J. and Peterman, A. (eds) *Gender in Agriculture: Closing the Knowledge Gap*. Food and Agriculture Organization of the United Nations, Rome and Springer, Dordrecht, pp. 187–208.

FLO (2011) Fairtrade International. Programmes. Gender. Available at: http://www.fairtrade.net/programmes/gender.html (accessed 20 April 2016).

Friedmann, H. (1999) Remaking 'traditions': how we eat, what we eat and the changing political economy of food. In: Brandt, D. (ed.) *Women Working the NAFTA Food Chain: Women, Food and Globalization.* Second Story Press, Toronto, pp. 36–60.

Goldsmith, R., Feygina, I. and Jost, J. (2013) The gender gap in environmental attitudes: a system justification perspective. In: Alston, M. and Whittenbury, K. (eds) *Research Policy: Addressing the Gendered Impacts of Climate Change.* Springer, Dordrecht, The Netherlands, pp. 159–171.

Gorman, M. (2006) Gender relations and livelihood strategies. In Bock, B. and Shortall, S. (eds) *Rural Gender Relations: Issues and Case Studies.* CABI International, Wallingford, UK, pp. 27–47.

Gray, L. (2014) *We Just Keep Running the Line: Black Southern Women and the Poultry Processing Industry.* LSU Press, Baton Rouge, Louisiana.

Guptill, A. and Welsh, R. (2014) The declining middle of American agriculture: a spatial phenomenon. In: Bailey, C., Jensen, L. and Ransom, E. (eds) *Rural America in a Globalizing World: Problems and Prospects for the 2010s.* West Virginia Press, Morgantown, West Virginia, pp. 36–50.

Hassanein, N. (1999) *Changing the Way America Farms: Knowledge and Community in the Sustainable Agriculture Movement.* University of Nebraska Press, Lincoln, Nebraska.

Hernandez, R., Reardon, T. and Guan, Z. (2010) Rural nonfarm employment and agricultural modernization and diversification in Guatemala. Paper prepared for presentation at the *Agricultural & Applied Economics Association 2010 AAEA, CAES, & WAEA Joint Annual Meeting,* Denver, Colorado, 25–27 July 2010.

Hoppe, R.A. and Korb, P. (2013) *Characteristics of Women Farm Operators and Their Farms. Economic Information Bulletin Number 111.* Economic Research Service, US Department of Agriculture, Washington, DC. Available at: https://www.ers.usda.gov/webdocs/publications/43749/37012_eib111.pdf?v=41390 (accessed 22 February 2017).

Hoppe, R.A., MacDonald, J.M. and Korb, P. (2010) *Small Farms in the United States: Persistence Under Pressure.* Economic Research Service, United States Department of Agriculture, Washington, D.C.

ILO (2010) *Agricultural Value Chain Development: Threat or Opportunity for Women's Employment. Gender and Rural Employment Policy Brief #4.* International Labour Organization Geneva, Switzerland. Available at: http://www.ilo.org/wcmsp5/groups/public/---ed_emp/documents/publication/wcms_150833.pdf (accessed 12 November 2016).

Jarosz, L. (2011) Nourishing women: toward a feminist political ecology of community supported agriculture in the United States. *Gender, Place & Culture* 18(3), 307–326. Available at: http://dx.doi.org/10.1080/0966369X.2011.565871 (accessed 22 February 2017).

Jiggins, J., Samanta, R.K. and Olawoye, J.E. (1998) Improving women farmers' access to extension services. In: Swanson, B.E., Bentz, R.P. and Sofranko, A.J. (eds) *Improving Agricultural Extension: A Reference Manual.* Food and Agriculture Organization of the United Nations, Rome. Available at: http://www.fao.org/docrep/w5830e/w5830e0b.htm#chapter9 (accessed 16 November 2016).

Jones, E., Smith, S. and Wills, C. (2012) Women producers and the benefits of collective forms of enterprise. *Gender & Development* 20(1), 13–32. Available at: http://dx.doi.org/10.1080/13552074.2012.663640 (accessed 22 February 2017).

Kasimis, C., Papadopoulos, A.G. and Zacopoulou, E. (2003) Migrants in rural Greece. *Sociologia Ruralis* 43(2), 167–184. Available at: http://dx.doi.org/10.1111/1467-9523.00237 (accessed 22 February 2017).

Kelly, R. and Shortall, S. (2002) 'Farmers' wives': women who are off-farm breadwinners and the implications for on-farm gender relations. *Journal of Sociology* 38(4), 327–343. Available at: http://dx.doi.org/10.1177/144078302128756714 (accessed 22 February 2017).

Kidder, T., Mapandi, Z. and Ortega, H. (2014) Not 'women's burden': how washing clothes and grinding corn became issues of social justice and development. *Gender & Development* 22(3), 495–513. Available at: http://dx.doi.org/10.1080/13552074.2014.963324 (accessed 22 February 2017).

Labrianidis, L. and Sykas, T. (2009) Migrants, economic mobility and socio-economic change in rural areas: the case of Greece. *European Urban and Regional Studies* 16(3), 237–256. Available at: http://dx.doi.org/10.1177/0969776409104691 (accessed 22 February 2017).

Le Mare, A. (2012) 'Show the world to women and they can do it': Southern Fair Trade Enterprises as agents of empowerment. *Gender & Development* 20(1), 95–109. Available at: http://dx.doi.org/10.1080/13552074.2012.663608 (accessed 22 February 2017).

Lee-Smith, D. and Trujillo, C.H. (1992) The struggle to legitimize subsistence: women and sustainable development. *Environment and Urbanization* 4(1), 77–84. Available at: http://dx.doi.org/10.1177/095624789200400108 (accessed 22 February 2017).

Liepins, R. (1998) The gendering of farming and agricultural politics: a matter of discourse and power. *Australian Geographer* 29(3), 371–388. Available at: http://dx.doi.org/10.1080/00049189808703230 (accessed 22 February 2017).

Little, J. (2002) Rural geography: rural gender identity and the performance of masculinity and femininity in the countryside. *Progress in Human Geography* 26(5), 665–670. Available at: http://dx.doi.org/10.1191/0309132502ph394pr (accessed 22 February 2017).

Lyon, S., Bezaury, J.A. and Mutersbaugh, T. (2010) Gender equity in fairtrade–organic coffee producer organizations: cases from Mesoamerica. *Geoforum* 41(1), 93–103. Available at: http://dx.doi.org/10.1016/j.geoforum.2009.04.006 (accessed 22 February 2017).

MacDonald, J.M., Korb, P. and Hoppe, R.A. (2013) *Farm Size and the Organization of US Crop Farming. Economic Research Report Number 152.* Economic Research Service, US Department of Agriculture, Washington, DC. Available at: https://www.ers.usda.gov/publications/pub-details/?pubid=45110 (accessed 22 February 2017).

Martin, P. and Jackson-Smith, D. (2013) *Immigration and Farm Labor in the US. National Agricultural and Rural Development Policy Center, Policy Brief 4.* Available at: http://nardep.info/uploads/Brief_FarmWorker.pdf (accessed 1 February 2016).

McAreavey, R. (2012) Resistance or resilience? Tracking the pathway of recent arrivals to a 'new' rural destination. *Sociologia Ruralis* 52(4), 488–507. Available at: http://dx.doi.org/10.1111/j.1467-9523.2012.00573.x (accessed 22 February 2017).

McMichael, P. (2009a) A food regime analysis of the world food crisis. *Agriculture and Human Values* 26, 281–293. Available at: http://dx.doi.org/10.1007/s10460-009-9218-5 (accessed 22 February 2017).

McMichael, P. (2009b) A food regime genealogy. *The Journal of Peasant Studies* 36(1), 129–169. Available at: http://dx.doi.org/10.1080/03066150902820354 (accessed 22 February 2017).

Ostrom, M. (2014) She's my farmer: exploring the role of gender in the farmers' market movement. Paper presented at *Annual Meeting of the Association for the Study of Food and Society and the Agriculture, Food and Human Values Society*, Burlington, Vermont, 18–22 June.

Otero, G., Pechlaner, G. and Gurcan, E.C. (2013) The political economy of 'food security' and trade: uneven and combined dependency. *Rural Sociology* 78(3), 263–289. Available at: http://dx.doi.org/10.1111/ruso.12011 (accessed 22 February 2017).

Pechlaner, G. and Otero, G. (2010) The neoliberal food regime: neoregulation and the new division of labor in North America. *Rural Sociology* 75(2), 179–208. Available at: http://dx.doi.org/10.1111/j.1549-0831.2009.00006.x (accessed 22 February 2017).

Peter, G., Bell, M.M., Jarnagin, S. and Bauer, D. (2000) Coming back across the fence: masculinity and the transition to sustainable agriculture. *Rural* Sociology 65(2), 215–233. Available at: http://dx.doi.org/10.1111/j.1549-0831.2000.tb00026.x (accessed 22 February 2017).

Preibisch, K.L. and Encalada Grez, E. (2010) The other side of *el Otro Lado*: Mexican migrant women and labor flexibility in Canadian agriculture. *Signs* 35(2), 289–316. Available at: http://dx.doi.org/10.1086/605483 (accessed 22 February 2017).

Razavi, S. (2009) The gendered impacts of liberalization: towards 'embedded liberalism'? In: Razavi, S. (ed.) *The Gendered Impacts of Liberalization: Towards 'Embedded Liberalism'?* Taylor & Francis, New York, pp. 1–36.

Rubin, D. and Manfre, C. (2014) Promoting gender-equitable agricultural value chains: issues, opportunities, and next steps. In: Quisumbing, A.R., Meinzen-Dick, R., Raney, T.L., Croppenstedt, A., Behrman, J. and Peterman, A. (eds) *Gender in Agriculture: Closing the Knowledge Gap.* Food and Agriculture Organization of the United Nations, Rome, pp. 287–313.

Rye, J.F. (2014) The Western European countryside from an Eastern European perspective: case of migrant workers in Norwegian agriculture. *European Countryside* 6(4), 327–346. Available at: http://dx.doi.org/10.2478/euco-2014-0018 (accessed 22 February 2017).

Rye, J.F. and Andrzejewska, J. (2010) The structural disempowerment of Eastern European migrant farm workers in Norwegian agriculture. *Journal of Rural Studies* 26(1), 41–51. Available at: http://dx.doi.org/10.1016/j.jrurstud.2009.06.003 (accessed 22 February 2017).

Sachs, C. (1991) Women's work and food: a comparative perspective. *Journal of Rural Sociology* 7(1–2), 49–55. Available at: http://dx.doi.org/10.1016/0743-0167(91)90039-U (accessed 22 February 2017).

Sachs, C. (2015) Gender and the international political economy of agri-food. In: Bonanno, A. and Busch, L. (eds) *The Handbook of International Political Economy of Agriculture and Food*. Edward Elgar, Northampton, Massachusetts, pp. 344–356.

Sachs, C. and Patel-Campillo, A.P. (2014) Feminist food justice: crafting a new vision. *Feminist Studies* 40(2), 396–410. Available at: http://www.jstor.org/stable/10.15767/feministstudies.40.2.396 (accessed 28 February 2017).

Sachs, C., Barbercheck, M.E., Brasier, K., Kiernan, N.E. and Terman, A.R. (2016) *The Rise of Women Farmers in Sustainable Agriculture*. University of Iowa Press, Iowa City, Iowa.

Saugeres, L. (2002a) Of tractors and men: masculinity, technology and power in a French farming community. *Sociologia Ruralis* 42(2), 143–159. Available at: http://dx.doi.org/10.1111/1467-9523.00207 (accessed 22 February 2017).

Saugeres, L. (2002b) The cultural representation of the farming landscape: masculinity, power and nature. *Journal of Rural Studies* 18(4), 373–384. Available at: http://dx.doi.org/10.1016/S0743-0167(02)00010-4 (accessed 22 February 2017).

Shortall, S. (2016) Gender and identity formation. In: Shucksmith, M. and Brown, D.L. (eds) *The Routledge International Handbook of Rural Studies*. Routledge/Falmer, London and New York, pp. 400–412.

Shortall, S. (2014) Farming, identity and well-being: managing changing gender roles within Western European farm families. *Anthropological Notebooks* 20(3), 67–81.

Shortall, S. (2010) *Women Working on the Farm: How to Promote their Contribution to the Development of Agriculture in Rural Areas in Europe*. Directorate General for Internal Policies, Policy Department B: Structural and Cohesion Policies, Agriculture and Rural Development, Brussels.

Smith, C. and Stevens, L. (1988) Farming and income generation in the female-headed small holder household: the case of a Haya village in Tanzania. *Canadian Journal of African Studies* 22(3), 552–566.

Southern Poverty Law Center (2010) *Injustice on our Plates: Immigrant Women in the US Food Industry*. Southern Poverty Law Center, Montgomery, Alabama.

Stalford, H., Baker, H. and Beveridge, F. (2003) *Children and Domestic Violence in Rural Areas: A Child-focused Assessment of Service Provision*. Save the Children, London.

Trauger, A., Sachs, C., Barbercheck, M.E., Brasier, K. and Kiernan, N.E. (2010) 'Our market is our community': women farmers and civic agriculture in Pennsylvania, USA. *Agriculture and Human Values* 27(1), 43–55. Available at: http://dx.doi.org/10.1007/s10460-008-9190-5 (accessed 22 February 2017).

Tronto, J.C. (2012) Democratic care politics in an age of limits. In: Razabi, S. and Staab, S. (eds) *Global Variations in the Political and Social Economy of Care: Worlds Apart*. Routledge/UNRISD Research in Gender and Development, New York, pp. 29–40.

USDA (2012) *2007 Census of Agriculture: Women Farmers*. US Department of Agriculture, National Agricultural Statistics Service, Washington, DC. Available at: http://www.agcensus.usda.gov/Publications/2007/Online_Highlights/Fact Sheets/Demographics/women.pdf (accessed 28 October 2015).

USDA (2014) *2012 Census Highlights: Farm Demographics – US Farmers by Gender, Age, Race, Ethnicity, and More*. US Department of Agriculture, Census of Agriculture, Washington, DC. Available at: https://www.agcensus.usda.gov/Publications/2012/Online_Resources/Highlights/Farm_Demographics/#fewer_women (accessed 12 October 2015).

USDA (2016) *Farm Labor: Background*. US Department of Agriculture, Economic Research Service, Washington, DC. Available at: https://www.ers.usda.gov/topics/farm-economy/farm-labor/background.aspx (accessed 22 February 2017).

USDA-NASS (2009) *2007 Census of Agriculture: United States Summary and State Data. Volume 1, Geographic Area Series, Part 51, AC-07-A-51*. US Department of Agriculture, National Agricultural Statistics Service, Washington, DC. Available at: https://www.agcensus.usda.gov/Publications/2007/Full_Report/usv1.pdf (accessed 12 February 2016).

US Department of Labor (2014) *Highest and Lowest Paying Occupations for Women*. US Department of Labor, Women's Bureau, Washington, DC. Available at: http://search.usa.gov/search?utf8=%E2%9C%93&affiliate=WB&query=graders+sorters&x=15&y=7 (accessed 12 November 2016).

US Department of Labor and US Bureau of Labor Statistics (2005) *Highlights of Women's Earnings in 2004*. US Department of Labor, Washington, DC. Available at: http://www.bls.gov/opub/reports/womens-earnings/archive/womensearnings_2004.pdf (accessed 12 November 2016).

Waugh, M.I. (2010) Examining the sexual harassment experiences of Mexican immigrant farm-working women. *Violence Against Women* 16(3), 237–261. Available at: http://dx.doi.org/10.1177/1077801209360857 (accessed 22 February 2017).

Weinberger, K. and Lumpkin, T.A. (2007) Diversification into horticulture and poverty reduction: a research agenda. *World Development* 35(8), 1464–1480. Available at: http://dx.doi.org/10.1016/j.worlddev.2007.05.002 (accessed 22 February 2017).

Wittman, H., Desmarais, A. and Wiebe, N. (2010) The origins and potential of food sovereignty. In: Wittman, H., Desmarais, A. and Wiebe, N. (ed) *Food Sovereignty: Reconnecting Food, Nature and Community*. Fernwood Publishing, Halifax, Nova Scotia, pp. 1–14.

World Bank (2007) *World Development Report: Agriculture for Development*. The World Bank, Washington, DC.

Wright, W. and Annes, A. (2016) Farm women and the empowerment potential in value-added agriculture. *Rural Sociology* 81(4), 545–571. Available at: http://dx.doi.org/10.1111/ruso.12105 (accessed 22 February 2017).

Xiao, C. and McCright, A.M. (2015) Gender differences in environmental concern: revisiting the institutional trust hypothesis in the USA. *Environment and Behavior* 47(1), 17–37. Available at: http://dx.doi.org/10.1177/0013916513491571 (accessed 22 February 2017).

Zelezny, L.C., Chua, P.P. and Aldrich, C. (2000) New ways of thinking about environmentalism: elaborating on gender differences in environmentalism. *Journal of Social Issues* 56(3), 443–457. Available at: http://dx.doi.org/10.1111/0022-4537.00177 (accessed 22 February 2017).

20 Glocal Networking for Gender Equality and Sustainable Livelihoods

W. Harcourt*

IISS, Erasmus University, The Hague, The Netherlands

Introduction

In this chapter a feminist political ecology (FPE) framework is elaborated and applied to an analysis of glocal networking for gender equality and sustainable livelihoods. The chapter examines case studies of rural women's organizations in India, Tanzania and Italy in order to look at the interface between the 'sidestreams' of women's knowledge and experiences of their environment and the 'mainstream' gender and sustainable development policy.

The chapter is divided into five sections. The first section reviews the literature informing FPE approaches. The second section explains the methodology of the chapter and in what way an FPE approach will be applied to the cases. The third section sets out three cases looking at: gender and sustainable livelihoods in rural India; women mobilizing for sustainable community livelihoods in Tanzania; and women's place-based global struggles for food sovereignty in Italy. The fourth section discusses the findings that emerge from the analysis of the cases, exploring: the contours of rural women's glocal networking; the role of interlocutors; lived feminist political ecology; and the connection between grassroots experiences to dominant development discourses. The fifth section proposes how the findings on women's glocal networking could contribute to the sustainable development policy framework (i.e. the Sustainable Development Goals (SDGs)).

Analytical Framework: Feminist Political Ecology

The chapter builds on the FPE framework that emerged in the 1990s in order to broaden political ecology to include feminist critiques of how gender relations shape power and knowledge (Rocheleau, 2015). The following literature review highlights the main concepts that are being developed in contemporary FPE research. FPE brings together scholars and activists working in a variety of fields – geography, anthropology, geography, gender studies, critical development studies, environmental justice and agrarian studies (Nirmal and Rocheleau, 2015). FPE comes from both academic endeavours and a stream of feminist activism that emerged in the wake of the United Nations' Conference on Environment and Development (UNCED) held in September 1992 in Rio de Janeiro, popularly known as 'Rio'. In the preparations for Rio, a global coalition of women's networks was formed to bring feminist issues on the environment

* Corresponding author e-mail: harcourt@iss.nl

to the governmental debates. These converged in the Women's Action Agenda 21 that came out of a global women's meeting held in Miami – 'World's Women for a Healthy Planet' (Harcourt, 1994; Dankelman, 2011) – and the women's tent 'Planeta Femea' in the forum for non-governmental organizations (NGOs) held alongside the government deliberations. The Women's Action Agenda 21 critiqued the mainstream approach of the environment and development agenda and argued for a new ethics for economics and nature and for sustained livelihoods (Harcourt, 1994; Wichterich, 2015). This advocacy-based analysis merged with gender and development environment scholarship in a series of publications in the early 1990s (Braidotti *et al.*, 1993; Harcourt, 1994). In response to these activist debates post-UNCED and the spate of academic publications, the first FPE text (Rocheleau *et al.*, 1996) set out a critique of the women and environment debates in policy and academe. In this classic 1996 text, the FPE opens up political ecology to include gendered power relations. It underlines the structuring nature of gender in environmental knowledge and women and men's differential access to, and control over, resources. It also points to the importance of understanding social movements as part of political struggles at the community level to (re) gain resource control and environmental protection in interventionist development processes. FPE, in looking at the political as well as the technical, presents ecology as inherently bound up in political processes that structure nature and gender relations. FPE sees its own knowledge production as part of a political process and undertakes research in order to find alternatives to environmental policy that it perceives as not taking into account adequately gender power relations (Harcourt and Nelson, 2015).

Building on that first FPE text learning and engaging in different women's environment movements, FPE in the 2000s explored more deeply how gender relations are shaped by interactions with ecological, technological and political-economic processes (Di Chiro, 2008). FPE research inspired by Judith Butler (2006), to look at how gender is 'performed', examined how meanings of gender are embedded and contested in people's actions and experiences in the environment. Gender was perceived as entangled in landscapes, bodies and livelihood strategies (Elmhirst, 2011, 2015; Hawkins and Ojeda, 2011). In these studies of how gendered power relations are performed in ecological, technological and political-economic processes, FPE looked at how gender ideologies and identities are produced, contested and employed in the governance of livelihoods and environments. FPE in the last years, under the loose banner of 'new feminist political ecology', has deepened these studies to explore the materiality of environmental practices and the emotions that produce gendered subjectivities, ideologies and identities (Elmhirst, 2011; Nightingale, 2011). In this new FPE research, a 'lived feminist political ecology' approach examines gendered 'embodied interactions and labours' along with the 'emotional and affective relations' (Harris, 2015, p. 158) in different environments and natures in different places, networks and political processes. Lived feminist political ecology looks at issues of care, emotion, bodily health and environmental health (Hawkins and Ojeda, 2011; Harcourt and Nelson, 2015). This interest in the well-being of the body, community and culture in environmental struggles is described as 'the interconnectedness of all life and the relevance of power relations including gender relations in decision making about the environment' (Nirmal and Rocheleau, 2015).

Complementing this brief review on how FPE has evolved, the chapter now takes up four main concepts that are currently being explored in the FPE framework: place, glocality, sustainable livelihoods and gender equality, which also require review in order to understand their importance in the analysis of the case studies that follow.

Place

The feminist geographer Doreen Massey defines place as inevitably inflected by the global (Massey, 1991). Understanding place as 'meeting places' of different cultures, knowledges and practices enables us to analyse place as fluid and changing where the lived, experienced, felt glocal realities co-constitute place and people as local contexts are a manifestation of transnational connectivity. As J.K. Gibson Graham suggests:

> [p]lace is a site of becoming and the ground of a global economic politics of local transformations . . . [and the] possibility of understanding local economies as places with highly specific economic identities and capacities rather than simply as nodes in a global capitalist system.
> (Gibson-Graham, 2007, p. 39)

J.K. Gibson-Graham contributed to an FPE research project from 2000 to 2004 on 'Women and Politics of Place' (WPP) (Harcourt and Escobar, 2005) that examined how women enter into political struggles through different levels of place – around the body, home, local environs and community. WPP aims to analyse the interrelations created in women's struggles around diverse embodiments, diverse ecologies and diverse economies (Escobar and Harcourt, 2005). The WPP work has led to several case studies that have examined women's role in place-based struggles in defending the safety and well-being of their community and environment (Florez *et al.*, 2016; Gloss, 2016). These and other scholars have looked at how struggles in different places have been linked through networks in what has been identified as 'place-based globalism' (Osterweil, 2005). These struggles are for the right to live and work in a healthy environment and the right to take responsibility to protect ecologies, livelihoods and natural resources from overuse, extraction and destruction. In these struggles 'in place', women experience how global forces of neoliberal capitalism are deeply impacting their lives. The WPP research focuses on describing and analysing how gendered political struggles against neoliberal global capital are anchored in the bodies, environments and economies of communities. The focus of the research has been on rural women's (and indigenous women's) political organizing in marginalized, economically poor, neocolonial 'developing' nations in the global South. Although there are diverse impacts specific to place, there are also larger global processes that connect women through their shared connections and political actions. WPP has looked at how these women's diverse struggles are connected in ways that allow rural women to challenge interventionist development discourses and practice. Working with the women's movements it looks at how to re-imagine and connect different forms of lifeworlds within and outside hegemonic neoliberal capitalism (Gibson-Graham, 2006).

Glocality

The term 'glocality' emerged in the WPP project to describe how the global and local co-constitute each other as women act 'in place' to protect and conserve, enhance and modify livelihoods and create connections with other places (Dirlik, 1998). Glocality helps to describe a lived feminist political ecology as rural women live through changes in agriculture practices that impact on their livelihoods not in isolation but connected to other environments and places. A glocality is formed by networks of relations among rural women's groups in their own place and environment that, via different interlocutors, stretch connections to other women's groups and places. The term 'interlocutor' is used here to describe the individuals or 'actants' who address social dilemmas via 'networked relationships' in transnational social movement advocacy groups (Fowler, 2016, p. 53).

Loosely speaking, the glocal networking by women working on gender and livelihood issues in the context of development processes builds on the webs of connection that have been formed since the 1992 Rio Conference. These connections can be traced in the monographs and edited collections on women local and global organizing in environmental politics (Braidotti *et al.*, 1993; Seager, 1993; Harcourt, 1994, 2012; Basu, 1995; Wanyeki, 2003; Ricciutell *et al.*,

2004; Desai, 2009; Dankelman, 2011). They are also evident in the 'grey literature' of the background documents and reports that make up official United Nations (UN) policy documents (UNEP, 2008; FAO, 2011; UN, 2015).

Sustainable livelihoods

The term 'sustainable livelihoods' also has a history in the FPE research via debates on mainstream sustainable development policy (Dankelman, 2011; Krishna, 2012b). Sustainable livelihoods has seen many permeations since the 1970s (Kothari and Harcourt, 2004).[1] Sustainable livelihoods has been an important concept for FPE because it takes a community- and people-centred approach to environment and resource management and the economy. Sustainable livelihoods views environmental concerns from a political perspective with a focus on the lives and resources of rural and marginal urban people, perceived as the poor and vulnerable objects of development programmes. In the 1990s it emerged both as a framework of analysis and an advocacy strategy by researchers, UN agencies and NGOs.[2] Sustainable livelihoods aimed to counter the mainstream sustainable development focus on economic growth with a focus on the participation of people and communities in their own economic and environmental management. Robert Chambers, a world-renowned champion of both sustainable livelihoods and participatory development, defined livelihoods as, 'the means of gaining a living, including tangible assets (resources and stores), intangible assets (claims and access), and livelihood capabilities' (Chambers and Conway, 1991, p. 5). What FPE adds to this definition is that sustainable livelihoods is also concerned with gendered knowledge, power and politics, engaging with women's and men's local knowledges in order to understand and support community coping strategies in the face of 'the violence of development' (Kothari and Harcourt, 2004).

Gender equality

As stated above, gender is central to FPE as a social category that informs and shapes societal and relations in nature. Men and women have different 'experiences of, responsibilities for, and interests in "nature" and environments' which 'vary by culture, class, race, and place and are subject to individual and social change' (Rocheleau *et al.*, 1996, p. 3). Gender relations determine the different productive and reproductive tasks and responsibilities between men and women and their different knowledges of nature and cultural relations.

In the 1990s and 2000s as FPE was developing analytically in policy, gender equality and women's empowerment became the central strategic approach to development and environment policy concerned with gender equality. Gender equality and women's empowerment was enshrined in the Beijing Fourth World Conference on Women in 1995. For the last 20 years the Beijing Declaration and Platform for Action of 1995 is seen as 'the current source of guidance and inspiration to realize gender equality and the human rights of women and girls, everywhere' (UN Women, 2015a).[3] In this post-Beijing era, gender and development policy examines how gender power relations affect men and women differently and determines who does what and how resources are allocated (Cornwall, 2007). The *World Development Report 2008: Agriculture for Development* (World Bank, 2008) recognizes the importance of women's role and the problems of gender inequality in agricultural practice and policy (Rao, 2012; Tsikata and Awetori Yaro, 2014; Verma, 2014). The UN report on sustainability states that, 'policies should ensure women's effective participation in and equal benefit from sustainable development projects and actively address entrenched discriminatory stereotypes and inequalities' (UN Women, 2015b, p. 29).

In academic research, the analysis of gender power relations has intersected with sustainable livelihoods through the analysis of gendered decision making in family and communities, as well as looking directly at power and conflict around sustainable livelihoods. The Indian economist Bina Agarwal (2003) has looked at how gender power relations inform the struggle of men and women to sustain ecologically viable livelihoods and community well-being. Agarwal's studies of rural women's gender power struggles in India illustrate how environmental rights and justice are deeply enmeshed in and determined by gender relations.

These four concepts – glocality, place, sustainable livelihoods and gender equality – help us to analyse how rural women have been engaged in different actions to protect their community's rights through women's organizations engaged in political struggles over ecological health, landscapes and livelihoods, and environmental decision making across the planet.

Methodology

The main methodological approach of the chapter is to use the FPE conceptual framework reviewed above to explore rural women's organizing in three geographical contexts: India, Tanzania and Italy. These examples have been selected from a multi-country study on women's community organizing on environmental politics at the International Institute of Social Studies (ISS) of Erasmus University in The Hague.[4] The study is researching women's community organizing in local and global arenas in order to map out different facets of environmental politics from a gender perspective, looking at the knowledge and power relations that determine women's access to and control over environmental resources.

This chapter uses an FPE framework in order to explore four questions:

1. How are glocalities being formed by women's groups connecting within and across environments and across places and scales?
2. How do the interlocutors (researchers who are recording and supporting rural women's responses and political organizing to changes of traditional resource management, ecosystems, rural community cultures and landscapes) make these connections across place and scale via 'networked relationships' linking community groups to transnational social movements and NGO advocacy groups?
3. What does a lived feminist political ecology approach reveal about the emotional and affective relations of communities with their environments?
4. How do these women's experiences and engagement in glocal political actions feed into dominant development discourses and practice around gender and the environment?

The research uses feminist standpoint and feminist participatory action research methodology. Feminist standpoint theory examines the experiences of women as a way to understand knowledge production and power relations, beginning with the lives of women on the margins of political and economic power. Such research describes and analyses the effects of power on women's understanding, emotions and lived experience of the environment while at the same time advocating ways for change. Feminist participatory action research aims to do research that in itself is taking action to improve the situation being studied. Like feminist standpoint theory, it reflects on the nature of knowledge and power and sees experience as a legitimate form of knowledge that can lead to new practices and create possibilities for social change. The interlocutors who undertake the research live and work with the local women's community organizations and are engaging in the research as part of their political engagement and commitment to the community. Their role as interlocutor contains power and influence as they are themselves engaged in struggles for social, gender and ecological justice. The narratives that are analysed in this chapter focus on processes that are aiming to produce societally relevant research that supports local communities' partners and feeds directly into policy debates around gender equality and sustainable development.

Research Findings: Three Case Studies

Gender and sustainable livelihoods in rural India

The first study is on the impact of the liberalization of the Indian economy on women's livelihoods in north-eastern and central India and the change in traditional resource management systems and practices (Krishna, 2012b, 2015). Feminist scholars have pointed to the increasing feminization of agriculture (Srivastava and Srivastava, 2010). A case study by Sumi Krishna describes how:

> Iniquitous resource exploitation . . . geared for speedy growth and productivity, have encouraged private and corporate exploitation of natural resources, with minerals and water now being the most contentious. The brunt of this is borne by women because of the patriarchal division of labour and responsibilities related to family provisioning.
>
> (Krishna, 2012b, p. 137)

In her research she documents how women in Apatani villages in Arunachal Pradesh have responded to these developments by searching for livelihood alternatives through collective self-organizing. The goal of her study is to recognize how these 'side-streams' connect and shape the 'wider socio-political and cultural contexts' of rural economies (Krishna, 2012b, p. 126).

The study looks at how grassroots women's groups, whose livelihoods have been threatened by economic change, are supported by women's NGOs to form small community collectives that connect different rural villages, and then to state-wide initiatives. Among the examples in the study are women in forest-dwelling communities who breed indigenous Aseel poultry. With the support of ANTHRA, an NGO of women veterinarians, the rural women work together to rebuild their livelihoods in a collective, *Tholakari Adivasi Mahila Vedika*, that links women in 80 villages. The activities not only create new markets in which to sell the poultry but also importantly give value to the women's knowledge of handling chickens and established the possibility for the women to give support to one another, as, in the process of organizing the collectives, the women leave their homes and villages, forming new networks of care and support. According to Krishna, these activities have shifted gender power relations in the home with an increase in women's own sense of self-worth and decision-making capacity with the men in the family and community. Being able to work collectively and to move from village to village has created a wider community of support that challenges traditional patriarchal attitudes towards women.

Another example is the rural *Jhansirani* Women's Self Help Group, aided by the M S Swaminathan Research Foundation, that supports grassroots women's groups to form new bio-enterprises (such as paper making) that allow village women to take the lead in the enterprise and benefit both from the money raised and again from the connections with other women, as well as self-value in their capabilities. What is important in the organizing of these collective networks is not so much the economic gain but the ability to tackle 'the socio-cultural practices and gender ideologies that restrict women's autonomy, mobility and capacity to participate in decisions' that enables an 'unravelling of the gendered structure of patriarchal attitudes and institutions' (Krishna, 2012b, p. 138).

The responses of the women's communities to major economic changes reveal a layered nature of glocal organizing. It is through interlocutors such as the NGOs that the lives of the rural women are changing. Through participatory action research by people like Krishna, the lives of these rural women become known and linked into other collective struggles in the social justice movement in India and further afield.[5]

A second very different example of glocal networking in India is the NGO Navdanya. Navdanya means 'nine crops', symbolizing biological and cultural diversity and the seeds as commons. The NGO was founded in 1984 by the Indian activist and eco-feminist Vandana Shiva. The NGO promotes biodiversity and organic farming and seed saving in order to protect biological and cultural diversity. As well as creating community seed banks across the country, Navdanya centres provide training for women (and men) farmers in sustainable agriculture along with direct marketing and fair trade organic network. Its main learning centre for biodiversity conversation and organic farming run by local women farmers is in Uttranchal in North India.[6]

As in the earlier study by Krishna, Navadanya connects women's grassroots groups horizontally within the same geographic context. In addition, Navadanya works on a national and global scale. It is linked to several global networks via the 'diverse women for diversity' network (known at the local level as *Mahila Anna Swaraj* (Women's Food Sovereignty)). In these connections the productive and reproductive work of peasant Indian women farmers is linked to the social movement Earth Democracy and other groups such as La Via Campesina (see below).

For the last 30 years Vandana Shiva has engaged in environmental politics as a writer and as an activist and interlocutor at the global level. Shiva and those who work with her have done pioneering advocacy work on biodiversity, food sovereignty, seeds and farmers' rights in UN international meetings, expanding glocal connections among rural communities. There are many examples of how such glocalities have been able to intervene in UN debates. For example, in 2013, Shiva along with other activists brought the issue of women's food sovereignty to the UN Conference on Nutrition. Illustrating her argument from her knowledge of the lives of rural women farmers in India, Shiva argued that nutrition has deep implications on people's rights, livelihoods and health. The outcome of the debates was the 'CSO Vision Statement on Nutrition' which proposed a vision of rural sustainable livelihoods based on small-scale and family food production systems – food sovereignty, women's empowerment, biodiversity, ecological principles and economic self-sufficiency.[7]

Women mobilizing for sustainable community livelihoods in Tanzania

The second case study is the Tanzania Gender Networking Programme (TGNP), one of the earliest feminist networks in Tanzania (Mbilinyi, 2015).[8] Since the 1990s, it has brought together academics and activists who have questioned the inequality of gender and economic relations of neoliberal capitalism as experienced in Tanzania's transition from a socialist economy. Majorie Mbilinyi, one of the founders and activists of the TGNP, has written about the glocal organizing of the TGNP for over two decades (Mbilinyi, 2015). She describes how TGNP actively participated in the Fourth World Conference on Women in Beijing in 1995. In following up the government promises, TGNP saw that there was little space for small-scale women farmers in Tanzania's post-Beijing gender and development programmes. Mbilinyi explains that TNGP decided to develop with rural women farmers a transformative and locally grounded set of activities. The Network worked with rural women's grassroots organizations in order to create exchange programmes among rural women's self-organized collectives. They shared their activities and analysis beyond Tanzania to other African women's groups through the quarterly newsletter, *Ulingo wa Jinsia*, published in English and Kiswahili. The Network sponsored a 'transformative feminist movement building study' in 2008, leading to a campaign for 'economic justice: making resources work for marginalized women' based on a series of workshops held in rural communities across Tanzania (Mbilinyi, 2015, p. 511).

The TGNP facilitated political and economic literacy workshops that provided the space for rural women to vocalize their experiences of gender inequality, social oppression and economic hardship. In the workshop dialogues the knowledge of the everyday lives and vision for grassroots women's future is merged with that of TNGP interlocutors. The participatory methodology used by the TNGP facilitators is for the rural women to assess their own situations using animation tools that support the women's critical assessment of the conditions determining their livelihoods. From this analysis, the workshops undertake problem-solving exercises set by the participants. One of the most powerful tools, according to Mbilinyi, is the use of a pictorial code based on rural women's analysis of their dreams, what they see as the root causes and what actions they imagine in order to overcome them. This process of participative creative listening led to the exposure of different problems ranging from land grabbing/investment to the high level of corruption and incompetence among government workers. Another powerful tool is visits and exchanges of different rural women's groups, connecting women at the ward and village level across the country.[9] Such glocal networking based on the knowledge of grassroots women and horizontal networking and movement building is upscaled to national and continental-wide campaigns that challenge state and agro-industry land grabbing as well as international NGOs, donor governments and UN organizations. Mbilinyi operates as an interlocutor in this process, connecting the work of TNGP with various donors (e.g. Oxfam, UK Department for International Development) and UN processes at the Food and Agriculture Organization of the United Nations and UN Women.[10]

Women's place-based global struggles for food sovereignty in Italy

The last case study is on different movements for food sovereignty in Italy. The global Slow Food movement was founded in Rome in 1986 to protest at the opening of an outlet of the international fast-food chain, McDonald's, in Rome. Slow Food is now an international association and describes itself as 'a global, grassroots organization with supporters in 150 countries around the world who are linking the pleasure of good food with a commitment to their community and the environment'.[11] It is a quintessential glocal network which connects local knowledge with this broader networking in order to 'counter the rise of fast food and fast life, the disappearance of local food traditions and people's dwindling interest in the food they eat, where it comes from, how it tastes and how our food choices affect the rest of the world'.[12]

Slow Food aims to connect producers with socially invested consumers. The political aim of Slow Food is to protect and support small producers, and to link them with those who are making decisions high up in the food production chain – the consumers, educational institutions, chefs and cooks, and agricultural research institutions.[13] Glocal networking includes bringing people from around the world to Turin, Italy, in a biannual global meeting Terra Madre (Mother Earth) that since 2004 has brought together the slow food network of food communities and producers. Since 2007 there are now regional Terra Madre meetings in Ireland, Tanzania, Brazil, Spain, Argentina, South Korea, Armenia, Russia and Japan. The aim is to assist local farmers and 'farming families' by creating a network of self-sufficient local economies, connecting local producers and consumers through this glocal network. Many women are part of the local associations of Terra Madre and Slow Food, where women's role in family life rather than women's empowerment and gender equality is promoted by the movements' activities (Counihan, 2014).

La Via Campesina is another global network that is mobilizing the resistance to a global bioeconomy. La Via Campesina is made up of 164 local and national organizations in 73 countries from Africa, Asia, Europe and the Americas. It claims to represent 200 million farmers (Shawki, 2014). A group of farmers' representatives – women and men – founded La Via Campesina in 1993 in Mons, Belgium when agricultural policies and agribusiness were becoming globalized and small farmers needed to develop and struggle for a common vision.[14] La Via Campesina brings together peasant and indigenous movements in the struggle for territorial recognition and rights, addressing inequities in access, ownership and control of resources. Its statements call for food sovereignty, or the right of self-determination of local communities to produce their own food in their own territory and to govern, manage and care for their ecosystems and natural wealth. La Via Campesina is an important example of a network that recognizes women's agency and the need for rural women to advance their economic, cultural and social claims to equality. In its activities, La Via Campesina has promoted a 'female peasant' political identity linked to land, food production and the defence of food sovereignty in opposition to the current agribusiness model. The political ecological struggle of women is situated on two levels: defending their rights as women within organizations and society in general; and the struggle as peasant women, together with their colleagues, against the neoliberal model of agriculture.

The women of La Via Campesina see their politics also as cultural when they ask for a revalorization of their traditional wisdom regarding the production of food, the selection and management of seeds, the breeding of animals, and the care of the earth and nature. In a highly symbolical gesture at La Via Campesina events, women perform the opening *mística* – a ceremonial act that recognizes peasant and indigenous values and ideals, seen as nourishing the spirit of struggle. The preparation of the *mística* is the tangible expression of peasant culture and represents a challenge to and defiance of the modern economic systems that oppress and marginalize such traditions.

La Via Campesina in Italy is an example of how glocal networking links small farmer and rural communities throughout regions in Italy, Europe and globally. One of the local rural communities in Lazio, Italy connected to La Via Campesina is Punti di Vista (PdiV), a small community-based organization based in Bolsena.[15] Founded in 1992, PdiV runs training sessions and workshops on local and global environmental issues in consumption and ecologically responsible waste reduction and management; gender and globalization; and sustainable local economies. The organization engages with the local La Via Campesina group in nearby Orvieto as well as with international, European and national organizations engaged in food, agriculture, social justice and gender issues.

PdiV connects family-run enterprises, such as goat herders, beekeepers and sheep cheese producers, many of whom came to the area in the 1990s with a vision of sustainable farming practices and a healthier, better quality of life for their families. PdiV represents a new generation of Italian women who, with their partners, are challenging agribusiness and globalization processes in their aim to reconnect to nature and find clean spaces for themselves and their children. PdiV runs a campaign around organic soap in an effort to halt environmental degradation around Bolsena, where waste and chemicals from agricultural production over the years are poisoning the lake and, more recently, soils added to the lake area to allow for new housing for tourists have changed the watershed.

The connection to La Via Campesina is important in the vision of new values in agricultural life reflecting on degrowth and alternative lifestyles based on environmental politics that put care for the Earth ahead of economic gain. The members of PdiV see connecting to other rural communities as important in the search for less-extractive forms of agricultural and consumer lifestyles in Italy and the broader human responsibility for the global commons.

Discussion

The contours of rural women's glocal networking

The case studies show how, through glocal networking among very different contexts, rural women's groups are connected within and across environments and across places and scales. The examples illustrate how local conditions are informed by global impacts but not in a simple top-down resistance of authority, governmental or industrial power. Rather there is a co-constituting of the global and the local in these glocal resistances. The organizing of the women at local levels is not only about sustainable livelihood strategies for economic survival, but also about finding opportunities to change social, environmental and cultural relations. The responses of the village women in north-eastern and central India in the case study by Krishna show how livelihood strategies were able to shift traditional patriarchal practices that confined women to their homes and created networks that could benefit the women and also link to advocacy for change at a state political level. The detailed studies of village women provided evidence on the impact of agricultural policies and how the lives of rural women need to be taken into account for sustainable livelihoods and state development. On another scale, glocal networks that are formed around Navdanya link the knowledge of organic farming by local women farmers in rural Uttranchal to global-level campaigns, knowledge sharing and advocacy on biodiversity. La Via Campesina operates as a huge glocality that links the work of peasant farmers around the world as the global interaction of flows of information support local mobilizing. The example of Italy shows how both La Via Campesina and the Slow Food campaign begun in Rome are part of these glocal changes feeding into changes around environmental and gendered knowledge within it. Such glocal networking for food sovereignty has contributed to changes in high-level food consumption in restaurants around the world as well as promoting alternative agricultural practices such as permaculture and organic food production to counter water pollution. The case of Tanzania is an illustration of how a highly political glocal networking process over time has linked grassroots women's survival strategies to national and continental-wide campaigns to end state and agro-industry land grabbing via national and international NGOs, which use the stories of the rural women to inform policy demands at government and UN level.

These examples represent just a few of the many nodes of rural women's organizing that, in a meshwork of connections, creates a strong glocal political presence in the sustainable development discourse. From an FPE perspective they show how place-based organizing cannot be confined or explained by looking only at the local environmental, political and social context. The larger global context in which groups are operating also informs and shapes the local conditions. As the diverse stories travel from place to place they make visible connections of rural women's lives to policy changes in both planned but also incidental ways.

The role of interlocutors

Key to the process of connections and therefore possibilities for change is the role of the interlocutor. Interlocutors are part of the landscape of change, who, as in the examples given here, deliberately undertake research with the women in order to bring about action. The interlocutors play an important role in recording and writing about the knowledge of the environment, the production of place and the lives of rural women. In adopting participatory action research methods, the interlocutors foster the pathways through which rural women's responses to changes of traditional resource management, ecosystems, rural community cultures and landscapes can be noted and acted upon.

The role and responsibility as well as the research ethics of the interlocutors are constantly under scrutiny, particularly if they are outsiders and when, as can often happen, it is the interlocutor rather than the rural women's group who is recognized or rewarded. In the case studies discussed here, the interlocutors are consciously making connections across place and scale via 'networked relationships' linking community groups to transnational social movements and NGO advocacy groups. Vandana Shiva stands out as a the most well-known example of an interlocutor who has personally connected many rural women's groups, campaigns and advocacy concerns in international multilateral settings as well as in transnational social movement fora and academic meetings. Her writings, digital presence on social media and speeches in international arenas have championed visible rural women's livelihoods, struggles and concerns. Similarly, Majorie Mbilinyi has written widely and engaged in many policy debates on gender power relations in industrial agricultural policy and in particular the recent land grabs (Leach, 2015). FPE recognizes the role of these interlocutors as part of the knowledge–power nexus that is informing our understanding not only of gender and environment practices, but also of how power operates through writings and advocacy and personal commitments to change. The work of these interlocutors is not only descriptive but also makes visible the power relations in gendered political ecological struggles of which the interlocutors are also embedded.

A lived feminist political ecology

As explained above, the concept of lived FPE aims to draw attention to processes that are often not commented upon in research and politics. By naming the role of interlocutors, the importance of networking, the personal commitments, the way research informs policy and is about action, FPE sees the researcher as closely connected to the political struggles being described. FPE sees the research process itself as part of the lived and emotional responses to the struggles, resistance and resilience of rural women who are often perceived as living in marginal or out-of-the-way places. It brings to the centre the knowledge and practices of women which are often not recognized as political. For example, analysing the *mística* ceremony of La Via Campesina as important to the political identity of the organization, helping to build community and redefining political behaviour takes into account the gendered role of women as spiritual leaders. Political research could well not notice or dismiss such practices as ornamental or irrelevant to the real power plays. Such performative practices are important in gender power relations and are part of the emotional and affective relations of communities with their environments. The use of art and theatre performances of the Tanzanian women's groups acknowledges and works with the emotions they feel at the changes in their lives. Recognizing the rural women's analysis of their dreams and what they see as the barriers, helps to create strategic knowledge that combined with the knowledge and connections of the leaders of the TGNP can build a stronger resistance and possibilities of change for the rural women. This is not to romanticize either *la mística* or the difficulties of change for rural women; rather it is to acknowledge these types of performances and emotions as part of the process of change. Lived FPE can also be seen in the creation of Terra Madre as a celebration of slow food at a multicultural and transnational level that connects many different lives in an emotional appeal to the Earth as mother and the need for recognition of place-based indigenous knowledge in the production of a fair food culture. Such a celebrated well-funded event does suggest some troubling questions around the power differentials among the people participating and who ultimately benefits. But it also illustrates how emotions and celebrations around food and indigenous cultures inform the lived politics of food sovereignty and how, in the face of fast food, such events raise awareness of the importance of preserving diverse food cultures where rural women are playing an increasingly visible part.

Connecting to dominant development discourses

The glocal networking of Vandana Shiva around biodiversity, Slow Food events and the food sovereignty campaigns of La Via Campesina are all ways in which glocalities are operating to connect rural women's experiences into development policies on gender and the environment. Such popular events, known through traditional and increasingly social media (websites, blogs,

Twitter, Facebook, etc.), feed into policy debates. Research such as the case studies discussed here, along with the so-called grey literature by NGOs, provide the analysis and knowledge taken up by government officials who are engaged in the policy debates around gender and environment. In addition, there are the professional channels where scholars are invited to give advice in expert panels, write background papers for UN reports and present at social movement discussions. These knowledge production processes at the popular, academic and policy levels are how narratives from the sidestreams merge into mainstream environment and development practice. The work of FPE is done in these different exchanges of information – in academic hallways, in UN and government rooms, in NGO events, in social movement campaigns and marches. These activities have become so much a part of broader political knowledge production that they are normalized and as a result not scrutinized enough as pathways for influencing and changing dominant discourses. These changes and influences are hard to measure directly, but the recognition of gender power relations and rural women's lives and the work in the current discourse on SDGs indicate that shifts have occurred.

Conclusion

Glocal networking for gender equality and sustainable livelihoods, then, is part of the gender and environment debate informing the SDGs launched in September 2015 (UN Women, 2015a). The huge question remains about how these sidestreams via these glocal networks will ultimately transform gender inequalities and sustain livelihoods. As the above suggests, there are no clear rules by which rural women's networks and interlocutors engage in policy arenas, or how promises made by the state are carried out. To be blunt, just pointing out there are unsustainable production and consumption patterns, unfair agricultural and trade agreements, gender-blind and unjust policy is not enough for change to come about. The current policy approach of eco-economic management, based on a commodification of nature and geopolitical inequalities, North and South, does not seem a likely a space for alternatives to flourish and regenerate emerging diverse place-based practices of rural women's cooperatives and organizations.

Nevertheless, after many rounds of consultations with members of governments, academia and civil society, the SDGs recognize the need to 'investigate the structural causes of poverty and inequality challenging the neoliberal economic model' (Easterly, 2015; McCloskey, 2015). They also speak of the need for a fair, sustainable economy where gender equality is one of the 17 goals and part of the other economic, social and environmental targets of the remaining 16 goals (UN Women, 2015a). At least in terms of the agreements, the SDGs provide a global mandate for gender equality, sustainable livelihoods and social justice. What can glocal networking add to these discussions?

Glocal networking presents a different way of understanding how change happens compared with the evidence- and goals-based SDGs. Glocal networking points to the connections across place and scales that build on people's lives, emotions and needs that cannot easily be measured or captured in more rational and bureaucratic discourses. Understanding the different ways glocalities are operating and mesh together requires continual attention to context and to connections that might not be able to be repeated in other places or be upscaled. This means there cannot be blueprints for sustaining livelihoods and gender equality, but there can be processes that set out to listen to and work with these different knowledges and ensure that the sidestreams are in conversation with these mainstreams. The process around the SDGs is a step towards trying to recognize different knowledges and experiences, as the breadth of the 17 goals suggests. The SDGs offer a particular political moment, but as the case studies in this chapter suggest, it is important to recognize there are different forms of political networking that are not directly linked to governance structures but are part of civic-driven change that can interlink. These civic innovation processes are defining livelihoods, food sovereignty, changing gender power relations and environmental strategies in small and large glocal events that do not always interweave with official processes. Those steering the SDG process have to work out how best to recognize and learn from these sidestream events, which for many civil society groups are where their experiences, voices and knowledge are best heard. The role of the interlocutor is important in this process as people who are able to move between different places translating across the borders of the UN, academia, social movements and community-based organizations. For example, the role of FPE interlocutor is evident in the *World Survey on the Role of Women in Development 2014: Gender Equality and Sustainable Development* (UN Women, 2015b) which drew on feminist writers, researchers and activists in the preparation of the report. Even if all the conversations could be reflected in the restrictions required of a UN report, it does recognize the many different pathways to sustainable development for rural women.

This chapter captures only partially what glocal networking is being done around the broad themes of sustainable livelihoods and gender equality. It is written as one of the many conversations that FPE is engaging in about how to bring about change where sidestreamed women's knowledge is at the centre, not on the peripheries. It aims to suggest rather than empirically prove how these glocal networking practices are informing the mainstream sustainable development policy. More detailed research is required that will involve deeper conversations based on further engagement and experiences, ethnographic studies undertaken over time and self-reflection. Such research will be able to map out more systematically the types of glocal processes that interweave rural women's organizations, glocal networking practices and political organizing to governmental and UN environmental policies and agricultural practices.

Notes

[1]For an extensive description of the history of the term sustainable livelihoods, see Krishna (2012a).

[2]*Agenda 21*, the non-binding 'Plan of Action' of the Rio UNCED, first mentions sustainable livelihoods: 'While managing resources sustainably, an environmental policy that focuses mainly on the conservation and protection of resources must take due account of those who depend on the resources for their livelihoods. Otherwise it could have an adverse impact both on poverty and on chances for long-term success in resource and environmental conservation' (UNEP, 1992).

[3]Sustainable livelihoods was not a high priority in the Declaration although it is mentioned in relation to women and poverty and women and economics. Declaration Chapter A on 'Women and poverty' asks, in paragraph 60(g), for governments to 'create an enabling environment that allows women to build and maintain sustainable livelihoods'. In Chapter F on 'Women and economics', paragraph 156 underlines the need for women 'to achieve economic autonomy and to ensure sustainable livelihoods for themselves and their dependents'. In Chapter K on 'Women and the environment', article 246 states: 'Women have an essential role to play in the development of sustainable and ecologically sound consumption and production patterns and approaches to natural resource management' (UN Women, 2015a).

[4]As part of the ISS civic innovation research initiative is a network called 'Women's environmental global/local organizing for community health and well-being' (WeGO). The Civic Innovation Research Initiative (CIRI) is one of the four research themes at the ISS exploring academic and activist concerns around citizenship struggles in relation to generational and gender equality, democratic rights, economic justice and sustainable development.

[5]See for example the study of Soma Kishore Parthasarathy (2012).

[6]See http://www.navdanya.org for more information (accessed 5 February 2016).

[7]See Shiva's article and other discussions in *Development* 57(2) special issue on Nutrition (http://link.springer.com/journal/41301/57/2/page/1, accessed 24 February 2017), produced in the lead-up to the Second International Conference on Nutrition, which explores the relationship between nutrition, food security and sustainable agriculture.

[8]For more information on TNGP, see http://www.tgnp.org (accessed 23 February 2016).

[9]Tanzania is divided into regions, districts, divisions, wards and then in rural areas into villages, see https://en.wikipedia.org/wiki/Subdivisions_of_Tanzania (accessed 23 February 2016).

[10]Information gathered from personal conversations and correspondence with TNGP (2010–2015).

[11]See http://www.slowfood.it (accessed 20 October 2015).

[12]See the description of Slow Food International at http://www.slowfood.com/ (accessed 20 October 2015).

[13]Slow Food's motto is that 'eating is an agricultural act and producing is a gastronomic act' (http://www.slowfood.com/, accessed 20 October 2015).

[14]http://viacampesina.org/en/index.php/organisation-mainmenu-44/what-is-la-via-campesina-mainmenu-45 (accessed 30 October 2015).

[15]The section on PdiV is based on the author's three-year study, the results of which are published in Harcourt (2016).

References

Agarwal, B. (2003) Gender and land rights revisited: exploring new prospects via the state, family and market. *Journal of Agrarian Change* 3(1–2), 184–224. Available at: http://dx.doi.org/10.1111/1471-0366.00054 (accessed 23 February 2017).

Basu, A. (ed.) (1995) *The Challenge of Local Feminism: Women's Movements in Glocal Perspective.* Westview Press, Boulder, Colorado.

Braidotti, R.E., Charkiewicz, S., Hausler, S. and Wieringa, S. (eds) (1993) *Women, The Environment and Sustainable Development.* Zed Books, London.

Butler, J. (2006) *Gender Trouble: Feminism and the Subversion of Identity.* Routledge, New York.

Chambers, R. and Conway, G.R. (1991) *Sustainable Rural Livelihoods: Practical Concepts for the 21st Century. IDS Discussion Paper 296.* Institute for Development Studies, Brighton, UK. Available at: https://www.ids.ac.uk/files/Dp296.pdf (accessed 5 February 2015).

Cornwall, A. (2007) Revisiting the gender agenda. *IDS Bulletin* 38(2), 69–78. Available at: http://dx.doi.org/10.1111/j.1759-5436.2007.tb00353.x (accessed 23 February 2017).

Counihan, C. (2014) Women, gender, and agency in Italian food activism. In: Counihan, C. and Siniscalchi, V. (eds) *Food Activism, Agency, Democracy and Economy.* Bloomsbury Press, London, pp. 61–75.

Dankelman, I. (2011) *Gender and Climate Change: An Introduction.* Routledge, London.

Desai, M. (2009) *Gender and the Politics of Possibilities.* Rowman and Little, Lanham, Maryland.

Di Chiro, G. (2008) Living environmentalisms: coalition politics, social reproduction, and environmental justice. *Environmental Politics* 17(2), 276–298. Available at: http://dx.doi.org/10.1080/09644010801936230 (accessed 23 February 2017).

Dirlik, A. (1998) Globalism and the politics of place. *Development* 41(2), 7–13.

Easterly, W. (2015) The SDGs Should Stand for Senseless, Dreamy, Garbled. Available at: http://sdg.earthsystemgovernance.org/sdg/publications/sdgs-should-stand-senseless-dreamy-garbled (accessed 7 July 2017).

Elmhirst, R. (2011) Introducing new feminist political ecologies. *Geoforum* 42(2), 129–132. Available at: http://dx.doi.org/10.1016/j.geoforum.2011.01.006 (accessed 23 February 2017).

Elmhirst, R. (2015) Feminist political ecology. In: Coles, A., Gray, L. and Momsen, J. (eds) *The Routledge Handbook of Gender and Development.* Routledge, London, pp. 58–66.

Escobar, A. and Harcourt, W. (2005) Practices of difference: introducing 'Women and the Politics of Place'. In: Harcourt, W. and Escobar, A. (eds) *Women and the Politics of Place.* Kumarian Press, Bloomfield, Connecticut, pp. 1–19.

FAO (2011) *The State of Food and Agriculture 2010–11. Women in Agriculture: Closing the Gender Gap in Development.* Food and Agriculture Organization of the United Nations, Rome. Available at: http://www.fao.org/docrep/013/i2050e/i2050e.pdf (accessed 24 February 2016).

Florez, J.F., Lara Veloza, G., Veloza Torres, P., Cardozo García, M. and Espejo, C. (2016) Politics of place at the Women's School of Madrid: experiences around bodies and territory. In: Harcourt, W. (ed.) *Bodies in Resistance: Gender and Sexual Politics in the Age of Neoliberalism.* Palgrave, London, pp. 25–56.

Fowler, A. (2016) Change actors and civic innovators: who triggers change? Systematizing the role of interlocutors in civic innovation processes. In: Biekart, K., Harcourt, W. and Knorringa, P. (eds) *Exploring Civic Innovation for Social and Economic Transformation.* Routledge, London, pp. 41–66.

Gibson-Graham, J.K. (2006) *A Postcapitalist Politics.* University of Minnesota Press, Minneapolis, Minnesota.

Gibson-Graham, J.K. (2007) *Economic Imaginaries,* web edn. Available at: http://www.communityeconomies.org/site/assets/media/KatherineGibson/Venice-gibson-graham.pdf (accessed 24 October 2016).

Gloss, D.M. (2016) Reclaiming the right to become other-women in other-places: the politics of place of the ecologist women of La Huizachera Cooperative, Mexico. In: Harcourt, W. (ed.) *Bodies in Resistance: Gender and Sexual Politics in the Age of Neoliberalism.* Palgrave, London, pp. 57–78.

Harcourt, W. (ed.) (1994) *Feminist Perspectives on Sustainable Development.* Zed Books, London.

Harcourt, W. (ed.) (2012) *Women Reclaiming Sustainable Livelihoods: Spaces Lost, Spaces Gained.* Palgrave, London.

Harcourt, W. (2016) Gender and sustainable livelihoods: linking gendered experiences of environment, community and self. *Agriculture and Human Values,* first published online 22 November 2016, pp. 1–13. Available at: http://dx.doi.org/10.1007/s10460-016-9757-5 (accessed 23 February 2017).

Harcourt, W. and Escobar, E. (eds) (2005) *Women and the Politics of Place.* Kumarian Press, Bloomfield, Connecticut.

Harcourt, W. and Nelson, I.L. (eds) (2015) *Practicing Feminist Political Ecology: Moving Beyond the Green Economy.* Zed Books, London.

Harris, L. (2015). Hegemonic waters and rethinking natures otherwise. In: Harcourt, W. and Nelson, I.L. (eds) *Practicing Feminist Political Ecologies: Moving Beyond the Green Economy.* Zed Books, London, pp. 157–181.

Hawkins, R. and Ojeda, D. (2011) Gender and environment: critical tradition and new challenges. *Environment and Planning D: Society and Space* 29(2), 237–253. Available at: http://dx.doi.org/10.1068/d16810 (accessed 25 February 2017).

Kothari, S. and Harcourt, W. (2004) Introduction: The violence of development. *Development* 47(1), 1–4.

Krishna, S. (2012a) Redefining sustainable livelihoods. In: Harcourt, W. (ed.) *Women Reclaiming Sustainable Livelihoods: Spaces Lost, Spaces Gained.* Palgrave, London, pp. 12–20.

Krishna, S. (2012b) Gender and sustainable livelihoods in India: sidestream/mainstream. In: Harcourt, W. (ed.) *Women Reclaiming Sustainable Livelihoods: Spaces Lost, Spaces Gained.* Palgrave, London, pp. 125–141.

Krishna, S. (2015) Women's transformative organizing for sustainable livelihoods: learning from Indian experiences. In: Baksh, R. and Harcourt, W. (eds) *Oxford Handbook on Transnational Feminist Movements.* Oxford University Press, Oxford, UK and New York, pp. 837–854.

Leach, M. (ed.) (2015) *Gender Equality and Sustainable Development (Pathways to Sustainability).* Routledge, London.

Massey, D. (1991) A global sense of place. *Marxism Today* (June), 24–29. Available at: http://www.unz.org/Pub/MarxismToday-1991jun-00024 (accessed 11 November 2016).

Mbilinyi, M. (2015) Transformative feminism in Tanzania: animation and grassroot women's struggles for land and livelihoods. In: Baksh, R. and Harcourt, W. (eds) *Oxford Handbook on Transnational Feminist Movements.* Oxford University Press, Oxford, UK and New York, pp. 507–530.

McCloskey, S. (2015) From MDGs to SDGs: we need a critical awakening to succeed. *Policy & Practice: A Development Education Review* 20(Spring), 186–194. Available at: https://www.developmenteducationreview.com/issue/issue-20/mdgs-sdgs-we-need-critical-awakening-succeed (accessed 24 February 2017).

Nightingale, A. (2011) Beyond design principles: subjectivity, emotion and the (ir)rational commons. *Society & Natural Resources* 24(2), 119–132. Available at: http://dx.doi.org/10.1080/08941920903278160 (accessed 25 February 2017).

Nirmal, P. and Rocheleau, D. (2015) Feminist political ecologies: grounded, networked and rooted on earth. In: Baksh, R. and Harcourt, W. (eds) *Oxford Handbook on Transnational Feminist Movements.* Oxford University Press, Oxford, UK and New York, pp. 793–814.

Osterweil, M. (2005) Placed-based globalism: locating women in the alternative globalization movement. In: Harcourt, W. and Escobar, A. (eds) *Women and the Politics of Place.* Kumarian Press, Bloomfield, Connecticut, pp. 174–189.

Parthasarathy, S.K. (2012) 'The dream merchants have no clothes': women's rights and empowerment in the microfinance regime. In: Harcourt, W. (ed.) *Women Reclaiming Sustainable Livelihoods: Spaces Lost, Spaces Gained.* Palgrave, London, pp. 142–161.
Rao, A. (2012) Global agricultural policies, institutional interventions and women's livelihoods. In: Harcourt, W. (ed.) *Women Reclaiming Sustainable Livelihoods: Spaces Lost, Spaces Gained.* Palgrave, London, pp. 83–100.
Ricciutell, L., Miles, A. and McFadden, M.H. (2004) *Feminist Political Activism and Vision: Local and Global Challenges.* Zed Books, London.
Rocheleau, D. (2015) A situated view of feminist political ecology from my networks, roots and territories. In: Harcourt, W. and Nelson, I.L. (eds) *Practicing Feminist Political Ecology: Moving Beyond the Green Economy.* Zed Books, London, pp. 29–66.
Rocheleau, D., Thomas-Slayter, B. and Wangari, E. (1996) *Feminist Political Ecology: Global Perspectives and Local Experiences.* Routledge, London.
Srivastava, N. and Srivastava, R. (2010) Women, work and employment outcomes in rural India. *Economic & Political Weekly* XLV(28), 52–63. Available at: http://environmentportal.in/files/Women%20Work%20and%20Employment%20Outcomes.pdf (accessed 11 November 2016).
Seager J. (1993) *Earth Follies: Feminism, Politics and the Environment.* Earthscan, London.
Shawki, N. (2014) New rights advocacy and the human rights of peasants: La Via Campesina and the evolution of new human rights norms. *Journal of Human Rights Practice* 6(2), 306–326. Available at: http://dx.doi.org/10.1093/jhuman/huu009 (accessed 24 February 2017).
Tsikata, D. and Awetori Yaro, J. (2014) When a good business model is not enough: land transactions and gendered livelihood prospects in rural Ghana. *Feminist Economics* 20(1), 202–226. Available at: http://dx.doi.org/10.1080/13545701.2013.866261 (accessed 24 February 2017).
UN (2015) *Transforming Our World: the 2030 Agenda for Sustainable Development.* United Nations, New York. Available at: https://sustainabledevelopment.un.org/post2015/transformingourworld (accessed 11 November 2016).
UNEP (1992) *Agenda 21 Plan of Action.* United Nations, New York. Available at: https://sustainabledevelopment.un.org/content/documents/Agenda21.pdf (accessed 11 November 2016).
UNEP (2008) *The Gender Plan of Action Under the Convention on Biological Diversity.* United Nations Environment Programme, Bonn. Available at: https://www.cbd.int/programmes/cross-cutting/gender/cop-09-inf-12-rev1-en.pdf (accessed 24 February 2016).
UN Women (2015a) *Monitoring Gender Equality and the Empowerment of Women and Girls in the 2030 Agenda for Sustainable Development.* UN Women, New York. Available at: http://www.unwomen.org/en/digital-library/publications/2015/9/indicators-position-paper (accessed 11 November 2016).
UN Women (2015b) *World Survey on the Role of Women in Development 2014: Gender Equality and Sustainable Development.* UN Women, New York. Available at: http://www.unwomen.org/en/digital-library/publications/2014/10/world-survey-2014 (accessed 11 November 2016).
Verma, R. (2014) Land grabs, power, and gender in East and Southern Africa: so, what's new? *Feminist Economics* 20(1), 52–75. Available at: http://dx.doi.org/10.1080/13545701.2014.897739 (accessed 24 February 2017).
Wanyeki, L.M. (ed.) (2003) *Women and Land in Africa.* Zed Books, London.
Wichterich, C. (2015) Contesting green growth, connecting care, commons and enough. In: Harcourt, W. and Nelson, I.L. (eds) *Practicing Feminist Political Ecologies: Moving Beyond the Green Economy.* Zed Books, London, pp. 67–100.
World Bank (2008) *The World Development Report 2008: Agriculture for Development.* The World Bank, Washington, DC. Available at: https://siteresources.worldbank.org/INTWDR2008/Resources/WDR_00_book.pdf (accessed 11 November 2016).

21 Sugar and Gender Relations in Malawi

M. Canning*
The University of Edinburgh, Edinburgh, UK

Introduction

Africa has been identified as 'the region of female farming par excellence' (Boserup, 1970, p. 16) and it is estimated that more than 60% of all food produced on the continent is grown by women (Maman and Tate, 2001, p. 127). A rich literature on gender and agriculture in Africa has documented women's contributions to domestic food production, cash-cropping and informal marketing of foodstuffs, and many scholars have framed gender as a central category shaping hierarchies in production and reproduction. Comparatively less, however, has been written about gender and plantations in Africa, with a few notable contributions forming a narrow literature (for example, Vaughan and Chipande, 1986; Konings, 1998, 2012; Norris and Worby, 2012). Indeed, the global scholarship on gender and plantations is very small when compared with a more substantial literature on gender and labour in other sectors of production such as export manufacturing. Jain and Reddock's (1998) edited collection on women plantation workers explores the ways in which gender relations, norms and the division of labour manifest differently in a variety of countries and historical periods, and ethnographic contributions from Asia (such as Chatterjee, 2001; Gunewardena, 2010) provide rich descriptions of the gendered nature of plantation labour in different cultural contexts. However, the need is great for nuanced empirical research focusing on the particularities of women's and men's experiences, and the gender dynamics between women and men in African contexts.

In Malawi, a country where 64% of the country's labour force is engaged in agriculture (NSO, 2014) – and where the plantation system has dominated agriculture-led development policy throughout the colonial period as well as the five decades since independence (Chirwa, 2004), the importance of understanding the impact of this economic model on rural gender relations is critical. Plantations are a powerful mechanism through which the incursion of global capital into rural African contexts is mobilized, and can bring about structural changes which strongly shape gender dynamics in plantation enclaves. The current political climate in Malawi has encouraged such incursions, as, in recent years, the country has adopted a number of outward-facing economic policies that encourage the expansion of its plantation sector[1] (Chinsinga *et al.*, 2013, pp. 1066–1069; DFID, 2013; Patel *et al.*, 2014). One such example of recent growth is that of Malawi's sugar industry, where industry expansion through the establishment of new plantations, cooperatives and outgrower schemes linking smallholders to sugar estates, as well as the enlargement of existing

* Corresponding author e-mail: s1264507@sms.ed.ac.uk

 Gender and Rural Globalization (eds B.B. Bock and S. Shortall) plantations, has helped cement sugar as Malawi's second most important export commodity after tobacco.[2] The gender implications of sugar cultivation are enormous, where production in Malawi draws on a predominantly male workforce, and thus understanding the gender implications of such expansion is important.

It is against this backdrop that this chapter addresses the following key questions: What is the impact of the commercial sugar production on rural gender relations in Malawi, both in the home and on the plantation? And, to what extent are the industry's economic processes, structures and values articulated through gender? The chapter draws on ethnographic research undertaken on a Malawian sugar plantation at a site of recent industry expansion and traces the experiences of women and men drawn into plantation wage work, many for the first time. A gender and development approach is employed that considers the experiences of both women and men and how production shapes relationships between them, and the analysis is contextualized with empirical and theoretical scholarship from within the fields of feminist economics and gender and development studies. First, the chapter provides an overview of the history of sugar production in Malawi, followed by a description of the methodology and selection of the field site. Next, an outline of the production of sugar and the ways in which labour is organized on a sugarcane plantation is provided. Then the discussion turns to the gender division of labour and how gender norms shape the sugar labour market in Malawi. Next, the chapter analyses the ways in which inclusion in the plantation workforce impacts gender relations in the home, looking at changes to bargaining power, time allocation and the gender division of labour in worker households. Following this, attention turns to the ways in which gender manifests in relationships of power between workers and management on the plantation. The changes experienced as workers move into the plantation workforce are analysed, the importance of patriarchal power in the control of labour is examined, and gendered forms of worker exploitation are identified. Ultimately the chapter concludes that plantation agriculture has a differential impact on rural gender relations; challenging local gender norms and opening up opportunities for increased bargaining power in some instances, but also increasing women's vulnerability and subordination vis-à-vis men by exposing them to new forms of patriarchal control.

Background and Context: The History of Sugar Production in Malawi

The history of Malawi is inextricably linked with the development of its plantation economy, as the production of cotton, coffee, tobacco and tea formed the basis of early settler engagement in this former British colony (McCracken, 2012). Capitalist penetration into pre-colonial subsistence and trading communities was the catalyst for the emergence of new patterns of labour, livelihoods and land tenure in the country (Mandala, 1990; McCracken, 2012). The competition of these early plantations with peasant agriculture – including rivalry for the land and labour to support each – led to the formation of three distinct sub-economies in Malawi: the plantation economy, the peasant cash-cropping economy and the labour reserve economy (Kydd and Christiansen, 1982, p. 356). These patterns of production shaped gender roles and gender relations in rural communities, as plantation work in Malawi was almost the total preserve of men (Vaughan, 1987, p. 128), and in some areas nearly 60% of able-bodied men were absent from home as migrant labourers (McCracken, 2012, p. 185). The legacy of this economic history can still be seen in Malawi to this day and agricultural policies have favoured plantations throughout each decade in the country's lifespan, despite the majority of Malawians engaging in smallholder production (Chirwa, 2004).

Sugar was introduced to Malawi relatively late and was an artefact not of colonialism, but of direct foreign investment by a British multinational corporation at the time of independence (Amer and MacGregor, 1966). The first sugar plantation and factory were established in Nchalo in the Chikwawa District in southern Malawi in 1965, with a second plantation and factory being built in Dwangwa in Nkhotakota District in central Malawi in 1976. Both sites were part of the Sugar Corporation of Malawi Ltd (Sucoma), owned by Lonrho, before being bought over by Illovo Sugar, Africa's leading sugar producer. For years, these plantation sites and the outgrowers' schemes which fed into them encapsulated almost the entire commercial sugar sector in Malawi. However, in recent years, industry expansion has seen the foundation of several new sugar companies in Malawi, including the establishment of smaller plantations, cooperatives and outgrower schemes, as well as expansion of Illovo's cane fields at both their Nchalo and Dwangwa sites (Chinsinga *et al.*, 2013). Today, sugar is one of Malawi's most lucrative export crops and has overtaken tea to become Malawi's second most important export commodity after tobacco (NSO, 2012).

Despite the importance of sugar to Malawi's economy, little attention has been given to Malawian sugar production by social scientists. Indeed, almost all scholarship on plantation agriculture in Malawi, across any commodity, focuses on the colonial period. Of these limited publications, very few have included gender as an important consideration in their analysis and only a single publication addresses the experiences of women plantation workers in Malawi (Vaughan and Chipande, 1986). The author's own research represents the first academic analysis of gender and sugar production in the country. However, plantations in Malawi are deserving of much greater attention from gender scholars, not only because they remain an established economic model, but because the mode of production which relies on the labour of thousands of people who live and work in a way dictated by plantation management gives rise to particular social, economic and cultural conditions that strongly shape gender dynamics in plantation enclaves. Plantations are defined by a hierarchal, centrally controlled, class-structured and most often male-dominated system of labour (Reddock and Jain, 1998, p. 4) and gender is an important factor in how labour is organized and controlled. Gender norms (which intersect with race, class, ethnicity, nationality and age) still strongly shape the plantation labour market to this day. The sugar industry in Malawi is an interesting case study for gender analysis because of the masculine mode of production involving the employment of a mostly male labour force, the hierarchal and 'factory-like' organization of labour within the plantation workforce, the unique context of the industry as a mostly 'tropical' export commodity produced by developing countries[3] and the significant implications for rural gender relations for workers and communities in sugar enclaves.

Methodology

This chapter draws on the author's ethnographic research conducted between February and April 2014 on a commercial sugar plantation in Malawi. The research included semi-structured interviews with 42 sugar workers, focus group discussions with 63 community members, and semi-structured interviews with community representatives including four Village Headmen, one Traditional Authority, one Malawian MP, two representatives of the Direct Labour Office, and senior management of the sugar company. Throughout, the research also involved non-participatory observation on and off the plantation site. The author's vantage point on company premises and within neighbouring communities allowed the process of sugar production to be observed first-hand, as well as the behaviour of the workers in relation to their work, managers and each other. The choice of field site was influenced primarily by the ability to obtain permission to conduct research within company premises. However, the recent industry expansion at the selected site afforded the perfect case study to analyse changes brought about for workers and communities as a direct result of sugar production.

The research methodology employed a gender and development approach which moves beyond the study of women, or the study of gender, to conceptualize the study of gender relations – between and among women and men – to understand the gender dynamics that exist in the social and economic structures, processes and values manifested in the commercial production of sugar in Malawi. Scholars note that the hallmark of a gender and development approach is that gender-aware conceptualization is built into the core of the analysis (Elson, 1995, p. 11) in an effort to 'build gender into the analysis of production rather than making the position of women the central focus' (Guyer, 2005, p. 110). A mix of women and men were selected as informants and the interaction between them was observed as part of the methodology; the position and experiences of both are considered throughout the discussion in this chapter.

Interviews with sugar workers were staged in a sheltered resting area deep within the cane fields. Interviews were undertaken with company permission, during working hours and without recourse to workers' salaries. Despite the interviews being held on company ground, the relatively secluded nature of the outdoor shelter facilitated a candid exchange. The workers interviewed were all field labourers working in the same position, for the same pay. The gender breakdown of informants was 26 men and 16 women, varying in age from 18 to 60 years. The gender ratio of informants was not reflective of the gender ratio of workers in the field, but rather a randomized selection of workers available for interview on any given day, in accordance with their excusal from work for this purpose. The company mediated the selection of sugar workers on a daily basis so as not to disrupt the highly integrated task system at work on the plantation. Despite the company's role in the selection process, no restrictions were experienced in speaking with key employees whom the author requested to interview. During the interviews, a detailed account was taken of people's livelihood strategies and questions asked about the social and economic impacts of the sugar industry to individuals and communities, the impact of economic restructuring on patterns of land use, gender norms and expectations, the valuation of women's and men's work at home and in paid and unpaid labour, the felt experience of work, the complexities of gendered power hierarchies in spaces of production, and the aspirations of workers and their families. In contrast to worker interviews, elite interviews took place off site, in the private homes and offices of the individuals concerned.

To capture the perspectives of those outside the plantation workforce, community focus group discussions were organized in three different villages adjacent to the plantation. Groups were organized with permission from Village Headmen, all of whom extended the invitation to conduct the research in or directly outside their homes as the recognized meeting place for community discussions. While this compromised privacy, the author was warmly welcomed by the communities who participated actively. Each group was asked questions about the impact of sugar production in their area, local gender norms, gendered power at work and home, and the impact which work and income has on personal relationships and community standing. Sixty-three community members formally participated in nine different groups, comprising a total of 38 men and 25 women of various ages. However, given the semi-public location of the discussion groups, many more contributed in passing as they were called away to attend other responsibilities, particularly women whose participation was sometimes interrupted by their role in preparing lunch for their families. Initially, groups were separated into men's and women's groups, but the participation in passing of additional individuals led to the adoption of mixed groups for the remaining five discussions. It was pleasing that the mixed groups were the most successful, and methodologically consistent with a gender and development approach that allowed gender relations between participants to be analysed as they spoke.

The work was supported throughout by a female research assistant who helped with translation. The choice was made not to record the interviews, and instead utilize the gaps in between translation to allow enough time to take detailed notes and capture quotes verbatim. Where the informant was comfortable using English, the interviews could be conducted without the help of the assistant. The author's positionality was positively affected by the pairing: where the assistant's elevated position as an older educated woman helped when speaking with senior officials, our gender as women and the author's younger age made interaction with younger female informants possible, particularly when discussing sensitive topics. Although an outsider to the community, the author was able to draw on pre-existing knowledge of the local language, customs and dress garnered from previous fieldwork in order to demonstrate respect in presentation and demeanour, which were felt to aid rapport. Further, the author's residence in a neighbouring village as well as basic use of the local rather than national language was a talking point met with positive affirmation. The assistant's position as a member of a neighbouring but unconnected village was an advantage as her perceived neutrality, but also inclusion in the wider community, helped gain trust and access to key informants. This positionality was particularly relevant during this research because it expedited the quality and quantity of interviews in an otherwise short space of time.

Research Findings

The production of sugar

Most of the employment generated by a sugarcane plantation happens in the field and thousands of workers may be employed at a single site. The level of mechanization in the field varies across global contexts (Frank, 1965; Coote, 1987; Abbott, 1990; Haraksingh, 1988; Bosma and Knight, 2004; Gunewardena, 2010); however, in Malawi, production is organized not through the use of technology, but rather a carefully integrated task system that exploits the most efficient use of workers' manual labour. Workers are assigned a regimented task of clearing, planting, weeding, fertilizing or cutting which they perform using simple hand tools such as hoes and machetes; much in the way that sugarcane has been cultivated for hundreds of years (Dunn, 1972). Labour is organized according to the principles of discipline, time management and specialization, and task work is integrated across shifts of workers in a way 'associated more with industry than with agriculture' (Mintz, 1985, p. 47). Cane cutting is the most demanding task and workers require great upper-body strength to make a clean diagonal cut through the dense cane. Even weeding, considered to be the least strenuous job, requires workers to position themselves bent over their tools for hours at a time.

The gender division of labour

Sugar production in Malawi remains predominantly male. The upper echelons of corporate management, skilled factory positions, watchmen and operators of machinery are almost invariably men, and recruitment for such posts is often predicated on formal qualifications such as a driver's licence, completion of secondary school and other types of formal training; all of which a woman in Malawi is less likely to have attained than a man. However, a strong preference for male labour also exists in the recruitment of field workers and cane cutters, the latter of whom largely work seasonally during the annual nine-month harvesting period beginning in April each year. Women are drawn into sugar work in the lowest-paid positions as field workers, as well as the lower levels of office administration; however, office jobs are minimal.

Gender norms in Malawi strongly influence the labour market in the sugar industry. In interviews, male sugar workers and male company management cited the physical nature of sugar cultivation as necessitating a male labour force, and community members too described sugarcane as a 'man's crop' due to the strength required to cut and carry it. Managers also employed local gender norms in the assignment of gender-specific tasks, comparing cane cutting to slashing (clearing); considered to be 'men's work' in the gender division of labour employed in Malawian smallholder production. It is notable however, that literature from other global contexts, such as the Philippines and Sri Lanka, shows women to be very active in all aspects of fieldwork in sugar production, including cane cutting (Coote, 1987; Gunewardena, 2010). In some global contexts, where better-paid work is available outside the agricultural sector, fieldwork on sugar plantations has been shown to be almost entirely female (see Gunewardena, 2010, who writes about gender and sugar in Sri Lanka). Of the women workers interviewed, the few who were included in the labour force were just as active as men in all aspects of fieldwork and received the same pay for their work. However, the sugar industry is still inherently male in the Malawian context and this highly gendered labour market shapes the gender dynamics on sugar plantations, the opportunities for wage work and the way income is distributed between women and men.

Wage work, income and decision-making power in the home

Access to income is an important factor shaping gender relations in sugar enclaves in Malawi. In much of the country, women play an active role in food production; and particularly in the matrilineal–matrilocal areas of southern and central Malawi where land traditionally passes through the female line, women have comparably greater security over land resources than women in other parts of southern Africa (Peters, 2010; Berge *et al.*, 2014). However, sugar plantations put considerable pressure on available land resources in adjacent areas and where people have had to give up their land for the purpose of plantation expansion, alternative income-earning opportunities are of paramount importance where there is limited access to land to grow food to supplement a sugar wage. Some of the workers interviewed balanced wage labour with smallholder farming and other off-farm income-generating activities, but many others such as landless or migrant workers and their families were totally dependent on the sugar wage alone. Where a sugar wage is synonymous with a male wage, this puts the majority of all income earned on a sugar plantation into the hands of men, which strongly shapes power dynamics in worker households.

Feminist economists have argued that the welfare of a household cannot be judged solely on the tracking of male income, as it may be unlikely to 'trickle down' to others in the house (Elson, 1995, p. 11). This is especially true in African contexts, where gender scholars have problematized constructions of African households as cohesive economic and social units that collectively pool resources in the way often assumed by Western economists and social scientists (Spring, 1995, p. 1; Benería, 2003; Cornwall, 2005, p. 7; Vaughan, 2005, pp. 119–120; Behrman *et al.*, 2012, p. 51). Spring (1995, p. 16) notes that in most parts of sub-Saharan Africa women and men operate separate income streams, and found in Malawi that while women were responsible for purchasing items for the family such as food, household items and clothing, men rarely contributed to such household staples and instead spent money on productive investments, entertainment and items for their own use. Examined in the context of a Malawian sugar plantation, the implications for gender relations are significant, where women are placed structurally in a position of economic dependence on male wage-earners and culturally have limited bargaining power in their bid to gain a share of that income.

Of the male workers interviewed, almost all retained full control over the decision of how much of their salary to give to their wives, even for those men who engaged in joint household decision making about how the family share should be spent. One man explained openly that he regularly lied to his wife about the value of his earnings so that he could retain a larger portion of his wage for himself, using the vagaries of the casual labour system as his explanation for the regular discrepancy in his monthly salary. Many workers said they never disclosed the value of their income to their wives at all, preferring instead to allot them a lump sum to manage for the family. Most felt that family money should be spent on food, groceries, school fees, health care, transport and agricultural inputs for subsistence plots, such as fertilizer and *ganyu* (casual) labour. Personal (male) money could be spent at will and while few men would speak openly about their own purchase of intoxicants such as alcohol and tobacco, everyone cited the prevalence of these substances among 'other men', as well as paying for transactional sex with girlfriends and sex workers.

This contrasts greatly with the experience of the women sugar workers interviewed, where access to their own formal-sector income had a transformative impact on their autonomy and decision-making power. Of the male workers spoken with, few were strangers to wage work, as many had dipped in and out of formal-sector employment throughout their lifetime. However, the majority of the women workers interviewed had only entered wage employment for the first time due to recent industry expansion in their immediate community and identified empowering changes in their relationships at home as a direct result of their new income. Women described how their salary conferred more power in discussing how that money should be spent and changed the nature of the exchange between themselves and their partners:

> Whenever a woman is working and earns money, nobody can come and control that money. But the money a man is earning, sometimes he can come and not even give a single tambala. Some men know the wife has been paid and force the woman to just give the money. But the woman they can feel independent. A woman whenever she is working is different than a man who is working. A man can say, 'I've got the money, you need to do this, this, this, this.' But when women gets the money, they sit down together to discuss.
>
> (Woman sugar worker)

In other cases, the opportunity to accumulate capital enabled women's financial independence from men altogether. Some young women cited the ability to support themselves as the catalyst for leaving unhappy or abusive marriages. For such women, regular formal-sector income was the determinant in ensuring they could run houses of their own rather than move back in with parents or rush into second marriages.

The limits to the liberating effects of capital accumulation

While wage work in the sugar industry may enable some degree of liberation for women able to independently support themselves without the necessity of a male kinship tie, and while income may confer a greater share in household decision making, the women interviewed still maintained a subordinate position vis-à-vis men. Even where women earned more than men, men still identified themselves as 'breadwinners' and household heads. And while women's income guaranteed them a greater share of decision making, men's wages, in contrast, did not necessitate the same concept of joint ownership. Only in female-headed households did women retain total control over their earnings.

Understanding the limits of the liberating effects of capital accumulation in this context is an important empirical contribution to a broad theoretical concern by feminist scholars around the extent to which wage work can challenge patriarchal oppression (Lim, 1983, p. 70). Ester Boserup (1970) first theorized that women's subordination in Africa stemmed from their exclusion from opportunities for capital accumulation, as European development initiatives favoured men in wage work, cash-cropping schemes and urbanization models. However, when examined through an Africanist lens, later empirical work has shown a greater variety of gendered patterns of labour and production across Africa than the simple binary theorized by Boserup; and this scholarship has rendered her understanding to be generalizing (Guyer, 2005). Benería and Sen (1997) note that while capital accumulation may have enabled some men in Africa to transcend their social position in indigenous class hierarchies, women were never privy to the same upward mobility. They argue that women's loss of status results not from the difficulties of accumulating capital in isolation, but from a complex interweaving of class and gender relations. A study of the Malawian sugar industry shows that while women's employment may be a catalyst which opens up some opportunities and alters gender relations in more equitable ways, these changes are limited and, critically, represent the minority of women's experiences. Many more are thrust into a position of structural inequality and financial dependence on men in areas shaped almost entirely by male-only employment prospects.

Further limits to the liberating effects of wage work for women are framed by the seasonal nature of employment required in the cultivation of sugarcane. Women have historically dominated the classes of casual and seasonal workers in commercial agriculture across global contexts, justified on the supposition of their competing responsibilities in subsistence agriculture and household activities (Coote, 1987; Orton *et al.*, 2001; Andrade *et al.*, 2009; Gunewardena, 2010; O'Laughlin and Ibraimo, 2013). This has been just as true in Malawi: plantations have historically shown a strong preference for male labour and called upon the labour of women, children and immigrants (from Mozambique) only to make up a casual workforce when it suited a company's needs, often at the lowest possible pay (Vaughan and Chipande, 1986; Vaughan, 1987; Chirwa, 1993, 1994). For the families met in Malawi who had fashioned their lives around the guarantee of a sugar income, seasonal employment and tenuous labour contracts provided a great deal of anxiety and instability in the planning of household livelihood strategies. This was especially true for many of the women workers interviewed, whose separation from ex-husbands and rejection of unwanted marriage proposals were secured on the financial assurance of their monthly salary.

Elson and Pearson (1997, p. 201) note that if a woman loses her job after reshaping her life on the basis of a wage income her only alternative may be sex work; a reality confirmed by one worker: 'if we men lose our jobs we become thieves, when women lose their jobs they become prostitutes' (man sugar worker). This is also a key finding for Norris and Worby (2012), who note that widespread lay-offs on Tanzanian sugar estates as a result of privatization have generated a new sexual economy where women exchange unprotected sex for money. Feminist scholars have identified a 'gender paradox of economic restructuring', where wage work facilitates opportunities for increased bargaining power and economic independence in relation to men, but increasingly places them at the mercy of global market forces (Elson, 1999, p. 618).

The production of sugarcane as a monoculture crop, which continues producing new canes each year through the process of ratooning, thus ensures that production may be tied up for a very long time. However, the industry is still vulnerable to global market forces as well as variable local conditions that can have a perverse effect on production, and world sugar prices are notoriously volatile (Abbott, 1990). The position of plantation workers at the very bottom of the value chain means that employment opportunities depend on the success of sugar companies in turning a profit in a global market. As Haraksingh (1988, p. 279) remarks in the context of Trinidad:

> for what eventually went to market was sugar, not cane [. . .] and the worker could do little about froghopper attacks, or transportation problems arising from

> inclement weather, or a breakdown in factory machinery, or conversion ratios of cane to sugar, or still less about poor management practices.

Thus the plantation system itself denies workers control over the means of production of sugar, rendering their labour contingent on factors far removed from the cane fields. As Elson (1999, p. 618) so perfectly states:

> it will be a hollow victory for women if greater equality within labour markets is achieved only to find that the real power is in the board rooms of multinational companies and on the trading floors of international capital markets, and that this is beyond their reach.

Time allocation and the division of labour in the home

Another change which employment in the sugar industry brings to the lives of workers is a change to the allocation of their time. Workers spend between six and eight hours a day, six days a week, labouring on the plantation. Notably, however, this does not result in a restructuring of the household division of labour between partners. Men who engaged in plantation work left the running of the household to their wives; a practice consistent with gender norms across Malawi which leave the responsibility of household tasks such as cooking, collecting water and fuelwood, looking after children, washing and cleaning to women. Women engaged in plantation wage work, however, lengthened and intensified their total labour time between work and home to meet their new demands.

In a study of Ghana's cocoa industry, Asare found that women's increased participation in commercial agriculture imposed three types of full-time chores on women as they balanced the responsibilities of employment, food production and domestic duties (Asare, 1995, p. 112). The author's own research revealed a fourth dimension in the marketing of goods and foodstuffs in the informal economy. In the Malawian context, many of the women workers interviewed had allocated every hour of their week to the balance of these four competing productive and reproductive responsibilities, sacrificing not only the opportunity for active participation in their communities and social contact with their families, but also rest and sleep after the gruelling experience of plantation task work. Floro (1995, p. 1921) highlights the difficult choice with which many poor working women are faced: either work to the biological limit of the length of their working day, or cut back on one or more areas of production and risk severe and even life-threatening poverty.

Floro (1995, p. 1925) also notes women may end up compromising the social and educational needs of their children in exchange for their assistance in meeting household demands. Indeed, the present research found significant changes not in the division of labour between partners, but in the outsourcing of *ganyu* (casual) labourers to fill in for family members (particularly men) on subsistence plots, as well as the utilization of children's labour in family labour systems. Girls especially were expected to take over much of the cooking, cleaning and childcare required in households where their mothers were working on the plantation, and were often expected to maintain small businesses and trade in informal markets in the afternoons where plantation work kept their parents from marketing. What emerges is a picture of children raising children and the increased absenteeism of girls from school to offset the new time constraints faced by their mothers due to their incorporation into a new realm of productive labour. Thus a contradiction emerges, where women's employment in the sugar industry may result in greater repercussions to their welfare and the well-being of younger girls in the household than that of men's. However, the significance of the sugar wage to patterns of decision making, independence and livelihood strategies may be much more important for women due to the other changes wage work can bring to their lives.

Changing relations of power on the plantation

In addition to creating changes in the home, plantation wage labour is also an important structural change which shapes gender relations at work. In some ways, traditional gender roles are redefined on the plantation as women and men become part of one proletarian workforce: united by their class as workers in a particular taskforce, rather than separated by their gender in the traditional division of labour found in the home and in subsistence agriculture. Many of the women workers interviewed took pride in being able to manage the same task work as their male colleagues, and male workers felt that women and men were equals on the plantation where they performed equal tasks for equal pay (despite women only being active in the lowest positions as field workers).

The positioning of women and men in the same productive capacities is a phenomenon completely different from the division of labour at home and in subsistence agriculture, where scholars have argued that gender divisions in African agriculture have historically been defined by a distinct but coordinated 'reciprocity of effort' (Sudarkasa, 2005, p. 28). In Malawian subsistence agriculture, the gendered organization of separate but complementary tasks offers women a relatively neutral space from which to escape the gendered power relations experienced in the home and other forms of production. In both the single-sex and mixed-sex focus group discussions, both women and men characterized the household hierarchy as dominated by men but had very different feelings about relations of power on family fields. One man said:

> At home, I am the bwana [boss] but in my garden we can say that we are both the bwana. In fact my wife she may be more, but mostly, we are just working side by side, me making ridges and she planting
>
> (Man sugar worker)

One woman joked in response: 'The woman is the bwana! You men are just resting!' (woman sugar worker), to which laughter ensued as women and men teased each other about the value of their work and position.

Such playful exchanges within the focus groups starkly contrasted with discussions of power hierarchies in the interviews held on the plantation, where people spoke in hushed tones, looking around them cautiously in fear their confession might be overheard by a boss or co-worker. Aside from the significant physical toll that task work exacts on workers' well-being, the emotional experience of social isolation and the construction of new forms of gendered power in the hierarchy of plantation labour widely differentiate these two forms of production. One worker stated:

> I'm always happy when we are working together in the [home] garden. We can tell one another different stories. We can say, 'no my wife, you must do that,' and she can say, 'no my husband, you must do that', joking. Just chatting together. Now these days, I'm spending time by myself at work on hard tasks. Always afraid the boss will come, shouting, shouting, shouting. It's no good. It's better to buzz off here [on the plantation] so we can spend good time together in our gardens.
>
> (Man sugar worker)

The power relations extolled in the plantation management hierarchy – characterized by the giving of orders by a male authority and the carrying out of orders by newly constituted subordinates – is a sharp contrast to the relative equality inscribed in the power relations on subsistence plots.

Patriarchal power and the control of labour

On the plantation, the management structure extends from senior manager, to field supervisor, to foreman (all of whom are men) to workers (most of whom are men, but some of whom are women). The system is a relatively sparse chain of command given the large number of workers employed. On modern plantations, it is the wage system, rather than the primacy of supervision, which characterizes the control of labour (Gibbon, 2011, pp. 39–40). In practice, however, it was found at the Malawi field site that the power vested in the plantation foreman through the giving and signing off of tasks cements supervision as a primary mode of control. The foremen reserved the right to give any task, to any person, at any time, and it is only after his personal satisfaction has been guaranteed that workers receive their pay.

The root of the authority vested in the plantation management system stems from the confluence of power and capital present in the production of a patriarchal model of control effective in ensuring the successful execution of the capitalist labour process. The authoritarian role of a male supervisor serves as a powerful tool in maximizing productivity (Gunewardena, 2010, pp. 387–388). Failure to control workers leading to a disruption in the production process has implications not only for profits, but also in the deterrence of prospective investors (Frank, 1965, p. 10). For a country like Malawi, competitive advantage in the sugar industry is lost by the import costs of chemicals, metals, fertilizers and factory parts compared with competitors that manufacture these components domestically. Competitive advantage is gained either through attaining a high sucrose content in the cane, or the systematic control of workers through the task system. The authoritarian control over workers is thus justified by the vital timing of cane operations (Gunewardena, 2010, p. 388).

Equally important is the reproduction of traditional power hierarchies in the context of the modern plantation. The history of sugar production and the plantation system in general has its roots in slavery and exploitation, associated most strongly with the colonization of the Caribbean (Dunn, 1972; Coote, 1987; Mintz, 1985, 1998; Abbott, 1990). The patterns of control and dependency which continue to structure supervisor/worker relations in African agriculture to this day are in many ways the legacy of hundreds of years of legalized inequality through the institution of slavery and colonialism (Orton *et al.*, 2001, pp. 473–474). In the Malawian context, the institutional operation of power is not just a post-colonial reproduction of these historical labour relations, but an indigenous repackaging of the foreman's role as village headman, or 'chief'; utilizing traditional instantiations of power to confer authoritarian command in the cane fields. One foreman interviewed characterized his management strategy as a reproduction of the power vested in traditional Malawian authorities:

> Should a manager meet resistance he must know how to conduct himself. I can tell you I practice dictatorship in the fields. As the chief, I must dictate. We cannot do democracy in this work. You may as well employ a reverend, a collared man. The job won't happen. Supervisors must take control over their workers. Yes my manager can decide the field strategy, but I'm the one who tells him, 'so and so is fit to be capital, this one no. So and so is fit to be watchman, this one no.' I inform him, as I know the workers. I am the most important in the field who knows his people best.
>
> (Management interview)

Elson and Pearson (1997, p. 199) note a similar phenomenon in global factory production, where multinational corporations may deliberately utilize traditional forms of patriarchal power in the control of workers (in particular women) on factory floors. In Malawi, traditional power is patriarchal power, where women are positioned in the lowest levels of the social hierarchy at home, in the community, in local and national government, and in the global economy.

Gendered forms of worker exploitation

The reproduction of patriarchal power in spaces of global capital renders women particularly vulnerable to super-exploitation (Elson and Pearson, 1997, p. 197). Benería and Sen (1997, pp. 47–48) argue that the process of capitalist development exploits existing gender hierarchies to place women in subordinate positions at every level of interaction between gender and class. Thus, markets are structured by gendered practices that are not only bearers of gender, but reinforcers of gender inequalities (Elson, 1999). The power manifested in the relationship between management and worker flows through the conduit of gender and the plantation becomes the site in which the potency of patriarchal control is fully mobilized to facilitate the financial and sexual extortion of workers, not only on the basis of their poverty, but also of their sex. Elson and Pearson (1997, p. 202) conclude, 'for women, it is gender subordination which is primary, capitalist exploitation secondary and derivative'.

In a system of control where favouritism and punishment are the mechanism through which power is mediated – a practice witnessed widely during ethnographic observation on the plantation site – workers must realize their labour-power outside the capitalist labour process. At the field site, workers who brought bananas and gifts were often excused early; men who could pay 3000mk (US$7.60) were promoted as watchmen. Some workers reported they had been contracted to cultivate rice in management's personal gardens at the company's expense. One male worker said to me:

> That boss is practising dictatorship here, whether one likes it or not. To my side, dictatorship should be prohibited. He says, 'do you want your job? Buy your place. If you want a job, come with a chicken, or a duck.' This is very, very bad and very sad to the community. Suppose you don't have a duck, or even a *kwacha*. Can you have a job? No you cannot.
>
> (Man sugar worker)

Yet this relationship of control is gendered:

> Women are being harassed here because they are women. When it comes to recruiting, they say, 'before I write your name, may I know you privately?' Then they go to work. To some women this is an advantage, but to some they are also victims.
>
> (Man sugar worker)

The interviews revealed that such harassment continued throughout the period of employment with implications for both women and men, in different ways. While women struggled to avoid unwanted advances, men complained they had been marked 'absent' when one particular foreman assumed they were having relationships with the women he was targeting. Women said they were vulnerable to the foreman grabbing their breasts, but unable to speak up for fear of losing their jobs. Women's subordination to the lowest realms of the plantation hierarchy was also reinforced through the act of begging when the burden became too great:

> When we are given tasks, we work all day, and when we wake our whole bodies ache. Sometimes tasks take two days to complete, our body is tired and aching and we feel pain. When we are sick we are afraid to stop because we are marked absent. This one [points to scar on her leg] came from cutting, but I didn't go to the clinic. I just worked. Some men are always talking to say 'no, this is the big task, we cannot do this' to the foreman. Then they are given something else to do, less work. When we make a complaint, we must sit down to the boss and say 'Bwana, please, this is too much. Please, think of us.' Like we are begging. A man can just tell the boss he can't do something. Women, it is like we must beg to him. Some women do stand up, but then they are absent, even though they have worked the whole day.
>
> (Woman sugar worker)

For women who do gain employment in the sugar industry, the institutional operation of power and capital in plantation management hierarchies exposes them to new risks, challenges and gender dynamics that deepen gender inequalities and create conditions under which women are vulnerable to exploitation and sexual assault. These are social experiences which shape women's lives and their experiences of wage work, just as men's experiences of exploitation are gendered in different ways. In interviewing officials at the District Labour Office, a representative said that abuses such as these were very common in Malawi, but few workers positioned at the bottom of the labour hierarchy were aware of their rights, and reporting carried the risk of dismissal for workers already teetering on the margins of economic survival. For men, but particularly for women whose economic position is uniquely precarious on a sugar plantation with few other opportunities for income, reporting such abuses may be an impossible choice.

Conclusion

Plantation systems are a powerful tool through which global capital shapes the social and economic lives of workers and communities in rural areas. Production, organized around the principles of discipline, efficiency and scale, gives rise to particular social, economic and cultural conditions which strongly shape gender relations in plantation enclaves. In some ways, traditional gender roles are redefined on the plantation as women and men work in the same productive capacities as part of one proletarian workforce. Opportunities for wage work are an important catalyst for women in securing a greater share of bargaining power at home, altering power dynamics between husbands and wives. For some women, their sugar wage was a key determinant in achieving financial independence from men altogether.

However, global capital is also articulated through gender relations, and the processes of sugar production are fused with gender ideologies which exploit gender stereotypes and exacerbate existing gender inequalities. The wage labour market in Malawi's sugar industry still remains heavily biased towards men, reinforcing the notion of men as 'breadwinners' and household heads. The gendered nature of income distribution within Malawian households results in minimal 'trickle-down' of male income to women and other dependants; and thus an industry grounded on a predominantly male workforce places women in a position of structural inequality and financial dependence on a massive scale. While wage work did enable some degree of liberation for the few women who were able to secure work, they still maintained a subordinate position to men overall. Income did not result in a restructuring of the household division of labour or time allocation between partners; women instead lengthened and intensified their total labour time to meet their new demands. Further, on the plantation, the root of the authority consigned in the plantation management stems from the union of power and capital present in the production of a patriarchal model of control. The authoritarian role of a male supervisor serves as a powerful tool in maximizing productivity, and while both women and men were exposed to new risks and challenges as they joined the plantation workforce, the reproduction of patriarchal power in spaces of global capital rendered women particularly vulnerable to super-exploitation and sexual assault.

In the context of the Malawian sugar industry, global processes impact local gender relations and norms, and gender relations and norms in turn shape these global processes. Expansion of the industry into rural spaces has brought about critical changes to the lives of workers and an altering of gender relations, which has both opened up opportunities in some ways and exacerbated gender inequalities in others. By placing gender at the core of an analysis of the impact of industry expansion, we are better able to understand the gendered power dynamics that are embedded in the industry's economic processes, structures and values which so strongly impact many thousands of lives.

Notes

[1]The Land Reform Program Implementation Strategy from 2003–2007, the Greenbelt Initiative in 2010 and the New Alliance for Food Security and Nutrition in 2013 are all policy initiatives designed to open up borders to global agribusiness for large-scale agriculture.

[2]Sugar and tea have flip-flopped over the last decade in the fight to become Malawi's second commodity, but most recent statistics place export revenue from sugar on top (NSO, 2012).

[3]Sugarbeet produced in the global North is also cultivated for sugar; however, sugarcane can only be grown in areas with high enough temperatures, rainfall and hours of sunshine – conditions found mostly in the developing world.

References

Abbott, G. (1990) *Sugar*. Routledge, London and New York.

Amer, J. and MacGregor, H. (1966) *The Nchalo Sugar Estate: A Major Agricultural Development in Malawi*. The Malawi Information Department, Zomba, Malawi.

Andrade, X., Cristiano, I., Casimiro, I. and Almeida, I. (2009) *Empowering Women through Access and Control over the Land in Context of Gender-biased Green-revolution Policies: Action Research Project in Manhiça District*. Forum Mulher Report, Maputo, Mozambique.

Asare, B. (1995) Women in commercial agriculture: the cocoa economy of southern Ghana. In: James, V. (ed.) *Women and Sustainable Development in Africa*. Praeger, London and Westport, Connecticut, pp. 101–112.

Behrman, J., Meinzen-Dick, R. and Quisumbing, A. (2012) The gender implications of large-scale land deals. *The Journal of Peasant Studies* 39(1), 49–79. Available at: http://dx.doi.org/10.1080/03066150.2011.652621 (accessed 24 February 2017).

Benería, L. (2003) *Gender, Development, and Globalisation: Economics as if All People Mattered*. Routledge, New York.

Benería, L. and Sen, G. (1997) Accumulation, reproduction and women's role in economic development: Boserup revisited. In: Visvanathan, N., Duggan, L. and Nisonoff, L. (eds) *The Women, Gender and Development Reader*. Zed Books, London, pp. 42–51.

Berge, E., Kambewa, D., Munthali, A. and Wiig, H. (2014) Lineage and land reforms in Malawi: do matrilineal and patrilineal landholding systems represent a problem for land reforms in Malawi? *Land Use Policy* 41, 61–69. Available at: http://dx.doi.org/10.1016/j.landusepol.2014.05.003 (accessed 24 February 2017).

Boserup, E. (1970) *Women's Role in Economic Development*. George Allen and Unwin Ltd, London.

Bosma, U. and Knight, R. (2004) Global factory and local field: convergence and divergence in the international cane-sugar industry, 1850–1940. *International Review of Social History* 49(1), 1–25. Available at: http://dx.doi.org/10.1017/S0020859003001342 (accessed 24 February 2017).

Chatterjee, P. (2001) *A Time for Tea: Women, Labor and Post/Colonial Politics on an Indian Plantation*. Duke University Press, Durham, North Carolina.

Chinsinga, B., Chasukwa, M. and Zuka, S. (2013) The political economy of land grabs in Malawi: investigating the contribution of Limphasa Sugar Corporation to rural development. *Journal of Agriculture and Environmental Ethics* 26(6), 1065–1084. Available at: http://dx.doi.org/10.1007/s10806-013-9445-z (accessed 24 February 2017).

Chirwa, W.C. (1993) Child and youth labour on the Nyasaland plantations, 1890–1953. *Journal of Southern African Studies* 19(4), 662–680. Available at: http://dx.doi.org/10.1080/03057079308708378 (accessed 24 February 2017).

Chirwa, W.C. (1994) Alomwe and Mozambican immigration labor in colonial Malawi, 1890–1945. *The International Journal of African Historical Studies* 27(3), 525–550.

Chirwa, E.W. (2001) *Access to Land, Growth and Poverty Reduction in Malawi. Report submitted to the North-South Institute, Canada. Wadonda Consult Working Paper*. Chancellor College, University of Malawi, Zomba, Malawi.

Coote, B. (1987) *The Hunger Crop: Poverty and the Sugar Industry*. Oxfam, Oxford, UK.

Cornwall, A. (2005) Introduction: perspectives on gender in Africa. In: Cornwall, A. (ed.) *Readings in Gender in Africa*. The International African Institute, London, pp. 1–19.

DFID (2013) *New Alliance for Food Security and Nutrition: 2013 Progress Report Summary*. Department for International Development, London.

Dunn, R. (1972) *Sugar and Slaves: The Rise of the Planter Class in the English West Indies, 1624–1713*. The University of North Carolina Press, Chapel Hill, North Carolina.

Elson, D. (1995) Male bias in the development process: an overview. In: Elson, D. (ed.) *Male Bias in the Development Process*. Manchester University Press, Manchester, UK, pp. 1–28.

Elson, D. (1999) Labor markets as gendered institutions: equality, efficiency and empowerment issues. *World Development* 27(3), 611–627. Available at: http://dx.doi.org/10.1016/S0305-750X(98)00147-8 (accessed 24 February 2017).

Elson, D. and Pearson, R. (1997) The subordination of women and the internationalization of factory production. In: Visvanathan, N., Duggan, L. and Nisonoff, L. (eds) *The Women, Gender and Development Reader.* Zed Books, London.

Floro, M. (1995) Economic restructuring, gender and the allocation of time. *World Development* 23(11), 1913–1929. Available at: http://dx.doi.org/10.1016/0305-750X(95)00092-Q (accessed 24 February 2017).

Frank, C. (1965) *The Sugar Industry in East Africa: An Analysis of Some Problems and Policy Questions Relating to the Expansion of the Sugar Industry in a Developing Economy.* East African Publishing House, Nairobi, Kenya.

Gibbon, P. (2011) *Experiences of Plantation and Large-Scale Farming in 20th Century Africa. DIIS Working Paper 2011:20.* Danish Institute for International Studies, Copenhagen, Denmark.

Gunewardena, N. (2010) Bitter cane: gendered fields of power in Sri Lanka's sugar economy. *Signs* 35(2), 371–396. Available at: http://dx.doi.org/10.1086/605481 (accessed 24 February 2017).

Guyer, J. (2005) Female farming in anthropology and history. In: Cornwall, A. (ed.) *Readings in Gender in Africa.* The International African Institute, London, pp. 103–111.

Haraksingh, K. (1988) Sugar, labour and livelihood in Trinidad, 1940–1970. *Social and Economic Studies* 37(1–2), 271–291.

Jain, S. and Reddock, R. (1998) *Women Plantation Workers: International Experiences.* Berg, Oxford, UK.

Konings, P. (1998) Women plantation workers and economic crisis in Cameroon. In: Jain, S. and Reddock, R. (eds) *Women Plantation Workers: International Experiences.* Berg, Oxford, UK, pp. 151–165.

Konings, P. (2012) *Gender and Plantation Labour in Africa: The Story of Tea Pluckers' Struggles in Cameroon.* Langaa Research and Publishing Common Initiative Group, Bamenda, Cameroon.

Kydd, J. and Christiansen, R. (1982) Structural change in Malawi since independence: consequences of a development strategy based on large-scale agriculture. *World Development* 10(5), 355–375. Available at: http://dx.doi.org/10.1016/0305-750X(82)90083-3 (accessed 24 February 2017).

Lim, L. (1983) Capitalism, imperialism, and patriarchy: the dilemma of third-world women workers in multinational factories. In: Nash, J. and Fernandez-Kelly, M. (eds) *Women, Men, and the International Division of Labor.* State University of New York Press, Albany, New York.

Maman, M. and Tate, T. (2011) *Women in Agriculture: A Guide to Research.* Routledge, New York and London.

Mandala, E. (1990) *Work and Control in a Peasant Economy: A History of the Lower Tchiri Valley in Malawi, 1859–1960.* University of Wisconsin Press, Madison, Wisconsin.

McCracken, J. (2012) *A History of Malawi 1859–1966.* James Currey, Woodbridge, UK.

Mintz, S. (1985) *Sweetness and Power: The Place of Sugar in Modern History.* Penguin Group, New York.

Mintz, S. (1998) Preface. In: Jain, S. and Reddock, R. (eds) *Women Plantation Workers: International Experiences.* Berg, Oxford, UK, pp. ix–xiv.

Norris, A. and Worby, E. (2012) The sexual economy of a sugar plantation: privatisation and social welfare in northern Tanzania. *American Ethnologist* 30(2), 354–370. Available at: http://dx.doi.org/10.1111/j.1548-1425.2012.01369.x (accessed 24 February 2017).

NSO (2012) *Statistical Yearbook: 2012.* National Statistical Office, Zomba, Malawi.

NSO (2014) *Malawi Labour Force Survey 2013: Key Findings Report.* National Statistical Office, Zomba, Malawi.

O'Laughlin B. and Ibraimo Y. (2013) *The Expansion of Sugar Production and the Well-being of Agricultural Workers and Rural Communities in Xinavane and Magude. IESE Caderno No. 12E/2013.* Instituto de Estudos Sociais e Económicos, Maputo.

Orton, L., Barrientos, S. and McClenaghan, S. (2001) Paternalism and gender in South African fruit employment: change and continuity. *Women's Studies International Forum* 24(3–4), 469–478. Available at: http://dx.doi.org/10.1016/S0277-5395(01)00166-2 (accessed 24 February 2017).

Patel, R., Bezner Kerr, R., Shumba, L. and Dakishoni, L. (2014) Cook, eat, man, woman: understanding the New Alliance for Food Security and Nutrition, nutritionism and its alternatives from Malawi. *The Journal of Peasant Studies* 42(1), 21–44. Available at: http://dx.doi.org/10.1080/03066150.2014.971767 (accessed 24 February 2017).

Peters, P. (2010) 'Our daughters inherit our land, but our sons use their wives' fields': matrilineal–matrilocal land tenure and the New Land Policy in Malawi. *Journal of East African Studies* 4(1), 179–199. Available at: http://dx.doi.org/10.1080/17531050903556717 (accessed 24 February 2017).

Reddock, R. and Jain, S. (1998) Plantation women: an introduction. In: Jain, S. and Reddock, R. (eds) *Women Plantation Workers: International Experiences*. Berg, Oxford, UK, pp. 1–5.

Spring, A. (1995) *Agricultural Development and Gender Issues in Malawi*. University Press of America, Lanham, Maryland.

Sudarkasa, N. (2005) The 'status of women' in indigenous African societies. In: Cornwall, A. (ed.) *Readings in Gender in Africa*. The International African Institute, London, pp.25–31.

Vaughan, M. (1987) *The Story of an African Famine: Gender and Famine in Twentieth-Century Malawi*. Press Syndicate of the University of Cambridge, Cambridge, UK.

Vaughan, M. (2005) Which family? Problems in the reconstruction of the history of the family as an economic and cultural unit. In: Cornwall, A. (ed.) *Readings in Gender in Africa*. The International African Institute, London, pp. 119–121.

Vaughan, M. and Chipande, G.H.R. (1986) *Women in the Estate Sector of Malawi: The Tea and Tobacco Industries. World Employment Programme Research, Working Paper WEP 10/WP.42*. International Labour Organization, Geneva, Switzerland.

22 The Role of Gender Indicators in Rural Development Programmes

F.A. Centrone[1,*], B.B. Bock[2,3], A. Mosso[1] and A. Calvo[1]

[1]Turin University, Grugliasco, Italy; [2]Wageningen University, Wageningen, The Netherlands; [3]Groningen University, Groningen, The Netherlands

Introduction

This chapter contributes to the debate among policy makers, development practitioners and researchers regarding the use of gender indicators in rural settings in the global South. It highlights both the main obstacles and the most recent attempts to develop novel and more insightful indicators.

Feminist researchers have repeatedly underlined the need for reliable gender-specific statistics to represent the position, relations and real involvement of women and men within rural development programmes. Currently, this area has also become a focal point for policy makers, who must demonstrate the necessity for targeted interventions and build alliances to support a specific approach and programme (because 'what gets measured gets managed'). Gender indicators may simultaneously be useful for development practitioners in different stages of project management, such as for determining the feasibility of a project, monitoring its progress and evaluating its impacts.

Indicators are commonly defined as tools that summarize substantial amounts of data with the purpose of representing a dynamic and complex situation and assessing (the direction of) change. Investigating the lives of rural women, in terms of status and opportunities, comprises a complex issue that requires the analysis of several empowerment factors, such as access to health, education, governance, knowledge and technology, the influence of which often varies considerably across different contexts and geographical scales (Calvo, 2013). Moreover, these variables are interdependent and interrelated in complex ways that are difficult to represent in an objective and understandable manner. Indicators may be useful in this context, particularly the most recent generation of multidimensional indices (originally applied to the more generic concepts of poverty and development). Indeed, these indices are more valuable than ordinary statistics by virtue of their potential to reflect this complexity, summarize various information in a single number and thus increase the range of analysis beyond a unilateral interpretation.

Many gender-sensitive indices have been developed in recent years. Several indices have been abandoned due to technical and other substantial limitations (e.g. the lack of gender-disaggregated data at regional and local levels, range of analysis, critical and difficult processing, lack of access to information, costs of the data-gathering process). New attempts have

* Corresponding author e-mail: alicentrone@gmail.com

incorporated these experiences and lessons learned from previous failures, providing a better understanding of what an indicator cannot achieve and how to complement it with other instruments. These limits (but also the improvements) are particularly important when elaborating gender indicators for rural areas, where there is an evident lack of information and recognition of women as 'actual farmers' and economic actors and where the difficulty for women of participating effectively in the formulation and implementation of agricultural and rural development processes is most evident (Bock, 2016).

Multidimensional gender indicators are typically developed at the national level to support international comparisons. However, they provide an average image of national gender relations, which hides potentially quite substantial diversities at the local level. Several studies (Narayan, 2005 in Vaz *et al.*, 2016) have stressed that contextual factors often matter much more for women's empowerment than individual differences (Malhotra and Schuler, 2005) (e.g. particularly in relation to marital violence, microfinance, fertility choices and political participation; see Mason and Smith, 2000; Koenig *et al.*, 2006; Desposato and Norrander, 2009). Depending on the different contexts, a woman may be 'empowered' in a particular dimension of her life but not in another; moreover, according to each particular status, origin and culture, 'what empowers one woman might not empower another' (Cornwall, 2016, p. 344). This statement indicates that contexts differ with respect to the relevance of empowerment criteria, and studies may reach different conclusions depending on the aspects of empowerment and relative indicators examined (Haile *et al.*, 2012).[2] Thus, it is important to develop indicators that are sensitive and adaptable to differences in context.

This chapter discusses the importance of gender indicators based on experiences in three rural regions in Senegal, an area in which we developed multidimensional gender indicators to support practitioners, particularly small and local actors, in project management issues and to inform policy makers. This chapter raises the following main question: how can we translate the gender indicators applied at the (inter)national level into context-sensitive indicators that are useful for rural development programmes at the regional level? First, a general overview of the major multidimensional gender indicators is provided; these clearly reflect the changing ideas regarding gender issues in rural development. Second, our attempt to develop indicators that can support rural gender projects at the local level is presented and discussed. Subsequently, the next steps in the process are stressed; in particular, the need to test these proposed indicators in the field, as well as to discuss the empowerment criteria and indicators that are most relevant to include with local women, practitioners and beneficiaries. A brief reflection is ultimately provided that highlights the main implications, constraints and challenges to overcome for the future development of rural gender indicators.

Gender Indicators: An Overview

The history of gender indicators began approximately 20 years ago, on the occasion of the Fourth World Conference on Women, which was convened by the United Nations (UN) in September 1995 in Beijing, China. Since that time, gender indicators have become more complex and manifold. Moreover, it continues to be difficult to transpose questions into indicators; for example, subjects such as the contribution of women to so-called 'care and reproductive activities' typically remain outside statistics. Table 22.1 provides a brief overview of several of the most relevant gender indicators proposed to date and specifies their scale unit (national or intra-national) and the investigated area (urban or rural). Far from being a completely exhaustive compendium, Table 22.1 indicates the sub-indicators that comprise each indicator, aggregated according to nine selected empowerment dimensions. Furthermore, a footnote provides a synthetic list that indicates the main concepts

Table 22.1. Summary of the most relevant gender indicators since the 1990s. (Authors' elaboration.)

							Empowerment dimension[a]								
Index	Year	Developed by	National	Regional	Urban	Rural	Gender	Life expectancy	Education	Labour	Health	Political participation	Resources access	Decision making	Civil rights
HDI	1990	UNDP	Y[b]	N[b]	N	N	N	Y	Y	Y	N	N	N	N	N
GDI	1995	UNDP	Y	N	N	N	Y	Y	Y	Y	N	N	N	N	N
GEM	1995	UNDP	Y	N	N	N	Y	N	N	Y	N	Y	N	Y	N
RSW	2000	Erasmus University	Y	N	N	N	Y	Y	Y	Y	N	N	N	N	N
AGDI	2004	UNECA	Y	N	Y	Y	Y	N	Y	Y	Y	Y	Y	Y	Y
GEI	2007	Social Watch	Y	N	N	N	Y	N	Y	Y	N	Y	N	Y	N
GGI	2005	WEF	Y	N	N	N	Y	N	Y	Y	N	Y	N	Y	N
SIGI	2009	OECD	Y	N	N	N	Y	N	N	N	Y	Y	Y	Y	Y
GHI	2006	IFPRI	Y	N	N	N	N	Y	N	N	Y	N	N	N	N
GII	2010	UNDP	Y	N	N	N	Y	N	Y	Y	Y	Y	N	N	N
MPI	2010	UNDP	Y	Y	N	Y	Y	N	Y	Y	Y	Y	Y	Y	N
WEAI	2012	USAID, OPHI, IFPRI	Y	Y	N	Y	Y	N	Y	Y	Y	Y	Y	Y	N

[a]Labour=earned income, paid employment, labour force participation, professional and technical work. Health=physical integrity, nutrition, maternal and child mortality, violence against women. Resources access=energy, water, sanitation, technology, land, loans, property, assets. Decision making=economic, professional and political decision making, access to means of production.
[b]Y=yes (included); N=not (included).

considered within some of the empowerment dimensions. Many of these indices were elaborated using databases supplied by UN agencies (e.g. International Labour Organization, World Bank, etc.), as well as by other national and statistical agencies.

With respect to the initial indicators, which may not be specifically gender-oriented, the Human Development Index (HDI) is also included in this list because it represents the starting point for elaborating a more accurate and gender-inclusive human development indicator. The inventors of this indicator (Anand and Sen, 1995, 2000) also contributed to the development of a greater gender vision by promoting the study of the Gender Development Index (GDI), with a focus on highlighting gender inequalities in the achievement of human development goals. The GDI was based on the same sub-indicators as the HDI (life expectancy, education and earned income); however, it was divided by sex.

In 1996, the United Nations Development Programme (UNDP) published the Gender Empowerment Measure (GEM), which specifically referred to the participation level of women and men in the economic and political life of selected countries (UNDP, 1996). The GEM used three variables to indicate the participation of women in terms of political presence, access to professional opportunities and bargaining power.

The beginning of the 2000s saw several new proposals for gender indicators. In 2000, Dijkstra and Hammer introduced the Relative Status of Women (RSW), which was calculated as the arithmetic average of the ratios between the female and male indices for education, life expectancy and returns to labour (Dijkstra, 2006).

In 2004, the United Nations Economic Commission for Africa (UNECA) processed the African Gender Development Index (AGDI), which aimed to estimate the size of the welfare inequality between African women and men (UNECA, 2011).

In 2005, Social Watch published the Gender Equity Index (GEI) to stress gender differences in political, economic and cultural power (Social Watch, 2005).

Between 2005 and 2006, the World Economic Forum (WEF), in collaboration with Harvard University and the London Business School, introduced another indicator, the Gender Gap Index (GGI), which was based on four domains: economic participation and opportunity, realization of those goals related to education, political empowerment and health (Lopez-Claros and Zahidi, 2005; WEF, 2006).

In 2009, the Organisation for Economic Co-operation and Development (OECD) produced the Social Institutions and Gender Index (SIGI) to indicate how social institutions can influence gender inequality. This indicator also introduced a civil rights component, which substantially influenced the status and empowerment of women (Branisa *et al.*, 2009). In particular, it considered the sub-indicators of different social domains: family, physical integrity, civil liberties and ownership rights (Jütting *et al.*, 2008).

Another important indicator, despite the absence of gender in its calculation, introduced in 2006 by the International Food Policy Research Institute (IFPRI) and still in use, is the Global Hunger Index (GHI); it is calculated using data collected by international agencies at the governmental level (IFPRI *et al.*, 2012b). Although the mention of 'gender' is not explicit in its reports, the GHI represents a crucial indicator for inclusion in gender analysis as at times the unpredictable current global climate and market (e.g. as a result of floods, droughts and the improper use of land for growing biofuel or products for the exports-oriented market) may cause different types of shocks that affect the crops of both women (typically in the case of environmental crises) and men (more often a result of market instabilities) (Carr, 2008).

Since 2010, within the most recent UNDP Human Development Reports, new concepts have been introduced in relation to gender disparities, such as vulnerability, resilience and sustainability. In particular, the most recent UNDP reports no longer refer to the GDI (Gender Development Index, described earlier); instead, the reports cite the Gender Inequality Index (GII). Basically, the GII of a nation reflects the disadvantages caused by deprivation in three domains: reproductive health, empowerment and the labour market (Seth, 2009).

Another apparently gender-blind indicator is the Multidimensional Poverty Index (MPI), which was launched by the Oxford Poverty & Human Development Initiative (OPHI) and the George Washington University in 2010 and was designed to analyse the 'poverty' level of a country and overcome the limits of the previous Human Poverty Index (HPI). The MPI measures the most serious forms of deprivation in the dimensions of health, education and standards of living, with respect to both the number of disadvantaged people and the intensity of their deprivation. Despite the lack of gender specificity, mostly because of the unavailability of gender-disaggregated data as described by several authors (see, for instance, Alkire and Santos, 2014), the MPI deserves to be included in this overview. In the presence of deprivations in education, health and public service access, particularly in several countries in the global South, women are often the main disadvantaged subjects.

In 2012, during the 56th session of the UN Commission on the Status of Women, the Women's Empowerment in Agriculture Index (WEAI) was presented; this index was promoted by the US Government's Feed the Future Initiative, led by the US Agency for International Development (USAID), in cooperation with the OPHI and IFPRI. This complex but exhaustive indicator measures the role, representativeness, empowerment and inclusion of women in agriculture, taking into account five sub-domains: production, access to and decision making over economic resources, control over income, leadership and time management (IFPRI *et al.*, 2012a; Alkire *et al.*, 2013). The WEAI is one of the few indices that assess the rural gender dimension; however, because it requires a robust methodology and detailed local surveys, which entail strong investments in human and financial resources, it may be relatively difficult for small local development actors to maintain or apply it.

From this long list, it can be deduced that gender empowerment comprises a difficult and complex concept not only to define, interpret (see Syed, 2010; Kabeer, 2011; Cornwall, 2016) and implement in practice, but also to measure and assess (Kabeer, 1999; Dijkstra, 2002; Charmes and Wieringa, 2003; Malhotra and Schuler, 2005; Schüler, 2006; Klasen and Schüler, 2011). Difficulties particularly increase when the empowerment analysis refers to the most immaterial aspects of life, such as decision-making power, autonomy over important life choices, self-confidence and awareness of women's rights (Cornwall, 2016). However, it is clear that many efforts and studies over the previous 20 years have improved the quality and scope of gender indicators, thereby making them more precise and allowing them to address and measure multiple aspects of human – and women's – lives. Nevertheless, there are several questions remaining. First, there is a lack of intra-national studies that propose gender indicators, which would be helpful for promoting local specific interventions (such as development programmes managed by small non-governmental organizations or similar). Second, there is an insufficient consideration of rural areas. If gender biases are present at the urban level, they are often more evident in rural settings, which, after all, are the most favoured spaces of many international development interventions.

These constraints open the door to our specific research question: how can we overcome these problems by proposing rural gender indicators that can be applied at the local level by small development actors?

Two different options may be viable:

1. Via the use of specific/ad hoc surveys, applied case by case. However, these surveys have several limits: they are not universally applicable, they are relatively expensive in terms of the economic and human resources involved, and they change over time.

2. Via the use of available local and/or regional databases to elaborate ad hoc indicators. This second possibility also presents several limitations related to the lack of completeness, which is predominantly a result of missing/unavailable data and the lack of qualitative aspects.

Therefore, we elaborated three rural gender indicators applicable at the local level. As models and starting points for these indicators, we use some of the previously described indicators proposed for international comparisons, as well as data originating from national statistical surveys.

This research aimed to demonstrate the usefulness of contextual regional indicators, the possibility of calculating them based on national data, and the ability to translate international and national data into regional information while ensuring that they are context sensitive. The overall idea is to propose regional gender indicators appropriate for multiple purposes: first, as a baseline assessment; and then as a tool to track and evaluate progress and changes in empowerment over time and during the implementation phases of international rural development programmes. Furthermore, these indicators may be viewed as a means to foster and potentially (indirectly) promote inclusive rural development projects (through a greater effective participation of women).

Background and Context: The Study Area

The research was developed in Senegal, within an international development project that aimed to improve the food security and agricultural production levels (through the use of drip-irrigation systems and technological tools) in three specific regions of Senegal.

The three research regions are situated in the central area of Senegal. Two regions (Fatick and Thies) are geographically and economically more advantaged (with access to the sea), whereas the third region (Diourbel) is more inland, drought-stricken and lacking an efficient infrastructure. Fatick comprises the more rural region (86% of the total population), whereas in Diourbel, a substantial portion of the population (64%) resides in urban areas. In contrast, the Thies region is characterized by a more uniform distribution of the population between rural and urban areas (ANSD, 2013a). These different pre-conditions are particularly important in sub-Saharan Africa, as well as in Senegal, where there are frequent, strong imbalances (Sahn and Stifel, 2003; Omigbodun *et al.*, 2010), as well as interconnections (Tacoli, 1998), between rural and urban contexts; these are even more evident when examining women's empowerment.

In these three regions, women are numerically slightly higher than men; however, it is not possible to determine whether they are more prevalent in rural areas, as might be supposed, because of the unavailability of specific gender-disaggregated data. In the Diourbel region, the proportion of women of working age in the total population is higher than that of men (77% of the active population). One potential explanation could be the substantial number of male migrants in this region (2.5% of the total), which is greater than the national average (2.3%). Fatick has an increased rate of teenage fertility (153 births per 1000 women aged 15–19 years), which is likely a result of the increased density of the population in rural areas (86% of the total) compared with the other two regions (ANSD, 2013a).

Methodology

The gender indicator process

The selected development project involved approximately 3000 women. The project goal for the research was to identify significant and easily obtainable regional and rural gender indicators that focused on three main empowerment dimensions strictly related to the project objectives:

1. Nutrition and food security;
2. Health and political representation;
3. Access to resources (with a specific distinction between rural, urban and agricultural settings).

First, on the basis of the wide selection of gender indicators developed to date (Table 22.1), and in accordance with our specific goals and available tools, we identified the most suitable indicators for calculation at the regional level as the following:

- Global Hunger Index (GHI);
- Gender Inequality Index (GII);
- Gender Economically Qualified Presence in Agriculture Index (GEQPAI).

Two of these indicators (the GHI and GII) have previously been proposed at the international level and are generally used to perform international comparisons between countries, whereas the third indicator (GEQPAI) was an original elaboration made by the authors. This latter indicator was conceived based on the research purpose, data availability and the observed constraints, but it was particularly designed to specifically address the gap in gender indicators at the rural and regional levels, as previously discussed.

Second, we analysed and selected the main national statistical sources (predominantly the surveys conducted by the National Statistical Service of Senegal – Agence Nationale de la Statistique et de la Démographie du Sénégal (ANSD)), which enabled us to calculate the selected indicators at the regional and rural levels. We subsequently elaborated three indicators, one indicator for each identified dimension of empowerment, for each sample region, with necessary adaptations to allow comparison with the initially intended indicators (see the following paragraphs). Below, we discuss the main outputs, criticisms and potentialities of these regional gender indicators and the next steps to be taken.

The Undernourishment Index (UI)

The initial intent was to calculate the GHI for each sample region. The GHI highlights the successes and failures in hunger reduction strategies and ranks countries on a 100-point scale, where 0 is the best score (no hunger) and 100 is the worst. The GHI combines three sub-indicators, giving equal weight to each (the percentage of undernourished individuals, the percentage of underweight children under 5 years old and the mortality rate in children under 5 years old) (IFPRI *et al.*, 2012b). However, in the current study, it was not possible to calculate the GHI because the proportion of undernourished individuals at the regional level was not provided by the national statistical agencies. Thus, the body mass index[1] (BMI), which is a parameter used to discriminate the undernourishment of a population, was included. Specifically, BMI values of 18.5, 17.0 and 16.0 kg/m^2 were proposed as universally valid thresholds below which an individual (male or female) could be suffering from mild, moderate or severe nutrition problems, respectively (Bailey and Ferro-Luzzi, 1995). For our specific elaboration, we selected the following scale: if $18.4\ kg/m^2 < BMI < 17\ kg/m^2$, chronic energy deficiency is present; whereas if $BMI < 17\ kg/m^2$, the malnutrition is considered severe. We used the percentage of the population of each region with $BMI < 18.4\ kg/m^2$, starting with data provided by the *Demographic and Health Survey with Multiple Indicators* report published in 2012 (ANSD, 2012).

For the GHI, our UI comprises a simple average of its sub-indicators: it ranges between 0 and 100 and indicates the percentage of the population that may suffer from nutritional and environmental problems (see Table 22.2).

The Adjusted Gender Inequality Index (AGII)

The second intent was to calculate the GII at the regional level according to the same criteria adopted by the 2011 Human Development Report (UNDP, 2011). It was nearly feasible, and we used the same method proposed by the UNDP (2011); however, we were again obliged to make an adjustment. Instead of using the proportion of women who occupy seats in parliament (information not provided within the national statistical surveys), we considered the proportion of women who occupy seats in local institutions; these data were obtained from alternative statistical sources (see the subsequent description). Therefore, we refer to this index as the Adjusted Gender Inequality Index (AGII). Similar to the GII, the AGII is a measure that captures the loss in achievements as a result of gender disparities in the dimensions of reproductive health, empowerment and labour force participation. The values range from 0 (perfect equality) to 1 (total inequality). The AGII is a composite indicator that involves two sub-indicators to assess women's reproductive health (the maternal mortality ratio and the adolescent fertility rate), two sub-indicators to evaluate gender empowerment (educational attainment of secondary level or above and the presence of women and men in local institutional posts) and the gender labour force participation sub-indicator (Table 22.3).

Index of Gender Economically Qualified Presence in three different contexts (agricultural, rural and urban: GEQPAI, GEQPRI and GEQPUI)

At the regional level, the UI and AGII do not indicate gender differences in terms of active (and economic) presence in the countryside or in terms of access to resources. Thus, beginning with the available information provided by Senegalese national statistical surveys[2] (ANSD, 2007, 2013a,b) and divided by sex, an indicator of the

Table 22.2. Sub-indicators and parameters for calculation of the Undernourishment Index (UI) at national and regional levels, Senegal. (Authors' elaboration.)

	National		Thies		Fatick		Diourbel	
Sub-indicator	Men	Women	Men	Women	Men	Women	Men	Women
Undernourished children, age < 5 years (%)	18		15		11		19	
Child mortality rate, age < 5 years (%)	0.87		0.53		0.88		1.04	
Population with BMI < 18.4 kg/m^2 (%)	28.2	22.0	24.2	20.8	25.3	16.1	40.9	29.5
UI (%)	15.6	13.5	13.4	12.2	12.5	9.5	20.3	16.5

Table 22.3. Sub-indicators and parameters for calculation of the Adjusted Gender Inequality Index (AGII) at national and regional levels, Senegal. (Authors' elaboration.)

	National		Thies		Fatick		Diourbel	
Sub-indicator	Men	Women	Men	Women	Men	Women	Men	Women
Maternal mortality ratio (deaths per 100,000 live births)	–	392	–	544	–	381	–	296
Adolescent fertility rate (births per 1000 women aged 15–19 years)	–	106	–	100	–	153	–	110
Seats in institutions (%)	71	29	75	25	85	15	96	4
Population with at least secondary level of education (%)	13.7	9.6	9.8	7.4	6.9	4.2	4.0	2.6
Active population (%)	78.6	57.0	67.0	50.0	86.0	67.0	74.0	77.0
AGII	0.557		0.577		0.611		0.624	

significant presence of women and men in agriculture was calculated for each study region. On the basis of two types of constraints, limited data availability and the desire to include only the most influential factors affecting the different levels of access to resources, we considered three sub-indicators. These sub-indicators (all divided by sex) used to calculate the main indicator (referred to as the Gender Economically Qualified Presence in Agriculture Index (GEQPAI)) are as follows: (i) the female/male adult literacy rate; (ii) the female/male ratio of the economically active population in agriculture; and (iii) the female/male proportion of resource managers in agriculture with respect to the women/men who are economically active in agriculture.

The same indicator was calculated at the rural (GEQPRI) and urban (GEQPUI) levels, with the aim of investigating and highlighting as much as possible the local/contextual gender differences within each sample region. In addition to other multidimensional indicators (for example, the HDI), the GEQPAI (the GEQPRI and the GEQPUI) are averages that range between 0 and 1. In particular, indicator values closer to 1 indicate smaller differences between women and men as qualified individuals in agricultural, rural and urban areas.

Research Findings

In relation to the empirical elaboration of the gender indicators, the main output of the UI concerns the Diourbel region (Fig. 22.1), in which the UI is approximately 20% for men and 17% for women; thus, it is consistently compared with the GHI of Senegal, 13.8% (IFPRI *et al.*, 2012), with worse values for both women and men. This finding indicates substantial nutritional problems in this particular region. With respect to the sub-indicators (Table 22.2) as a means to understand the main factors that affect the final value of the UI, the indicators of child undernutrition (19%) and child mortality (1.04%) are the most influential/explicatory elements with regard to the Diourbel region.

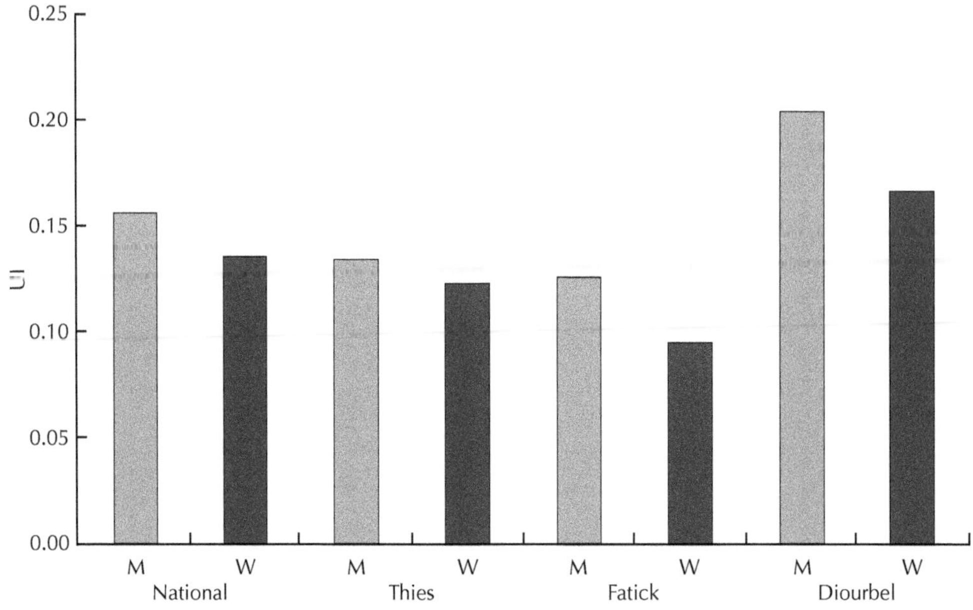

Fig. 22.1. Undernourishment Index (UI) of women (W) and men (M) at national and regional levels, Senegal. (Authors' elaboration.)

Moreover, a distinctive feature is related to the greater value of the UI exhibited by men compared with women. This finding, which has been reported previously by other studies in reference to Ethiopia (Bailey and Ferro-Luzzi, 1995; Hadley *et al.*, 2011), holds nationally and in all three regions of the sample, with the most dramatic differences between women and men being observed in the Diourbel region (almost 4%). In addition, nutritional and health studies (Macia *et al.*, 2010; Duboz *et al.*, 2012) have demonstrated significantly different prevalences of underweight (BMI $<18.4\,\text{kg/m}^2$) and obesity (BMI $\geq 30\,\text{kg/m}^2$) between women and men. In-depth and more disaggregated analyses should be conducted at the local level to specifically understand at which ages, where and why these differences are present. Specifically, investigation should focus on local factors (distinguishing between rural and urban levels and making a cross-comparison between different age groups) and other factors (such as education and marital status) that may have greater effects on the nutrition of the target individuals.

Turning next to the second indicator (the AGII), the findings indicate that the regions exhibit values similar to those found nationally, which enhances the applicability and reliability of the AGII. There is a small difference in relation to the Diourbel region, in which the AGII confirms the worst situation (0.624), with increased inequalities compared with the indicator for the entire country (0.57) and the other sample regions (Fig. 22.2).

Furthermore, as shown in Table 22.3, we can observe that 'our' AGII calculated at the national level (0.557) is slightly higher than the GII (0.54) reported in the Human Development Report of 2013 (UNDP, 2013); our value is moderately higher in the Thies region (0.577) and more so in the Fatick (0.611) and Diourbel (0.624) regions (Table 22.3). An upward shift of the AGII indicates greater diversity and inequality in terms of opportunities for women compared with men. In the Diourbel region, however, the proportion of women of working age in the total population is slightly higher than that of men; this represents a potential source of workers and suggests eligible targets (refer to 'Active population (%)', Table 22.3). One potential explanation is the greater number of male migrants in this region, 2.5% of the total population (ANSD, 2013a), compared with the other regions. Focusing on educational attainment, the main insight originates again from the Diourbel region, which exhibits a

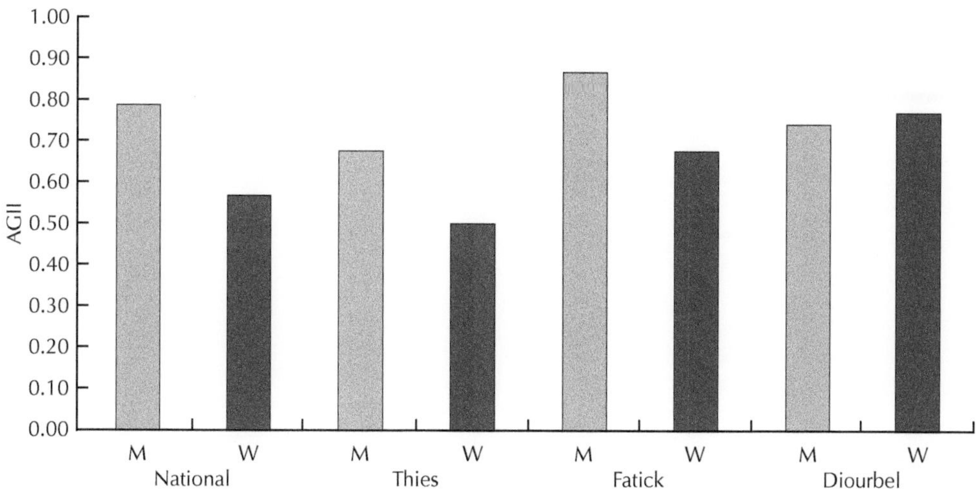

Fig. 22.2. Adjusted Gender Inequality Index (AGII) of women (W) and men (M) at national and regional levels, Senegal. (Authors' elaboration.)

very low education level (approximately one-third of the national average) compared with the other regions.

The higher rate of teenage fertility in Fatick is likely justified by the increased density of the population in rural areas (86%; ANSD, 2013) compared with the other two regions. Finally, the Thies region globally appears to be a relative 'best' example, particularly in terms of education and political representation. However, despite these positive results, this region exhibits the highest maternal mortality ratio (0.544), as well as the lowest proportion of economically active women (50%).

Concerning our third indicator (Index of Gender Economically Qualified Presence in agriculture (GEQPAI), rural (GEQPRI) and urban (GEQPUI) contexts), in both the rural and urban contexts the differences between men and women with paid work (who are therefore economically recognizable and consequently 'qualified') are low (see Fig. 22.3); however, in all cases, the index of the economically qualified presence of women is always lower than that of men. The worst situation, in terms of economic aspects, is again associated with Diourbel, with the exception of the index related to the rural sector (the GEQPRI) which indicates that women in this region appear to be slightly more qualified and economically active in this dimension.

This GEQPAI clearly highlights the agricultural and rural vocation of the three sample regions in relation to both women and men. The proportion of women who appear to have increased possibilities to be economically qualified in agriculture is highest in the Fatick region, which comprises the most agriculturally oriented region in the sample. This positive result is likely explained by the heavy weight of the sub-indicator of agricultural resource managers (which may include access to and management of financial resources, skills, inputs, technical training, extension services and technology) in relation to the total indicator. In particular, in an independent analysis of the sub-indicator related to resource managers in agriculture (Fig. 22.4), the differences between women and men are drastically evident.

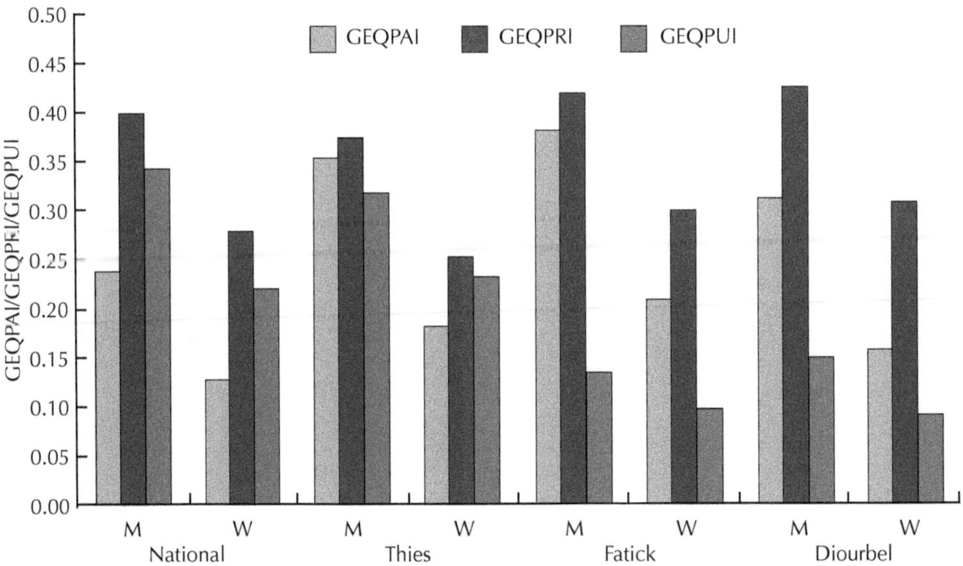

Fig. 22.3. Indices of Gender Economically Qualified Presence in three different contexts (agricultural, rural and urban: GEQPAI, GEQPRI and GEQPUI) of women (W) and men (M) at national and regional levels, Senegal. (Authors' elaboration.)

As clearly indicated in Fig. 22.4, in all three sample regions (despite their reciprocal general differences), women are largely under-represented in resource management. Women are considered to be economically active in agriculture (Table 22.3), but they are not actually responsible for the resources; this indicates that women may act more as a general workforce than as skilled entrepreneurs. This reality evidently entails dependence and contributes to the randomness and unsustainability of a female presence that is economically active today, but may not be tomorrow. The resource management shortcoming implies several remaining questions. Is it a problem of land access and ownership or a failure of access and availability of inputs, machineries, skills and resources for farming activities? Or both? And to what extent? The multidimensional gender indicators may serve as a tool to achieve a relatively quick overview of particular aspects of empowerment and their relative differences; however, to adequately answer these open questions, further in-depth investigations (also more local participatory indicators) are required.

Discussion

This research promotes the fundamental importance of applying contextual gender development indicators to specific intervention areas. The investigated regional indicators enable an explicit understanding of how and how much the three sample regions differ from each other, particularly in terms of nutrition, empowerment and access to resources. In our case, the role played by regional indicators is particularly evident when we examine the most vulnerable areas (nutrition and general empowerment) and those areas with lower gender competences to manage agricultural resources, such as the Diourbel region, where few women manage paid work. The first direct implication in terms of policies and project management issues is related to the necessity of designing punctual interventions that can fill the specific gaps and deficiencies measured at the local level. This finding indicates for example, that in the most critical contexts that result from gender institutional and infrastructural failures, prior to implementing 'advanced' agricultural development projects, it should

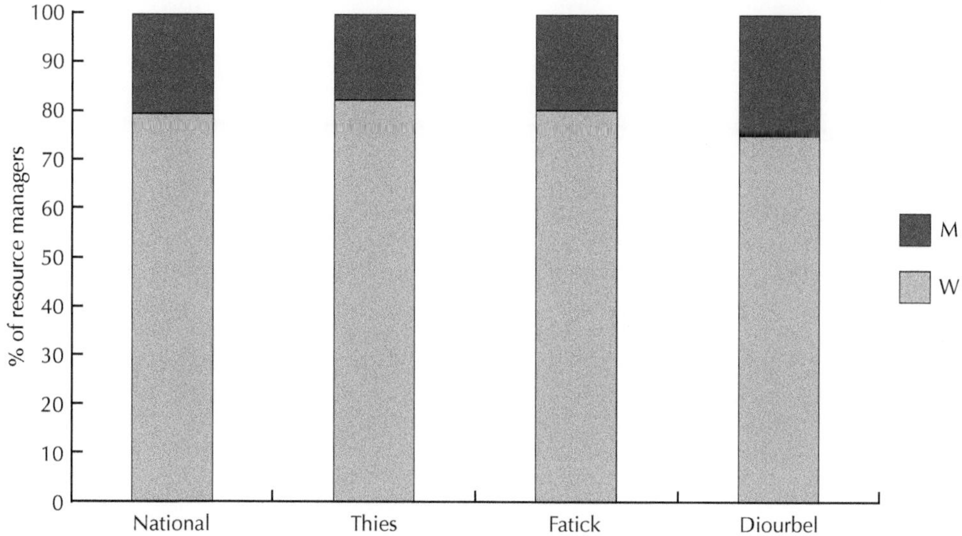

Fig. 22.4. Distribution of resource managers (women (W) and men (M)) in agriculture at national and regional levels, Senegal. (Authors' elaboration.)

first be necessary to enhance the empowerment and education levels of women and girls and only then to progressively enforce the most appropriate economic and technological aspects. This recommendation is particularly valid with regard to the real sustainability and legacy of rural development projects. In contrast, the areas in which the skills, potentialities and presence of women are more widespread and recognized (such as the regions of Fatick and Thies in the sample) appear to comprise more suitable contexts for the effective implementation of more productive and technologically oriented agricultural development programmes. We argue that in these latter areas, it may likely be easier for women to adopt enhanced technologies in order to efficiently improve the quality and quantity of their production. In any case, we suggest that it is necessary to carefully select the goals, targets and activities of 'generalized' food security projects, conducting local surveys to understand whether, where and the extent to which gender discrepancies are present. Otherwise, development practitioners risk implementing the same 'package' of gender-oriented project activities in territories with extremely different starting conditions.

As indicated by methodologists (see, for instance, Metso and Le Feuvre, 2006; Ritchie *et al.*, 2013), in most cases the initial theoretical research design is subordinated to specific contextual and practical findings that require modification and adaptation of the initial methodological intent (for indicator elaboration) along the pathway. Therefore, conscious of the limits and the necessary adaptations throughout the research process, as well as on the basis of available data and final outcomes, we suggest that our translation of international gender indicators to the regional scale may positively contribute to the better discovery of local differences in terms of gender empowerment achievements (or lack thereof).

The main evidence for the importance of *in situ* gender indicators arises from the case of the Diourbel region, which exhibits the most different values or deep deficiencies in terms of empowerment (such as access and equality regarding education, inputs, socio-politic representation and financial resources) compared with the national situation. Furthermore, this study demonstrates that national statistical databases allow the deduction and development of regional indicators with values very similar to the national indicators proposed by international agencies and institutions. Thus, these indicators may also be attainable for 'small' international development actors, whose aim is to obtain an 'immediate', global, comparable and easily attainable picture of gender equality at the local and regional levels. Consequently, although some scientific value and reliability may be lost, regional gender indicators can play an important role in supporting, from the beginning, the shape and implementation of most adapted rural development projects. Their use implies a decrease in the high costs required for the monitoring and evaluation phases and contributes to the most effective 'operationalization' of gender empowerment in countries of the global South.

Conclusions

In answer to our initial question regarding how and whether it is suitable to adapt international indicators to regional and local contexts, we can state that yes, this attempt is useful because it highlights punctual differences (i.e. 'what, where and whom') in terms of gender empowerment achievements and/or failures with respect to specific contexts. This knowledge may facilitate the better design and implementation of gender-oriented interventions within rural development and agricultural programmes. In particular, the translation of indicators from an international to a regional scale is effective and can be used as an intermediate (and more easily achievable) monitoring/evaluation solution between international comparison operations and household surveys. Simultaneously, even if this exercise has been valuable, additional information is still needed (such as that related to the most economic and productive aspects of agriculture, e.g. access to inputs, technical training, extension services and decision making) to fulfil the technical requirements of the applied gender indicators. This type of attempt may also contribute to overcoming the diffused lack of data, to the achievement of a more comprehensive expertise on gender indicators, and to increased exchange, knowledge and awareness of gender issues and methodologies between international, national and local stakeholders.

Furthermore, as a result of this research on gender indicators, we have become increasingly conscious of the necessity of considering and analysing the complete set of mechanisms and processes that underlie the elaboration of these tools. This implies knowing and documenting not only the data collected but also where the data originate, the scale at which the data are collected, who collects the data, and how the data are interpreted and presented (Chant, 2006). As Sen highlighted (Sen, 1987), there are often substantial discrepancies between subjective perceptions of empowerment, equality and well-being, and empowerment as measured by 'objective' indicators (Jackson, 1996), such as those previously discussed. Therefore, we are conscious of the future necessity of testing these indicators in the field through the use of more qualitative (see Pradhan, 2003) and participatory approaches or 'self-rating' poverty exercises (Chant, 2006), such as Participatory Poverty Assessments (PPAs), as well as on the basis of previous positive attempts of field-based and participatory indicators (see Lilja and Dixon, 2008; Njuki *et al.*, 2008; Xiaoyun and Remenyi, 2008). We are simultaneously aware of the controversial incidences behind some participation struggles, including the Participatory Rural Appraisal (PRA) (Mosse, 1994; Mayoux, 1995; Cornwall, 2003). Through the research, it is also possible, and in some cases potentially dangerous, that the role of indicators as (relatively) new governance and power tools (Merry, 2009; Davis *et al.*, 2012) will become progressively more substantial. They may be viewed as a 'top-down' construction, an expression of external and personal views and perceptions regarding the gender equality definition of the authors. Moreover, the increased attention towards impact/effect quantification and the subsequent (wished) better management of the development (and research) project risks critically influencing the answers of the 'assessed' individuals. This may occur through both their desire (and also potential fear) to comply (or not) with the expectations of the researchers and as a result of the risk/threat related to not obtaining further investments. These factors may 'frustrate' the final outputs and the desired greater subjectivity and quality of the collected information.

In the general debate regarding gender indicators, a main key point includes how to combine the attention to local values, perceptions and characteristics with the need for universal and shared tools. These two different directions have similar purposes; however, they are accomplished through different means, including the following: producing more effective and inclusive rural development processes, particularly in Southern contexts; and providing a specific and appropriate weight to the whole actors' objectives and outcomes. In practice, this implies proposing, discussing and applying new interactive ways to achieve real 'data for the people by the people', which provides the expected 'empowered' individuals a meaningful role in the process and returns to them the importance and usefulness of the collected information (in an understandable and accessible manner). Moreover, it also captures their perceptions, level of interest and awareness to fill the gap between theory and practice, framework and field, and counted and unexpected.

Acknowledgements

This work was conducted in the framework of the PAPSEN project in Senegal, an international development project funded by the General Direction of Development Cooperation (DGCS) of the Italian Ministry of Foreign Affairs and jointly managed with the Senegalese Ministry of Agriculture. The authors thank the DGCS, the director and the members of the Local Technical Unit (UTL) of Dakar (Senegal) and the PAPSEN staff, as well as the ANSD and all official national and international representatives met during the several field studies, conducted between March 2013 and May 2015, for their support, availability and cooperation.

Notes

[1]According to the World Health Organization, BMI is defined as an individual's weight in kilograms divided by the square of the individual's height in metres (kg/m^2); it is an attempt to quantify whether an individual is underweight, normal weight, overweight or obese.

[2]Surveys consulted were: Situation économique et sociale du Sénégal en 2011 (SES 2011) (ANSD, 2013a), prémière Enquête de Suivi de la Pauvreté au Sénégal (ESPS-I, 2005–2006) (ANSD, 2007) and deuxième Enquête de Suivi de la Pauvreté au Sénégal (ESPS-II, 2011) (ANSD, 2013b).

References

Alkire, S. and Santos, M.E. (2014) Measuring acute poverty in the developing world: robustness and scope of the Multidimensional Poverty Index. *World Development* 59, 251–274. Available at: http://dx.doi.org/10.1016/j.worlddev.2014.01.026 (accessed 25 February 2017).

Alkire, S., Meinzen-Dick, R., Peterman, A., Quisumbing, A., Seymour, G. and Vaz, A. (2013) The Women's Empowerment in Agriculture Index. *World Development* 52, 71–91. Available at: http://dx.doi.org/10.1016/j.worlddev.2013.06.007 (accessed 25 February 2017).

Anand, S. and Sen, A.K. (1995) *Gender Inequality in Human Development: Theories and Measurement. Human Development Report Office Occasional Paper No. 19.* United Nations Development Programme, New York.

Anand, S. and Sen, A.K. (2000) The income component of the human development index. *Journal of Human Development* 1(1), 83–106. Available at: http://dx.doi.org/10.1080/14649880050008782 (accessed 25 February 2017).

ANSD (2007) *Prémière Enquête de Suivi de la Pauvreté au Sénégal (ESPS I, 2005–2006). Rapport National.* Agence Nationale de la Statistique et de la Démographie du Sénégal, Dakar. Available at: http://www.ansd.sn/ressources/rapports/Rapport_ESPS.pdf (accessed 4 March 2017).

ANSD (2012) *Enquête Démographique et de Santé à Indicateurs Multiples (EDS-MICS) 2010–2011. Rapport final (Demographic and Health Survey with Multiple Indicators).* Agence Nationale de la Statistique et de la Démographie du Sénégal and ICF International, Dakar and Calverton, Maryland. Available at: http://www.ansd.sn/ressources/rapports/EDS-MICS_2010-11_Rapport_final.pdf (accessed 4 March 2017).

ANSD (2013a) *Situation économique et sociale du Sénégal en 2011 (SES 2011). Version Définitive.* Agence Nationale de la Statistique et de la Démographie du Sénégal, Dakar. Available at: http://www.ansd.sn/index.php/citations/3 (accessed 7 July 2017).

ANSD (2013b) *Deuxième Enquête de Suivi de la Pauvreté au Sénégal (ESPS-II, 2011). Rapport Definitif.* Agence Nationale de la Statistique et de la Démographie du Sénégal, Dakar. Available at: http://anads.ansd.sn/index.php/catalog/17/download/349 (accessed 4 March 2017).

Bailey, K.V. and Ferro-Luzzi, A. (1995) Use of body mass index of adults in assessing individual and community nutritional status. *Bulletin of the World Health Organization* 73(5), 673–680.

Branisa, B., Klasen, S. and Ziegler, M. (2009) *New Measures of Gender Inequality: The Social Institutions and Gender Index (SIGI) and Its Subindices. Poverty, Equity and Growth Discussion Paper No. 10.* Courant Research Centre, Goettingen, Germany.

Bock, B.B. (2016) The rural. In: van der Tuin, I. (ed.) *Macmillan Interdisciplinary Handbooks: Gender.* Vol. 2: *Nature.* Macmillan, New York, pp. 199–216.

Calvo, A. (2013) Technology and education in rural women entrepreneurship: approaches and problems in some case studies in Mali and in Italy. *Scientific Journal of Agricultural Economics* 2(1), 1–16.

Carr, E.R. (2008) Men's crops and women's crops: the importance of gender to the understanding of agricultural and development outcomes in Ghana's Central Region. *World Development* 36(5), 900–915. Available at: http://dx.doi.org/10.1016/j.worlddev.2007.05.009 (accessed 25 February 2017).

Chant, S. (2006) Re-thinking the 'feminization of poverty' in relation to aggregate gender indices. *Journal of Human Development* 7(2), 201–220. Available at: http://dx.doi.org/10.1080/14649880600768538 (accessed 25 February 2017).

Charmes, J. and Wieringa, S. (2003) Measuring women's empowerment: an assessment of the Gender-related Development Index and the Gender Empowerment Measure. *Journal of Human Development* 4(3), 419–435. Available at: http://dx.doi.org/10.1080/1464988032000125773 (accessed 25 February 2017).

Cornwall, A. (2003) Whose voices? Whose choices? Reflections on gender and participatory development. *World Development* 31(8), 1325–1342. Available at: http://dx.doi.org/10.1016/S0305-750X(03)00086-X (accessed 25 February 2017).

Cornwall, A. (2016) Women's empowerment: what works? *Journal of International Development* 28(3), 342–359. Available at: http://dx.doi.org/10.1002/jid.3210 (accessed 25 February 2017).

Davis, K.E., Kingsbury, B. and Merry, S.E. (2012) Indicators as a technology of global governance. *Law & Society Review* 46(1), 71–104. Available at: http://dx.doi.org/10.1111/j.1540-5893.2012.00473.x (accessed 25 February 2017).

Desposato, S. and Norrander, B. (2009) The gender gap in Latin America: contextual and individual influences on gender and political participation. *British Journal of Political Science* 39(1), 141–162. Available at: http://dx.doi.org/10.1017/S0007123408000458 (accessed 25 February 2017).

Dijkstra, A.G. (2002) Revisiting UNDP's GDI and GEM: towards an alternative. *Social Indicators Research* 57(3), 301–338. Available at: http://dx.doi.org/10.1023/A:1014726207604 (accessed 25 February 2017).

Dijkstra, A.G. (2006) Towards a fresh start in measuring gender equality: a contribution to the debate. *Journal of Human Development* 7(2), 275–283. Available at: http://dx.doi.org/10.1080/14649880600768660 (accessed 25 February 2017).

Duboz, P., Chapuis-Lucciani, N., Boëtsch, G. and Gueye, L. (2012) Prevalence of diabetes and associated risk factors in a Senegalese urban (Dakar) population. *Diabetes & Metabolism* 38(4), 332–336. Available at: http://dx.doi.org/10.1016/j.diabet.2012.02.011 (accessed 25 February 2017).

Hadley, C., Belachew, T., Lindstrom, D. and Tessema, F. (2011) The shape of things to come? Household dependency ratio and adolescent nutritional status in rural and urban Ethiopia. *American Journal of Physical Anthropology* 144(4), 643–652. Available at: http://dx.doi.org/10.1002/ajpa.21463 (accessed 25 February 2017).

Haile, H.B., Bock, B. and Folmer, H. (2012) Microfinance and female empowerment: do institutions matter? *Women's Studies International Forum* 35(4), 256–265. Available at: http://dx.doi.org/10.1016/j.wsif.2012.04.001 (accessed 25 February 2017).

IFPRI, OPHI and USAID (2012a) *Women's Empowerment in Agriculture Index.* International Food Policy Research Institute, Oxford Poverty & Human Development Initiative, and US Agency for International Development Feed the Future Initiative, Washington, DC.

IFPRI, Concern WorldWide, Welthungerhilfe and Green Scenery (2012b) *2012 Global Hunger Index. The Challenge of Hunger: Ensuring Sustainable Food Security Under Land, Water, and Energy Stresses.* International Food Policy Research Institute, Concern WorldWide, Welthungerhilfe and Green Scenery, Bonn, Washington, DC and Dublin.

Jackson, C. (1996) Rescuing gender from the poverty trap. *World Development* 24(3), 489–504. Available at: http://dx.doi.org/10.1016/0305-750X(95)00150-B (accessed 25 February 2017).

Jütting, J.P., Morrisson, C., Dayton-Johnson, J. and Drechsler, D. (2008) Measuring gender (in)equality: the OECD gender, institutions and development data base. *Journal of Human Development* 9(1), 65–86. Available at: http://dx.doi.org/10.1080/14649880701811401 (accessed 25 February 2017).

Kabeer, N. (1999) Resources, agency, achievements: reflections on the measurement of women's empowerment. *Development and Change* 30(3), 435–464. Available at: http://dx.doi.org/10.1111/1467-7660.00125 (accessed 25 February 2017).

Kabeer, N. (2011) *Contextualising the Economic Pathways of Women's Empowerment: Findings from a Multi-Country Research Programme. Pathways Policy Paper.* Pathways of Women's Empowerment RPC, Brighton, UK.

Klasen, S. and Schüler, D. (2011) Reforming the Gender-related Development Index and the Gender Empowerment Measure: implementing some specific proposals. *Feminist Economics* 17(1), 1–30. Available at: http://dx.doi.org/10.1080/13545701.2010.541860 (accessed 25 February 2017).

Koenig, M.A., Stephenson, R., Ahmed, S., Jejeebhoy, S.J. and Campbell, J. (2006) Individual and contextual determinants of domestic violence in North India. *American Journal of Public Health* 96(1), 132–138. Available at: http://dx.doi.org/10.2105/AJPH.2004.050872 (accessed 25 February 2017).

Lilja, N. and Dixon, J. (2008) Operationalising participatory research and gender analysis: new research and assessment approaches. *Development in Practice* 18(4–5), 467–478. Available at: http://dx.doi.org/10.1080/09614520802181202 (accessed 25 February 2017).

Lopez-Claros, A. and Zahidi, S. (2005) *Women's Empowerment: Measuring the Global Gender Gap.* World Economic Forum, Geneva, Switzerland.

Macia, E., Duboz, P. and Gueye, L. (2010) Prevalence of obesity in Dakar. *Obesity Reviews* 11(10), 691–694. Available at: http://dx.doi.org/10.1111/j.1467-789X.2010.00749.x (accessed 25 February 2017).

Malhotra, A. and Schuler, S.R. (2005) Women's empowerment as a variable in international development. In: Narayan, D. (ed.) *Measuring Empowerment: Cross-Disciplinary Perspectives.* World Bank, Washington, DC, pp. 71–88.

Mason, K. and Smith, H.L. (2000) Husbands' versus wives' fertility goals and use of contraception: the influence of gender context in five Asian countries. *Demography* 37(3), 299–311. Available at: http://dx.doi.org/10.2307/2648043 (accessed 25 February 2017).

Mayoux, L. (1995) Beyond naivety: women, gender inequality and participatory development. *Development and Change* 26(2), 235–258. Available at: http://dx.doi.org/10.1111/j.1467-7660.1995.tb00551.x (accessed 25 February 2017).

Merry, S.E. (2009) Measuring the world: indicators, human rights, and global governance. *Proceedings of the Annual Meeting (American Society for International Law)* 103, 239–243.

Metso, M. and Le Feuvre, N. (2006) *Quantitative Methods for Analysing Gender, Ethnicity and Migration.* The University of York, York. Available at: https://www.york.ac.uk/res/researchintegration/Integrative_Research_Methods/Metso%20and%20Le%20Feuvre%20Quantitative%20Methods%20April%202007.pdf (accessed 27 February 2017).

Mosse, D. (1994) Authority, gender and knowledge: theoretical reflections on the practice of participatory rural appraisal. *Development and Change* 25(3), 497–526. Available at: http://dx.doi.org/10.1111/j.1467-7660.1994.tb00524.x (accessed 25 February 2017).

Narayan, D. (ed.) (2005) *Measuring Empowerment: Cross-Disciplinary Perspectives.* World Bank, Washington, DC.

Njuki, J., Mapila, M., Kaaria, S. and Magombo, T. (2008) Using community indicators for evaluating research and development programmes: experiences from Malawi. *Development in Practice* 18(4–5), 633–642. Available at: http://dx.doi.org/10.1080/09614520802181913 (accessed 25 February 2017).

Omigbodun, O.O., Adediran, K.I., Akinyemi, J.O., Omigbodun, A.O., Adedokun, B.O. and Esan, O. (2010) Gender and rural–urban differences in the nutritional status of in-school adolescents in south-western Nigeria. *Journal of Biosocial Science* 42(5), 653–676. Available at: http://dx.doi.org/10.1017/S0021932010000234 (accessed 25 February 2017).

Pradhan, B. (2003) Measuring empowerment: a methodological approach. *Development* 46(2), 51–57. Available at: http://dx.doi.org/10.1057/palgrave.development.1110445 (accessed 25 February 2017).

Ritchie, J., Lewis, J., Nicholls, C.M. and Ormston, R. (eds) (2013) *Qualitative Research Practice: A Guide for Social Science Students and Researchers.* SAGE Publications Ltd, London.

Sahn, D.E. and Stifel, D.C. (2003) Urban–rural inequality in living standards in Africa. *Journal of African Economies* 12(4), 564–597. Available at: http://dx.doi.org/10.1093/jae/12.4.564 (accessed 25 February 2017).

Schüler, D. (2006) The uses and misuses of the Gender-Related Development Index and Gender Empowerment Measure: a review of the literature. *Journal of Human Development* 7(2), 161–181. Available at: http://dx.doi.org/10.1080/14649880600768496 (accessed 25 February 2017).

Sen, A. (1987) The standard of living: lecture I, concepts and critiques. In: Sen, A. (ed.) *The Standard of Living.* Cambridge University Press, Cambridge, UK, pp. 1–19.

Seth, S. (2009) Inequality, interactions, and human development. *Journal of Human Development and Capabilities* 10(3), 375–396. Available at: http://dx.doi.org/10.1080/19452820903048878 (accessed 25 February 2017).

Social Watch (2005) No country treats its women the same as its men: the Gender Equality Index – a new perspective. In: *Social Watch Roars and Whispers. Gender and Poverty: Promises Versus Action*. Available at: http://www.socialwatch.org/sites/default/files/pdf/en/gei2005_eng.pdf (accessed 4 March 2017).

Syed, J. (2010) Reconstructing gender empowerment. *Women's Studies International Forum* 33(3), 283–294. Available at: http://dx.doi.org/10.1016/j.wsif.2010.03.002 (accessed 25 February 2017).

Tacoli, C. (1998) Rural–urban interactions; a guide to the literature. *Environment and Urbanization* 10(1), 147–166.

UNDP (1996) *Human Development Report 1996. Economic Growth and Human Development*. United Nations Development Programme, New York. Available at: http://hdr.undp.org/en/content/human-development-report-1996 (accessed 25 February 2017).

UNDP (2011) *Human Development Report 2011. Sustainability and Equity: A Better Future for All*. United Nations Development Programme, New York. Available at: http://hdr.undp.org/en/content/human-development-report-2011 (accessed 25 February 2017).

UNDP (2013) *Human Development Report 2013. The Rise of the South: Human Progress in a Diverse World*. United Nations Development Programme, New York. Available at: http://hdr.undp.org/sites/default/files/reports/14/hdr2013_en_complete.pdf (accessed 4 March 2017).

UNECA (2011) *The African Gender and Development Index*. United National Economic Commission for Africa, Addis Ababa.

Vaz, A., Pratley, P. and Alkire, S. (2016) Measuring women's autonomy in Chad using the Relative Autonomy Index. *Feminist Economics* 22(1), 264–294. Available at: http://dx.doi.org/10.1080/13545701.2015.1108991 (accessed 25 February 2017).

WEF (2006) *The Global Gender Gap Report 2006*. World Economic Forum, Geneva, Switzerland.

Xiaoyun, L. and Remenyi, J. (2008) Making poverty mapping and monitoring participatory. *Development in Practice* 18(4–5), 599–610. Available at: http://dx.doi.org/10.1080/09614520802181699 (accessed 25 February 2017).

23 Beneficial for Women? Global Trends in Gender, Land and Titling

S. Jacobs*
Manchester Metropolitan University, Manchester, UK

Introduction

This chapter explores the impact of agricultural land titling for women. Land titling has taken place in the context both of neoliberal policy making and of growth of demands for gender equality. 'Neoliberal globalization' is a complex phenomenon with contradictory implications for the lives and livelihoods of many women. This is particularly so for rural women, who are disproportionately among the poorest sectors of many populations, often relying mainly or solely on agriculture and petty trading to survive. In many societies, the question of land is thus crucial to women's lives, livelihoods and social standing (Agarwal, 1994; Mudege, 2008; Archambault and Zoomers, 2015a).

Globalization's dominant aspect is, arguably, a deep economic restructuring, enacted via the spread of versions of capitalism which eschew social welfare state provision and increase social inequality (Harvey, 2011). This restructuring also entails an increased emphasis on individualization, which, in the context of 'reforms' of land tenure, often takes the form of strengthening statutory land tenure for individuals while simultaneously increasing pressures to dismantle common property regimes.

While being in many cases a negative development for women's livelihoods, globalizing trends can nevertheless have more positive outcomes, particularly within social movements (Keck and Sikkinck, 1998). Transnational women's and feminist movements have used the technological and political opportunities provided by globalization to organize, especially since the 1990s. Thus, in recent years, non-governmental organizations (NGOs), women's movements and, at times, global bodies have attended to some of the inequalities faced by rural women, urging that agricultural land be titled jointly in both husbands' and wives' names or else individually.

The background to such discussions is now well explored. Women, globally, tend to have less direct access to land or else poorer ownership rights where such legal entitlements apply. This is particularly marked in contexts of patrilineal kinship systems. Moreover, women's land is usually of lower quality than land owned by men (FAO, 2011). Even where kinship is bilateral, as in much of Europe and the Americas, women none the less still often experience inferior rights and access (Brandth, 2002; Varley, 2010; Jacobs, 2014a).

Spurred on in part through neoliberal global policy, the main recent international trend in land and agriculture has been the

* Corresponding author e-mail: s.jacobs@mmu.ac.uk

move towards privatization and titling of land. This trend, however, does not necessarily affect men and women in the same ways. Where land is privatized and owned outright, women are significantly less likely to possess it formally (FAO, 2011; Doss *et al.*, 2014c). Where land is held communally, as in much of sub-Saharan Africa and among some indigenous peoples elsewhere, women's rights are usually mediated through a husband or father (Budlender and Alma, 2011). Customarily in sub-Saharan Africa, wives do have the right to hold a small plot allocated for food crops, and this and its products should be at her disposal. Where a woman is divorced or single with children, she should have access to her father's lineage land to feed her family. However, in many patrilineal African systems,[1] women's rights are usually contingent, even when customarily they should be secure. Budlender (2013) thus writes that the received wisdom concerning African land rights for women of different marriage statuses (i.e. married/single/widowed) is that they can usually access land but may lack rights of control over it. (Widows are meant to hold land in trust for sons, so do have rights of control, albeit temporarily.) Nevertheless, negotiation of such entitlements can be arduous, with the rights often insecure in practice (Thorley, 2015).

According to the Food and Agriculture Organization of the United Nations (FAO), a distinction should be made between land use rights, control over land and rights of transfer (FAO, 2002). Adding further nuance to the difference between ownership and control, Doss *et al.* (2014b) classify property rights as: access (the right to be on the land); withdrawal (the right to take something from the land, such as firewood or water); management (including cultivation); exclusion (the right to prevent use by others); and alienation (the right to lease, bequeath or sell the land). Using this schema, women may have rights of access or withdrawal, and sometimes management, but they are less likely to have rights to exclude others or to sell or bequeath land.

African customary systems do nevertheless usually acknowledge the rights of all lineage *members* (and sometimes their dependants, including wives) to access land for food cultivation (Kingwill, 2014; Thorley, 2015). Given current world trends of land 'grabbing',[2] this safeguard should not be taken for granted. Nevertheless, women are likely to lack full decision-making powers, to be poorer than men in their communities and to enjoy only marginal control over land. They may also be subject to intimate violence when asserting rights (Daley, 2011; Jacobs, 2015).

With this background in mind, the remainder of this chapter explores some international trends concerning land, land tenure and gendered impacts. The discussion is divided into two main sections. The first outlines current trends in land titling led by global agencies and focuses particularly on the significant role of World Bank gendered policies. The second section summarizes selected recent research pertaining to the impacts of both joint and individual land titling. This is followed by a concluding discussion and some brief recommendations on increasing the role of states in protecting women's land rights.

Land Titling: Background, Institutions and Debates

Land titling (i.e. the formalization of land tenure law and practice) forms an important aspect of the economic liberalization of economies generally, and of the spread of capitalist relations into agriculture in particular. Titling is a type of formalization that can give full or partial property rights to users, superseding customary tenure practices and thereby giving statutory law authority over property. Titling usually strengthens the new owner's rights, including the right to exclude others (Stein and Cunningham, 2015). Although titling often occurs in conjunction with privatization of land rights, it does not always lead automatically to full privatization. For instance, states can retain ownership while granting permits for use (often, these are long-term) and restricting rights to buy and sell (see below).

Attempts to formalize land tenure have taken place in sub-Saharan African countries (see e.g. Manji, 2006), but also in South and East Asia and within former state socialist societies (Jacobs, 2010; Spoor, 2012). This process has been promoted by international financial institutions (IFIs), and particularly the World Bank, for some years. More recently, bilateral 'deals' involving tenure formalization have been enacted by the G8 and other bodies (see below).

Rationales for titling

This thinking has been influenced especially by de Soto (2000), who saw formalized property rights as the main cause of Western economic growth and development. At least four main political and economic benefits of land titling have been put forward (Hall, 2013). First, titling is meant to promote security by protecting land from predation or seizure by various social actors, including the state itself (Stein and Cunningham, 2015). Second, titling is thought to stimulate investment on farms, since cultivators have a greater stake with secure tenure. Third, titling makes it possible to use land as collateral when seeking loans. Fourth and relatedly, titling makes land much easier to buy and sell, in theory raising its financial value. (Indicating the deeply economic inflections of these arguments, improving women's rights is usually mentioned only as a secondary benefit of titling.) Such arguments are emphasized in varying degrees by different proponents. Combined, however, they constitute a powerful discourse and central pillar of contemporary neoliberal development policy, which is claimed to prompt growth and reduce poverty (FAO, 2011; USAID, 2013). Yet, through this process the meaning of 'land reform' also changes from its original sense of state-backed redistribution of land – particularly to the poorest people (Jacobs, 2013) – to the individualization of property rights and fundamental land law reorganization.

The World Bank and gender issues

Since the 1980s, the World Bank has been one of the most significant organizations urging changes in land tenure and titling policy (see World Bank, 2008, 2009). Gill and Cutler (2014) term this development a 'world market revolution'; that is, a global project to 'liberalize' economies and to open them to investment. Nevertheless, alongside this economic agenda, the World Bank has also incorporated an explicit gender focus in recent years in its writings and policy briefs. One of the earliest mentions by a major global institution of improved rights for women as an outcome of titling appears in a World Bank report (Deininger, 2003), while the 2000/2001 and 2006 *World Development Reports* (WDRs) (World Bank, 2000, 2006a) highlighted the close relationship between gender equality, economic growth and women's paid employment (Elson, 2012; Mahon, 2012; Razavi, 2012).

One rationale for concern over women's land rights might relate to the World Bank's less clearly articulated ambition to help form better functioning and more egalitarian families, albeit solely *heterosexual* families (Bedford, 2009). For instance, although World Bank policies encourage men to engage more with housework and childcare, Bedford (2009) notes that this ignores other possibilities that would involve public funding, most notably nurseries. However, the main justification claimed for women's land rights has been instrumental. The FAO estimates that the inequality or 'gap' in rural women's access to land and other resources causes a 20–30% reduction in yields on woman-managed farms (FAO, 2011). It is argued by the World Bank and other organizations that granting women land titles will address this issue by prompting investment in agriculture, leading to improved farming practices (World Bank, FAO and IFAD, 2009; FAO, 2011). The World Bank termed such business-focused approaches to women's control over agricultural (and other) land 'smart economics', claiming that it was capable of raising productivity, reducing poverty and improving family welfare simultaneously

(World Bank, 2006b, cited in Wisborg, 2014; see Chant and Sweetman, 2012). The smart economics approach was taken up with reference to agriculture and gender in the *Gender and Agriculture Sourcebook* (World Bank, FAO and IFAD, 2009) and particularly in the FAO's 2011 report on Women in Agriculture. Additionally, it has been acknowledged by the FAO (2011) that women are more likely to use income for the benefit of households, and particularly children within those households.

The World Bank's WDR 2012, *Gender Equality and Development*, makes some advances on the, at times, simplistic claims of 'smart economics' (World Bank, 2012). The Global Social Policy symposium, which evaluated the contents of WDR 2012, thus welcomed this report's acknowledgment of the *intrinsic* as well as instrumental importance of gender equality. WDR 2012 also recognizes that many initiatives continue to rely on women's unpaid labour, and highlights continued discrimination in family and property law, plus the need for greater investment in water and sanitation – without which legal reforms remain ambitious but impeded in practice (Elson, 2012; Razavi, 2012).

Despite serious gaps and a primarily instrumental approach to women's rights, the World Bank has at least paid some limited attention to gender equity. A new draft strategy paper covering 2016–2023, for instance, offers a broader approach than WDR 2012, acknowledging the obstacles posed by women's unpaid labour (including caring work), their lack of sexual and reproductive rights, and gender-based violence (*Bretton Woods Observer*, 2016, p. 6). However, although such policy shifts are welcome, much is left out; and most notably the role of the World Bank itself (Elson, 2012).

Extent of titling

Land titling is most likely to occur where land is becoming a scarce resource and where it attracts commercial interest, either from corporations or from the state and/or national elites. However, this is not a new programme. An early land titling project was enacted in Thailand from 1984 (tellingly, for the globalized nature of this phenomenon, with the help of an external organization – the Australian aid agency). Reflecting the exponential nature of land titling, this project saw the number of title deeds issued increase from 4 million in 1984 to 13 million in 1998 (Hall, 2013, p. 120). Land titling has since gathered pace, however. By 1996, such projects were widespread, covering many countries and world regions.

In sub-Saharan Africa, the predominance of customary law means that titling has given rise to much discussion and opposition. The idea that land is an 'asset' rather than a place to dwell, earn a livelihood or cultivate food – not to mention often a site of identity rooted in ancestral or kinship ties – runs counter to many customary systems. Opposition to titling thus prompted, at least for a time, a grassroots reaffirmation that customary systems could be efficient in terms of production (Bruce and Migot-Adholla, 1994). There have also been moves to go further and codify customary law, although this is not the dominant trend.[3] Communal tenure is often seen as both a barrier to investment and a site of traditional values, however, making 'reform' a particularly fraught issue.

Despite the risks, pressures to title land in sub-Saharan Africa have increased in recent years. Not only IFIs but also state development agencies such as the US Agency for International Development (USAID), the G8, the Australian Agency for International Development (AusAID) and the German Agency for Technical Cooperation (GTZ) have promoted titling projects. Stein and Cunningham (2015) further note that, following the 2012 G8 summit, held at Camp David, President Obama's New Alliance for Food Security and Nutrition prompted the G8 member states to launch a number of African 'cooperation frameworks', which also included commitments to ease access for private corporations. Subsequently, and not by coincidence, the G8 focused on strengthening property rights in Africa. Bilateral partnerships were then established in Senegal (with France), in South Sudan and Niger (with the EU), in Namibia (with Germany) and in Tanzania and Nigeria (with the UK's Department for International Development) (Stein and Cunningham, 2015, p. 6).

The impact of titling differs, of course, where land is already privatized. Customary arrangements often exist alongside formal legislation (Sikor and Lund, 2009); yet small farmers are likely to lack the legal documentation necessary to use such laws. Where communal land tenure does not obtain, and ownership is (as is often the case) held in informal arrangements, powerful individuals – who are more able to exploit the legal system – will often play key roles in regulating access (Wily, 2012; Hall, 2013; Wisborg, 2014). Hence, those with greater social power and capital are likely to prevail in land negotiations and disputes.

Pros and cons of land titling for women

A great deal of literature has critiqued the potential impact of titling in general and for women in particular. Three key strands of critique which emerge from this work are now considered.

Different perspectives exist concerning the potential impacts of land titling for women in agrarian households. Land titling might empower women by securing their rights and livelihoods given that most rural women are poor and face serious discrimination within families and communities. Property holding may also enhance women's status within communities (Agarwal, 2003; USAID, 2013) leading to indirect, beneficial impacts on their lives. However, instances where titling has resulted in the security, investment and productivity envisaged by proponents are probably most likely where land is already held privately.

Land titling is intended to encourage privatization and the spread of capitalist relations in agriculture (see Mandel, 1968) and is often successful in doing so. Like other commercializations, this process tends to benefit those who are better-off and therefore able to acquire more land, leaving the poor to face marginalization, accompanied often by loss of their land. Such losses particularly affect women, who are disproportionately poor. Moreover, such effects seem inherent to the process itself; poor landowners are encouraged to use their property as collateral for loans in order to increase productivity, yet if the borrower is unable to repay, the land is lost (Fortin, 2005).

Foreign land acquisitions have gained much attention. This is understandable, given that such acquisitions, or 'grabs', are reminiscent of colonial predations. However, many if not most acquisitions are by *local* actors, often linked with the state. This includes corporations, local elites and government agencies (Park, 2015). Be they local or global, however, those with more wealth and power are most likely to gain control over property of any kind, including land. This effect is exacerbated by the remoteness of many rural communities, plus the high costs of litigation and conflict resolution, both of which increase the advantages held by those with power and/or capital (Askew *et al.*, 2013). In particular, rural women who may need urgently to access litigation are unlikely, in most cases, to be able to do so.

A second critique of the gendered effects of land titling concerns loss of secondary rights. Although in *most* societies women's property rights are often greatly inferior to men's, many women do have some 'secondary' claims on land through marriage or kinship links, or else through community membership. Nevertheless, commercialization and redistributive land reforms have usually been based around an automatic assumption that husbands or fathers are 'household heads' who enjoy sole decision-making rights. Distributing titles or land based on this assumption has further reduced wives' power within households and over agriculture (Deere and León de Leal, 1987; Lastarria-Cornhiel, 1997; Jacobs, 1983, 2010). Titling can entail erasure of their secondary rights, as well as seasonal rights (Chimhowu and Hulme, 2006). Some attempts have been made to offset this effect. In Rwanda, for example, the Succession Law (1999) does presume common property ownership within marriage (Polavarapu, 2011), while the Organic Land Law (2005) ensures that women's interests in land and agricultural decision making be respected. Nevertheless, Rwandan marriage law continues to consider men as household heads, undermining the egalitarian aspects of land law (Polavarapu, 2011).

This problem also occurs in other parts of the world, since the formulation of laws, and their enactment in policy, often reflects male bias (Elson, 1995). In Madagascar, for instance, civil servants, along with many rural people, assume a model of the male-headed 'unitary household' (Widman, 2015). A related phenomenon concerns contradictions between codified marriage laws and land laws. In Madagascar, marriage law assumes a partial common property regime. However, at present customary law in marriage contradicts land law. Widman (2015) thus argues that mandatory joint titling for couples would harmonize these two sets of laws.

A third stream of critique questions the ability of titling to ensure security. Some of the problematic assumptions of land titling are clearly political rather than to do with the workings of financialized capitalism per se. Most significantly, the beneficial outcomes of titling are predicated upon the existence of a state which is able to enforce laws and to make fair adjudications. However, the absence of such state capacity is one of the problems that land titling is meant to solve (Hall, 2013; Ossome, 2014).

Recent Research on Gender and Land Titling

The above critiques point to useful directions for analysis. However, the exact impacts and results of policy changes require detailed empirical investigation. Hence, the chapter now turns to a discussion of several recent case studies.

Interest in, and research on, gender and land has greatly increased in recent years, with much of this material being sparked off by debates on land titling and its effects. An indication of the volume of recent research is now given. Google Scholar returns approximately 150 papers, articles, chapters and reports on 'gender AND land'[4] published in 2014 alone. This represents an exponential increase over previous years. Part of the explanation for this upsurge is that research in this area has won some notable support from major global institutions, particularly the World Bank, the FAO and UN Women, as well as from aid agencies such as USAID, Norad and Danida. Research institutions, such as the International Food Policy Research Institute (IFPRI) and the International Center for Research on Women (ICRW), plus NGOs such as Action Aid, Oxfam and Mokoro, have also contributed to the debate, often offering critical views. The global alliance, the International Land Coalition (ILC), which is notable for its ability to combine activism and research, has taken up gender issues, including holding difficult discussions over gender relations within family farming.

However, despite this increased attention, detailed empirical studies of the impacts of land titling are fairly recent and are few. The next subsection provides an outline of such studies, particularly those which focus on gender. Although not a comprehensive review, this provides a flavour of the key trends in current research. Some studies on this topic do not explore gendered relational processes but do address linked issues. For instance, Newman *et al.* (2015) discuss effects of joint titling on productivity in Vietnam; the study finds that there exist no detrimental effects that could counteract an 'equity' argument.

Customary law and 'traditional' beliefs

Many studies have emphasized the difficulty of obtaining women's land rights, particularly where customary law predominates. For instance, in post-genocide Rwanda, civil law has been used to attempt vigorously to promote equal access to land. Yet the persistence of customary law remains an impediment to this objective: since most disputes are brought to community courts, it is *customary* law that applies (Daley *et al.*, 2010). Similarly in Vanuatu, although women enjoy property rights under the constitution, nevertheless customary law is primarily recognized as the basis for land ownership (Nagarajan and Parashar, 2013). This places the effective control of land with fathers and husbands. Less often, customary law may form the basis of more egalitarian gender relations and decision making, as found in Wiig's (2013) study of highland Peru.

Discrimination against women within customary land and marriage law is often accompanied by social attitudes and beliefs that curtail their access to land and ability to farm. In Limpopo, South Africa, for example, despite women's widespread participation in farming, Murugani *et al.* (2014) find strong negative perceptions of women's agricultural capabilities. Furthermore, a number of studies report that wives who assert land rights are seen as disloyal, thereby threatening their marriage (see Tripp, 2004; Widman, 2015). Murugani *et al.* (2014) similarly find women emphasizing the need to maintain 'harmonious' marital relationships in order to retain access to land (see also Thorley, 2015, on the Acholi of northern Uganda).

That the impact of customary attitudes and beliefs can be considerably stronger than measures introduced to equalize land access is illustrated by the case of Orissa, India. Doss *et al.* (2014c) found that very strong normative prohibitions against any public displays of female authority meant that formal land rights in Orissa had had little impact on women's ability to take decisions concerning farming. Thus, customary and traditional law, practice and norms may form a powerful barrier, even where more egalitarian laws and norms have been proposed or introduced.

Women hold less land

While women usually hold much less land than do men, this is not only the case where married women are customarily forbidden from holding land on their own account, but also where such strictures do not apply. In Madagascar, according to the country's Land Certification database, 21% of plots are held by women and 0.5% are held jointly by couples (Widman, 2015, p. 118; see also Widman, 2014). Less than 3% of plots acquired by couples were certified as 'joint' holdings. In Vietnam, despite a strongly patrilineal culture, national laws since 2008 have attempted to provide for more equal gendered access to land. A large-scale Vietnamese study of two sites, in the north and the south, indicated that most women still suffered disadvantage (albeit with geographical differences in this effect) (Alvarado *et al.*, 2015). In Vietnam, land continues to be held by the state, with cultivators holding Land Use Rights Certificates (LURCs). In the southern province of Long An, only 1% of plots were registered in both names of couples, whereas over 70% of plots were held by men and 25% of plots were registered in women's names. In contrast, in the northern site, Hung Yen, 35% of plots were registered as belonging to couples. Men held 53% of the remainder while women (alone or with other women) held 9–10% of plots (Alvarado *et al.*, 2015, p. 10). The study found, additionally, that the plots held by men were larger than those held by women or couples. Thus, rural women seeking rights usually begin from a disadvantaged base in terms of landholding.

Law and policy

A number of practical and legal obstacles may inhibit women's ability to make land claims, including lack of knowledge, male-biased laws, contradictions between sectors of law and lack of implementation. Rural women, and particularly those who live in remote communities, may be unware of changes in legislation. This can be the case even in a society with strong official reach, such as Vietnam. Alvarado *et al.* (2015) thus found that fewer Vietnamese women than men knew how to access LURCs. In their sample of some 841 households, there was general understanding that women could inherit houses and land, and that land could not be alienated without a wife's consent. However, few people knew that widows should not lose land if they remarried, that husbands and wives were able to hold their own property or LURCs within marriage, or that childless women had the right to land and houses.

A further problem is that egalitarian laws may simply not be implemented. In Vietnam, despite land laws enacted in 2003 and 2013, it remains commonplace for only the husband's name to appear on an LURC. Moreover, women almost never acquire LURCs or shares of joint property upon divorce (Alvarado *et al.*, 2015, p. 12).

Gaining land

A third strand of discussion concerns how women gain rights to land where these are available. In at least some contexts, married women may gain partial security within patrilineal customary systems, so long as their marriages are successful according to traditional expectations. In Murugani *et al.*'s (2014) Limpopo study, wives in marriages of longer duration and with a number of children felt more secure in terms of access to land (see also Jacobs and Kes, 2015; Thorley, 2015). In many societies, however, maintaining 'harmonious' marriage relations often entails wives behaving compliantly and avoiding appearing to be demanding. Such 'security' is clearly precarious.

In South Africa, a number of women have avoided the dilemmas of customary law within marriage by 'refusing' to wed, or at least considering this option. In the Limpopo study, 36% of women in one site and 12% in another felt that marriage weakened their claims to land and that they would have stronger rights remaining on their father's property (Murugani *et al.*, 2014, p. 215). This echoes Claassens' (2013) research on the declining popularity of marriage among rural women in South Africa, in which marriage was reported by many participants to offer numerous risks and few advantages.

Examining the institutional arena rather than individual agency, the most usual way for women to gain, or to be granted, land rights is 'from above', through mandatory legal or other changes enacted by a state. In China, for instance, land rights were granted to peasant women in 1950 following the 1949 Communist Revolution (Frenier, 1983). Yet, elsewhere, such rights have remained remarkably uncommon until recently. (An exception is Peru, where policies mandated by the authoritarian state have provided women with much stronger claims to land (Wiig, 2013).)

In the absence of state implementation, in some instances NGOs have attempted to link joint land or property titling to their funding grants (Widman, 2015), albeit with variable success. In other cases, there have been efforts to refer to international human rights instruments, particularly the Convention on the Elimination of all Forms of Discrimination against Women (CEDAW). This was attempted in Vanuatu by rights activists. However, as Nagarajan and Parashar (2013, p. 101) note, 'most women's lives are lived in spaces that the CEDAW does not reach'. The existence of women's movements can also be of importance. Women in Budlender's (2013) South African study thus felt that the reasons that single, divorced and widowed women were more able to access and control land in recent years was due to increasing 'democracy' and 'women's rights'. Conversely, absence of progress in ensuring women's land rights is sometimes linked to lack of feminist movements (see Widman, 2015).

Individual versus joint titling: three studies

Two studies compare the impacts of joint versus individual titling. Doss *et al.* (2014c) compare the impacts of joint versus individual land titling in three different African countries – Mali, Malawi and Tanzania – finding that both joint and individual titling within couples had positive impacts in each case. However, the effects for individual ownership were strongest, especially concerning wives' ability to make agricultural decisions.[5] Individual landownership also gave women more influence over food consumption, cash-crop farming and decisions regarding agricultural inputs and marketing, although somewhat less so over non-agricultural decisions (Doss *et al.*, 2014c, p. 13). Jacobs' and Kes's (2015) study of couples in central Uganda and Kwa-Zulu Natal province, South Africa, produced more ambiguous data. Spouses often disagreed over the meaning of 'joint titling', with married women more likely to assert joint ownership than were their husbands. However, joint ownership was weakly associated[6] with coupled women being able to make transactions over land and housing, while the existence of documentation which included a woman's name meant that husbands were more likely to recognize their wives' claims to land.

A third study, conducted by Wiig (2013) in highland Peru, was able to report a clearer and more positive impact of joint titling. Peru is unusual both because joint titling has been strictly enforced from above for two decades, and also because gender norms among the indigenous Quechua are relatively egalitarian. Wiig (2013) compared 'old' redistributive land reform dating from the 1960s with more recent titling reforms within the same area. The survey of some 1280 households indicated a statistically significant 'empowerment' effect of joint titling for wives; again, particularly regarding agricultural decision making.

Conclusion

Land remains a crucial subsistence resource for many. Several points emerge from the recent literature that provides gender-based contextualization to this need. First, where customary land law is adhered to, its influence is powerful, often due to patrilineal or otherwise male-dominated contexts. However, customary law does afford some relative security for women, particularly with regard to rights of access and food cultivation. Yet such customs are often breached (Murugani *et al.*, 2014), particularly outside marriage, but also within it, at times making it difficult for women to activate their traditional rights. Another theme arising from the literature is that laws governing marriage and land can be contradictory with regard to women's rights. Where marriage or land law accords legal dominance to a husband or father, it generally remains the case that this will prevail. Moreover, women's assertion of land rights within wedlock is often seen as a sign of 'disloyalty' or lack of commitment to the marriage.

Recent studies indicate that women within couples do receive *some* benefit from joint titling, for example improving household bargaining power. Individual titling or ownership gave coupled women clearer rights; although accessing such rights risked 'payback' in terms of how women were viewed by the family and/or community. More advantaged women, who faced fewer obstacles, or who were more able to weather community disapproval, were more likely to be able to assert ownership or titling claims. Most women still remain at the 'bottom' in terms of income and assets, even within wealthier countries. Aggravating this problem, enforcement of gender-beneficial law entails recourse to formal courts. Only an elite of women are likely to be able to access such litigation.

Nevertheless, there remains a preference for individual titling by many rural people (Hall, 2013), including women (Agarwal, 2003; Wanyeki, 2003; Jacobs, 2004; Widman, 2015). This may indicate a general distrust among women of the protections supposedly entailed in customary land tenure.

Much of the current research notes that both joint and individual titling has some positive impact on women's lives – for example, improving household bargaining power. However, joint land titling is a recent development. The studies cited are not longitudinal, hence it is not currently possible to assess whether future market processes will lead to women losing land gained through titling. Is land titling a vehicle that can secure rights for women? Or alternatively, does it serve as a 'trojan horse' for further privatization (Monsalve Suárez, 2006), so that a vehicle for democratization of land rights may instead result in loss of the land itself through foreclosure and land sales?

How rural women's land rights fare within the context of the major contemporary trend of increasing state and corporate land acquisition[7] has been the subject of some recent studies (see e.g. Daley, 2011; Behrman *et al.*, 2012; Wisborg, 2014; Park, 2015). Although the evidence remains sparse, there is at least some worrying indication that, rather than raising women's rights to the level of men's, rural women and rural men are both in danger of losing their existing and secondary rights.

Despite some positive developments, the pressure to introduce land titling for women can often be traced back to the spreading influence of IFIs. These institutions, in pursuing neoliberal agendas, often undermine or destroy rural people's livelihoods (e.g. Bryceson, 2002; Gwynne and Kay, 2004; Li, 2010; Jacobs, 2014a).

It is thus likely to be *states* which hold the key to improving women's land rights. The most effective and positive improvements in women's rights have previously occurred only through 'top-down' government action. Neoliberal policies undermine states, particularly in functions supporting health, education, labour rights and social equality. And male-dominated local communities have rarely been quick to reform women's rights, either in terms of land or more generally. State bodies, with the power to enforce transformation, may be able to mediate between the opposing forces of patriarchal forms of customary law and untrammelled market forces, neither of which is likely to be positive for land and decision-making rights. The role of women's rights organizations in this transformation will remain key, as governments will likely remain pressurized by both external financial action and internal cultural reaction.

As the matter stands, significant obstacles remain. Much testimony indicates how difficult it is to put women's land rights into practice, even once gained on paper. For instance, even the relatively straightforward change of including a space for two people's signatures on contracts, thereby permitting easier joint titling, has proved notably intractable (Deere, 2003; Doss *et al.*, 2014a; Widman, 2015). There exists no simple solution to this, given the dual challenge of external acquisitions that threaten to expropriate land and the 'internal' problem of the 'family–land–sexuality' nexus (Jacobs, 2014b). One response could be to make women's rights much more visible. For instance, Wiig (2013) notes that, in Peru, legitimacy (and potential cultural change) were gained by the public witnessing of the signing of land deeds. Campaigns against gendered violence (such as that of La Vía Campesina), whether within organizations or on peasant/smallholder farms (Jacobs, 2015; Park *et al.*, 2015), can assist women to gain a measure of security and autonomy.

In sum, the research reviewed in this chapter indicates that there have been some positive recent achievements for women's land rights, not least that they are being highlighted in contemporary discussion and debates. In this complex terrain it is important to build on such achievements, including changes in community attitudes, and, perhaps most significantly, to resist the dismantling of state and other public institutions. These are the very bodies that might help to encourage, or where needed *enforce*, new rights for women 'on the ground'.

Acknowledgement

Many thanks are extended to Dr John David Jordan for assistance with and suggestions for editing.

Notes

[1]Matrilineal systems also exist, particularly in the central 'belt' of Africa, and also in parts of South and South-East Asia, and these do afford greater rights for women, as they hold land through lineage membership. However, the number of matrilineal societies has for some time been in decline.

[2]Land 'grabs' or land transactions are equally important international trends and have given rise to a large literature. See, for instance, Borras and Franco (2012) and Archambault and Zoomers (2015b). Space does not allow discussion of land grabs and their gendered impacts here.

[3]Thus, the World Bank partially 'backtracked' over the insistence that all land be titled or privatized. There have been related moves to codify customary law – such as in Mozambique – and this has sometimes provided community protection (Hanlon, 2004; Sikor and Lund, 2009; Archambault and Zoomers, 2015b). However, this has not been the main direction of change.

[4]The search used 'gender OR women and land' as keywords. Some caveats and explanation are in order. The data accessed are by no means comprehensive, but do give some indication of publications that are relevant to gender, women and questions of agriculture and land tenure. However, many 'results' are irrelevant (for instance, 'coronary disease among *women* in New Zea*land*'), while a larger number were only of marginal relevance. Examining notices over a 2-year period (June 2013–May 2015), lists of publications were sent on average nine or ten times monthly, with approximately eight or nine publications in each. Of these, usually two or three items contained some relevant information or data.

[5]Doss *et al.* (2014c) caution that increased female decision making does not equate with wives having the 'final say'; in most instances this is a male prerogative.

[6]Jacobs and Kes (2015) note that, in their study, clear and consistent advantages for women could not be demonstrated statistically.

[7]Both men and women are likely to lose land through processes of marketization and/or acquisition. However, women are likely to lose land disproportionately as they start out as weaker social actors, without the 'assets' demanded by the market.

References

Agarwal, B. (1994) *A Field of One's Own*. Cambridge University Press, Cambridge, UK.

Agarwal, B. (2003) Gender and land rights revisited: exploring new prospects via the state, family and market. *Journal of Agrarian Change* 3(1–2), 184–224. Available at: http://dx.doi.org/10.1111/1471-0366.00054 (accessed 26 February 2017).

Alvarado, G., Kuat, T.H., Zayid, D., Nguyen, T.V.A., Nguyen, T.P.T., Mukasa, S., Trinh, T.N.D. and Vu, X.T. (2015) *Women, Land and Law in Vietnam*. International Center for Research on Women, Washington, DC. Available at: http://www.icrw.org/wp-content/uploads/2016/10/WomenLandLaw.pdf (accessed 26 February 2017).

Archambault, C. and Zoomers, A. (2015a) Introduction. In: Archambault, C. and Zoomers, A. (eds) *Global Trends in Land Tenure Reform: Gender Impacts*. Routledge, London and New York, pp. 1–12.

Archambault, C. and Zoomers, A. (2015b) From *de facto* to *de jure*. In: Archambault, C. and Zoomers, A. (eds) *Global Trends in Land Tenure Reform: Gender Impacts*. Routledge, London and New York, pp. 71–73.

Askew, K., Maganga, F. and Odgaard, R. (2013) Of land and legitimacy: a tale of two lawsuits. *Africa* 83(1), 120–141. Available at: http://dx.doi.org/10.1017/S0001972012000745 (accessed 26 February 2017).

Bedford, K. (2009) *Developing Partnerships: Gender, Sexuality and the Reformed World Bank*. University of Minnesota Press, Minneapolis, Minnesota.

Behrman, J., Meinzen-Dick, R. and Quisumbing, A. (2012) The gender implications of large-scale land deals. *The Journal of Peasant Studies* 39(1), 49–79. Available at: http://dx.doi.org/10.1080/03066150.2011.652621 (accessed 26 February 2017).

Borras, S.M. Jr and Franco, J.C. (2012) Global land grabbing and trajectories of agrarian change: a preliminary analysis. *Journal of Agrarian Change* 12(1), 34–59. Available at: http://dx.doi.org/10.1111/j.1471-0366.2011.00339.x (accessed 26 February 2017).

Brandth, B. (2002) Gender identity in European family farming: a literature review. *Sociologia Ruralis* 42(3), 181–200. Available at: http://dx.doi.org/10.1111/1467-9523.00210 (accessed 26 February 2017).

Bretton Woods Observer (2016) World Bank Releases New Gender Strategy. *Bretton Woods Observer* (Winter), 6. Available at: http://www.brettonwoodsproject.org/2016/02/world-bank-releases-gender-strategy/ (accessed 26 February 2017).

Bruce, J. and Migot-Adholla, S.E. (eds) (1994) *Searching for Land Tenure Security in Africa.* Kendall Hunt Publishers, Dubuque, Iowa.

Bryceson, D.F. (2002) The scramble in Africa: reorienting rural livelihoods. *World Development* 30(5), 725–739. Available at: http://dx.doi.org/10.1016/S0305-750X(02)00006-2 (accessed 26 February 2017).

Budlender, D. (2013) Women, marriage and land: findings from a three-site survey. *Acta Juridica* 28(1), 28–48.

Budlender, D. and Alma, E. (2011) *Women and Land: Securing Rights for Better Lives.* International Development Research Centre, Ottawa. Available at: https://www.idrc.ca/en/book/infocus-women-and-land-securing-rights-better-lives/10625/47431/1/IDL-47431.pdf (accessed 7 July 2017).

Chant, S. and Sweetman, C. (2012) Fixing women or fixing the world? 'Smart economics', efficiency approaches and gender equality in development. *Gender & Development* 20(3), 517–529. Available at: http://dx.doi.org/10.1080/13552074.2012.731812 (accessed 26 February 2017).

Chimhowu, A. and Hulme, D. (2006) Livelihood dynamics in planned and spontaneous resettlement in Zimbabwe: converging and vulnerable. *World Development* 34(4), 728–750. http://dx.doi.org/10.1016/j.worlddev.2005.08.011 (accessed 26 February 2017).

Claassens, A. (2013) Recent changes in women's land rights and contested customary law in South Africa. *Journal of Agrarian Change* 13(1), 71–92. Available at: http://dx.doi.org/10.1111/joac.12007 (accessed 26 February 2017).

Daley, E. (2011) *Gendered Impacts of Commercial Pressures on Land.* International Land Coalition, Rome.

Daley, E., Dore-Weeks, R. and Umuhoza, C. (2010) Ahead of the game: land tenure reform in Africa and the process of securing women's rights. *Journal of East African Studies* 4(1), 131–153. Available at: http://dx.doi.org/10.1080/17531050903556691 (accessed 26 February 2017).

Deere, C.D. (2003) Women's land rights and social movements in the Brazilian agrarian reform. *Journal of Agrarian Change* 3(1–2), 257–288. Available at: http://dx.doi.org/10.1111/1471-0366.00056 (accessed 26 February 2017).

Deere, C.D. and León de Leal, M. (1987) *Rural Women and State Policy: Feminist Perspectives on Latin American Agricultural Development.* Westview Press, Boulder, Colorado.

Deininger, K. (2003) *Land Policies for Growth and Poverty Reduction: World Bank Land Policy Research Report.* World Bank, Washington, DC.

De Soto, H. (2001) *The Mystery of Capital: Why Capitalism Triumphs in the West and Fails Everywhere Else.* Black Swan, London.

Doss, C., Summerfield, G. and Tsikata, D. (2014a) Land, gender and food security. *Feminist Economics* 20(1), 1–23. Available at: http://dx.doi.org/10.1080/13545701.2014.895021 (accessed 26 February 2017).

Doss, C., Meinzin-Dick, R. and Bomuhangi, A. (2014b) Who owns the land? Perspectives from rural Ugandans and implications for large-scale land acquisitions. *Feminist Economics* 20(1), 76–100. Available at: http://dx.doi.org/10.1080/13545701.2013.855320 (accessed 26 February 2017).

Doss, C., Kim, S.M., Njuki, J., Hillenbrand, E. and Miruka, M. (2014c) *Women's Individual and Joint Property Ownership: Effects on Household Decision-making. Discussion Paper 01347.* International Food Policy Research Institute/CGIAR, Washington, DC. Available at: https://landportal.info/library/resources/landwiserecord2459item2492/womens-individual-and-joint-property-ownership-effects (accessed 7 July 2017).

Elson, D. (ed.) (1995) *Male Bias in the Development Process*, 2nd edn. Manchester University Press, Manchester, UK.

Elson, D. (2012) Review of *World Development Report 2012*: gender equality and development. *Global Social Policy* 12(2), 178–183. Available at: http://dx.doi.org/10.1177/1468018112443673b (accessed 26 February 2017).

FAO (2011) *The State of Food and Agriculture 2010–11. Women in Agriculture: Closing the Gender Gap for Development.* Food and Agriculture Organization of the United Nations, Rome. Available at: http://www.fao.org/docrep/013/i2050e/i2050e.pdf (accessed 26 February 2017).

Fortin, E. (2005) Reforming land rights: the World Bank and the globalization of agriculture. *Social and Legal Studies* 40(2), 147–177. Available at: http://dx.doi.org/10.1177/0964663905051217 (accessed 26 February 2017).

Frenier, M.D. (1983) The effects of the Chinese Communist land reform on women and their families. *Women's Studies International Forum* 6(1), 41–55.

Gill, S. and Cutler, C. (2014) *New Constitutionalism and World Order?* Cambridge University Press, Cambridge, UK.

Gwynne, R. and Kay, C. (2004) *Latin America Transformed: Globalization and Modernity*, 2nd edn. Routledge, London.

Hall, D. (2013) *Land*. Routledge, London and New York.

Hanlon, J. (2004) Renewed land debate and the 'cargo cult' in Mozambique. *Journal of Southern African Studies* 30(3), 603–626. Available at: http://dx.doi.org/10.1080/0305707042000254128 (accessed 26 February 2017).

Harvey, D. (2011) *The Enigma of Capital*. Profile Publishers, London.

Jacobs, K. and Kes, A. (2015) The ambiguity of joint asset ownership: cautionary tales from Uganda and South Africa. *Feminist Economics* 21(3), 23–55. Available at: http://dx.doi.org/10.1080/13545701.2014.926559 (accessed 26 February 2017).

Jacobs, S. (1983) Women and land resettlement in Zimbabwe. *Review of African Political Economy* 10(27–28), 33–50. Available at: http://dx.doi.org/10.1080/03056248308703545 (accessed 26 February 2017).

Jacobs, S. (2004) Livelihoods, security and needs: gender and land reform in South Africa. *Journal of International Women's Studies* 6(1), 1–19. Available at: http://vc.bridgew.edu/cgi/viewcontent.cgi?article=1478&context=jiws (accessed 26 February 2017).

Jacobs, S. (2010) *Gender and Agrarian Reforms. Routledge International Studies of Women and Place*. Routledge, London and New York.

Jacobs, S. (2013) Agrarian reforms. *Current Sociology* 61(5–6), 862–885. Available at: http://dx.doi.org/10.1177/0011392113486660 (accessed 9 March 2017).

Jacobs, S. (2014a) Women's land rights: tenure, organisational issues, the local and the global. Special Issue: 'Materialist Feminisms against Neoliberalism'. *Politics and Culture*, 10 March. Available at: https://politicsandculture.org/2014/03/10/womens-land-rights-tenure-organisational-issues-the-local-and-the-global-by-susie-jacobs/ (accessed 26 February 2017).

Jacobs, S. (2014b) Gender, land and sexuality: exploring connections. In: Jacobs, S. and Klesse, C. (eds) Special Issue: 'Gender, Sexuality and Political Economy'. *International Journal of Politics, Culture and Society* 27(2), 173–190. Available at: http://dx.doi.org/10.1007/s10767-013-9156-5 (accessed 26 February 2017).

Jacobs, S. (2015) Gender, food sovereignty and agrarian reforms. In: Trauger, A. (ed.) *Food Sovereignty in International Context*. Routledge/Earthscan, Abingdon, UK and New York, pp. 176–197.

Keck, M. and Sikkinck, K. (1998) *Activists beyond Borders*. Cornell University Press, Ithaca, New York.

Kingwill, R. (2014) Papering over the cracks: an ethnography of land title in the Eastern Cape. *Kronos* 40(1), 241–268.

Lastarria-Cornhiel, S. (1997) The impact of privatization on gender and property rights in Africa. *World Development* 25(8), 1317–1333. Available at: http://dx.doi.org/10.1016/S0305-750X(97)00030-2 (accessed 26 February 2017).

Li, T.M. (2010) To make live or let die? Rural dispossession and the protection of surplus populations. *Antipode* 41(1), 66–93. Available at: http://dx.doi.org/10.1111/j.1467-8330.2009.00717.x (accessed 26 February 2017).

Mahon, R. (2012) Introduction: the World Bank's new approach to gender equality? *Global Social Policy* 12(2), 173–174. Available at: http://dx.doi.org/10.1177/1468018112443673 (accessed 26 February 2017).

Mandel, E. (1968) *Marxist Economic Theory*. Merlin, London.

Manji, A. (2006) *The Politics of Land Reform in Africa: From Communal Tenure to Free Markets*. Zed Books, London.

Monsalve Suárez, S. (2006) Gender and land. In: Rosset, P., Patel, R. and Courville, M. (eds) *Promised Land: Competing Visions of Agrarian Reform*. Food First, Oakland, California, pp. 192–207.

Mudege, N. (2008) Gender roles in agricultural knowledge in a land resettlement context: the case of Mupfurudzi, Zimbabwe. *Development Southern Africa* 25(4), 455–468. Available at: http://dx.doi.org/10.1080/03768350802318597 (accessed 26 February 2017).

Murugani, V., Thamaga-Chitja, J., Koanisi, U. and Shimelis, H. (2014) The role of property rights in rural women's land use security and household food security for improved livelihood in Limpopo Province. *Journal of Human Ecology* 46(2), 205–221.

Nagarajan, V. and Parashar, A. (2013) Space and law, gender and land: using CEDAW to regulate for women's rights to land in Vanuatu. *Law and Critique* 24(1), 87–105. Available at: http://dx.doi.org/10.1007/s10978-012-9116-7 (accessed 26 February 2017).

Newman, C., Tarp, F. and van den Broeck, K. (2015) Property rights and productivity: the case of joint land titling in Vietnam. *Land Economics* 91(1), 91–105. Available at: http://dx.doi.org/10.3368/le.91.1.91 (accessed 26 February 2017).

Ossome, L. (2014) Can the law secure women's rights to land in Africa? Revisiting tensions between culture and land commercialization. *Feminist Economics* 20(1), 155–177. Available at: http://dx.doi.org/10.1080/13545701.2013.876506 (accessed 26 February 2017).

Park, C.M.Y. (2015) New 'webs of power' and agrarian transformations in Cambodia: where are the women? BICAS (BRICS Initiative in Critical Agrarian Studies) Working Paper 8. Available at: https://www.tni.org/files/download/bicas_working_paper_8_park.pdf (accessed 10 June 2015).

Park, C.M.Y., White, B. and Julia (2015) We are not all the same: taking gender seriously in food sovereignty discourse. *Third World Quarterly* 36(3), 584–599. Available at: http://dx.doi.org/10.1080/01436597.2015.1002988 (accessed 12 March 2017).

Polavarapu, A. (2011) Procuring meaningful land rights for women of Rwanda. *Yale Human Rights and Development Journal* 14(1), 105–154. Available at: http://digitalcommons.law.yale.edu/cgi/viewcontent.cgi?article=1100&context=yhrdlj (accessed 10 July 2015).

Razavi, S. (2012) WDR 2012: too little, too late and already a little out of date? *Global Social Policy* 12(2), 193–197. Available at: http://dx.doi.org/10.1177/1468018112443673 (accessed 26 February 2017).

Sikor, T. and Lund, C. (2009) Access and property: a question of power and authority. *Development and Change* 40(1), 1–22. Available at: http://dx.doi.org/10.1111/j.1467-7660.2009.01503.x (accessed 26 February 2017).

Spoor, M. (2012) Agrarian reform and transition: what can we learn from the East? *Journal of Peasant Studies* 39(1), 175–194. Available at: http://dx.doi.org/10.1080/03066150.2011.652949 (accessed 26 February 2017).

Stein, H. and Cunningham, S. (2015) *Land Grabbing and Formalization in Africa: A Critical Enquiry. ASC Working Paper 124.* African Studies Centre, Leiden, The Netherlands. Available at: https://openaccess.leidenuniv.nl/bitstream/handle/1887/37024/ASC-075287668-3722-01.pdf?sequence=2 (accessed 1 February 2016).

Thorley, L. (2015) Holding on: gender relations, food security and women's options and strategies for maintaining access to land in the Acholi region of northern Uganda. PhD thesis, University of Bradford, Bradford, UK.

Tripp, A.M. (2004) Women's movements, customary law and land rights in Africa; the case of Uganda. *Africa Studies Quarterly* 7(4), 1–19. Available at: http://asq.africa.ufl.edu/files/Tripp-Vol-7-Issue-4.pdf (accessed 5 July 2015).

USAID (2013) *Land Tenure, Property Rights and Gender. USAID Issue Brief.* US Agency for International Development, Washington, DC. Available at: https://www.land-links.org/wp-content/uploads/2016/09/USAID_Land_Tenure_Gender_Brief_06124-1.pdf (accessed 7 July 2017).

Varley, A. (2010) Modest expectations: gender and property rights in urban Mexico. *Law & Society Review* 44(1), 67–100. Available at: http://dx.doi.org/10.1111/j.1540-5893.2010.00396.x (accessed 26 February 2017).

Wanyeki, M. (ed.) (2003) *Women and Land in Africa.* Zed Books, London.

Widman, M. (2014) Land tenure insecurity and formalising land rights in Madagascar: a gender perspective on the certification program. *Feminist Economics* 20(1), 130–154. Available at: http://dx.doi.org/10.1080/13545701.2013.873136 (accessed 26 February 2017).

Widman, M. (2015) Joint land certificates in Madagascar: the gendered outcome of a 'gender-neutral' policy. In: Archambault, C. and Zoomers, A. (eds) *Global Trends in Land Tenure Reform: Gender Impacts.* Routledge, London and New York, pp. 110–126.

Wiig, H. (2013) Joint titling in rural Peru: impact on women's participation in household decision-making. *World Development* 52, 104–119. Available at: http://dx.doi.org/10.1016/j.worlddev.2013.06.005 (accessed 26 February 2017).

Wily, L.A. (2012) Looking back to see forward: the legal niceties of land theft in land rushes. *Journal of Peasant Studies* 39(3–4), 751–775. Available at: http://dx.doi.org/10.1080/03066150.2012.674033 (accessed 26 February 2017).

Wisborg, P. (2014) Transnational land deals and gender equality: utilitarian and human rights approaches. *Feminist Economics* 20(1), 24–51. Available at: http://dx.doi.org/10.1080/13545701.2013.862341 (accessed 26 February 2017).

World Bank (2000) *World Development Report 2000/2001. Attacking Poverty.* World Bank, Washington, DC. Available at: https://openknowledge.worldbank.org/handle/10986/11856 (accessed 7 July 2017).

World Bank (2006a) *World Development Report 2006. Equity and Development.* World Bank, Washington, DC. Available at: http://www-wds.worldbank.org/servlet/WDSContentServer/WDSP/IB/2005/09/20/000112742_20050920110826/Rendered/PDF/322040World0Development0Report02006.pdf (accessed 26 February 2017).

World Bank (2006b) *Gender Equality as Smart Economics: A World Bank Group Gender Action Plan (Fiscal years 2007–10).* World Bank, Washington, DC. Available at: http://siteresources.worldbank.org/INTGENDER/Resources/GAPNov2.pdf (accessed 26 February 2017).

World Bank (2008) *World Development Report 2008. Agriculture for Development.* World Bank, Washington, DC. Available at: https://siteresources.worldbank.org/INTWDR2008/Resources/WDR_00_book.pdf (accessed 26 February 2017).

World Bank (2009) *Awakening Africa's Sleeping Giant: Prospects for Commercial Agriculture in the Guinea Savannah Zone and Beyond.* World Bank, Washington, DC. Available at: http://siteresources.worldbank.org/INTARD/Resources/sleeping_giant.pdf (accessed 1 August 2015).

World Bank (2012) *World Development Report 2012: Gender Equality and Development.* World Bank, Washington, DC. Available at: https://siteresources.worldbank.org/INTWDR2012/Resources/7778105-1299699968583/7786210-1315936222006/Complete-Report.pdf (accessed 26 February 2017).

World Bank, FAO and IFAD (2009) *Gender and Agriculture Sourcebook.* World Bank, Washington, DC. Available at: http://siteresources.worldbank.org/INTGENAGRLIVSOUBOOK/Resources/CompleteBook.pdf (accessed 26 February 2017).

24 Conclusions – Future Directions

S. Shortall[1,*] and B.B. Bock[2,3]

[1]Newcastle University, Newcastle upon Tyne, UK; [2]Wageningen University, Wageningen, The Netherlands; [3]Groningen University, Groningen, The Netherlands

Introduction

When we were asked to do an updated version of *Rural Gender Relations*, we thought it an important task but were more excited by the idea of doing a new book. This book is new, in that all of its chapters are original rather than updates of the 2006 chapters, although some of the same contributors have participated in both volumes. Doing a new book has allowed us to extend the global reach of the contributors as we were particularly keen to include the global South and developing countries. The way in which we organized this book is the same as the previous one. It was very useful to have a number of thematic sections. As before, each of the four thematic components of this volume – gender and mobility, gender and agriculture, rurality and gender identity, and gender and international development – addresses topical debates within rural studies and the social sciences more generally. All of the components begin with a theoretical overview that also provides a summary of the chapters, and to do this again in this concluding chapter would be repetitious. Instead, we have decided in this short conclusion to reflect on what we view as the key issues for rural gender studies and some of the issues that need to be addressed going forward. We focus particularly on two issues that we believe are key issues for future consideration: greater reflexivity about gender research in rural studies and how gender is represented in professional rural academia; and comparative research that allows us to grasp the increasing global interconnectedness of gender relations across rural and urban places and across and within the global North and South that results from the globalization of society and social relations in general.

Positioning Gender in Rural Academia

Ten years ago we said, and it remains the case, that rural gender relations research is a well established specialist field within, among other disciplines, rural sociology, rural geography and rural economics. It has flourished over the last decade and has developed further in both theoretical and empirical sophistication. We quoted Whatmore, who warned that the study of farm women was:

> in danger of becoming isolated as a specialist field of research interest in a way which obscures the broader significance of gender and the analysis of farm women to key debates in rural sociology about the

* Corresponding author e-mail: sally.shortall@newcastle.ac.uk

> survival of family farming in advanced capitalist societies and the nature of the farm labour process.
>
> (Whatmore, 1988, p. 239)

Twenty-seven years later, we still have this concern. Rural gender studies remain an insulated research area within rural studies, and even more so within mainstream sociology, geography, economics, politics, and even mainstream women/gender studies. We have in the past commented on the insular nature of the gender streams at rural conferences. While scholars doing research on gender and governance, power relations, agriculture, identity, social exclusion and so on were keen to go to the streams on the more general question, there were few from the sessions on governance, agriculture or social exclusion coming to hear the gender perspective on these more general questions. In many respects, we return to the age-old feminist question of whether exclusionary or inclusive strategies are most appropriate and effective. If we wish to demonstrate the broader significance of gender to rural studies, is it best achieved through an established sub-discipline of rural gender studies or through integrating aspects of rural gender studies with rural research questions? Our concern is that rural gender relations and those who do research in this sub-discipline are seen as 'doing gender', which is interpreted as being relevant only to research questions on gender and not to the wider programme of rural research. This problem though is not confined to rural studies, which in itself is a sub-area of a number of disciplines. It is our view that not only does research on rural gender relations need to be well informed about general debates in the social sciences, but the study of rural gender relations can also lend theoretical and empirical insights to more general debates.

Previously we discussed the more general integration of gender as a factor in research programmes. This is currently encouraged by EU research programmes that oblige research teams to consider gender issues. However this is often addressed in a nominal way, by increasing the number of female researchers involved in the research team, rather than a comprehensive integration of gender analysis into the research question. The Research Council of Norway is exemplary in this regard. Regardless of the quality of the science, if gender issues are not adequately addressed, the proposal is off the table. There is an understanding of the importance of gender perspectives and a commitment to ensure they are embedded in the research it funds.

We have also reflected more recently on gender within rural academia. We are currently the President of the European Society for Rural Sociology (ESRS) (Shortall) and editor of *Sociologia Ruralis* (Bock), and in these roles have discussed the gender of rural organizations and awareness of gender more generally. The gender balance of the ESRS is quite good now, but has not always been, and it varies across the regional networks. We were surprised to see a very unequal gender balance at the International Rural Sociology Association's (IRSA) conference, with some networks nominating all men to the IRSA Council. When we raised this as an issue, some people commented that it was not an issue for IRSA but rather the networks. However when we raised it at the Council, it was agreed that in future IRSA would advise networks to encourage a balance of gender, diversity and stage of career when nominating members. We do not believe there is a deliberate attempt to exclude women, but there is a gender blindness – it is not noticed unless it affects you and in this way it is systemically perpetuated. It is not enough to have a gender stream or a gender section in a book but remain blind to gender in other ways. We must reflect on the gender composition of panels, the gender balance of invitees to meetings, of our networks and of leadership positions as all of these send subliminal messages about the extent to which we are genuinely committed to gender equality.

Comparative Research

In some respects it is quite disheartening that some of our observations from 10 years ago remain true. We noted then that rural gender studies have tended to rely heavily on qualitative, regionally specific studies and consequently comparative (quantitative and qualitative) research remains an underdeveloped area within the field. This is still generally the case; nevertheless, this volume presents several examples of comparative research. Caoimhe Ni Dhonaill evaluates the responsiveness of social initiatives to elderly men's needs and interests in urban and rural places (Chapter 16), and also Donatello Greco and Chiara Zanetti compare the experiences of migrant women in more and less remote rural areas as well as cities (Chapter 3). Several authors in this volume present research that is internationally comparative. Majda Černič Istenič and Chrysanthi Charatsari assess the effect of farm women's participation in education and extension programmes in Greece and Slovenia (Chapter 10), whereas Janet Momsen reflects on the effect of migration on rural gender relations in Mexico and the Caribbean (Chapter 6). Three chapters even cross-cut the borders between countries in the global North and South: Wendy Harcourt does so when looking into glocal initiatives for gender equality (Chapter 20), Susie Jacobs when studying the gendered effects of land titling programmes (Chapter 23), and also Carolyn Sachs and Elisabeth Garner approach their analysis of the gendered effects of changing agro-food systems from a global perspective (Chapter 19). These chapters are important in adding to our understanding of the construction of gender relations and the circumstances under which more equality may be achieved, economically, socially and politically. They point to the fact that similar changes develop in similar ways in different parts of the world and often in parallel to responses to global forces, such as increased mobility, privatization of land or changing agro-food systems. In doing so, they reconfirm the increasing interconnectedness of our societies which also includes the rural.

There are more chapters in this volume which unravel the growing interrelation and interdependence of places, such as that of Donatello Greco and Chiara Zanetti (Chapter 3), which demonstrates the dependency of elderly rural citizens in Italy on care provided by migrants from Ukraine. Or the chapter about Uzbekistan by Nargiza Nizamedinkhodjayeva, Bettina Bock and Peter Mollinga which identifies the importance of migration experiences for shifting power relations according to gender and generation (Chapter 4). The chapter by Jessica Duncan and Monika Agarwal provides testimony of the opposite: rising incorporation in global industries, and the resulting immobilization of rural men and women, may result in the deterioration of women's socio-economic position (Chapter 5). In addition, the chapters that discuss singular case studies add to our ability to understand recent developments from a comparative perspective, given the fact that their assemblage in the different sections of this volume facilitates the contemplation of differences and correspondences.

The book proves in our view the fruitfulness of comparative research and the comparative assemblage of case studies. Further development of comparative gender research is, hence, still important.

As existing research has shown, comparative research not only facilitates rigorous analysis of existing research questions, but also enables a more critical analysis of less developed research questions, such as the impact of geopolitics, ideology, historical trends and culture on rural gender relations. While this volume has a much broader global reach than the previous book and includes many more chapters from the global South and developing countries,[1] there is scope for further, deeper, comparative research. Difficulties accessing research funding has been one barrier. It will be interesting to see if the Global Challenges Research Fund (GCRF) announced in 2015 by the UK Government will change this landscape. The GCRF is a £1.5 billion fund announced by the UK Government to support cutting-edge research that addresses the challenges faced by developing countries. It aims to grow the research base in the UK and strengthen capacity overseas to address research challenges informed by the expressed needs of developing countries. Ten years ago we expected that within the EU, comparison between the 'new' and 'old' member states would emerge. This has not developed to any great extent. It remains our view that the most important form of comparative research that needs systematic development is between developed and developing countries. Many of our debates and areas of research are mirrored in the literature from developing countries. There is an elaborate and very thorough body of research on many of the research questions topical in Western studies: gender and agriculture, governance, gender budgeting, migration, education, and additional areas such as women's access to water and how they are treated by development agencies. For some research questions, developing countries have a far stronger body of knowledge and lead the field theoretically and empirically. This is particularly the case for gender budget analysis, participatory development and integrated resource management.

There is a need to develop both a framework and an agenda for comparative research. Because of the tradition of singular, local research projects, comparative frameworks are hardly developed when it comes to rural gender issues. We need a systematic analysis of existing knowledge on relevant gender issues, identifying gaps in knowledge and understanding. Following this the design of a comparative research framework could be considered, and this could then be used as a starting point for developing comparative research projects.

Ten years ago we noted that a serious obstacle to comparative research is the quality of baseline information on rural women. Many questionnaires for national surveys were designed 20 years ago and they provide information that does not take account of changed gender relations, work roles and economic situations. There is a need to 'write' gender into the collection of statistics, in order to provide information about rural women's employment patterns, participation rates, types of activities, educational levels, engagement with rural development programmes and other important indicators. We both made this point in reports commissioned by the European Parliament (Bock, 2010; Shortall, 2010). Without comparable international statistics, comparative research is practically impossible. As well as statistics, however, there needs to be recognition of what can be learned from research conducted on similar questions but in situations radically different from our own.

While baseline information and comparable statistics are essential tools to undertake comparative research, it would be naïve to present the difficulties as only relating to the quality of our information. Comparative research is difficult and needs considerable commitment to ensure accurate and useful research. The more different the places we wish to compare, the more problematic it is to compare them only by way of statistics or another quantitative methodology that may misrepresent the significance of differences. A comparative case study approach can allow us to examine macro-sociological processes that contribute to similarities and differences in rural gender relations. Yet we need to be realistic about the difficulties of doing good comparative qualitative research. It will require common frameworks and sets of definitions that are general enough to allow for comparison yet specific enough to be meaningful in a specific situation. There is a need to develop a sound qualitative comparative research methodology that maintains the insights of in-depth case studies, but yet generates knowledge that can be generalized.

So, going forward, there are patterns of change and continuity. Some of the challenges remain the same, and greater reflexivity as a profession may be one means of considering how to ensure gender issues and by extension gender studies are given the attention they need.

Note

[1]We are aware that these are contested terms but it is not a debate for this chapter.

References

Bock, B. (2010) *Social Situation of Women in Rural Areas. IP/B/AGRI/IC/2010_089*. European Parliament, Brussels.

Shortall, S. (2010) *Women Working on the Farm: How to Promote their Contribution to the Development of Agriculture and Rural Areas in Europe. IP/B/AGRI/IC/2010_090*. European Parliament, Brussels.

Whatmore, S. (1988) From women's roles to gender relations: developing perspectives in the analysis of farm women. *Sociologia Ruralis* 28(4), 239–247. Available at: DOI:10.1111/j.1467-9523.1988.tb00342.x (accessed 27 February 2017).

INDEX

Note: All United Nations agencies appear under UN. Gender indicators appear only under their full names, not their acronyms. Locators in **bold** refer to figures. Page numbers in *italics* refer to tables.